AF606464

BD 034035201 5

EXCITATORY AMINO ACID RECEPTORS
Design of Agonists and Antagonists

ELLIS HORWOOD SERIES IN PHARMACEUTICAL TECHNOLOGY
incorporating Pharmacological Sciences

Series Editor: M. H. RUBINSTEIN, Professor of Pharmaceutical Technology, The Liverpool Polytechnic

Armstrong & James	**UNDERSTANDING EXPERIMENTAL DESIGN AND INTERPRETATION IN PHARMACEUTICS**
Bloomfield et al	**MICROBIAL QUALITY ASSURANCE IN PHARMACEUTICALS, COSMETICS, TOILETRIES**
Broadley	**AUTONOMIC PHARMACOLOGY**
Cartwright & Matthews	**PHARMACEUTICAL PRODUCT LICENSING: Requirements for Europe**
Cartwright & Matthews	**INTERNATIONAL PHARMACEUTICAL PRODUCT REGISTRATION: Quality, Safety, Efficacy**
Clark & Moos	**DRUG DISCOVERY TECHNOLOGIES**
Cole	**PHARMACEUTICAL PRODUCTION FACILITIES: Design and Application**
Cole, Hogan, Aulton	**PHARMACEUTICAL TABLET COATING TECHNOLOGY**
Cook	**POTASSIUM CHANNELS: Structure, Classification, Function and Therapeutic Potential**
Craig & Newton	**DIELECTRIC ANALYSIS OF PHARMACEUTICAL SYSTEMS**
D'Arcy & McElnay	**PHARMACY AND PHARMACOTHERAPY OF ASTHMA**
Denyer & Baird	**GUIDE TO MICROBIOLOGICAL CONTROL IN PHARMACEUTICS**
Doods & Van Meel	**RECEPTOR DATA FOR BIOLOGICAL EXPERIMENTS: A Guide to Drug Selectivity**
Field & Goldthorpe	**DRUG RESISTANCE IN VIRUSES: Principles, Mechanisms and Clinical Perspectives**
Ford & Timmins	**PHARMACEUTICAL THERMAL ANALYSIS: Techniques and Applications**
Glasby	**DICTIONARY OF ANTIBIOTIC PRODUCING ORGANISMS**
Gould	**PHYSICOCHEMICAL PROPERTIES OF DRUGS: A Handbook for Pharmaceutical Scientists**
Hardy et al	**DRUG DELIVERY TO THE GASTROINTESTINAL TRACT**
Hider & Barlow	**POLYPEPTIDE AND PROTEIN DRUGS: Production, Characterisation, Formulation**
Junginger	**DRUG TARGETING AND DELIVERY: Concepts in Dosage Form Design**
Krosgaard-Larsen & Hansen	**EXCITATORY AMINO ACID RECEPTORS: Design of Agonists and Antagonists**
Kourounakis & Rekka	**STEROIDS, DRUG RESPONSE AND METABOLISM: Pharmacochemical Approach to Defensive Steroids**
Labaune	**HANDBOOK OF PHARMACOKINETICS: The Toxicity Asssessment of Chemicals**
Law	**IMMUNOASSAY PROCEDURES: A Practical Guide**
Martinez	**PEPTIDE HORMONES AS PROHORMONES**
Rainsford	**ANTI-RHEUMATIC DRUGS: Actions and Side Effects**
Ramabhadran	**PHARMACEUTICAL DESIGN AND DEVELOPMENT: A Molecular Biological Approach**
Ridgway Watt	**TABLET MACHINE INSTRUMENTATION IN PHARMACEUTICS: Principles and Practice**
Roth et al	**PHARMACEUTICAL CHEMISTRY VOLUME 1 Drug Synthesis**
Roth et al	**PHARMACEUTICAL CHEMISTRY VOLUME 2 Drug Analysis**
Rubinstein	**PHARMACEUTICAL TECHNOLOGY Controlled Drug Release Volume 1**
Rubinstein	**PHARMACEUTICAL TECHNOLOGY Tableting Technology Volume 1**
Rubinstein	**PHARMACEUTICAL TECHNOLOGY Drug Stability**
Rubinstein	**PHARMACEUTICAL TECHNOLOGY VOLUMES 5 AND 6**
Russell & Chopra	**UNDERSTANDING ANTIBACTERIAL ACTION AND RESISTANCE**
Rutherford	**PHARMACEUTICAL SPECIFICATIONS: Standards for Drugs**
Taylor & Kennewell	**MODERN MEDICINAL CHEMISTRY**
Theobald	**RADIOPHARMACEUTICALS: Using Radioactive Compounds in Pharmaceutics and Medicine**
Thomas & Thurston	**CHEMISTRY FOR PHARMACY, PHARMACOLOGY AND THE HEALTH SCIENCES**
Tweed	**CLINICAL TRIALS FOR THE PHARMACEUTICAL INDUSTRY**
Van Meel, Hauel, Shelley	**CARDIOTONIC AGENTS FOR THE TREATMENT OF HEART FAILURE**
Vergnaud	**CONTROLLED DRUG RELEASE OF ORAL DOSAGE FORMS**
Washington	**PARTICLE SIZE ANALYSIS IN PHARMACEUTICS AND OTHER INDUSTRIES**
Washington et al	**PHARMACOKINETIC MODELLING USING STELLA ON THE APPLE MACINTOSH (TM)**
Wells	**PHARMACEUTICAL PREFORMULATION**
Wells & Rubinstein	**PHARMACEUTICAL TECHNOLOGY Controlled Drug Release Volume 2**
Wilson & Washington	**PHYSIOLOGICAL PHARMACEUTICS BIOLOGICAL BARRIERS TO DRUG ABSORPTION**

The above is a complete list of all Ellis Horwood titles in the pharmaceutical and pharmacological sciences, both published and in preparation. Further details can be obtained from Ellis Horwood or from Simon & Schuster International Group.

EXCITATORY AMINO ACID RECEPTORS

Design of Agonists and Antagonists

Editors

P. KROGSGAARD-LARSEN D.Sc. Ph.D.

J. J. HANSEN M.Sc. Ph.D.

both of The Royal Danish School of Pharmacy
Copenhagen, Denmark

ELLIS HORWOOD

NEW YORK LONDON TORONTO SYDNEY TOKYO SINGAPORE

First published in 1992 by
ELLIS HORWOOD LIMITED
Market Cross House, Cooper Street,
Chichester, West Sussex, PO19 1EB, England

A division of
Simon & Schuster International Group
A Paramount Communications Company

Printed and bound in Great Britain

British Library Cataloguing-in-Publication Data

Excitatory amino acid receptors: design of agonists and antagonists. —
(Ellis Horwood series in pharmacological science)
I. Krogsgaard-Larsen, P. II. Hansen, J. J.
III. Series
615
ISBN 0-13-296716-2

Library of Congress Cataloging-in-Publication Data

Excitatory amino acid receptors: design of agonists and antagonists /
editors, P. Krogsgaard-Larsen, J. J. Hansen.
p. cm. — (Ellis Horwood series in pharmacological sciences)
Includes bibliographical references and index.
ISBN 0-13-296716-2
1. Excitatory amino acids — Agonists. 2. Excitatory amino acids — Antagonists.
3. Excitatory amino acid — Receptors.
I. Krogsgaard-Larsen, Povl. II. Hansen, Jan Jørn. III. Series.
RM666.E87E93 1992
615'.7-dc20 91-46025
CIP

Table of contents

Coloured illustrations appear together in a **colour section** between pages 160 and 161. A black and white reproduction of each coloured illustration appears at the appropriate place in the text.

Preface

(*S*)-Glutamic acid (Glu) and probably also (*S*)-aspartic acid (Asp) are the major excitatory amino acid (EAA) neurotransmitters in the mammalian central nervous system (CNS). The EAAs are neurotoxic, causing neuronal degeneration and, ultimately, cell death after local administration in brain tissues or on different types of cultured neurons. There is some degree of correlation between neuroexcitatory and neurotoxic potencies of EAA agonists, suggesting that EAA receptors play a key role in certain neurodegenerative processes ('excitotoxicity').

'Excitotoxic' mechanisms are likely to be a major causative factor in the severe neuronal injuries in ischaemia, anoxia, and hypoglycaemia. Furthermore, hyperactivity at EAA synapses has been associated with the aetiology of certain neurodegenerative disorders such as Huntington's chorea, epilepsy, and Alzheimer's disease. Hypoactivity at EAA synapses may play a role in schizophrenia. In the progression of Alzheimer's disease degeneration of EAA neurons is observed in certain brain areas.

Multiple EAA receptors exist in the CNS. At least five pharmacologically distinct receptors have been identified and characterized. On the basis of molecular cloning approaches, some of these receptor subtypes have been shown to represent families of structurally related receptor 'isoforms', and it is likely that all of the pharmacologically defined receptor subtypes are heterogeneous. There is growing evidence of functional and, perhaps, structural interaction between different receptor subtypes, notably between the (*RS*)-2-amino-3-(3-hydroxy-5-methylisoxazol-4-yl)propanoic acid (AMPA) and (2*S*,3*S*,4*S*)-kainic acid (KAIN) receptors. These ionotropic EAA receptors, which are receptor-gated ion channels, are probably hetero-oligomeric receptor complexes. It is possible that the complex pharmacology of these EAA receptors reflects to some extent the existence of receptors of more or less different subunit stoichiometry in different parts of the brain.

These aspects represent major therapeutic prospects and obvious challenges for neuroscientists in the EAA field, not least medicinal chemists involved in design and synthesis of EAA receptor ligands. In this complex research area it is of particular importance to stress that there is no acceptable alternative to specificity. In order to map out the physiological relevance and pharmacological importance of various receptor 'isoforms' and subunits, it is essential to develop ligands, which interact specifically with the individual receptor subunits or with well-defined combinations of them. The availability of such research tools is also a prerequisite for detailed studies of the mechanisms underlying the interactions between different receptor subtypes. The design of such receptor ligands represents important steps in the development of drugs with specific actions on unbalanced or derailed mechanisms in neurodegenerative disorders.

The first phases of this kind of drug design project are making very good progress. Thus, highly selective antagonists and agonists for the *N*-methyl-D-aspartate (NMDA) receptor subtype have been developed. Some of these ligands appear to be selective for NMDA 'isoreceptors', and further progress can be anticipated in this therapeutically important field. Similarly, highly selective agonists and antagonists for AMPA and/or KAIN receptors are available. In light of the probable heterogeneity of this/these receptor subtype(s) and their interactions, there is an urgent need for subunit-specific receptor ligands in this area. The pharmacological and therapeutic interests in metabotropic and L-AP4 receptors and in EAA uptake systems have accelerated drug design projects relevant to these EAA synaptic mechanisms.

The specific and highly selective EAA receptor ligands, so far available, have been developed using a broad range of classical and modern principles in drug design. Bioisosteric replacement of functional groups of Glu or Asp by structurally related groups or heterocyclic units showing different physicochemical and electronic properties have proved particularly effective in the design of specific EAA receptor ligands. Conformational restriction or virtual immobilization of the highly flexible molecule of Glu has led to a number of selective EAA receptor ligands. These results may reflect that Glu adopts different 'active conformations' at different receptor subtypes.

The molecule of Glu contains one chiral as well as prochiral centres, making it highly asymmetric during its binding to chiral receptor macromolecules. Consequently, stereochemistry is a factor of major importance in the design of specific EAA receptor ligands. In most cases, the biological effects of mono- or polychiral Glu bioisosteres reside in a single enantiomer. Enantiomers of chiral EAAs have been obtained using classical resolution, stereo-controlled synthetic reactions, enzymatic resolution procedures, or chiral chromatographic techniques.

Combined use of these principles and methods of drug design is likely to lead to ligands showing specific affinity for EAA receptor 'isoforms' or subunits. For different reasons, full agonists or antagonists may not be optimal therapeutic agents. Partial agonists at distinct 'isoforms' of receptors or 'partial agonists' showing agonist and antagonist effects at different subunits of the same receptor complex may be of particular therapeutic interest in, for example, Alzheimer's disease.

Since neuroexcitatory and neurotoxic mechanisms to some extent are different, a major goal in the EAA field is the development of neuroprotective receptor ligands showing little or, ideally, no effect on normal neuroexcitatory mechanisms. In any case, a prerequisite for the development of therapeutically useful agents with specific effects on EAA receptor mechanisms is intimate collaboration between medicinal chemists, molecular biologists, pharmacologists, and, ultimately, clinicians. The challenges and prospects in the EAA field are tremendous.

P. Krogsgaard-Larsen
J. J. Hansen

Abbreviations of names of compounds

α-AA: (*RS*)-2-aminoadipic acid (= (*RS*)-2-amino-5-carboxypentanoic acid)
ABPA: (*RS*)-2-amino-3-(3-hydroxy-5-bromomethylisoxazol-4-yl)propanoic acid
ACBC: 1-aminocyclobutane-1-carboxylic acid
ACBD (= MG): (1*RS*,3*RS*)-1-aminocyclobutane-1,3-dicarboxylic acid (= 'methanoglutamic acid'); see also CMG
ACPD (= ADCP): (1*RS*,3*RS*)-1-aminocyclopentane-1,3-dicarboxylic acid
ACRO: acromelic acid, especially acromelic acid A
ADCP: see ACPD
4-AHCP: (*RS*)-2-amino-3-(3-hydroxy-7,8-dihydro-6*H*-cyclohepta[1,2-*d*]isoxazol-4-yl)propanoic acid
AMAA: (*RS*)-2-amino-2-(3-hydroxy-5-methylisoxazol-4-yl)acetic acid
AMNH: (*RS*)-2-amino-3-(2-(3-hydroxy-5-methylisoxazol-4-yl)methyl-5-methyl-3-oxoisoxazolin-4-yl)propanoic acid
AMOA: (*RS*)-2-amino-3-(3-(carboxymethoxy)-5-methylisoxazol-4-yl)propanoic acid
AMPA: (*RS*)-2-amino-3-(3-hydroxy-5-methylisoxazol-4-yl)propanoic acid
AP3: (*RS*)-2-amino-3-phosphonopropanoic acid
AP4: (*RS*)-2-amino-4-phosphonobutanoic acid
AP5: (*RS*)-2-amino-5-phosphonopentanoic acid
AP7: (*RS*)-2-amino-7-phosphonoheptanoic acid
APPA: (*RS*)-2-amino-3-(3-hydroxy-5-phenylisoxazol-4-yl)propanoic acid
Asp: (*S*)-aspartic acid
ATPA: (*RS*)-2-amino-3-(3-hydroxy-5-*tert*-butylisoxazol-4-yl)propanoic acid
BMAA: (*S*)-2-amino-3-methylaminopropanoic acid (= *β*-methylamino-L-alanine)
BOAA: see *β*-ODAP
Br-HIBO: (*RS*)-2-amino-3-(3-hydroxy-4-bromoisoxazol-5-yl)propanoic acid
CA: (*RS*)-cysteic acid
CCG (= CPG): 2-(2-carboxycyclopropyl)glycine (= 'cyclopropylglutamic acid')
CGP 37849: (*RS*)-2-amino-4-methyl-5-phosphono-3-(*E*)-pentenoic acid
CGS 19755: (2*RS*,4*SR*)-4-(phosphonomethyl)piperidine-2-carboxylic acid (= (2*RS*,4*SR*)-(2-carboxypiperidin-4-yl)methylphosphonic acid)

7-Cl-KYNU (= 7-ClKYN): 7-chlorokynurenic acid
CMG: '*cis*'-1-aminocyclobutane-1,3-dicarboxylic acid (= '*cis*-methanoglutamic acid'); see also ACBD
CNQX: 6-cyano-7-nitroquinoxaline-2,3-dione
CPAA: *trans*-(2-carboxypyrrolidin-3-yl)acetic acid
CPG: see CCG
CPP: (*RS*)-4-(3-phosphonoprop-1-yl)piperazine-2-carboxylic acid (= (*RS*)-3-(2-carboxypiperazin-4-yl)propylphosphonic acid)
CPP-ene: (*RS*)-4-(3-phosphonoprop-2-(*E*)-en-1-yl)piperazine-2-carboxylic acid (= (*RS*)-(*E*)-3-(2-carboxypiperazin-4-yl)propenylphosphonic acid)
CSA: (*RS*)-cysteine sulphinic acid
CSS (= SSC): (*RS*)-cysteine-*S*-sulphonic acid (= *S*-sulpho-(*RS*)-cysteine)
γ-DGG (= DGG): γ-D-glutamylglycine
DNQX: 6,7-dinitroquinoxaline-2,3-dione
DOMO: (2*S*,3*S*,4*S*)-domoic acid
EAA: excitatory amino acid
GABA: 4-aminobutanoic acid
GAMS: γ-D-glutamylaminomethylsulphonic acid
GDEE: (*S*)-glutamic acid diethyl ester
Glu: (*S*)-glutamic acid or (*RS*)-glutamic acid
Gly: glycine
HCA: (*RS*)-homocysteic acid
HCSA: (*RS*)-homocysteine sulphinic acid
HIBO: (*RS*)-2-amino-3-(3-hydroxyisoxazol-5-yl)propanoic acid
HA-966: *N*-hydroxy-3-amino-2-pyrrolidone
4-HPCA: (*RS*)-3-hydroxy-4,5,6,7-tetrahydroisoxazolo[4,5-*c*]pyridine-4-carboxylic acid
5-HPCA: (*RS*)-3-hydroxy-4,5,6,7-tetrahydroisoxazolo[5,4-*c*]pyridine-5-carboxylic acid
7-HPCA: (*RS*)-3-hydroxy-4,5,6,7-tetrahydroisoxazolo[5,4-*c*]pyridine-7-carboxylic acid
HQUIN: homoquinolinic acid
IBO: ibotenic acid (= (*RS*)-2-amino-2-(3-hydroxyisoxazol-5-yl)acetic acid)
KAIN: (2*S*,3*S*,4*S*)-kainic acid
KYNU (= KYN): kynurenic acid
Me-HIBO: (*RS*)-2-amino-3-(3-hydroxy-4-methylisoxazol-5-yl)propanoic acid
MG: see ACBD and CMG
MK-801: 5-methyl-10,11-dihydro-5*H*-dibenzo[*a*,*d*]cyclohepten-5,10-imine (= dizocilpine)
MNQX: 5,7-dinitroquinoxaline-2,3-dione
NAAG: *N*-acetyl-(*S*)-aspartyl-(*S*)-glutamic acid
NBQX: 6-nitro-7-sulphamoylbenzo(*f*)quinoxaline-2,3-dione (= 2,3-dihydroxy-6-nitro-7-sulphamoylbenzo(*f*)quinoxaline)
NMDA: *N*-methyl-(*R*)-aspartic acid
NPC 451: (*RS*)-2-(2-phosphonoethyl)phenylalanine
γ-ODAB: (*RS*)-2-amino-4-(*N*-oxalylamino)butanoic acid
α-ODAP: 3-amino-(*S*)-2-(*N*-oxalylamino)propanoic acid

β-ODAP (= ODAP = BOAA): (*S*)-2-amino-3-(*N*-oxalylamino)propanoic acid
OPS: see SOP
OSS: see SOS
PCP: phencyclidine
PDA: piperidinedicarboxylic acid
PTZ: pentylenetetrazole
QUIN: quinolinic acid
QUIS: (*S*)-quisqualic acid
SKF 10047: *N*-allylnormetazocine
SOP (= OPS): (*RS*)-serine-*O*-phosphonic acid (= *O*-phospho-(*RS*)-serine)
SOS (= OSS): (*RS*)-serine-*O*-sulphonic acid (= *O*-sulpho-(*RS*)-serine)
SSC: see CSS
TCP: *N*-(1-(2-thienyl)cyclohexyl)piperidine
Trans-APPA: (*RS*)-2-amino-5-phosphono-3-(*E*)-pentenoic acid

1

Excitatory amino acids, excitotoxicity and neurodegenerative disorders

Dirk Sauer and **Graham E. Fagg**
Biology Research Laboratories, Pharmaceutical Division, Ciba-Geigy Ltd, CH-4002 Basel, Switzerland

ABSTRACT

From the pioneering efforts of three research groups in the 1950s, studies of excitatory amino acid receptor mechanisms have exploded in the past decade to encompass almost every facet of neuroscience. Amongst other specific functions, L-glutamate is accepted as the major excitatory neurotransmitter in the central nervous system and as the primary endogenous acidic amino acid toxin underlying a number of neurodegenerative disorders. Of the five L-glutamate receptors defined, the *N*-methyl-D-aspartate (NMDA)-preferring subtype is best characterized. Hyperactivation of the NMDA receptor by abnormally high extracellular glutamate levels leads to lethal Ca^{2+} influx (via the NMDA receptor ion channel) and neuronal death. Prolonged activation of 2-amino-3-(3-hydroxy-5-methylisoxazol-4-yl)propanoic acid (AMPA) and kainate receptors also can trigger neurodegeneration. Current evidence indicates that 'excitotoxic' mechanisms underlie the neuronal damage which follows acute brain insults such as cerebral ischaemia and trauma, and may participate in chronic degenerative diseases, such as Alzheimer's and AIDS-related dementias. In animal models, NMDA and AMPA receptor antagonists show therapeutic potential as anti-ischaemic agents. Studies with these antagonists will open the way to greater understanding of neurodegenerative mechanisms and disease processes, and will facilitate the rational design of new therapeutic agents for these debilitating conditions.

INTRODUCTION

The concept that L-glutamate can function both to transmit excitatory signals between brain neurons and to cause neurological dysfunction and cell death originates from experimental observations made more than thirty years ago. In a series of pioneering studies in the 1950s, L-glutamate was shown to increase neuronal firing when iontophoresed onto cat spinal neurons (Curtis *et al.* 1959), to induce convulsions when applied intracerebrally in dogs (Hayashi 1954) and to destroy retinal neurons when injected systemically in the immature mouse (Lucas and Newhouse 1957). Although these findings were not linked together at that time, research during the 1970s and 1980s has shown that acidic amino acids are major excitatory synaptic transmitters in the central nervous system (CNS), that they act via specific membrane receptors, and that, under abnormal circumstances, they operate through the same receptor mechanisms to elicit pathological phenomena such as epileptiform discharges and ischaemic neurodegeneration.

The aim of this chapter is to overview excitatory amino acid mechanisms and their role in the process of neurodegeneration. It is not our intention to review this topic exhaustively, but rather to provide a framework of biological significance for the more chemically weighted articles which follow. Selective receptor agonists and antagonists have played a crucial part in understanding excitatory amino acid physiology and pathology, and the development of potent, subtype-specific antagonists not only will facilitate future progress in this direction, but also may form the bases of new classes of therapeutic agents for neurological disorders.

EXCITATORY AMINO ACIDS AND THEIR RECEPTORS

Endogenous ligands

L-Glutamate is regarded as the major excitatory transmitter in the vertebrate CNS (Watkins and Evans 1981, Fagg and Foster 1983, Collingridge and Lester 1989) and as the primary candidate for an endogenous acidic amino acid neurotoxin (Olney *et al.* 1971, Choi 1988, 1991, Meldrum and Garthwaite 1990). Structurally analogous neuroexcitants found in brain tissue include L-aspartate, L-homocysteate, L-cysteine sulphinate, and the trytophan metabolite quinolinate. Whether any of these substances have roles as excitatory neurotransmitters or endogenous neurotoxins is largely unresolved. Homocysteate, for example, is a potent excitatory amino acid receptor agonist (Mayer and Westbrook 1987) which, on the basis of its presence in and its depolarization-evoked release from brain tissue (Do *et al.* 1986), has been proposed to serve a transmitter function in subsets of excitatory neurons; however, some of these data recently have been questioned following a re-evaluation of the analytical methods used to measure homocysteate in tissue samples (Kilpatrick *et al.* 1991). Quinolinate is a weak *N*-methyl-D-aspartate (NMDA) receptor agonist (see Collingridge and Lester 1989) but, since its release is not evoked by physiological stimuli (Stone and Burton 1988), is not considered to serve a neurotransmitter function; nevertheless, prolonged exposure to low concentrations of quinolinate leads to neuronal death (Whetsell and Schwarcz 1989), and this substance has been proposed to play a role in chronic neurodegenerative diseases such as Huntington's disease and AIDS-related dementia (Schwarcz and Meldrum 1986, Stone and Burton 1988, Heyes *et al.* 1989).

Receptor subtypes

Five distinct membrane receptors for excitatory amino acids are currently recognized (Table 1; Fagg and Massieu 1991). Evidence that more than one such receptor might

Table 1 — Acidic amino acid receptor subtypes[a]

	NMDA	AMPA	Kainate	L-AP4	Glu_G
Other receptor names	AA_1	AA_2 Quisqualate	AA_3	L-APB	Metabotropic Q_p
Selective agonists	*cis*-met-Glu NMDA	AMPA	Kainate Domoate	L-AP4	*trans*-ACPD (Quisqualate)
Selective antagonists	CGP 40116 D-CPPene MK-801[b] 7-Cl-kynurenate[b]	NBQX			
Effector mechanism	Cation channel, Ca^{2+}-permeable	Cation channel	Cation channel	Unknown	IP_3/Ca^{2+} mobilization

[a] Nomenclature according to a Symposium of the International Union of Pharmacology (Flims 1990).
[b] Not competitive antagonists. Abbreviations not defined in the text: *cis*-met-Glu, *cis*-2,4-methanoglutamate; *trans*-ACPD, *trans*-1-aminocyclopentane-1,3-dicarboxylic acid; CGP 40116 is the D-enantiomer of CGP 37849.

exist arose during the 1970s from observations that neurons in the dorsal and ventral horns of the cat spinal cord were differentially sensitive to the excitatory effects of acidic amino acid analogues; L-glutamate and kainate were shown preferentially to stimulate dorsal horn interneurons, and L-aspartate and NMDA to activate Renshaw cells (Johnston *et al.* 1974). The discovery of substances (D-α-aminoadipate (D-α-AA), L-glutamic acid diethylester (GDEE), and Mg^{2+}) able to antagonize the actions of acidic amino acids led to a clearer subdivision of excitatory neuronal responses, and by the end of the 1970s three excitatory amino acid receptor sybtypes had been proposed. These were characterized as (1) NMDA-preferring and D-α-AA-sensitive, (2) quisqualate-preferring and GDEE-sensitive and (3) kainate-preferring and D-α-AA- and GDEE-insensitive (Davies and Watkins 1979, McLennan and Lodge 1979, Ault *et al.* 1980); they became widely known as the NMDA, quisqualate and kainate receptor subtypes, respectively. More recently, the 'quisqualate' receptor has been renamed the 'AMPA' receptor, based on the non-selectivity of quisqualate and the greater selectivity of 2-amino-3-(3-hydroxy-5-methylisoxazol-4-yl)propanoic acid (AMPA) for this receptor subtype (Table 1; Watkins *et al.* 1990, Fagg and Massieu 1991).

Research during the 1980s confirmed the existence of these three excitatory amino acid receptors and yielded evidence for two additional subtypes (which should not be termed 'excitatory', since there is no evidence that they directly lead to neuronal depolarization and excitation). One of these is sensitive to low concentrations of the glutamate analogue L-2-amino-4-phosphonobutanoate (L-AP4), and appears to be located presynaptically on subpopulations of glutamate-releasing nerve terminals (Cotman *et al.* 1986), and the second is linked to phosphatidylinositol (PI) metabolism in the postsynaptic neuron (see Schoepp *et al.* 1990). These are referred to as the L-AP4 and Glu_G ('metabotropic') receptor subtypes, respectively (Table 1; Fagg and Massieu 1991).

Functionally, electrophysiological studies indicate that the NMDA and AMPA receptor subtypes gate cation channels in the postsynaptic membrane. These two are the most widely distributed excitatory amino acid receptor subtypes, and appear to mediate the excitatory actions of released glutamate at synapses throughout the CNS (see Mayer and Westbrook 1987, Monaghan *et al.* 1989, Collingridge and Lester 1989). The precise functions of other receptor subtypes remain to be elucidated. Recent investigations indicate that many of the excitatory actions of applied kainate may be mediated via the AMPA receptor, thereby raising questions about the significance of a separate 'kainate receptor' (Honoré *et al.* 1988); on the other hand, it is clear that high-affinity [^{3}H]kainate binding sites (Monaghan *et al.* 1989) and kainate 'receptors' (Agrawal and Evans 1986) with distinct pharmacological profiles and regional distributions exist, but that their physiological or pathological roles have not been clarified (see Fagg and Massieu 1991). The significance of the L-AP4 subtype is similarly unclear; as mentioned above, L-AP4 appears to act presynaptically to regulate glutamate release, but the distribution of L-AP4-sensitive synapses in the CNS is remarkably restricted (Collingridge and Lester 1989, Fagg and Massieu 1991). The Glu_G receptor has been the subject of extensive study in recent years, but its involvement in brain physiology and pathology is not clearly defined. At the biochemical level, Glu_G receptors have been reported to act in conjunction with AMPA receptors to stimulate the release of arachidonic acid from cultured striatal neurons (Dumuis *et al.* 1990), raising the possibility that they may participate in long-term synaptic changes or in neurodegeneration; Glu_G receptor activity has been observed to change during CNS maturation and following brain ischaemia or kindling (Schoepp *et al.* 1990).

Agonists and antagonists

Well-characterized pharmacological tools are of critical importance for evaluating the roles of defined receptor mechanisms in the physiology and pathology of the CNS. The NMDA receptor subtype is the most extensively studied of the excitatory amino acid receptors, and several sites of pharmacological or regulatory significance have been identified (Fig. 1). At the transmitter recognition site, *cis*-2,4-methanoglutamate is the most potent and selective agonist currently reported (Lanthorn *et al.* 1990), while the D-isomers of 2-amino-4-methyl-5-phosphono-3-pentenoate (CGP 37849)

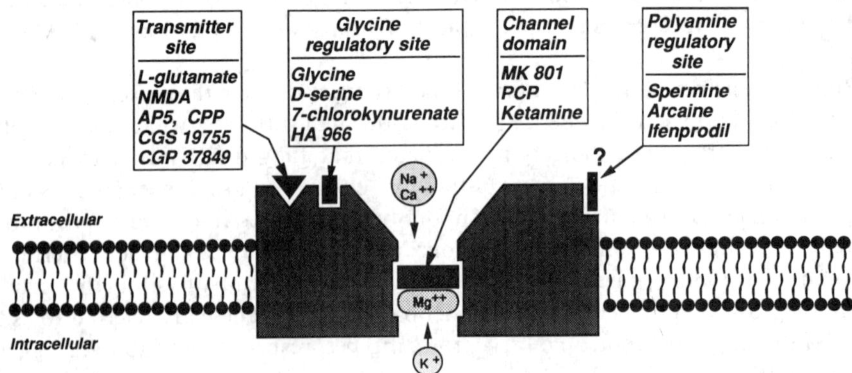

Fig. 1 — Schematic model of the NMDA receptor complex, illustrating major sites of pharmacological and regulatory significance.

(Fagg *et al.* 1990) and 3-(2-carboxypiperazin-4-yl)propenyl-1-phosphonic acid (CPPene) (Lowe *et al.* 1990) are the most potent and selective antagonists. Other well-documented antagonists of the NMDA receptor include uncompetitive (channel) blockers (e.g. phencyclidine (PCP), ketamine and MK-801; Kemp *et al.* 1987, Lodge *et al.* 1988) and non-competitive antagonists acting at the glycine (e.g. 7-chlorokynurenic acid, HA 966; Lodge and Johnson 1990, Palfreyman and Baron 1991) and polyamine regulatory sites (e.g. arcaine, ifenprodil, SL 82.0715; Carter *et al.* 1991).

Far fewer pharmacological agents are available for other excitatory amino acid receptors. AMPA itself is the agonist of highest selectivity for the AMPA receptor (Krogsgaard-Larsen *et al.* 1980), and a quinoxaline derivative, 6-nitro-7-sulphamoylbenzo[*f*]quinoxaline-2,3-dione (NBQX), the only selective antagonist (Sheardown *et al.* 1990). No antagonists are available for high-affinity kainate receptors; agonists and antagonists have not been defined at L-AP4 receptors; and only weak and non-selective antagonists of the Glu_G receptor have been reported (Schoepp *et al.* 1990). Comprehensive reviews of the pharmacological properties and structure–activity relationships of excitatory amino acid receptor subtypes are to be found in subsequent chapters of this volume.

EXCITOTOXICITY: PHENOMENON AND BASIC MECHANISMS

The term 'excitotoxicity' was coined by Olney and his colleagues in the 1970s as a semi-mechanistic description of the neuronal necrosis which follows exposure of CNS tissue to excitatory amino acids. These investigators observed that glutamate and its analogues, when administered to immature rodents, induced degenerative changes in neurons of the hypothalamus and other regions with poorly formed blood–brain barrier (Olney 1971, Olney *et al.* 1971). Subsequent investigations demonstrated that there was a good correlation between the neurotoxic and the neuroexcitatory potencies of a range of related amino acids, suggesting that there was a direct causal relationship between cellular activation and cell death (Olney 1978).

Neurons suffering excitotoxic damage show a characteristic cytopathology. In the acute phase damage is postsynaptic, and presynaptic terminals and axons are unaffected. Dendrites (where excitatory inputs and receptors are concentrated) are swollen, and perikaryal mitochondria and endoplasmic reticulum are dilated. Subsequently, the cytoplasm becomes condensed and vacuolated, and nuclear pyknosis is apparent (Olney and Sharpe 1969, Olney 1971, Meldrum and Garthwaite 1990). Cytopathological changes of this type are found in a number of degenerative conditions, including cerebral ischaemia and trauma (Brown and Brierly 1972, Simon *et al.* 1984a, van Reempts 1984, Faden *et al.* 1989) and status epilepticus (Evans *et al.* 1984), and represent one line of evidence that excitotoxic mechanisms play a primary role in the cerebral damage associated with these disorders.

NMDA receptors and excitotoxic cell damage

The mechanisms of excitotoxicity *in vitro* have been studied by several groups. Choi (1987) has shown that two separable events take place following application of glutamate to cortical cell cultures: (1) acute neuronal swelling, which requires extracellular Na^+ and Cl^-, can be induced by a variety of depolarizing agents and is reversible; and (2) delayed neuronal degeneration, which is dependent on extracellular

Ca^{2+} and is essentially irreversible (also see Choi 1988, 1991). Ca^{2+}-dependent, NMDA receptor-mediated neurodegeneration has been observed by other groups using cultured neurons or brain slices *in vitro* (Garthwaite *et al.* 1986, Rothman *et al.* 1987, Garthwaite and Garthwaite 1989), and it has been demonstrated that the accumulation of Ca^{2+} intracellularly is correlated with the subsequent neuronal degeneration (Kurth *et al.* 1989, Choi 1991). Although Ca^{2+} might in principle enter the cell by a number of routes (e.g., voltage-dependent Ca^{2+} channels, Na^{+}–Ca^{2+} exchange), pharmacological studies indicate that the NMDA receptor channel (which is permeable to Ca^{2+}; MacDermott *et al.* 1986) is primarily responsible for L-glutamate/excitotoxin-induced increases in intracellular Ca^{2+} under the experimental conditions employed (Choi 1987, Garthwaite and Garthwaite 1987). These investigations provide a rational and quantitative basis to support a critical role for Ca^{2+} in degenerative processes, and strengthen earlier observations on the accumulation of this cation in seizure- and ischaemia-damaged neurons *in vivo* (Griffiths *et al.* 1984, Simon *et al.* 1984b, Deshpande *et al.* 1987; also see Meldrum and Garthwaite 1990, Choi 1991).

The precise sequence of degenerative events initiated by high levels of intracellular Ca^{2+} is unknown, although activation of Ca^{2+}-dependent proteases, phospholipases or kinases, followed by the generation of cytotoxic free radicals or direct structural breakdown, are possibilities (Fig. 2; see Choi 1988, 1991, Siesjö 1989, Meldrum and Garthwaite 1990). Elucidation of the major intracellular biochemical pathways leading to neuronal death will form an important research focus in the coming years,

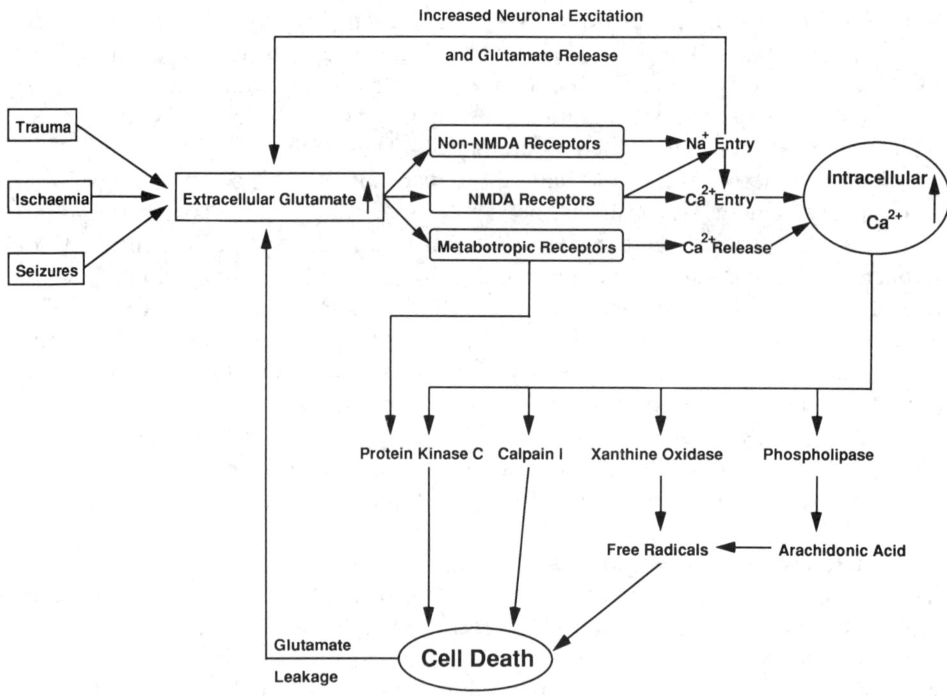

Fig. 2 — Putative mechanisms of excitotoxic cell damage (modified from Choi 1988).

since this may result in the identification of key mechanistic targets for the development of novel cerebroprotective therapies.

The features of NMDA receptor-mediated neurodegeneration *in vivo* have been examined following direct injection of receptor agonists (e.g. NMDA, quinolinate) into brain regions such as the hippocampus or striatum (Schwarcz *et al.* 1983, Foster *et al.* 1988, Fagg *et al.* 1991). As in the case of ischaemic neurodegeneration (see below), excitotoxic cell death is a delayed or protracted event, since neurons can be rescued by NMDA receptor antagonists even when administered several hours after injection of the excitoxin (Fig. 3; Foster *et al.* 1988). The mechanisms of delayed neuronal degeneration *in vivo* are not clear. One possibility is that, following exposure of a core of neurons to high concentrations of an NMDA receptor agonist, hyperactivation of NMDA receptors (and neuronal death) spreads outwards by a combination of excitatory synaptic transmission and glutamate leakage (see Fig. 2), and that drug-induced neuroprotection observed after some hours simply reflects blockade of the spread of damage. An alternative hypothesis is that prolonged stimulation of NMDA receptors is required to elicit neuronal degeneration, possibly because the effects of acute receptor hyperactivation can be limited by intrinsic (e.g. Ca^{2+}-buffering) or extrinsic (e.g. synaptic inhibition) neuronal mechanisms. These or other events (or a combination) may contribute to the time-course of neurodegeneration after excitotoxic insult (for additional discussion see Choi 1988, 1991, Foster *et al.* 1988, Meldrum and Garthwaite 1990). Clarification of the mechanisms underlying delayed excitotoxic degeneration is a prerequisite for the rational development of neuroprotective drugs which will be effective during the post-insult 'time window of opportunity'.

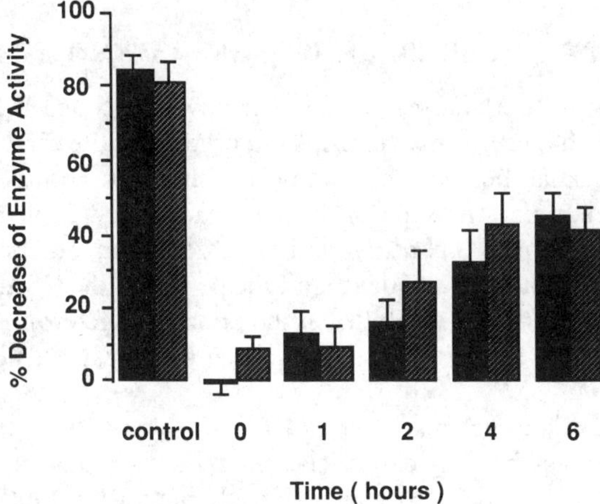

Fig. 3 — Competitive (CGP 37849, 50 mg/kg i.p., hatched bars) and uncompetitive NMDA receptor antagonists (MK-801, 2 mg/kg i.p., solid bars) prevent quinolinic acid-induced loss of striatal cholinergic neurons (choline acetyltransferase activity) when administered up to 6 hours after excitotoxic lesion. NMDA antagonists (controls received no drug) were administered at the indicated times after unilateral injection of 200 nmol quinolinic acid into the rat striatum, and enzyme activity was measured seven days later. (Data are from L. Massieu *et al.*, unpublished.)

Non-NMDA receptors and excitotoxic cell damage

Although NMDA receptor mechanisms have received most attention with regard to the excitotoxic process, activation of non-NMDA excitatory amino acid receptors has been shown to lead to neuronal death *in vitro* and *in vivo* (Frandsen *et al.* 1989, Koh *et al.* 1990, reviewed by Meldrum and Garthwaite 1990, Choi 1991). In general terms, agonists with distinct receptor selectivities display excitotoxic effects on different neuronal populations; examples of this 'selective vulnerability' are the reciprocal sensitivities of cultured NADPH-diaphorase-containing cortical neurons to the toxic effects of non-NMDA and NMDA receptor agonists (Koh and Choi 1988), and the high vulnerability of hippocampal CA3/4 pyramidal neurons to kainate (Nadler *et al.* 1978). In some cases, selective vulnerability appears to be a function of regional variations in receptor expression (e.g., CA3/4 neurons have a high density of kainate receptors; Monaghan *et al.* 1989) although other factors also may be involved (Meldrum and Garthwaite 1990, Choi 1991).

A feature of non-NMDA receptor-mediated neurodegeneration *in vitro* is the long exposure times required to initiate neuronal death. Whereas several minutes exposure to glutamate or NMDA is sufficient to trigger the death of cultured cortical neurons, several hours exposure to AMPA or kainate are required (Koh *et al.* 1990a). Although AMPA and kainate receptor channels mostly are not permeable to Ca^{2+}, this cation appears to be important for expression of the neurotoxic actions of both agonists, since Ca^{2+}-free medium attenuates kainate toxicity (Rothman *et al.* 1987) and nifedipine reduces AMPA-evoked degeneration (Weiss *et al.* 1990). The prolonged exposure time required for the neurotoxic effects of non-NMDA receptor agonists has led to suggestions that non-NMDA receptor mechanisms may play a role in chronic neurodegenerative diseases (Meldrum and Garthwaite 1990, Choi 1991).

EXCITOTOXICITY IN CEREBRAL ISCHAEMIA AND TRAUMA

Several distinct lines of evidence have led to the now widely accepted hypothesis that excitatory amino acid receptor mechanisms underlie the acute cerebral damage which follows brain ischaemia; these will be summarized in this section. Clinically, cerebral ischaemia occurs in situations such as stroke, head injury, asphyxia, subarachnoid haemorrhage, cardiac arrest, cardiovascular surgery and others. Cerebrovascular disease is the third leading cause of death in Europe and in the USA, and CNS trauma the leading cause of death and disability in the under-45 age group (McCulloch *et al.* 1991). Effective therapy of such disorders is a major clinical and social need.

Animal models of brain ischaemia

Ischaemic brain damage broadly can be classified into two major types: (1) acute and focal; and (2) delayed and diffuse (Meldrum and Garthwaite 1990). Experimentally, focal ischaemia (a model of human stroke) is usually elicited by permanent or transient occlusion of the MCA (Ginsberg and Busto 1989), resulting in localized infarction in those brain regions supplied by the occluded vessel. The infarct comprises a core of tissue which has little blood supply and depleted energy levels (and hence is destined to die), and a surrounding 'penumbra' zone, where blood flow and energy levels are partially maintained (and which potentially is rescuable;

Ginsberg and Busto 1989, Meldrum 1990, McCulloch *et al.* 1991). In contrast, global ischaemia (a model of cardiac arrest, hypotension or asphyxia) is induced by temporary occlusion of arteries (carotid/vertebral) supplying the entire cranium. Cerebral damage elicited in this manner is typically diffuse, affecting selectively vulnerable neuronal populations throughout the brain (e.g., neurons in hippocampal area CA1, striatum, cortical layers 3, 5 and 6, cerebellum) and occurring with a delay of one to two days (Kirino 1982, Pulsinelli *et al.* 1982, Smith *et al.* 1984).

Excitatory amino acid levels rise in cerebal ischaemia
In addition to the cytopathological evidence outlined above, studies of cerebral extracellular metabolites by microdialysis *in vivo* indicate that excitatory amino acids are involved in the generation of ischaemic brain damage. Focal and global ischaemia, percussion injury, subdural haematoma and hypoglycaemia have all been shown markedly to increase the extracellular concentrations of glutamate and aspartate in those regions subsequently exhibiting neuronal damage (Benveniste *et al.* 1984, Hagberg *et al.* 1985, Globus *et al.* 1988, Hillered *et al.* 1989, Faden *et al.* 1989, Butcher *et al.* 1990, Phillis *et al.* 1991, McCulloch *et al.* 1991). This appears to be a consequence of increased release under anoxic conditions and impaired reuptake by energy-depleted neurons and glia (Drejer *et al.* 1985, Nicholls and Attwell 1990). For glutamate, the magnitude of the increase ranges from 3.5–10-fold following global ischaemia, brain trauma and subdural haematoma (Benveniste *et al.* 1984, Globus *et al.* 1988, Faden *et al.* 1989, Phillis *et al.* 1991, McCulloch *et al.* 1991) to as much as 80-fold after focal ischaemia in the rat (Hillered *et al.* 1989). This places extracellular glutamate concentrations during ischaemia in the same range as those which have been shown to be neurotoxic *in vitro* (see Choi 1988, Meldrum 1990).

Removal of excitatory inputs attenuates ischaemic brain damage
Another link in the chain of events relating excitatory amino acid synaptic mechanisms and ischaemic neuronal death was provided by experiments involving hippocampal deafferentation. The hippocampus is not only highly vulnerable to global ischaemic damage but, by virtue of the organization of its excitatory pathways (which utilize acidic amino acid transmitters; Fagg and Foster 1983), is also a region amenable to lesions of discrete fibre tracts. Using this approach, Wieloch *et al.* (1985) first showed that destruction of the perforant path (excitatory afferents to CA1 pyramidal neurons) reduces damage to CA1 neurons induced by subsequent bilateral carotid occlusion. Other investigators, employing chemical or surgical lesions of distinct CA1 afferents, have made similar observations (Johansen *et al.* 1986, Onodera *et al.* 1986, Jorgensen *et al.* 1987; although Buchan and Pulsinelli (1990) found no neuroprotection when using the more severe four-vessel occlusion model). Benveniste *et al.* (1989) went one step further by demonstrating that destruction of CA3 neurons (from which the excitatory Schaffer collaterals to CA1 originate) not only protects CA1 neurons from subsequent ischaemic damage, but also prevents the increase in extracellular glutamate (but not GABA) induced during and after the ischaemic insult. Jorgensen *et al.* (1990) showed that this same lesion prevents the increase in CA1 glucose metabolism normally observed after global ischaemia. These data strengthen the hypothesis that, at least following global ischaemia, hyperactivation

(hypermetabolism) and degeneration of vulnerable neurons results from an increased release of glutamate, and that this is derived primarily from the terminals of excitatory amino acid afferents.

Excitatory amino acid receptor antagonists block ischaemic neurodegeneration
Possibly the most persuasive evidence that excitatory amino acid receptor mechanisms occupy a key position in the cellular events leading to ischaemic damage is that antagonists, especially of the NMDA receptor, have been shown in many laboratories to prevent post-ischaemic neuronal death *in vitro* and *in vivo* (see Choi 1988, 1991, Meldrum and Garthwaite 1990, Meldrum 1990, McCulloch *et al.* 1991).

Jørgensen and Diemer (1982) provided the initial stimulus for pharmacological investigations using glutamate receptor antagonists. Although their study showed no protection by GDEE (probably owing to the low potency and instability of this antagonist), success was soon achieved with the more potent and selective NMDA receptor antagonists which then were becoming available. Using a rat model of global ischaemia (bilateral carotid occlusion), Simon *et al.* (1984a) observed that intrahippocampal injection of the NMDA receptor antagonist AP7 reduced early morphological signs of damage to CA1 pyramidal neurons, and these observations were confirmed and extended by work from other groups using systemic administration of competitive and uncompetitive NMDA receptor blockers (e.g. Gill *et al.* 1987, Sauer *et al.* 1988, Swan and Meldrum 1990). However, positive results have not emerged from all studies. Several groups have been unable to demonstrate neuroprotective effects of NMDA receptor antagonists in models of global ischaemia (e.g. Wieloch *et al.* 1989, Buchan and Pulsinelli 1990). The reasons for these inconsistencies have been ascribed to the use of different animal models (two-vessel or four-vessel occlusion) and other methodological parameters between laboratories that influence brain temperature, insult severity and the resultant energy status of the ischaemic tissue (for an excellent discussion of this issue see McCulloch *et al.* 1991; also Buchan 1990, Meldrum 1990). At present, the consensus of opinion is that NMDA receptor antagonists prevent ischaemic neuronal damage in those models or brain regions in which energy levels are partially preserved, but that they are ineffective under conditions of total energy failure (Wieloch *et al.* 1989, Siesjö and Bengtsson 1989, McCulloch *et al.* 1991).

Clear and consistent evidence for the cerebroprotective efficacy of NMDA receptor antagonists, and a pointer to one of their primary therapeutic potentials, has come from experiments involving focal cerebral ischaemia in rodents and cats. Uncompetitive blockers, such as PCP, MK-801 and dextromethorphan (Ozyurt *et al.* 1988, Park *et al.* 1988, Steinberg *et al.* 1989, Tamura *et al.* 1988, Bielenberg 1989), the competitive antagonists 2-amino-7-phosphonoheptanoic acid (AP7), CGS 19755, D-CPPene and CGP 40116 (Figs 4 and 5; Roman *et al.* 1989, Simon and Shiraishi 1990, Bullock *et al.* 1990, Sauer *et al.* 1991) and the putative polyamine site antagonists ifenprodil and SL 82.0715 (Gotti *et al.* 1988, Carter *et al.* 1991) have all been shown markedly to reduce infarct volume following MCA occlusion. In all cases, protection is observed principally in the cortical zones of the infarct (the volume of cortical damage is reduced by up to 60%), with little effect in striatal regions (Fig. 4A, B). This has been attributed to the fact that MCA occlusion reduces blood flow to cortical regions less than it does to the striatum (i.e., cortical regions experience incomplete ischaemia, and represent the 'penumbral' zones of the infarct; Fig. 4C, see also

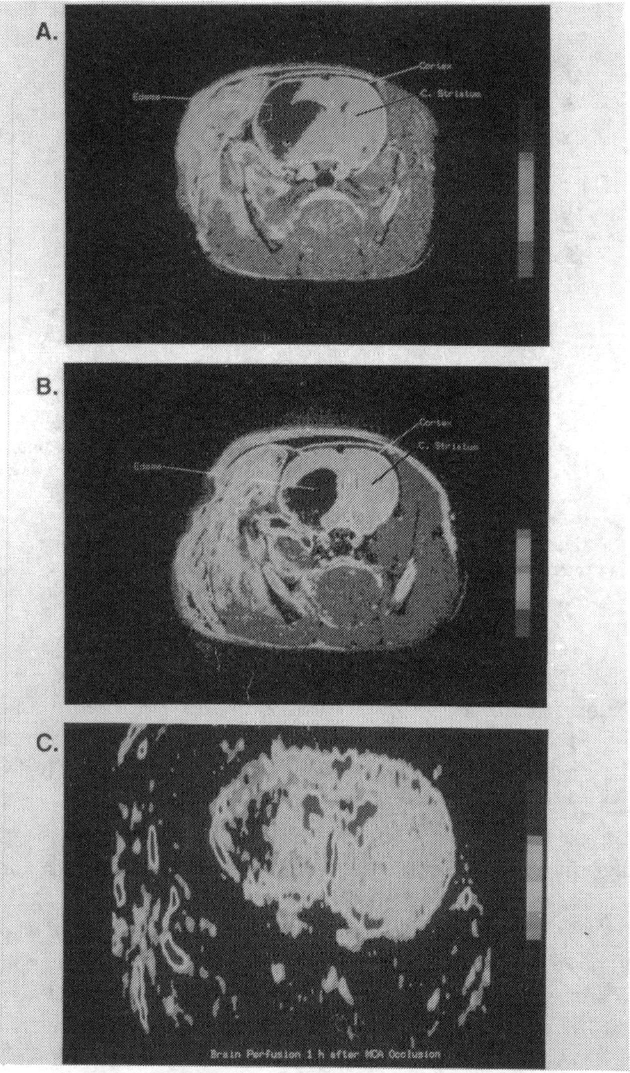

Fig. 4 — (**See colour section**.) Magnetic resonance images (T_2 weighted, coronal planes) taken two days after middle cerebral artery occlusion in Fischer 344 rats which received either (A) no treatment or (B) 20 mg/kg i.v. CGP 40116 immediately post-occlusion. Images were taken using a Spectrospin BMT 47/30 (4.7 T) instrument and colour-enhanced using a Bruker Aspect X32 workstation. Regions of oedema are colour-coded red; these regions were shown to be necrotic using histological techniques. Note the pronounced reduction in cortical oedema in the animal treated with CGP 40116. (C) Magnetic resonance subtraction image (before and after i.v. infusion of a gadolinium-based contrast agent) showing cerebral perfusion 1 hour following middle cerebral artery occlusion. Dark areas indicate regions of low vascular perfusion, and red high. Note the lack of perfusion in the striatum, and the moderate perfusion in cortical regions of the occluded side. (Courtesy of Dr P. Allegrini.)

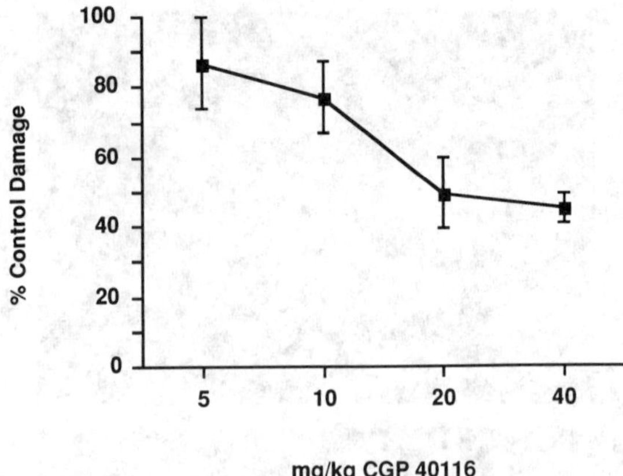

Fig. 5 — Dose-response curve describing the reduction of cortical infarct volume by CGP 40116 following middle cerebral artery occlusion in Fischer 344 rats. CGP 40116 (the D-enantiomer of CGP 37849) was administered i.v. immediately after occlusion and infarct volume was determined two days later by magnetic resonance imaging (see Fig. 4). Data are expressed relative to the infarct volume in untreated rats (= 100%). Curve-fitting (logistic equation) indicated that CGP 40116 maximally reduced cortical infarction by 62% with an ED_{50} of 11 mg/kg i.v. (Data are from D. Sauer *et al.*, unpublished.)

Ginsberg and Busto 1989, Shiraishi *et al.* 1989, McCulloch *et al.* 1991). Choi (1990) has recently summarized hypotheses to account for the predominance of NMDA receptor-mediated injury in the penumbra of an ischaemic infarct, and hence why focal ischaemic brain damage is a primary target for protection by NMDA receptor antagonists.

In addition to their ability to reduce focal ischaemic brain damage, NMDA receptor antagonists have been observed to reduce neuronal damage following hypoglycaemia *in vivo* and *in vitro* (Wieloch 1985, Monyer *et al.* 1989), traumatic injury *in vitro* (Tecoma *et al.* 1989) and subdural haematoma *in vivo* (Bullock *et al.* 1990b). They have additionally been reported to improve neurological outcome following head or spinal cord trauma in animal models *in vivo* (Faden *et al.* 1989).

In contrast to the properties of NMDA receptor antagonists, recent reports indicate that the AMPA receptor antagonist NBQX can prevent delayed neuronal death which follows transient complete global ischaemia (Sheardown *et al.* 1990), but is less effective against focal ischaemic damage (Gill *et al.* 1991). Although confirmation is required, these observations strengthen the concept that excitotoxic (albeit distinct) mechanisms are involved both in global and in focal ischaemic brain damage, and indicate that an optimal therapy of ischaemic CNS disorders in man will comprise either a 'cocktail' of antagonists targeted at different receptor subtypes or drugs with broad actions at several key molecular targets.

EXCITOTOXICITY IN EPILEPSY

The mechanistic rationale and potential therapeutic value of utilizing NMDA receptor antagonists as anti-epileptic agents has been documented extensively during the past decade (see Patel *et al.* 1988, Chapman 1991), and will not be reviewed here.

Instead, we shall focus on one form of epilepsy — status epilepticus — which is characterized by sustained seizure activity, and shall briefly overview the evidence that excitotoxic mechanisms underlie the neuronal death which occurs in this syndrome.

Prolonged limbic seizures elicited by chemical convulsants in animals have long been known to elicit a characteristic hippocampal pathology (notably a marked loss of CA3/4 pyramidal neurons) which resembles that found in post-mortem tissue of patients with status epilepticus (Nadler *et al.* 1978, Ben-Ari 1985, Olney 1986). As in the case of cerebral ischaemia (above), several converging lines of evidence indicate that the observed neuropathology is mediated by excitotoxic mechanisms:

(1) Increased glutamate release rapidly follows application of convulsant drugs to brain tissue *in vitro* (De Boer *et al.* 1982).
(2) Intracerebral injection of glutamate or prolonged stimulation of perforant path fibres results in selective damage to hippocampal CA3/4 pyramidal neurons (Sloviter 1983, Sloviter and Dempsey 1985).
(3) Acutely damaged neurons exhibit the same ultrastructural appearance following excitotoxic damage and status epilepticus (Evans *et al.* 1984, Olney 1986).
(4) Ca^{2+} is accumulated by damaged neurons (Griffiths *et al.* 1984).
(5) NMDA receptor antagonists prevent the neuronal damage which follows sustained convulsant-induced limbic seizures in animals (Fariello *et al.* 1989, Clifford *et al.* 1990, Fujikawa *et al.* 1991).

Although neuroprotective, NMDA receptor blockers do not reduce seizure activity in these models. Since, however, non-NMDA receptor antagonists effectively attenuate burst firing evoked in CA3 neurons *in vitro* (Chamberlin *et al.* 1990), it may be worthwhile evaluating the anticonvulsant and neuroprotective effects of this class of compound in animal models of status epilepticus, when appropriate agents become available.

EXCITOTOXICITY IN CHRONIC NEURODEGENERATIVE DISEASES

The hypothesis that excitatory amino acid receptor mechanisms underlie the pathogenesis of chronic neurodegenerative diseases originated primarily from knowledge of the excitotoxicity of acidic amino acids (reviewed above), together with observations that the pathological changes in diseases like Huntington's (HD) and Alzheimer's (AD) occur mainly in those regions receiving a high density of excitatory amino acid afferents. Although evidence in favour of this postulate is largely circumstantial, recent reports that the toxicity of excitatory amino acids is enhanced in the presence of abnormal protein products found in AD and AIDS-related dementia (see below) provide a basis for focused study. The following paragraphs provide a very brief summary of this area.

Huntington's disease

The main evidence that excitotoxic mechanisms are involved in the pathogenesis of HD is correlative. The principal pathology in HD is loss of GABAergic and cholinergic neurons in the caudate nucleus–putamen, with sparing of somatostatin-,

NADPH-diaphorase- and acetylcholinesterase-containing cells (Martin 1984, Ferrante *et al.* 1987). A similar pattern of selective neurodegeneration follows injection of the NMDA receptor agonist quinolinic acid into the rat striatum (Schwarcz *et al.* 1983, Beal *et al.* 1986; but see Davies and Roberts 1987). Autoradiographic experiments showing a preferential loss of NMDA receptors in post-mortem brain tissue from HD patients is consistent with the idea that NMDA receptor mechanisms mediate the neuronal loss (Young *et al.* 1988).

One interpretation of these observations is that the neuropathology associated with HD results from hyperactivation of the NMDA receptor by increased levels of an endogenous agonist. To date, however, no such increase has been detected. Although evidence of altered tryptophan metabolism has been found in post-mortem brain tissue from HD patients (Schwarcz *et al.* 1988), no changes in quinolinic acid levels have been found (Reynolds *et al.* 1988). However, Whetsell and Schwarcz (1989) have demonstrated that, in culture, prolonged exposure to submicromolar concentrations of quinolinic acid leads to neuronal death. One speculative hypothesis, therefore, is that individuals carrying the HD gene suffer an age-related change in metabolism that leads to small but sustained increases in quinolinic acid, and this ultimately results in the neurodegenerative and behavioural features characteristic of the disease. Experiments to determine mechanisms by which quinolinic acid elicits neuronal death at low concentrations will be valuable.

Alzheimer's disease

Investigations of pre- and postsynaptic markers of excitatory amino acid synaptic transmission have led to mixed and generally inconclusive results regarding a causative role for excitatory amino acids in AD (see Greenamyre and Young 1989; see also chapter 2). Evidence (albeit circumstantial) that excitotoxic mechanisms may be involved in the neuronal changes characteristic of this disease have arisen from other lines of study:

(1) Glutamate induces dendritic changes (Mattson 1988) and paired helical filaments (DeBoni and Crapper-McLachlan 1985) in cultured neurons similar to those typical of AD.
(2) The neurofibrillary tangles of AD share antigenic properties with those found in the amyotrophic lateral sclerosis–Parkinsonian–dementia syndrome of Guam (Guiroy *et al.* 1987), a disorder which has been linked tentatively to a plant-derived excitatory amino acid toxin (Spencer *et al.* 1987).
(3) Anatomically, the presence of neurofibrillary tangles and plaques in AD brains is correlated with the distribution of glutamate-using cortical association fibres (Pearson *et al.* 1985).

A recent intriguing study has shown that cultured cortical neurons are more vulnerable to excitatory amino acid-mediated damage following exposure to the abnormal β-amyloid protein found in AD plaques (Koh *et al.* 1990a). Increased vulnerability to toxic damage required prolonged ($>$ 1 day) exposure to β-amyloid protein, and was not receptor-specific, in that damage mediated by NMDA and kainate were both enhanced. These findings are important because they provide a concrete link between excitotoxic mechanisms and the disease process. They do not show that excitatory amino acid-mediated events are the primary cause of AD, but they do suggest a hypothesis in which, under conditions pertaining in the disease, facilitated excitotoxic mechanisms contribute to the neuropathology and, by extension, to the dementia. A corollary of this hypothesis is that dementia is the symptom of a secondary degenerative component of the disease, and that the primary disease state and the symptoms may be separable mechanistically and therapeutically. Studies to elucidate the cellular mechanisms by which β-amyloid protein augments excitotoxic processes will undoubtedly lead to further progress in understanding AD and in developing rational therapies for this disease.

AIDS-related dementia

Infection with HIV-1 (human immunodeficiency virus) leads in many patients secondarily to dementia and neurological dysfunction. Neuropathological examination indicates that the brains of such patients show atrophy, macrophage infiltration and neuronal loss (Navia *et al.* 1986). The mechanisms underlying the neuropathology are not clear. One possibility is that viral infection itself is directly responsible (Navia *et al.* 1986). Heyes *et al.* (1989) have proposed an alternative hypothesis, involving excitotoxic neurodegeneration resulting from increased levels of quinolinic acid in infected patients.

Heyes *et al.* (1989) collected cerebrospinal fluid (CSF) samples from patients with AIDS and from age-matched volunteers, and found that quinolinic acid concentrations in patients were significantly greater than those in controls; furthermore, CSF quinolinic acid levels were correlated with the degree of dementia in infected individuals (Heyes 1991). SIV infection or septicaemia in monkeys was similarly associated with sustained increases in the CSF levels of quinolinic acid, which attained high nanomolar to low micromolar concentrations (Heyes and Lackner 1991). The probable basis of this effect appears to be stimulation of quinolinic acid synthesis by γ-interferon-induced activation of indoleamine-2,3-dioxygenase in macrophages. Heyes and colleagues conclude that increases in CSF quinolinic acid may be a generalized phenomenon associated with immunological activation or with infectious diseases which affect the CNS.

The question arising from these data is whether the CSF levels of quinolinic acid in AIDS patients are sufficient to account for the neuropathological changes observed. Recent observations add weight to the hypothesis that they may be. Lipton *et al.* (1991) have reported that the HIV coat protein gp120 leads to increased intracellular Ca^{2+} and cell death in retinal ganglion cells *in vitro*, and that this is mediated via the NMDA receptor; gp120 appeared to sensitize neurons to NMDA receptor-mediated excitotoxicity, such that the micromolar levels of endogenous glutamate present in the medium became lethal. Coupled with the observations of Whetsell and Schwarcz (1989), that prolonged exposure to submicromolar concentrations of quinolinic acid can lead to neuronal damage in culture, these findings add weight to the hypothesis put forward by Heyes *et al.* (1989), that excitotoxic processes may contribute to the neuropathology (and dementia) associated with AIDS.

Interestingly, these findings with gp120 are analogous to those described for β-amyloid protein by Koh *et al.* (1990a) (see above). This brings to the forefront new questions on the aetiology of chronic neurodegenerative diseases: do ‘protein-sensitization' phenomena of this type play a general role in the pathogenesis of such disorders? What are the mechanisms whereby gp120 and β-amyloid protein sensitize neurons to the toxic effects of excitatory amino acids? Are they the same? Investigations directed towards these issues will likely uncover basic mechanisms of chronic neurodegenerative disease, clarify the part played by excitotoxic processes, and open up new therapeutic approaches to disease management.

CONCLUSIONS: PAST, PRESENT AND FUTURE

There is probably no better attestation to the significance attributed to any field of research than the speed with which it grows. Although the generalized concept of 'excitotoxicity' was proposed early in the 1970s, a key stimulus which focused the efforts of many groups were the publications in 1984 that excitatory synaptic activity (Rothman 1984), glutamate release (Benveniste *et al.* 1984) and NMDA receptor activation (Simon *et al.* 1984a) were involved in the pathogenesis of hypoxic–ischaemic brain damage. In the subsequent brief period of time, sufficient evidence has accumulated to establish beyond doubt that excitotoxic mechanisms are the primary mediator of neuronal death following events such as cerebral ischaemia, trauma, haemorrhage and status epilepticus.

Progress in this area would not have been possible without receptor antagonists. Most obviously, the range of potent and selective NMDA receptor antagonists available has enabled the function of this receptor type to be evaluated in a variety of experimental models of cerebral ischaemia and seizures, and the best of these compounds will likely be candidates for clinical evaluation in the coming years. Far fewer agents are available with which to examine the roles of non-NMDA receptors, although the recently described AMPA receptor antagonist NBQX is fuelling investigations to define mechanistic differences between global and focal brain ischaemia. Second-generation AMPA receptor antagonists may well play a role with NMDA receptor blockers in the therapy of ischaemic brain damage. Further medicinal chemistry studies to develop excitatory amino acid receptor antagonists which are potent, subtype-selective, water-soluble and which cross the blood–brain barrier are essential for continued progress.

Research into excitatory amino acid neuronal mechanisms and their roles in health and disease stands at the end of a decade of exceptional progress. Is it all over? We do not believe so. On the contrary, the successes of recent years have opened up many new possibilities for exploration at basic and pharmaceutical research levels. Receptors have been identified whose function is unknown; receptor cloning is under way and structure–function analyses will surely follow; and the roles of receptor subtypes in many physiological events, including neural development and plasticity, remain to be elucidated. In terms of excitotoxicity and neurodegeneration, major targets for the future are chronic degenerative diseases, including Alzheimer's and AIDS-related dementias. Do excitotoxic mechanisms contribute to the pathogenesis of these diseases? Whilst it seems unlikely that such mechanisms are the primary cause, recent investigations (outlined above) have provided hints in favour of a concept sketched by Choi (1988), that excitatory amino acid receptor-mediated toxicity might be operative in secondary stages of these diseases as a 'final common path' leading to neuronal death and dementia. Studies to evaluate the interactions of excitotoxic and other neurodegenerative disease mechanisms seem certain to bring reward.

REFERENCES

Agrawal, S. G. and Evans, R. H. (1986) *Br. J. Pharmacol.* **87** 345–355.

Ault, B., Evans, R. H., Francis, A. A., Oakes, D. J. and Watkins, J. C. (1980) *J. Physiol.* **307** 413–428.

Beal, M. F., Kowall, N. W., Ellison, D. W., Mazurek, M. F., Swartz, K. J. and Martin, J. B. (1986) *Nature* **321** 168–171

Ben-Ari, Y. (1985) *Neuroscience* **14** 375–403.

Benveniste, H., Drejer, J., Schousboe, A. and Diemer, N. H. (1984) *J. Neurochem.* **43** 1369–1374.

Benveniste, H., Jørgensen, M. B., Diemer, N. H. and Hansen, A. J. (1988) *Acta Neurol. Scand.* **78** 529–536.

Benveniste, H., Jørgensen, M. B., Sandberg, M., Christensen, T., Hagberg, H. and Diemer, N. H. (1989) *J. Cereb. Blood Flow Metab.* **9** 629–693.

Bielenberg, G. W. (1989) *J. Cereb. Blood Flow Metab.* **9** S298.

Brown, A. W. and Brierley, J. B. (1972) *J. Neurol. Sci.* **16** 59–84.

Buchan, A. M. (1990) *Cerebrovasc. Brain Metab. Rev.* **2** 1–26.

Buchan, A. M. and Pulsinelli, W. A. (1990) *Brain Res.* **512** 7–14.

Bullock, R., Graham, D. I., Chen, M.-H., Lowe, D. and McCulloch, J. (1990a) *J. Cereb. Blood Flow Metab.* **10** 668–674.

Bullock, R., McCulloch, J., Graham, D. I., Lowe, D., Chen, M.-H. and Teasdale, G. M. (1990b) *Stroke* **21** (Suppl. III) 32–36.

Butcher, S. P. Bullock, R., Graham, D. I. and McCulloch, J. (1990) *Stroke* **21** 1727–1733.

Carter, C., Benavides, J., Dana, C., Schoemaker, H., Perrault, G., Sanger, D. and Scatton, B. (1991). In: Meldrum, B. S. (ed.) *Excitatory Amino Acid Antagonists.* Blackwell Scientific Publications, Oxford, pp. 130–163.

Chamberlin, N. L., Traub, R. D. and Dingledine, R. (1990) *J. Neurophysiol* **64** 1000–1008.

Chapman, A. G. (1991) In: Meldrum, B. S. (ed.) *Excitatory Amino Acid Antagonists.* Blackwell Scientific Publications, Oxford, pp. 265–286.

Choi, D. W. (1987) *J. Neurosci.* **7** 369–379.

Choi, D. W. (1988) *Neuron* **1** 623–634.

Choi, D. W. (1990) *J. Neurosci.* **10** 2493–2501.

Choi, D. W. (1991). In: Meldrum, B. S. (ed.) *Excitatory Amino Acid Antagonists.* Blackwell Scientific Publications, Oxford, pp. 216–236.

Choi, D. W., Maulucci-Gedde, M. A. and Kriegstein, A. R. (1987) *J. Neurosci.* **7** 357–368.

Clifford, D. B., Olney, J. W., Benz, A. M., Fuller, T. A. and Zorumski, C. F. (1990) *Epilepsia* **31** 382–390.

Collingridge, G. L. and Lester, R. A. (1989) *Pharmacol. Rev.* **40** 143–210.

Cotman, C. W., Flatman, J. A., Ganong, A. H. and Perkins, M. N. (1986) *J. Physiol.* **378** 403–415.

Curtis, D. R., Phillis, J. W. and Watkins, J. C. (1959) *Nature* **183** 611.

Davies, J. and Watkins, J. C. (1979) *J. Physiol.* **297** 621–635.

Davies, S. W. and Roberts, P. J. (1987) *Nature* **327** 326–329.

DeBoer, T., Strof, C. and van Duijn, H. (1982) *Brain Res.* **253** 153–160.

De Boni, U. and Crapper-McLachlan, D. R. (1985) *J. Neurol. Sci.* **68** 105–118.

Deshpande, J. K., Siesjö, B. K. and Wieloch, T. (1987) *J. Cereb. Blood Flow Metab.* **7** 89–95.

Do, K. O., Herrling, P. L., Streit, P., Turski, W. A. and Cuénod, M. (1986) *J. Neurosci.* **6** 2226–2234.

Drejer, J., Benveniste, H., Diemer, N. H. and Schousboe, A. (1985) *J. Neurochem.* **45** 145–151.

Dumuis, A., Pin, J. P., Oomagari, K., Sebben, M. and Bockaert, J. (1990) *Nature* **347** 182–184.

Evans, M. C., Griffiths, T. and Meldrum, B. S. (1984) *Neuropathol. Appl. Neurobiol.* **10** 285–302.

Faden, A. J., Demedink, P., Panter, S. S. and Vink, R. (1989) *Science* **244** 798–800.

Fagg, G. E. and Foster, A. C. (1983) *Neuroscience* **4** 701–719.

Fagg, G. E., and Massieu, L. (1991). In: Meldrum, B. S. (ed.) *Excitatory Amino Acid Antagonists.* Blackwell Scientific Publications, Oxford, pp. 39–63.

Fagg, G. E., Baud, J., Pozza, M. F., Olpe, H.-R., Schmutz, M., Baumann, P., Bittiger, H., Allgeier, H., Heckendorn, R., Angst, C., Brundish, D. and Dingwall, J. G. (1990) *Br. J. Pharmacol.* **99** 791–797.

Fagg, G. E., Massieu, L., Schmutz, M., Pozza, M. F., Brugger, F. and Olpe, H.-R. (1991). In: Meldrum, B. S., Moroni, F., Simon, R. P. and Woods, J. H. (eds.) *Excitatory Amino Acids 1990.* Raven Press, New York, pp. 687–693.

Fariello, R. G., Golden, G. T., Smith, G. G. and Reyes, P. F. (1989) *Epilepsy Res.* **3** 206–213.

Ferrante, R. J., Beal, M. F., Kowall, N. W., Richardson, E. P. and Martin, J. B. (1987) *Brain Res.* **411** 162–166.

Foster, A. C., Gill, R. and Woodruff, G. N. (1988) *J. Neurosci.* **8** 4745–4754.

Frandsen, A., Drejer, J. and Schousboe, A. (1989) *J. Neurochem.* **53** 297–299.

Fujikawa, D. G. *et al.* (1991) *Neurosci. Abstracts* **17** 789.

Garthwaite, G. and Garthwaite, J. (1987) *Neurosci. Lett.* **83** 241–246.

Garthwaite, G. and Garthwaite, J. (1989) *Neurosci. Lett.* **97** 316–322.

Garthwaite, G., Hajos, F. and Garthwaite, J. (1986) *Neuroscience.* **18** 437–447.

Gill, R. and Lodge, D. (1991) *Br. J. Pharmacol.* (in press).

Gill, R., Foster, A. C. and Woodruff, G. N. (1987) *J. Neurosci.* **7** 3343–3349.

Ginsberg, M. D. and Busto, R. (1989) *Stroke* **20** 1627–1642.

Globus, M. Y.-T., Busto, R., Dietrich, W. D., Martinez, E., ValdeÆs, I. and Ginsberg, M. D. (1988) *J. Neurochem.* **51** 1455–1464.

Globus, M. Y.-T., Dietrich, W. D., Busto, R., ValdeÆs, I. and Ginsberg, M. D. (1989) *J. Cereb. Blood Flow Metab.* **9** S5.

Goldberg, M. P., Weiss, J. H., Pham, P.-C. and Choi, D. W. (1987) *J. Pharmacol. Exp. Ther.* **243** 784–791.

Goldberg, M. P., Visekul, V. and Choi, D. W. (1988) *J. Pharmacol. Exp. Ther.* **245** 1081–1087.

Gotti, B., Duverger, D., Bertin, J., Carter, C., Dupont, R., Frost, J., Gandilliere, B., Mackenzie, E. T., Rousseau, J., Scatton, B. and Wick, A. (1988) *J. Pharmacol. Exp. Ther.* **247** 1211–1221.

Greenamyre, J. T. and Young, A. B. (1989) *Neurobiol. Aging* **10** 593–602.

Griffiths, T., Evans, M. C. and Meldrum, B. S. (1984) *Neuroscience* **12** 557–567.

Guiroy, D. C., Miyazaki, M., Multhaup, G., Fischer, P., Garruto, R. M., Beyreuther, K., Masters, C. L., Simms, G., Gibbs, C. J. and Gajdusek, D. C. (1987) *Proc. Nat. Acad. Sci. USA* **84** 2073–2077.

Hagberg, H., Lehmann, A., Sandberg, M., Nystrom, B., Jacobson, I. and Hamberger, A. (1985) *J. Cereb. Blood Flow Metab.* **5** 413–419.

Hayashi, T. (1954) *Keio J. Med.* **3** 183–192.

Heyes, M. P. and Lackner, A. (1991) *J. Neurochem.* **55** 338–341.

Heyes, M. P., Rubinow, D., Lane, C. and Markey, S. P. (1989) *Ann. Neurol.* **26** 275–277.

Hillered, L., Hallström, A., Segersvärd, S., Persson, L. and Ungerstedt, U. (1989) *J. Cereb. Blood Flow Metab.* **9** 607–616.

Honoré, T., Davies, S. N., Drejer, J., Fletcher, E. J., Jacobsen, P., Lodge, D. and Nielsen, F. E. (1988) *Science* **241** 701–703.

Johansen, F. F., Jørgensen, M. B. and Diemer, N. H. (1986) *Brain Res.* **377** 344–347.

Johnston, G. A. R., Curtis, D. R., Davies, J. and McCulloch, R. M. (1974) *Nature* **248** 804–805.

Jørgensen, M. B. and Diemer, N. H. (1982) *Acta Neurol. Scand.* **66** 536–546.

Jørgensen, M. B., Johansen, F. F. and Diemer, N. H. (1987) *Acta Neuropathol.* **73** 189–194.

Jørgensen, M. B., Wright, D. C. and Diemer, N. H. (1990) *J. Cereb. Blood Flow Metab.* **10** 243–251.

Kemp, A., Foster, A. C. and Wong, E. H. F. (1987) *Trends Neurosci.* **10** 294–299.

Kilpatrick, I. C., Waller, S. J. and Evans, R. H. (1991) *Trends Pharmacol. Sci.* **12** 95–96.

Kirino, T. (1982) *Brain Res.* **239** 57–69.

Koh, J.-Y. and Choi, D. W. (1988) *J. Neurosci.* **8** 2153–2163.

Koh, J.-Y., Goldberg, M. P., Hartley, D. M. and Choi, D. W. (1990a) *J. Neurosci.* **10** 693–705.

Koh, J., Yang, L. L. and Cotman, C. W. (1990b) *Brain Res.* **533** 315–320.

Krogsgaard-Larsen, P., Honoré, T., Hansen, J. J., Curtis, D. R. and Lodge, D. (1980) *Nature* **284** 64–66.

Kurth, M. C., Weiss, J. H. and Choi, D. W. (1989) *Neurology* **39**, 217.

Lanthorn, T. H., Hood, W. F., Watson, G. B., Compton, R. P., Rader, R. K., Gaoni, Y. and Monahan, J. B. (1990) *Eur. J. Pharmacol.* **182** 397–404.

Lipton, S. A., Sucher, N. J., Kaiser, P. K. and Dreyer, E. B. (1991) *Neurology* **41** (Suppl. 1) 198.

Lodge, D. and Johnson, K. M. (1990) *Trends Pharmacol. Sci.* **11** 81–86.

Lodge, D., Aram, J. A., Church, J., Davies, S. N., Martin, D., Millar, J. and Zeman, S. (1988). In: Lodge, D. (ed.) *Excitatory Amino Acids in Health and Disease.* Wiley, Chichester, pp. 237–259.

Lowe, D. A., Neijt, H. C. and Aebischer, B. (1990) *Neurosci. Lett.* **113** 315–321.

Lucas, D. R. and Newhouse, J. P. (1957) *Arch. Opthalmol.* **58** 193–201.

MacDermott, A. B., Mayer, M. L., Westbrook, G. L., Smith, S. J. and Barker, J. L. (1986) *Nature* **321** 519–522.

Maragos, W. F., Greenamyre, J. T., Penney, J. B. Jr and Young, A. B. (1987) *Trends Neurosci.* **10** 65–68.

Martin, J. B. (1984) *Neurology* **34** 1059–1072.

Mattson, M. P. (1988) *Brain Res. Rev.* **13** 179–212.

Mattson, M. P., Guthrie, P. B., Hayes, B. C. and Kater, E. B. (1989) *J. Neurosci.* **9** 1223–1232.

Mayer, G. L. and Westbrook, G. L. (1984) *J. Physiol.* **354** 29–53.

Mayer, G. L. and Westbrook, G. L. (1987) *Prog. Neurobiol.* **28** 197–276.

McCulloch, J., Bullock, R. and Teasdale, G. M. (1991). In: Meldrum, B. S. (ed.) *Excitatory Amino Acid Antagonists.* Blackwell Scientific Publications, Oxford, pp. 287–326.

McLennan, H. and Lodge, D. (1979) *Brain Res.* **169** 83–90.

Meldrum, B. S. (1990) *Cerebrovasc. Brain Metab. Rev.* **2** 27–57.

Meldrum, B. S. and Garthwaite, J. (1990) *Trends Pharmacol. Sci.* **11** 379–387.

Monaghan, D. T., Bridges, R. J. and Cotman, C. W. (1989) *Annu. Rev. Pharmacol. Toxicol.* **29** 365–402.

Monyer, H., Goldberg, M. P. and Choi, D. W. (1989) *Brain Res.* **483** 347–354.

Nadler, J. V., Perry, B. W. and Cotman, C. W. (1978) *Nature* **271** 676–677.

Navia, B. A., Cho, E.-S., Petito, C. K. and Price, R. W. (1986) *Ann. Neurol.* **19** 517–524.

Olney, J. W. (1971) *J. Neuropathol. Exp. Neurol.* **30** 75–90.

Olney, J. W. (1978). In: McGeer, E. and Olney, J. W. (eds) *Kainic Acid as a Tool in Neurobiology.* Raven Press, New York, pp. 95–121.

Olney, J. W. (1986) *Adv. Exp. Med. Biol.* **203** 631–645.

Olney, J. W. and Sharpe, L. G. (1969) *Science* **166** 386–388.

Olney, J. W., Ho, O. L. and Rhee, V. (1971) *Exp. Brain Res.* **14** 61–76.

Onodera, H., Sato, G. and Kogure, K. (1986) *Neurosci. Lett.* **68** 169–174.

Ozyurt, E., Graham, D. I., Woodruff, G. N. and McCulloch, J. (1988) *J. Cereb. Blood Flow Metab.* **8** 757–762.

Palfreyman, M. G. and Baron, B. M. (1991). In: Meldrum, B. S. (ed.) *Excitatory Amino Acid Antagonists.* Blackwell Scientific Publications, Oxford, pp. 101–129.

Park, C. K., Nehls, D. G., Graham, D. I., Teasdale, G. M. and McCulloch, J. (1988) *Ann. Neurol.* **24** 543–551.

Patel, S., Chapman, A. G., Millan, M. H. and Meldrum, B. S. (1988). In: Lodge, D. (ed.) *Excitatory Amino Acids in Health and Disease.* Wiley, Chichester, pp. 353–378.

Pearson, R. C. A., Esiri, M. M., Hiorns, R. W., Wilcock, G. K. and Powell, T. P. S. (1985) *Proc. Nat. Acad. Sci. USA* **82** 4531–4534.

Phillis, J. W., Walter, G. A. and Simpson, R. E. (1991) *J. Neurochem.* **56** 644–450.

Pulsinelli, W. A., Brierley, J. B. and Plum, F. (1982) *Ann. Neurol.* **11** 491–498.

Reynolds, G. P., Pearson, S. J., Halket, J. and Sandler, M. (1988) *J. Neurochem.* **50** 1959–1960.

Rothman, S. M. (1984) *J. Neurosci.* **4** 1884–1891.

Rothman, S. M., Thurston, J. H. and Hanhart, R. E. (1987) *Neuroscience* **22** 471–480.

Sauer, D. Nuglisch, J., Rossberg, C., Mennel, H.-D., Beck, T., Bielenberg, G. W. and Krieglstein, J. (1988) *Neurosci. Lett.* **91** 327–332.

Sauer, D., Massieu, L., Allegrini, P. R., Amacker, H., Schmutz, M. and Fagg, G. E. (1991). In: Marangos, P. J. and Lal, H. (eds) *Emerging Strategies in Neuroprotection.* Birkhäuser Boston, Cambridge (in press).

Schoepp, D., Bockaert, J. and Sladeczek, F. (1990) *Trends Pharmacol. Sci.* **11** 508–515.

Schwarcz, R. and Meldrum, B. S. (1986) *Lancet* **ii** 140–143.

Schwarcz, R. Whetsell, W. O. and Mangano, R. M. (1983) *Science* **219** 316–318.

Schwarcz, R., Okuno, E., White, R. J., Bird, E. D. and Whetsell, W. O. (1988) *Proc. Nat. Acad. Sci. USA* **85** 4079–4081.

Sheardown, M. J., Nielsen, E. O., Hansen, A. J., Jacobsen, P. and Honoré, T. (1990) *Science* **247** 571–574.

Shiraishi, K., Sharp, F. R. and Simon, R. P. (1989) *J. Cereb. Blood Flow Metab.* **9** 765–773.

Siesjö. B. K. (1989) *Cerebrovasc. Brain Metab. Rev.* **1** 165–211.

Siesjö. B. K. and Bengtsson, F. (1989) *J. Cereb. Blood Flow Metab.* **9** 127–140.

Simon, R. and Shiraishi, K. (1990) *Ann. Neurol.* **27** 606–601.

Simon, R. P., Swan, J. H., Griffiths, T. and Meldrum, B. S. (1984a) *Science* **226** 850–852.

Simon, R. P., Griffiths, T., Evans, M. C., Swan, J. H. and Meldrum, B. S. (1984b) *J. Cereb. Blood Flow Metab.* **4** 350–361.

Sloviter, R. S. (1983) *Brain Res. Bull.* **10** 675–697.

Sloviter, R. S. and Dempsey, D. W. (1985) *Brain Res. Bull.* **15** 39–60.

Smith, M. L., Auer, R. N. and Siesjö, B. K. (1984) *Acta Neuropathol.* **64** 319–332.

Spencer, P. S., Nunn, P. B., Hugon, J., Ludolph, A. C., Ross, S. M., Roy, D. N. and Robertson, R. C. (1987) *Science* **237** 517–522.

Steinberg, G. K., Saleh, J., Kunis, D., DeLaPaz, R. and Zarnegar, S. R. (1989) *Stroke* **20** 1247–1252.

Stone, T. W. and Burton, N. R. (1988) *Prog. Neurobiol.* **30** 333–368.

Swan, J. H. and Meldrum, B. S. (1990) *J. Cereb. Blood Flow Metab.* **10** 343–351.

Tamura, A., Kirino, T., Sano, K., Tomukai, N., Hirakawa, M. and Narita, K. (1988) *J. Clin. Exp. Med.* **146** 131–132.

Tecoma, E. S., Monyer, H., Goldberg, M. P. and Choi, D. W. (1989) *Neuron* **2** 1541–1545.

Van Reempts, J. (1984) *Behav. Brain Res.* **14** 99–108.

Watkins, J. C. and Evans, R. H. (1981) *Annu. Rev. Pharmacol. Toxicol.* **21** 165–204.

Weiss, J. H., Goldberg, M. P. and Choi, D. W. (1986) *Brain Res.* **380** 186–190.

Weiss, J. H., Hartley, D. H., Koh, J. and Choi, D. W. (1990) *Science* **247** 1474–1477.

Whetsell, W. O., Jr and Schwarcz, R. (1989) *Neurosci. Lett.* **97** 271–275.

Wieloch, T. (1985) *Science* **230** 681–683.

Wieloch, T., Lindvall, O., Blomquist, P. and Gage, F. H. (1985) *Neurol. Res.*, **7** 24–26.

Wieloch, T., Gustafson, I. and Westerberg, E. (1989) *J. Cereb. Blood Flow Metab.* **9** S6.

Young, A. B., Greenamyre, J. T., Hollingsworth, Z., Albin, R., D'Amato, C., Shoulson, I. and Penney, J. B. (1988) *Science* **241** 981–983.

2

Excitatory amino acid receptors: multiplicity and ligand selectivity of receptor subtypes

Povl Krogsgaard-Larsen, Ulf Madsen, Bjarke Ebert and **Jan J. Hansen**
PharmaBiotec Research Centre, Department of Organic Chemistry, The Royal Danish School of Pharmacy, Copenhagen, Denmark

ABSTRACT

The classification of central excitatory amino acid (EAA) receptors is outlined. In the introductory section the apparent role of EAA receptors in neurodegenerative disorders, notably Alzheimer's disease, is discussed. On the basis of a simple model illustrating neuronal interactions pertinent to Alzheimer's disease, the relevance of different synaptic mechanisms as therapeutic targets in this disease is briefly discussed. The structural requirements for activation or blockade of different subtypes of EAA receptors are outlined. In the analyses of the relationship between structure and biological activity of subclasses of EAA agonists, the importance of structural, conformational, stereochemical, and steric factors are emphasized. It is concluded that the relative importance of these structural parameters for agonist potency at different EAA receptors varies considerably. The complex structure of the *N*-methyl-D-aspartate receptor is illustrated. The apparent identity/lack of identity of 2-amino-3-(3-hydroxy-5-methylisoxazol-4-yl)propanoic acid and kainic acid receptors in different tissues is discussed. Although virtually all of the compounds described have charged structures under physiological conditions they are, in most cases, depicted in the uncharged forms for the sake of clearness.

NEURODEGENERATIVE DISORDERS

Neurodegenerative disorders are characterized by progressive loss of certain types of neurons in different regions of the central nervous system (CNS) (Nappi *et al.* 1988). Whereas degeneration of dopamine neurons is a key factor in Parkinson's disease, degenerative loss of acetylcholine (ACh) and serotonin neurons is a dominating characteristic of Alzheimer's disease. In Parkinson's disease dopamine neurons in the basal ganglia are degenerating, whereas cortical and hippocampal ACh neurons are particularly vulnerable in the brains of Alzheimer patients (Wurtman *et al.* 1990).

The primary cause(s) of neurodegenerative diseases, including ischaemic and seizure-related brain damages, are far from being fully elucidated. Several factors may play important roles in the aetiology of these diseases, including free radical processes, genetic disposition and, perhaps, autoimmune mechanisms (Nappi *et al.* 1988). Studies in recent years have, however, been focused on the role of the central excitatory amino acid (EAA) neurotransmitters, (*S*)-glutamic acid (Glu) and (*S*)-aspartic acid (Asp), in the processes causing neuron injury and, ultimately, death (Rothman and Olney 1986, Lodge 1988, Bridges *et al.* 1988, Greenamyre and Young 1989) (see chapter 16). The view that hyperactivity of central Glu neurons is an important causative factor in neurodegenerative processes ('excitotoxicity') (Fig. 1) is supported by *in vitro* and *in vivo* studies in a variety of model systems (Lodge 1988) (see chapter 1). Thus, regioselective excitotoxic lesions in animal brains by Glu and other EAAs cause patterns of neurodegenerations similar to those observed in brains of patients suffering from neurodegenerative diseases.

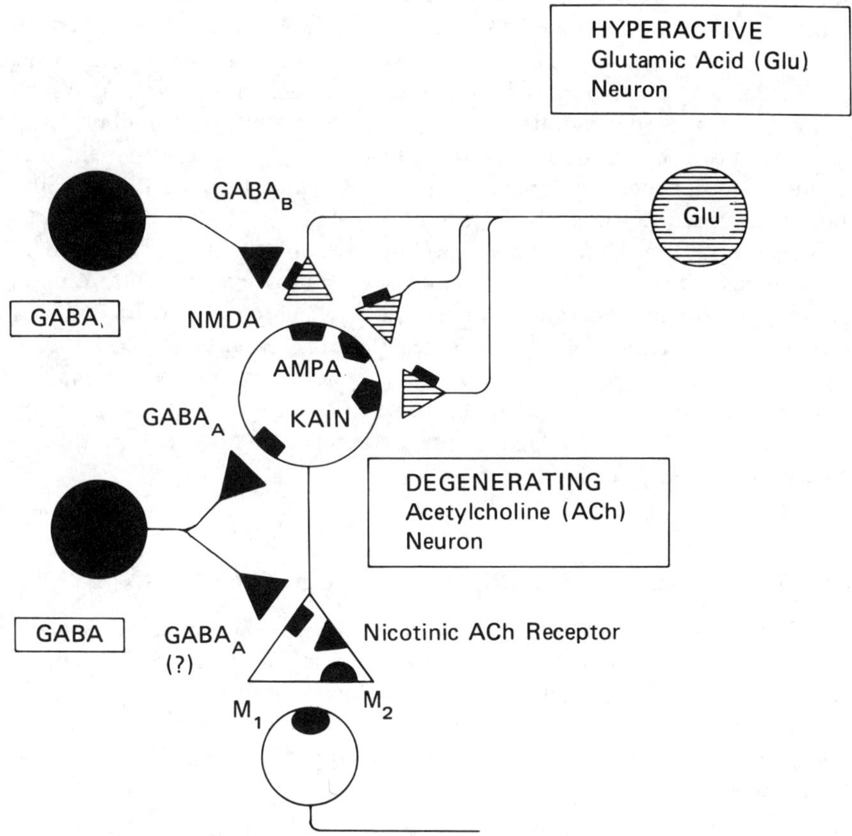

Fig. 1 — Schematic illustration of synaptic contacts between central acetylcholine (ACh), (*S*)-glutamic acid (Glu), and 4-aminobutanoic acid (GABA) neurons and their receptors.

SYNAPTIC MECHANISMS AS THERAPEUTIC TARGETS IN ALZHEIMER'S DISEASE

In the progression of Alzheimer's disease, loss of Glu neurons is observed in addition to loss of ACh and serotonin neurons (Greenamyre and Young 1989). Thus, hyperactive as well as hypoactive EAA neuronal mechanisms may be operative in Alzheimer's disease. The hypoactivity of central EAA neuronal pathways has recently been proposed to be associated with learning and memory deficits (Greenamyre and Young 1989, Bowen 1990), and there is some evidence that hypoactivity of central EAA system(s) may also be a key factor in schizophrenia (Deutsch *et al.* 1989).

Since hyperactivity as well as hypoactivity of EAA neuronal systems may be implicated in, for example, Alzheimer's disease, drugs capable of both protecting and activating EAA receptors may be of therapeutic interest. Thus, an obvious challenge is to design EAA antagonists or, perhaps, partial agonists with appropriately balanced agonist/antagonist profiles (Christensen *et al.* 1989, Huettner 1991).

Normal function of ACh neurons appears to be dependent on stimulation by Glu neurons (Fig. 1), and, therefore, it may be difficult to prevent Glu receptor-mediated degeneration of ACh neurons by Glu antagonists without reducing or blocking the excitatory input from Glu terminals to these neurons. Consequently, future therapies in Alzheimer's disease based on Glu antagonists probably have to be supplemented by concomitant treatment with appropriate ACh agonists, possibly muscarinic M_1 receptor agonists, as an attempt to maintain the function of muscarinic ACh synapses. Thus, agonists or partial agonists at M_1 receptors may be of major therapeutic interest, not only as drugs for symptomatic treatment of Alzheimer patients but also as essential components of neuroprotective treatments of such patients with EAA receptor antagonists (Krogsgaard-Larsen *et al.* 1989).

Other synaptic mechanisms may also be relevant therapeutic targets in Alzheimer's disease (Fig. 1). Thus, antagonists at muscarinic M_2 autoreceptors or agonists at presynaptic nicotinic ACh receptors, which appear to be involved in a positive feedback control of ACh release (Lapchak *et al.* 1989), may be future therapeutic agents in Alzheimer's disease. Furthermore, appropriate pharmacological manipulation of $GABA_A$ or $GABA_B$ receptors in order to stimulate ACh release or reduce Glu release, respectively, may have therapeutic prospects in this disease (Krogsgaard-Larsen *et al.* 1989).

EXCITATORY AMINO ACID RECEPTOR SUBTYPES — AN OVERVIEW

Until a few years ago, EAAs were thought to mediate their actions through three different classes of receptors (Watkins and Evans 1981, McLennan 1983) (Fig. 2). As the result of extensive neurochemical, pharmacological and, in recent years, molecular biological studies, central EAA receptors are now most conveniently subdivided into five main classes, some of which, if not all, are heterogeneous (Fig. 3) (Monaghan *et al.* 1989, Watkins *et al.* 1990, Hansen and Krogsgaard-Larsen 1990):

(1) NMDA receptors at which *N*-methyl-(*R*)-aspartic acid (NMDA), quinolinic acid (QUIN), and ibotenic acid (see Fig. 5) are selective agonists. NMDA receptors are competitively and very effectively blocked by a number of phosphono amino

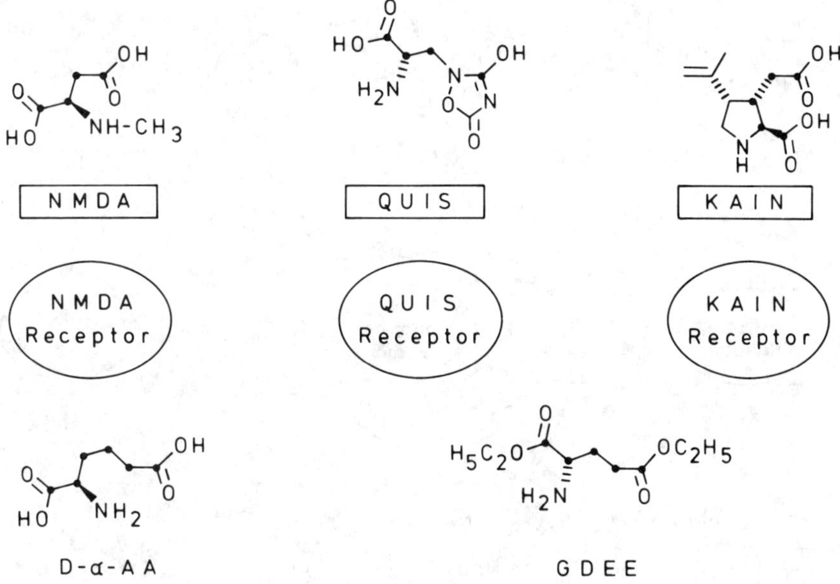

Fig. 2— Schematic illustration of the earlier classification of central EAA receptors based on the EAA agonists NMDA, QUIS, and KAIN and the antagonists D-α-aminoadipic acid (D-α-AA) and Glu diethyl ester (GDEE) showing some selectivity for NMDA and QUIS receptors, respectively.

acids, notably (*R*)-2-amino-5-phosphonopentanoic acid (D-AP5) (Fig. 3), (*RS*)-3-(2-carboxypiperazin-4-yl)propyl-1-phosphonic acid (CPP), NPC 451, and the piperidine analogue of D-AP5, CGS 19755 (Fig. 4) (Watkins and Collingridge 1989, Watkins *et al.* 1990) (see chapter 8). Evidence derived from electrophysiological and receptor binding studies strongly suggests heterogeneity of NMDA receptors (Monaghan *et al.* 1988, Young and Fagg 1990) (see chapter 6). Thus, NMDA receptors in the brain and the spinal cord show different agonist sensitivities (Lodge 1988). Furthermore, NMDA receptors in different brain regions show dissimilar pharmacological profiles consistent with the view that several isoforms of NMDA receptors may exist in the mammalian CNS (Stone and Burton 1988, Monaghan and Cotman 1989).

(2) AMPA receptors at which (*S*)-quisqualic acid (QUIS) is a non-selective and (*RS*)-2-amino-3-(3-hydroxy-5-methylisoxazol-4-yl)propanoic acid (AMPA) a highly selective agonist (Krogsgaard-Larsen *et al.* 1980, Krogsgaard-Larsen and Honoré 1983). A number of quinoxalinediones, notably 6-cyano-7-nitroquinoxaline-2,3-dione (CNQX) and 6,7-dinitroquinoxaline-2,3-dione (DNQX), are very potent, but non-selective, antagonists at AMPA receptors (Fig. 3) (Honoré *et al.* 1988, 1989) (see chapter 11). More recently, (*RS*)-2-amino-3-[3-(carboxymethoxy)-5-methylisoxazol-4-yl]propanoic acid (AMOA) has been shown to antagonize AMPA-induced excitation, showing less effect on excitations by QUIS or kainic acid (KAIN) (Krogsgaard-Larsen *et al.* 1991) (see Fig. 17).

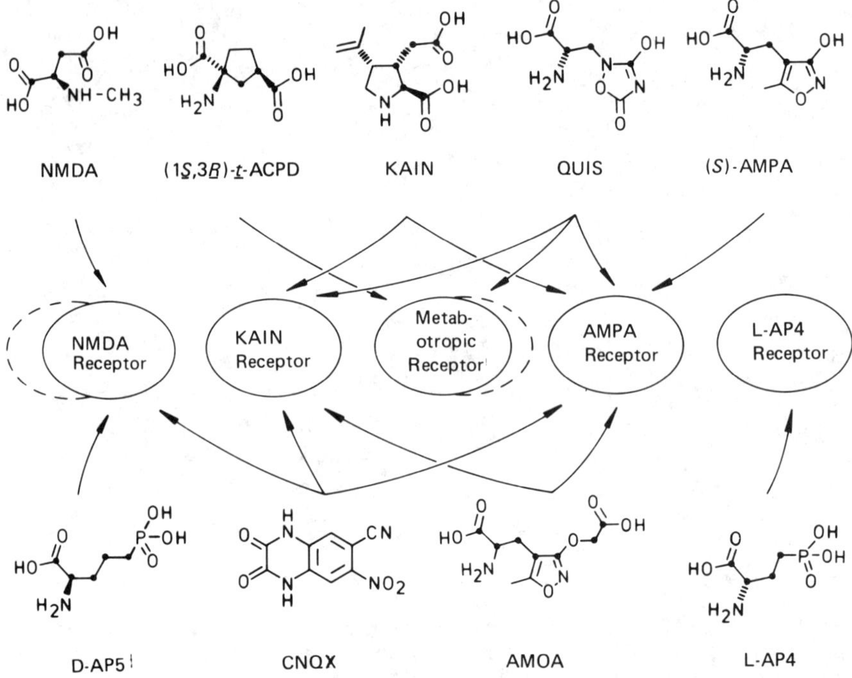

Fig. 3 — Schematic illustration of the multiplicity of central EAA receptors and the sites of action of a number of agonists (upper row) and antagonists (bottom row).

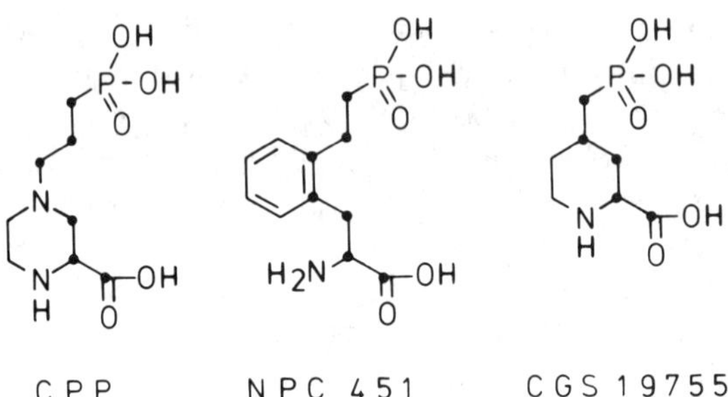

Fig. 4 — Structures of some NMDA antagonists.

(3) KAIN receptors, which are selectively activated by KAIN (Coyle 1983). A number of other naturally occurring amino acids show very potent KAIN agonist actions (Shinozaki 1988) (see later section and chapter 12). The 3-isoxazolol amino acid (*RS*)-2-amino-3-[2-(3-hydroxy-5-methylisoxazol-4-yl)methyl-5-methyl-3-oxoisoxazolin-4-yl]propanoic acid (AMNH), which is structurally related to AMOA (see Fig. 16), has recently been shown to be a selective, but relatively weak, KAIN receptor antagonist (Krogsgaard-Larsen *et al.* 1991).

(4) Recent investigations by a number of research groups have disclosed that, in addition to ion channel-linked receptors mediating fast depolarizing responses in the CNS, an EAA receptor, named the metabotropic receptor, exists that is coupled to inositol 1,4,5-trisphosphate (IP_3) and diacylglycerate turnover (Sladeczek *et al.* 1985, Monaghan *et al.* 1989, Schoepp and Johnson 1988). Whereas QUIS and ibotenic acid are non-selective agonists at this receptor, the former compound being the most active agonist so far described, the *trans* form of 1-aminocyclopentane-1,3-dicarboxylate (*t*-ACPD), notably the (1*S*,3*R*)-enantiomer (Fig. 5), appears to be a selective agonist at the metabotropic receptor (Watson *et al.* 1990, Schoepp *et al.* 1990). There is some evidence to suggest that two metabotropic receptors exist (Manzoni *et al.* 1990). Whereas neither receptor subtype is activated by AMPA, one is reported to be more sensitive to QUIS, whereas the other one is preferentially activated by ibotenic acid (Manzoni *et al.* 1990). None of the EAA receptor antagonists so far described interact specifically with the metabotropic receptor(s) (Watkins *et al.* 1990). However, (*R*)-2-amino-3-phosphonopropanoic acid (L-AP3) has recently been shown to reduce markedly the response to (1*S*,3*R*)-*t*-ACPD (Irving *et al.* 1990) (see chapter 15).

(5) L-AP4 receptors, through which (*S*)-2-amino-4-phosphonobutanoic acid (L-AP4) inhibits synaptic excitation (Johnson and Koerner 1988) (see chapter 14).

THE NMDA RECEPTOR COMPLEX

The NMDA receptor is actually a receptor complex coupled to ion channel(s) which, in contrast to those associated with AMPA or KAIN receptors, show a voltage-dependent blockade by magnesium (Lodge 1988, Barnard and Costa 1989, Watkins and Collingridge 1989, Reynolds 1990) (Fig. 6). Following membrane depolarization,

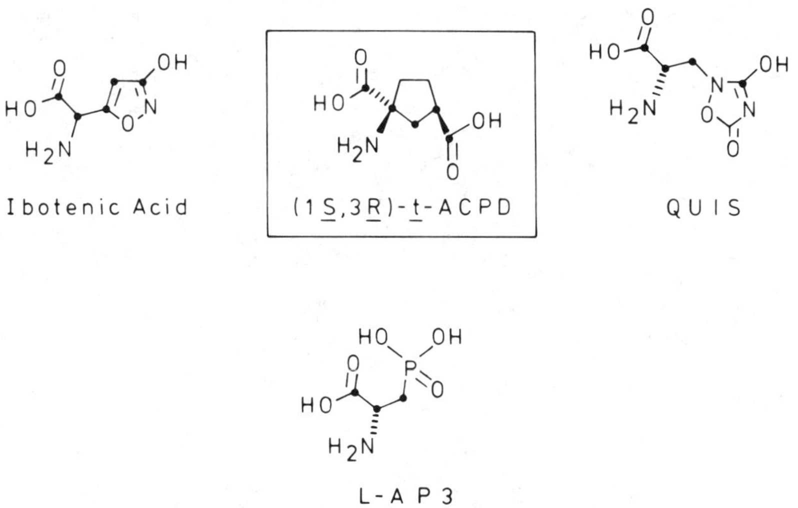

Ibotenic Acid

(1S,3R)-t-ACPD

QUIS

L-AP3

Fig. 5 — Structures of the selective metabotropic EAA receptor agonist (1*S*,3*R*)-*t*-APCD and the non-selective agonists ibotenic acid and QUIS. L-AP3 is an antagonist at this class of EAA receptors.

for example via activation of AMPA or KAIN receptors in the cell membrane, the magnesium blockade is released, causing further depolarization. Thus, the NMDA receptor may function as an amplification system capable of generating long-lasting physiological changes, and it has been associated with the phenomenon of long-term potentiation (LTP), which is a sustained increase in synaptic sensitivity following high-frequency stimulation (Collingridge and Singer 1990). LTP can last for weeks and may play a key role in memory and learning processes. This component of synaptic plasticity mediated by NMDA receptors may, however, also be involved in the processes causing neuronal damage and epileptiform activity. These aspects have brought the NMDA receptor complex into focus as a potential target for therapeutic attack in certain neurological diseases, including Alzheimer's disease (Monaghan *et al.* 1989) (see chapter 9).

The NMDA receptor complex also contains distinct binding sites for zinc ions, for polyamines, and for glycine, which are closely associated with the ion channel of this complex (Fig. 6) (Lodge and Johnson 1990, Chadwick and Choi 1990) (see chapter 9). The physiological roles of these additional sites are unknown, but activation of the glycine site, which is distinctly different from the strychnine-sensitive glycine receptor, strongly potentiates the responses to NMDA via allosteric/co-agonist interaction with the NMDA receptor site (Lodge and Johnson 1990, Huettner 1991).

The dissociative anaesthetics, such as ketamine and phencyclidine (PCP), selectively inhibit NMDA-associated excitatory effects without affecting excitation induced by agonists at AMPA, KAIN or metabotropic receptors (Lodge 1989, Lodge and Johnson 1990). The binding site for another non-competitive antagonist, MK-801, is identical to or closely associated with that for PCP. In analogy with the findings for the blockade by magnesium ions of the NMDA responses, PCP, MK-801, and other dissociative anaesthetics block NMDA responses in a voltage-dependent manner, indicating a close association between the binding sites concerned and the ion channel (Fig. 6) (Kemp *et al.* 1987). Furthermore, the blockade of the binding site(s) for PCP

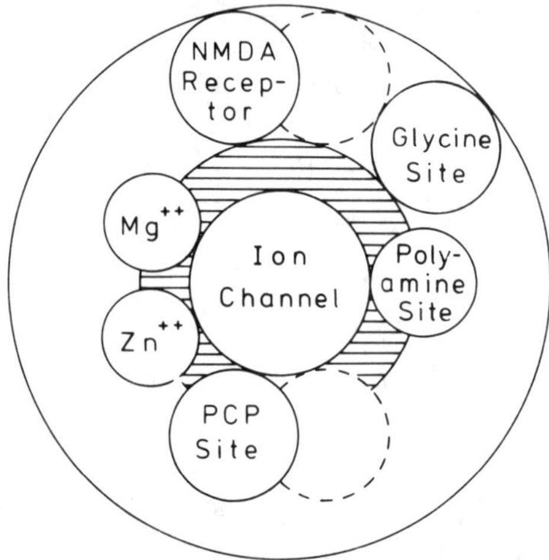

Fig. 6 — Schematic illustration of the NMDA receptor complex.

and MK-801 is use-dependent. Thus, a prerequisite for these antagonist effects is the presence of an NMDA receptor agonist, suggesting that opening of the ion channel uncovers these binding sites (Kemp *et al.* 1987).

In principle, compounds like MK-801 would be almost ideal drugs for treatment of neurodegenerative diseases caused by hyperactivation of NMDA receptors (Fig. 1). The use dependence of the antagonist effects of such compounds suggests that, for example, MK-801 would show limited effect on the EAA neurotransmission under normal conditions. Excessive release of Glu or other endogenous EAAs and, consequently, overstimulation of NMDA receptors would, however, be expected to disclose the binding site(s) for these non-competitive antagonists in the fully open ion channels. Unfortunately, psychotomimetic side-effects of this class of pharmacological agents may severely restrict their use in the human clinic (Watkins and Collingridge 1989).

NMDA, Glu AND SOME CYCLIC ANALOGUES AS NMDA AGONISTS

NMDA itself is the standard agonist for studies of the pharmacology of NMDA receptors (Watkins and Colingridge 1989, Hansen and Krogsgaard-Larsen 1990). However, in spite of its high potency, NMDA binds to NMDA receptor sites with very low affinity, and attempts to use radioactive NMDA as a ligand for binding studies have been unsuccessful (Lodge 1988, Watkins and Collingridge 1989).

From a structure–activity point of view, NMDA is unique (Fig. 7). Not only is NMDA the sole *N*-methylated acidic amino acid showing an NMDA receptor affinity similar to that of its unmethylated form, but it is also the only example of a D-amino acid, which is markedly more potent as an agonist than the corresponding L-form. Thus, (*R*)-Glu, *N*-methyl-Glu, and, in particular, *N*-methyl-(*R*)-Glu interact much less potently with NMDA receptor sites than does Glu (Watkins and Collingridge 1989). This structure–activity profile of NMDA is also different from that of the potent and highly selective NMDA agonist, AMAA (see subsequent section). These observations may reflect that NMDA does not interact directly with the NMDA recognition site of the receptor complex (Fig. 6). The precise mechanism of action of NMDA probably remains to be established.

The acidic amino acids piperidine-2,3-dicarboxylic acid (2,3-PDA) and 2,4-PDA are cyclic analogues of Asp/NMDA and Glu, respectively (Fig. 8). Whereas the *trans* forms of 2,3-PDA and 2,4-PDA are NMDA receptor agonists, *cis*-2,3-PDA and *cis*-2,4-PDA show NMDA antagonist properties (Watkins and Collingridge 1989). Using

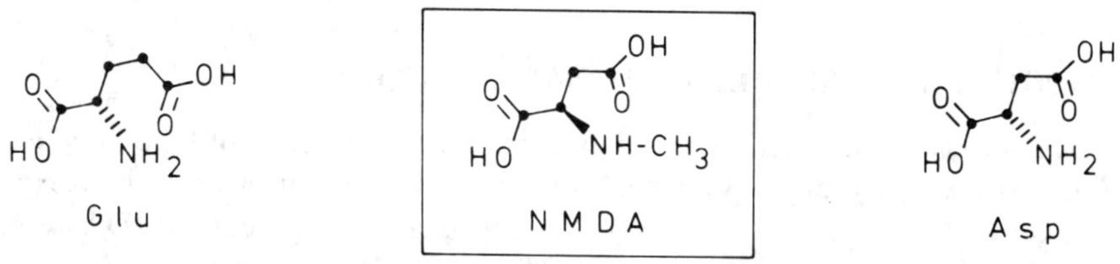

Fig. 7 — Structures of Glu, NMDA, and Asp.

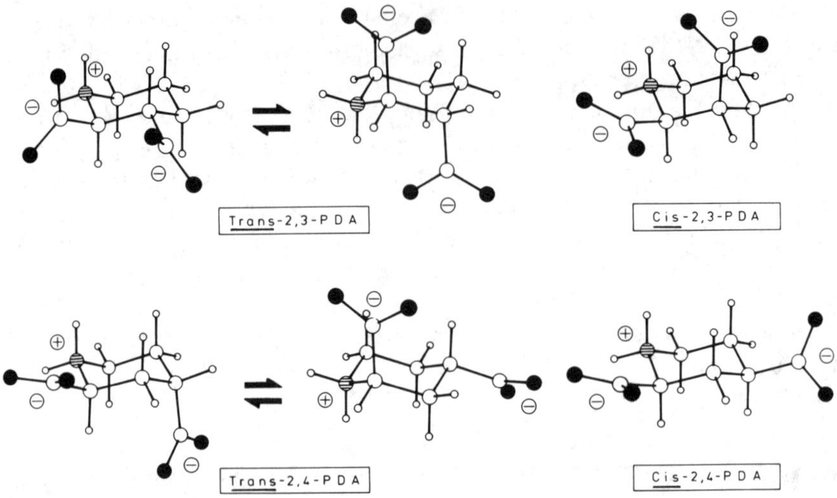

Fig. 8 — Approximate conformations in aqueous solutions of the NMDA antagonists *cis*-2,3-PDA and *cis*-2,4-PDA, and an illustration of the conformational mobility in aqueous solution of the NMDA agonists *trans*-2,3-PDA and *trans*-2,4-PDA. The conformational analyses were performed at pH 7.4 using ^{1}H-NMR spectroscopic techniques, and correlated with molecular mechanics calculations.

^{1}H-NMR spectroscopy and molecular mechanics calculations, the relationship between structure, conformational flexibility, and effects on NMDA receptors of these cyclic amino acids has been studied. Whilst the NMDA antagonist *cis*-2,4-PDA was shown to exist almost exclusively in one conformation in aqueous solution, the NMDA agonist *trans*-2,4-PDA exists in aqueous solution as an equilibrium mixture of two conformations. Similarly, only one conformation of the NMDA antagonist *cis*-2,3-PDA could be detected in aqueous solution, whereas the NMDA agonist *trans*-2,3-PDA, like *trans*-2,4-PDA, exists as an equilibrium mixture of two conformations (Fig. 8). The *cis* form of 2,5-PDA, which shows weak NMDA agonist effects, also exists as a mixture of two conformations in water (Madsen *et al.* 1990a).

These observations seem to indicate that a certain degree of conformational flexibility of Glu or Asp/NMDA analogues is an important, though perhaps not strictly necessary, condition for agonist activity at NMDA receptors. Alternatively, the active NMDA agonist conformation(s) are not accessible to, for example, *cis*-2,3-PDA or *cis*-2,4-PDA, which show NMDA antagonist profiles.

IBOTENIC ACID AND RELATED NMDA AGONISTS

The *Amanita muscaria* constituent ibotenic acid is an analogue of Glu, in which the terminal carboxylate group of Glu has been replaced by a bioisosteric 3-hydroxyisoxazole group (Fig. 9) (see chapter 10). Ibotenic acid is a potent NMDA receptor agonist, but it also interacts with AMPA and KAIN receptor sites and with metabotropic receptors (Krogsgaard-Larsen *et al.* 1980, Schoepp *et al.* 1990).

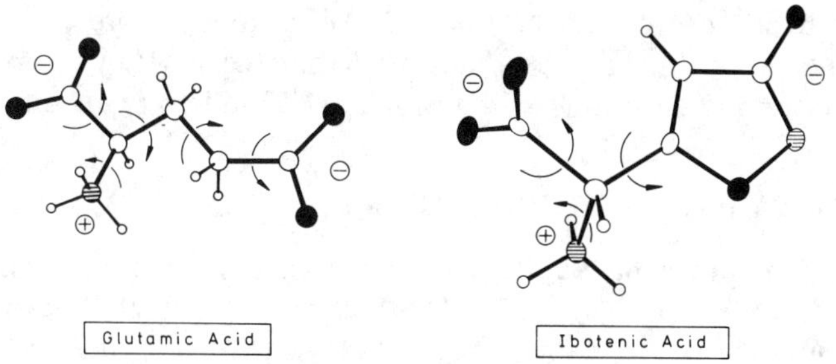

Glutamic Acid

Ibotenic Acid

Fig. 9 — Structures and relative conformational mobilities of the fully charged molecules of Glu and ibotenic acid.

Ibotenic acid is chemically unstable (Nielsen *et al.* 1985), and in an attempt to develop chemically stable compounds showing potent and specific NMDA agonist effects, a number of ibotenic acid analogues structurally related to NMDA have been synthesized and tested, including (*RS*)-2-amino-2-(3-hydroxy-5-methylisoxazol-4-yl)acetic acid (AMAA) (Fig. 10) and related mono- and bicyclic analogues (Madsen *et al.* 1990b). None of these compounds showed detectable affinity for AMPA, high- and low-affinity KAIN, or strychnine-insensitive glycine receptor sites (Fig. 6), and AMAA was shown to be more potent than NMDA as an NMDA agonist. *N*-Methylation of AMAA or cyclization to give (*RS*)-3-hydroxy-4,5,6,7-tetrahydroisoxazolo[4,5-*c*]pyridine-4-carboxylic acid (4-HPCA) were accompanied by a marked loss of activity, while *N*,*N*-dimethylation (Fig. 10) resulted in almost complete loss of activity. The neurotoxic effects of AMAA after striatal injections in rats are more

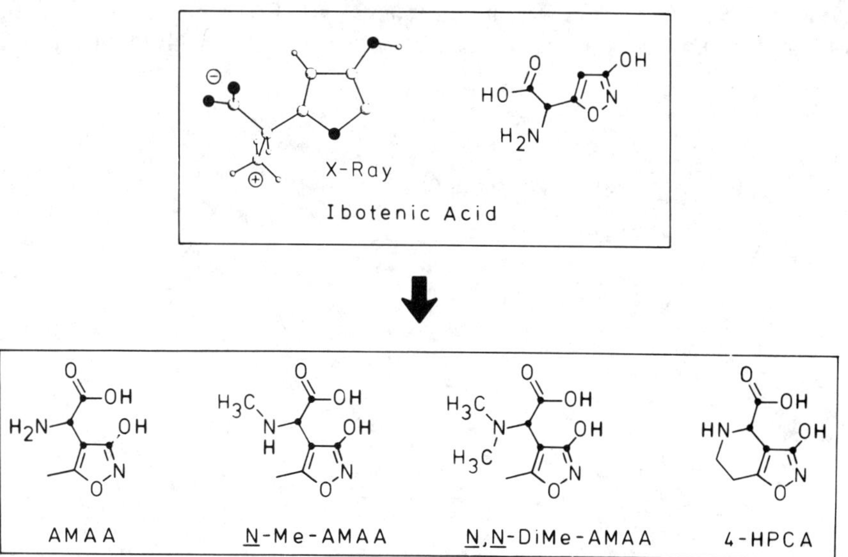

Fig. 10 — Structures of ibotenic acid and the specific NMDA agonist AMAA, an analogue of ibotenic acid related to Asp. The structures of some analogues of AMAA are shown.

pronounced than those of KAIN, and AMAA is much more toxic than QUIN (Madsen *et al.* 1990b). It has been suggested that the NMDA receptor affinity of AMAA resides primarily in the (*S*)-enantiomer (U. Madsen *et al.*, unpublished).

AMPA AND KAINIC ACID RECEPTORS — COMPLEX INTERACTIONS

Receptor binding studies have revealed the existence of both high- and low-affinity AMPA and KAIN receptor sites (Fig. 11) (Honoré *et al.* 1989, Watkins *et al.* 1990). The equilibrium between the two stages of AMPA receptors is facilitated by chaotropic isothiocyanate ions. Calcium ions, on the other hand, somehow block high-affinity KAIN receptor sites (Braitman and Coyle 1987, Krogsgaard-Larsen *et al.* 1991), but the underlying mechanism is not clear. The high-affinity binding sites for [^{3}H]AMPA and [^{3}H]KAIN in the mammalian CNS show different ligand binding characteristics (Krogsgaard-Larsen *et al.* 1980), but there is evidence to suggest that the low-affinity stages of these receptors are functionally important and that AMPA receptors mediate at least some of the excitatory effects of KAIN (Honoré 1989) (see chapters 10 and 11).

Although high-affinity AMPA and KAIN receptor sites show very different pharmacological characteristics, the distinction between AMPA and KAIN receptors is unclear (see chapters 3 and 10). Functional expression of four cloned cDNAs encoding EAA receptors in cultured mammalian neurons generated receptors showing AMPA-selective binding pharmacology (Keinänen *et al.* 1990). On the other hand, KAIN-evoked currents in these cells were markedly reduced not only by CNQX but also by AMPA. There is, however, other evidence to suggest that the agonist effects of AMPA and its inhibition of KAIN responses are mediated by different sites (Sugiyama *et al.* 1989).

In cerebellar granule cell cultures KAIN was shown to cause extensive cell death (Kato *et al.* 1991). These cells were not affected by AMPA or QUIS, but these EAAs, as well as DNQX, antagonized the neurotoxic effects of KAIN. Similarly, AMPA

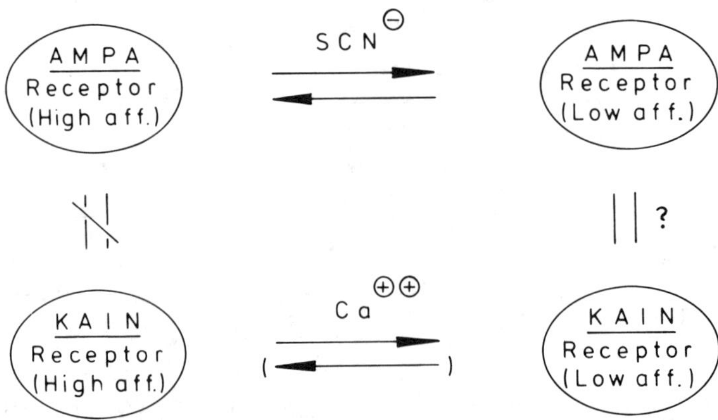

Fig. 11 — Apparent relationships between high- and low-affinity sites or stages of AMPA and KAIN receptors.

(Rassendren *et al.* 1989) and QUIS (Perouansky and Grantyn 1989) have been shown to inhibit responses by KAIN in other test systems. It was concluded (Perouansky and Grantyn 1989) that QUIS acts as a low-affinity competitive antagonist at the KAIN site and as a high-affinity agonist at its own receptor(s). These authors hypothesized that several EAA receptors could converge to one and the same protein complex capable of promoting different conduction states of the common ionic channel, in agreement with earlier observations (Ascher and Novak 1988). The pharmacology of KAIN and AMPA binding sites, co-purified from the *Xenopus* CNS, suggests that KAIN and AMPA recognize a single receptor protein or a protein complex, in which the two binding sites are identical or strongly interact (Henley *et al.* 1989, Ambrosini *et al.* 1990).

The nature and degree of interactions between AMPA and KAIN receptors in different parts of the CNS, as detected under different experimental conditions, are still far from being fully elucidated. Based on the pieces of information so far available, it may be postulated that non-NMDA receptors are hetero-oligomeric complexes comprising subunits with preferential affinities for the EAAs AMPA, KAIN, or QUIS and the non-NMDA antagonists CNQX or AMOA (Fig. 12). It is possible that the subunit composition of such receptors varies markedly in different parts of the CNS,

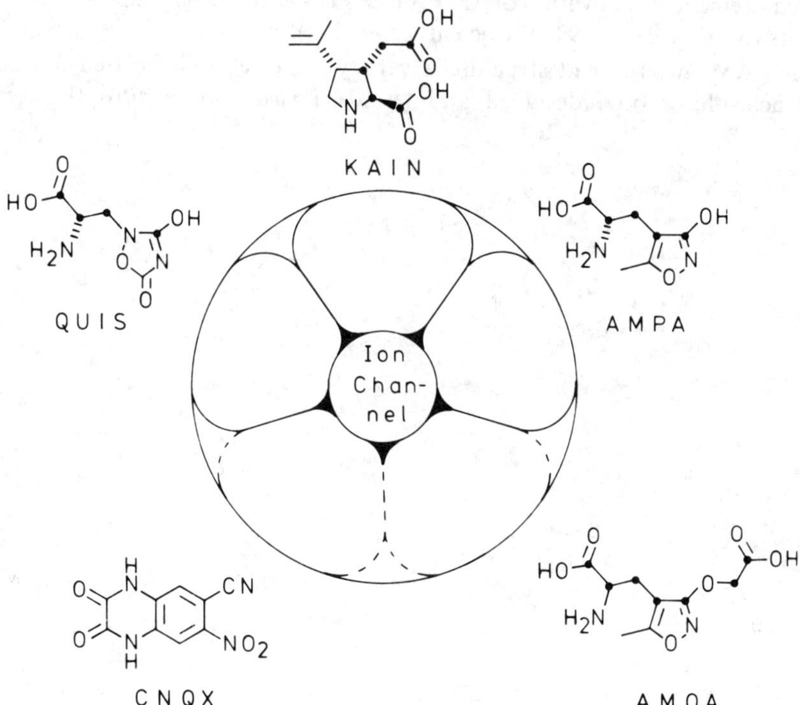

Fig. 12 — Hypothetical structure of a hetero-oligomeric non-NMDA receptor complex containing subunits with preferential affinities for the EAA agonists QUIS, KAIN, or AMPA and the antagonists CNQX or AMOA.

which could explain the dissimilar EAA sensitivity of different types of neurons and the tissue- and neuron-selective antagonist sensitivity of EAA responses. One or more of such receptor subunits may constitute the modulatory unit proposed to exist at AMPA receptors (Honoré 1989).

AMPA RECEPTOR AGONISTS AND PARTIAL AGONISTS

AMPA is a highly selective and very potent agonist at AMPA receptors, and radiolabelled AMPA has become the standard ligand for studies of this subtype of EAA receptors (Honoré *et al.* 1982, Murphy *et al.* 1987). Likewise (*RS*)-3-hydroxy-4,5,6,7-tetrahydroisoxazolo[5,4-*c*]pyridine-7-carboxylic acid (7-HPCA), a bicyclic analogue of ibotenic acid, and the isomeric bicyclic amino acid (*RS*)-3-hydroxy-4,5,6,7-tetrahydroisoxazolo[5,4-*c*]pyridine-5-carboxylic acid (5-HPCA), which is a conformationally restricted analogue of AMPA (Fig. 13), are selective agonists at AMPA receptors (Krogsgaard-Larsen *et al.* 1985) (see chapter 10). The preferred conformations of 5-HPCA in solution, as determined by ^{1}H-NMR spectroscopy (Nielsen *et al.* 1986), and of 7-HPCA in the solid state, as determined by X-ray crystallography (Krogsgaard-Larsen *et al.* 1985), are very similar: both of the compounds adopt almost planar conformations with the carboxylate groups in equatorial and pseudo-equatorial positions, respectively. The relatively rigid strutures of 5- and 7-HPCA suggest that a high degree of conformational mobility of agonists is not a requirement for activation of AMPA receptors as has been proposed for NMDA receptors (Madsen *et al.* 1990a) (see earlier section).

Using AMPA as a lead structure a variety of structurally related 3-isoxazolol amino acids have been designed and tested *in vivo* and *in vitro* (Hansen and

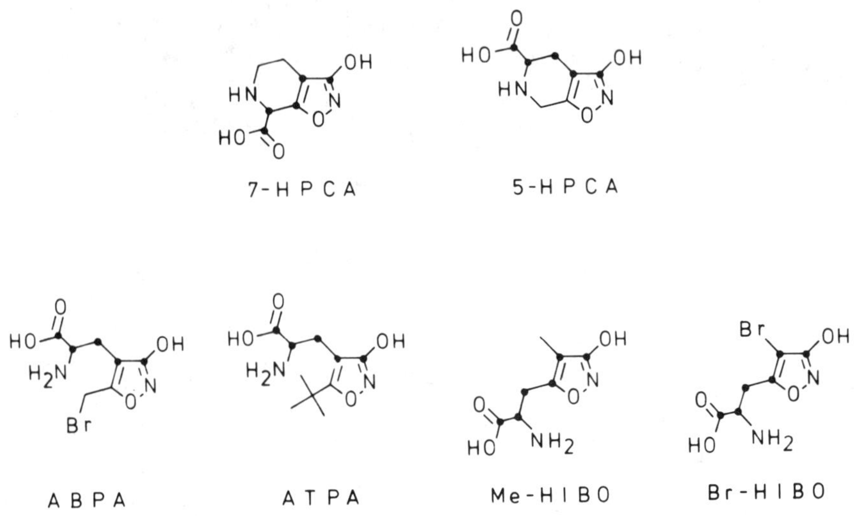

Fig. 13 — Structures of the specific AMPA receptor agonists 7-HPCA, 5-HPCA, ABPA, ATPA, (*RS*)-2-amino-3-(3-hydroxy-4-methylisoxazol-5-yl)propanoic acid (Me-HIBO), and (*RS*)-2-amino-3-(3-hydroxy-4-bromoisoxazol-5-yl)propanoic acid (Br-HIBO).

Krogsgaard-Larsen 1990). The *tert*-butyl analogue of AMPA, (*RS*)-2-amino-3-(3-hydroxy-5-*tert*-butylisoxazol-4-yl)propanoic acid (ATPA) (Fig. 13), is an agonist at AMPA receptors somewhat weaker than AMPA (Lauridsen *et al.* 1985) (Fig. 14). Interestingly, replacement of the bulky and spherical *tert*-butyl group of ATPA by the planar phenyl group affords a compound, (*RS*)-2-amino-3-(3-hydroxy-5-phenylisoxazol-4-yl)propanoic acid (APPA), which shows markedly lower potency and efficacy than ATPA as an AMPA receptor agonist (Fig. 14). APPA actually appears to be the first example of a partial agonist at AMPA receptors (Christensen *et al.* 1989).

Structure–activity studies on these and related AMPA analogues have led to a hypothetical model for the binding site of the AMPA receptor capable of accommodating bulky substituents of agonist molecules (Fig. 15). This hypothetical 'cavity' at the AMPA receptor site does not seem to be occupied during the binding of 5-HPCA.

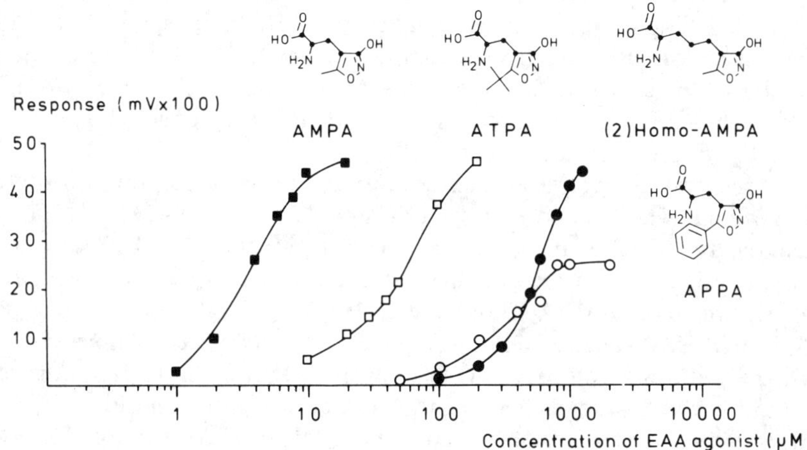

Fig. 14 — Excitatory responses (normalized) of AMPA, ATPA, (*RS*)-2-amino-5-(3-hydroxy-5-methylisoxazol-4-yl)pentanoic acid ((2)homo-AMPA), and APPA in the rat cortical slice preparation. (Data from Christensen *et al.* 1989.)

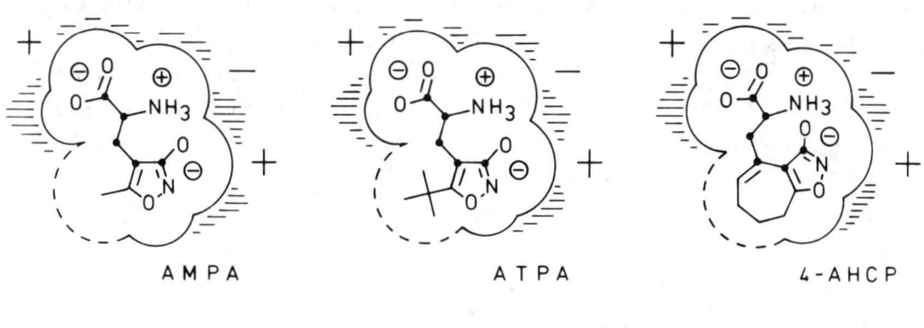

Fig. 15 — Illustration of the binding of the specific AMPA receptor agonists AMPA, ATPA, and 4-AHCP to a hypothetical model of the AMPA receptor assumed to contain a lipophilic 'cavity' capable of accommodating bulky substituents of the agonist molecules.

Structure–activity studies on (*RS*)-2-amino-3-(3-hydroxy-5-bromomethylisoxazol-4-yl)propanoic acid (ABPA) (Fig. 13), which is virtually identical with AMPA in terms of potency and specificity as an AMPA receptor agonist (Krogsgaard-Larsen *et al.* 1985), may support the existence of a 'cavity' at the AMPA receptor. Thus, in spite of the presence of a chemically reactive bromomethyl group in the molecule of ABPA, this AMPA agonist does not interact irreversibly with AMPA receptors *in vivo* or *in vitro* (Krogsgaard-Larsen *et al.* 1985; Nielsen *et al.* 1986). This inability of ABPA to alkylate nucleophilic groups at the AMPA receptor site may reflect that the reactive bromomethyl group fits into this 'cavity' during the receptor binding of ABPA.

The bicyclic homologue of AMPA, (*RS*)-2-amino-3-(3-hydroxy-7,8-dihydro-6*H*-cyclohepta[1,2-*d*]isoxazol-4-yl)propanoic acid (4-AHCP) (Fig. 15), has been shown to be an extremely potent agonist at AMPA receptors in the cat spinal cord (Krogsgaard-Larsen *et al.* 1984), and it binds to AMPA receptor sites in rat cortical membranes with an affinity comparable with that of AMPA (Lund *et al.* 1991). Molecular modelling studies have shown that 4-AHCP, in spite of its bicyclic structure, is a conformationally flexible compound, which can adopt a broad range of conformations. We envisage that the bulky seven-membered ring of 4-AHCP is embraced by the 'cavity', assumed to exist at the AMPA receptor site (Fig. 15), during its binding to this receptor (see chapter 10).

The stereochemical requirements for activation of AMPA receptors are at present being mapped out (Hansen and Krogsgaard-Larsen 1990). Like AMPA, 4-bromohomoibotenic acid (Br-HIBO) and 4-methylhomoibotenic acid (Me-HIBO) (Fig. 13) are potent and selective AMPA receptor agonists (Krogsgaard-Larsen *et al.* 1980). The (*S*)- and (*R*)-isomers of AMPA (Hansen *et al.* 1983), Br-HIBO (Hansen *et al.* 1989), and Me-HIBO (J. J. Hansen *et al.*, unpublished) have been synthesized using enzymatic procedures. In all cases, the GDEE- or CNQX-sensitive excitatory effects reside almost exclusively in the (*S*)-enantiomers, and there is a positive correlation between potency as neuroexcitants and affinity for AMPA receptor sites of the enantiomers of AMPA, Br-HIBO and Me-HIBO (see chapter 10).

AMINO ACID AMPA RECEPTOR ANTAGONISTS

The specificity of (*S*)-AMPA, (*S*)-Br-HIBO, and related heterocyclic Glu analogues is assumed to be associated with the rigid and planar structure of the acidic 3-isoxazolol unit of these compounds (Fig. 16) as established by X-ray crystallography (Krogsgaard-Larsen *et al.* 1985, Hansen *et al.* 1989, Brehm *et al.* 1988, 1990). These structural characteristics of AMPA and Br-HIBO are only to a limited extent shared by the structurally related heterocyclic amino acid QUIS (Jackson *et al.* 1989). The conformational flexibility and lack of coplanarity of the heterocyclic ring of QUIS may be of some importance for the ability of this amino acid to interact not only with AMPA receptors, but also with KAIN and metabotropic receptors (Fig. 3) and with EAA-associated enzyme and transport sites (Monaghan *et al.* 1989, Hansen and Krogsgaard-Larsen 1990, Watkins *et al.* 1990).

Using AMPA as a lead structure in attempts to design antagonists selective for non-NMDA receptors we have now synthesized and tested the two compounds AMOA and AMNH (Fig. 16) as potential EAA antagonists (Krogsgaard-Larsen *et al.* 1991).

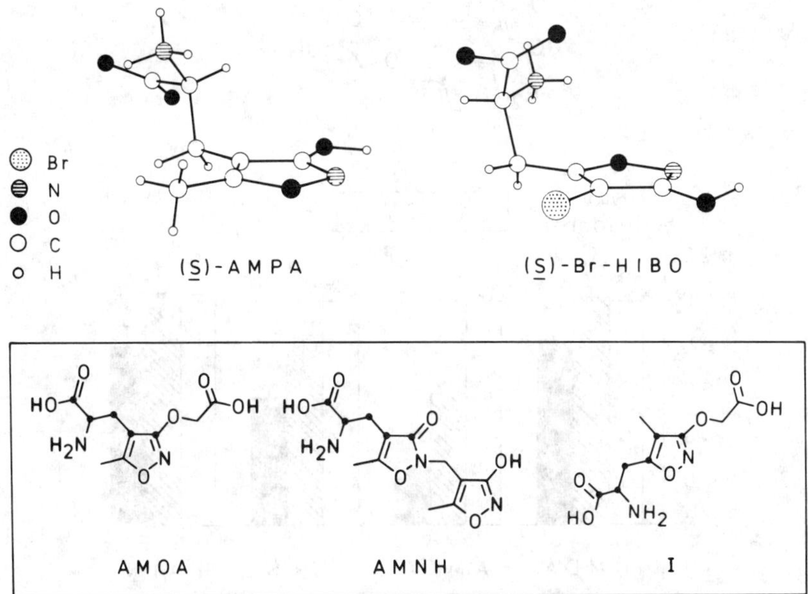

Fig. 16 — Structures of (*S*)-AMPA and (*S*)-Br-HIBO, as established on the basis of X-ray analyses, and of the non-NMDA antagonists AMOA and AMNH. Compound I, which is an isomer of AMOA, is inactive.

AMOA and AMNH were shown to be stable in aqueous solution at pH values close to physiological pH. Neither AMOA nor AMNH showed detectable affinities for the receptor, ion channel, or modulatory sites of the NMDA receptor complex. Quantitative receptor autoradiographic and conventional binding techniques were used to study the affinities of AMOA and AMNH for non-NMDA receptor sites. Both compounds were inhibitors of the binding of [^{3}H]AMPA (IC_{50} = 90 and 29 μM, respectively). AMNH as well as AMOA were very weak inhibitors of the high-affinity binding of [^{3}H]KAIN. AMNH, and to a much smaller extent AMOA, was, however, shown to inhibit low-affinity [^{3}H]KAIN binding (IC_{50} = 40 μM) in the presence of 100 mM calcium chloride (Krogsgaard-Larsen *et al.* 1991).

In the rat cortical slice preparation, AMOA was shown to antagonize competitively excitation induced by AMPA (Fig. 17) with some selectivity, whereas AMNH proved to be a weak but rather selective antagonist of KAIN-induced excitation (not illustrated). AMOA was a relatively weak antagonist of excitation by QUIS, whereas AMNH did not affect excitation by this non-selective AMPA receptor agonist. On cat spinal neurons, both AMOA and AMNH reduced AMPA- and KAIN-induced excitations, but, again, the excitatory effects of QUIS were much less sensitive (Krogsgaard-Larsen *et al.* 1991). Quite surprisingly, the carboxymethoxy analogue I (Fig. 16) of Me-HIBO (Fig. 13) is totally inactive as an EAA receptor antagonist (I.T. Christensen *et al.*, unpublished).

AMNH and, in particular, AMOA effectively protected rat striatal neurons against the neurotoxic effects of KAIN, whereas the toxic effects of AMPA were reduced only by AMOA (Table 1). Neither antagonist showed protection against the

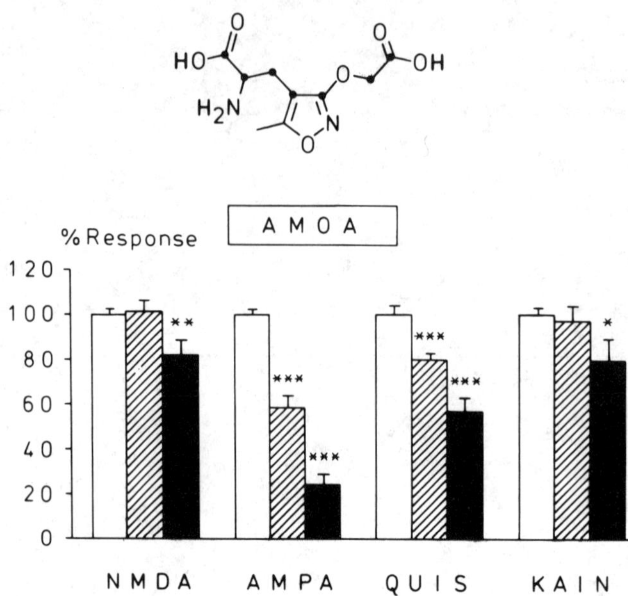

Fig. 17 — Effects of AMOA on the neuronal excitations induced by NMDA, AMPA, QUIS, or KAIN in the rat cortical slice preparation. Excitatory effects of NMDA (10 μM), AMPA (5 μM), QUIS (10 μM), or KAIN (5 μM) in the absence of antagonist are shown by the unshaded histograms. The effects of the same concentrations of agonists in the presence of AMOA (250 μM — hatched; or 1 mM — black). Percentage response (normalized) SEM, $n = 3$. (Data from Krogsgaard-Larsen *et al.* 1991.)

Table 1 — Neuroprotective effects of some EAA antagonists

Agonist	Antagonist	Remaining enzyme activity (%) CHAT	GAD
AMPA		53 ± 6	40 ± 12
—	+ AMOA	68 ± 7	63 ± 9
—	+ AMNH	53 ± 10	39 ± 7
—	+ CPP	55 ± 8	50 ± 13
QUIN		61 ± 8	53 ± 8
—	+ AMOA	52 ± 2	64 ± 11
—	+ AMNH	50 ± 12	49 ± 4
—	+ CPP	98 ± 3	99 ± 4
KAIN		34 ± 5	44 ± 8
—	+ AMOA	90 ± 7	102 ± 10
—	+ AMNH	87 ± 10	68 ± 4

Solutions (1 μl) containing KAIN (10 nmol), QUIN (300 nmol), or AMPA (100 nmol) with or without the test compounds, CPP (150 nmol), AMOA (150 nmol), or AMNH (150 nmol) were infused into the corpus striatum of rats. Following a three-day recovery period, animals were sacrificed and the percentage remaining activities of the marker enzymes choline acetyltransferase (CHAT) and glutamic acid decarboxylase (GAD) determined. (Data from Krogsgaard-Larsen *et al.* 1991.)

cell damage caused by intrastriatal injection of the NMDA agonist QUIN (Krogsgaard-Larsen *et al.* 1991). In cultured cortical neurons, AMOA effectively protected against KAIN-induced toxicity, whereas no significant protective effect of AMOA could be demonstrated when the cells were exposed to AMPA or NMDA (Frandsen *et al.* 1990a).

These comparative studies on AMOA and AMNH as non-NMDA receptor antagonists and as neuroprotective agents (Table 1) strongly suggest that the mechanisms underlying neuroexcitation and neurotoxicity are not identical. Thus, although AMOA is a selective AMPA receptor antagonist (Fig. 17), the neurotoxic effects induced by KAIN are much more effectively blocked by AMOA than those induced by AMPA.

KAINIC ACID RECEPTOR AGONISTS AND ANTAGONISTS

The prototypic agonist at KAIN receptors is KAIN itself (Figs 3 and 18), which is a natural product (Coyle 1983, Shinozaki 1988). The epimer of KAIN, β-KAIN (Fig. 18), is virtually devoid of affinity for KAIN receptor sites, whereas the naturally occurring amino acid, domoic acid, is at least as potent as KAIN as an agonist at KAIN receptors (Coyle 1983). More recently, the mushroom constituent acromelic acid A (Fig. 18), which is a heterocyclic analogue of domoic acid, has been shown to depolarize crayfish opener muscle fibres more potently than any of the KAIN receptor agonists described earlier (Shinozaki 1988) (see chapter 12).

A number of quinoxalinediones, including CNQX (Fig. 3), effectively reduce KAIN-induced neuronal excitation in addition to their antagonist effects at AMPA receptors (Honoré 1989, Honoré *et al.* 1988) (see chapter 11). Although AMNH only

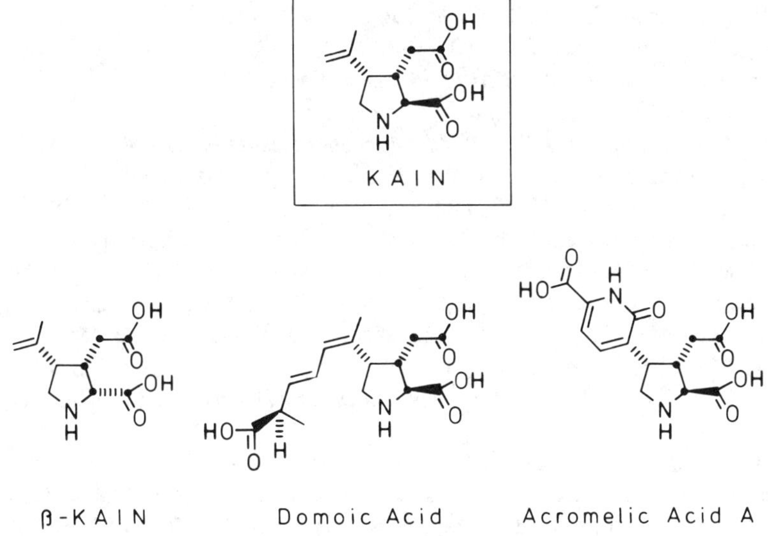

Fig. 18 — Structure of KAIN, some KAIN receptor agonists, and the biologically inactive analogue β-KAIN.

seems to be capable of reducing partially excitation by KAIN (Krogsgaard-Larsen *et al.* 1991), this compound and phenobarbitone (Frandsen *et al.* 1990b) appear to be the most selective KAIN receptor antagonists so far described.

CONCLUDING REMARKS

During the past years the central EAA receptors have been extensively studied. So far, the physiological, pharmacological, and medicinal chemical studies have been focused primarily on NMDA receptors. A wide variety of very effective and highly selective competitive as well as non-competitive NMDA antagonists have been developed. Some of these compounds have major therapeutic interest. There also is a growing interest in NMDA receptor agonists and/or partial agonists as potential therapeutic agents.

In recent years the pharmacology of non-NMDA receptors has been in a state of rapid development. In light of the availability of selective agonists and, in some cases, effective antagonists, pharmacological and therapeutic progress in this area of EAA receptor research is to be expected.

ACKNOWLEDGEMENTS

This work was supported by grants from the Danish Medical and Technical Research Councils and from the Lundbeck Foundation. The secretarial assistance of Mrs Birte Hare is gratefully acknowledged.

REFERENCES

Ambrosini, A., Henley, J. M. and Barnard, E. A. (1990) *Biochem. Soc. Trans.* **18** 401–402.

Ascher, P. and Novak, L. (1988) *J. Physiol.* **399** 227–245.

Barnard, E. A. and Costa, E. (eds) (1989) *Allosteric Modulation of Amino Acid Receptors: Therapeutic Implications.* Raven Press, New York.

Bowen, D. M. (1990) *Br. J. Psychiat.* **157** 327–330.

Braitman, D. J. and Coyle, J. T. (1987) *Neuropharmacology* **26** 1247–1251.

Brehm, L., Jørgensen, F. S., Hansen, J. J. and Krogsgaard-Larsen, P. (1988) *Drug News Perspect.* **1** 138–144.

Brehm, L., Christensen, I. T., Nielsen, B., Reinhardt, A., Ebert, B., Madsen, U., Jørgensen, F. S., Hansen, J. J. and Krogsgaard-Larsen, P. (1990) In: Jensen, B., Jørgensen, F. S. and Kofod, H. (eds) *Frontiers in Drug Research: Crystallographic and Computational Methods.* Munksgaard, Copenhagen, pp. 92–105.

Bridges, R. J., Geddes, J. W., Monaghan, D. T. and Cotman, C. W. (1988) In: Lodge, D. (ed.) *Excitatory Amino Acids in Health and Disease.* Wiley, New York, pp. 321–335.

Chadwick, W. C. and Choi, D. W. (1990) *J. Neurosci.* **10** 108–116.

Christensen, I. T., Reinhardt, A., Nielsen, B., Ebert, B., Madsen, U., Nielsen, E. Ø., Brehm, L. and Krogsgaard-Larsen, P. (1989) *Drug Des. Del.* **5** 57–71.

Collingridge, G. L. and Singer, W. (1990) *Trends Pharmacol. Sci.* **11** 290–296.

Coyle, J. T. (1983) *J. Neurochem.* **41** 1–11.

Deutsch, S. I., Mastropaolo, J., Schwartz, B. L., Rosse, R. B. and Morihisa, J. M. (1989) *Clin. Neuropharmacol.* **12** 1–13.

Frandsen, Aa., Krogsgaard-Larsen, P. and Schousboe, A. (1990a) *J. Neurochem.* **55** 1821–1823.

Frandsen, Aa., Quistorff, B. and Schousboe, A. (1990b) *Neurosci. Lett.* **111** 233–238.

Greenamyre, J. T. and Young, A. B. (1989) *Neurobiol. Ag.* **10** 593–602.

Hansen, J. J. and Krogsgaard-Larsen, P. (1990) *Med. Res. Rev.* **10** 55–94.

Hansen, J. J., Lauridsen, J., Nielsen, E. and Krogsgaard-Larsen, P. (1983) *J. Med. Chem.* **26** 901–903.

Hansen, J. J. Nielsen, B., Krogsgaard-Larsen, P., Brehm, L., Nielsen, E. Ø. and Curtis, D. R. (1989) *J. Med. Chem.* **32** 2254–2260.

Henley, J. M., Ambrosini, A., Krogsgaard-Larsen, P. and Barnard, E. A. (1989) *New Biologist* **1** 153–158.

Honoré, T. (1989) *Med. Res. Rev.* **9** 1–23.

Honoré, T., Lauridsen, J. and Krogsgaard-Larsen, P. (1982) *J. Neurochem.* **38** 173–178.

Honoré, T., Davies, S. N., Drejer, J., Fletcher, E. J., Jacobsen, P., Lodge, D. and Nielsen, F. E. (1988) *Science* **241** 701–703.

Honoré, T., Drejer, J., Nielsen, E. Ø. and Nielsen, M. (1989) *Biochem. Pharmacol.* **38** 3207–3212.

Huettner, J. E. (1991) *Biochem. Pharmacol.* **41** 9–16.

Irving, A. J., Schofield, J. G., Watkins, J. C., Sunter, D. C. and Collingridge, G. L. (1990) *Eur. J. Pharmacol.* **186** 363–365.

Jackson, D. E., Bycroft, B. W. and King, T. J. (1989) *J. Comput.-Aided Mol. Design* **2** 321–328.

Johnson, R. L. and Koerner, J. F. (1988) *J. Med. Chem.* **31** 2057–2066.

Kato, K., Puttfarcken, P. S., Lyons, W. E. and Coyle, J. T. (1991) *J. Pharmacol. Exp. Ther.* **256** 402–411.

Keinänen, K., Wisden, W., Sommer, B., Werner, P., Herb, A., Verdoorn, T. A., Sakmann, B. and Seeburg, P. H. (1990) *Science* **249** 556–560.

Kemp, J. A., Foster, A. C. and Wong, E. H. F. (1987) *Trends Neurosci.* **10** 294–298.

Krogsgaard-Larsen, P., Honoré, T., Hansen, J. J., Curtis, D. R. and Lodge, D. (1980) *Nature* **284** 64–66.

Krogsgaard-Larsen, P. and Honoré, T. (1983) *Trends Pharmacol. Sci.* **4** 31–33.

Krogsgaard-Larsen, P., Nielsen, E. Ø. and Curtis, D. R. (1984) *J. Med. Chem.* **27** 585–591.

Krogsgaard-Larsen, P., Brehm, L., Johansen, J. S., Vinzents, P., Lauridsen, J. and Curtis, D. R. (1985) *J. Med. Chem.* **28** 673–679.

Krogsgaard-Larsen, P., Jensen, B., Falch, E. and Jørgensen, F. S. (1989) *Drugs Fut.* **14** 541–561.

Krogsgaard-Larsen, P., Ferkany, J. W., Nielsen, E. Ø., Madsen, U., Ebert, B., Johansen, J. S., Diemer, N. H., Bruhn, T., Beattie, D. T. and Curtis, D. R. (1991) *J. Med. Chem.* **34** 123–130.

Lapchak, P. A., Araujo, D. M., Quirion, R. and Collier, B. (1989) *J. Neurochem.* **53** 1843–1851.

Lauridsen, J., Honoré, T. and Krogsgaard-Larsen, P. (1985) *J. Med. Chem.* **28** 668–672.

Lodge, D. (ed.) (1988) *Excitatory Amino Acids in Health and Disease.* Wiley, New York.

Lodge, D. (1989) *Drugs of Today* **25** 395–411.

Lodge, D. and Johnson, K. M. (1990) *Trends Pharmacol. Sci.* **11** 81–86.

Lund, T. M., Madsen, U., Ebert, B., Jørgensen, F. S. and Krogsgaard-Larsen, P. (1991) *Med. Chem. Res.* **1** 136–141.

Madsen, U., Brehm, L., Schaumburg, K., Jørgensen, F. S. and Krogsgaard-Larsen, P. (1990a) *J. Med. Chem.* **33** 374–380.

Madsen, U., Ferkany, J. W., Jones, B. E., Ebert, B., Johansen, T. N., Holm, T. and Krogsgaard-Larsen, P. (1990b) *Eur. J. Pharmacol. Mol. Pharmacol. Sect.* **189** 381–391.

Manzoni, O. J. J., Finiels-Marlier, F., Sassetti, I., Bockaert, J., le Peuch, C. and Sladeczek, F. A. J. (1990) *Neurosci. Lett.* **109** 146–151.

McLennan, H. (1983) *Prog. Neurobiol.* **20** 251–271.

Monaghan, D. T. and Cotman, C. W. (1989) In: Watkins, J. C. and Collingridge, G. L. (eds) *The NMDA Receptor.* Oxford University Press, Oxford, pp. 53–64.

Monaghan, D. T., Olverman, H. J., Nguyen, L., Watkins, J. C. and Cotman, C. W. (1988) *Proc. Nat. Acad. Sci. USA* **85** 9836–9840.

Monaghan, D. T., Bridges, R. J. and Cotman, C. W. (1989) *Annu. Rev. Pharmacol. Toxicol.* **29** 365–402.

Murphy, D. E., Snowhill, E. W. and Williams, M. (1987) *Neurochem. Res.* **12** 775–782.

Nappi, G., Hornykiewicz, O., Fariello, R. G., Agnoli, A. and Klawans, H. (eds) (1988) *Neurodegenerative Disorders: The Role Played by Endotoxins and Xenobiotics.* Raven Press, New York.

Nielsen, E. Ø., Schousboe, A., Hansen, S. H. and Krogsgaard-Larsen, P. (1985) *J. Neurochem.* **45** 725–731.

Nielsen, E. Ø., Madsen, U., Schaumburg, K., Brehm, L. and Krogsgaard-Larsen, P. (1986) *Eur. J. Med. Chem.* **21** 433–437.

Perouansky, M. and Grantyn, R. (1989) *J. Neurosci.* **9** 70–80.

Rassendren, F.-A., Lory, P., Pin, J.-P., Bockaert, J. and Nargeot, J. (1989) *Neurosci. Lett.* **99** 333–339.

Reynolds, I. J. (1990) *Life Sci.* **47** 1785–1792.

Rothman, S. M. and Olney, J. W. (1986) *Ann. Neurol.* **19** 105–111.

Schoepp, D. D. and Johnson, B. G. (1988) *J. Neurochem.* **53** 273–278.

Schoepp, D. D., Bockaert, J. and Sladeczek, F. (1990) *Trends Pharmacol. Sci.* **11** 508–515.

Shinozaki, H. (1988) *Prog. Neurobiol.* **30** 399–435.

Sladeczek, F., Pin, J.-P., Recasens, M., Bockaert, J. and Weiss, S. (1985) *Nature* **317** 717–719.

Stone, T. W. and Burton, N. R. (1988) *Prog. Neurobiol.* **30** 333–368.

Sugiyama, H., Watanabe, M., Taji, H., Yamamoto, Y. and Ito, I. (1989) *Neurosci. Res.* **7** 164–167.

Watkins, J. C. and Collingridge, G. L. (eds) (1989) *The NMDA Receptor.* Oxford University Press, Oxford.

Watkins, J. C. and Evans, R. H. (1981) *Annu. Rev. Pharmacol. Toxicol.* **21** 165–204.

Watkins, J. C., Krogsgaard-Larsen, P. and Honoré, T. (1990) *Trends Pharmacol. Sci.* **11** 25–33.

Watson, G. B., Monaghan, D. T. and Lanthorn, T. H. (1990) *Eur. J. Pharmacol.* **179** 479–481.

Wurtman, R. J., Corkin, S., Growdon, J. H. and Ritter-Walker, E. (eds) (1990). *Alzheimer's Disease*. Raven Press, New York.

Young, A. B. and Fagg, G. E. (1990) *Trends Pharmacol. Sci.* **11** 126–133.

3

Isolation and structural characterization of excitatory amino acid receptors

Graham Johnson
Bristol-Myers Squibb Company, Pharmaceutical Research Institute, Research Parkway, Wallingford, CT 06492, USA

ABSTRACT

Numerous biochemical, biophysical, electrophysiological and, more recently, molecular biological techniques have been used to investigate and characterize diverse aspects of excitatory amino acid receptor size, structure and function. Each of these studies has provided a small piece of the overall puzzle of how receptors are constructed and which structural and environmental features are critical to their efficient operation. In this review an attempt has been made to draw together this wealth of disparate information into a coherent if somewhat speculative picture of excitatory amino acid receptor structure.

INTRODUCTION

Over the last ten years tremendous strides have been made in defining the pharmacology of excitatory amino acid receptors. Through these studies, five major receptor types have been defined (*N*-methyl-(*R*)-aspartic acid (NMDA), (*RS*)-2-amino-3-(3-hydroxy-5-methylisoxazol-4-yl)propanoic acid (AMPA), kainate (KAIN), metabotropic (1-aminocyclopentane-1,3-dicarboxylate (ACPD)) and L-2-amino-4-phosphonobutanoic acid (L-AP4)) and their second messenger systems identified. However, our understanding of how these receptors are assembled and what tertiary structural features are responsible for their observed ligand specificity and electrophysiological properties is still rudimentary. As with many other fields, the application of molecular biological techniques has begun to shed light on some of these questions. However, even in the absence of these powerful techniques numerous biochemical observations have been made that afford insights into the structure of both NMDA and non-NMDA excitatory amino acid receptors. In this review an attempt has been made to assemble these individual biochemical, pharmacological and molecular biological observations into a mutually consistent picture of excitatory amino acid receptor structure.

RECEPTOR BIOCHEMISTRY

Glycosylation

The use of chemical and biochemical tools to probe the structural characteristics of excitatory amino acids has proven fruitful. Through binding with wheatgerm aglutinin (WGA) or concanavalin A, glycosylation of extracellular receptor protein domains has been established for both NMDA and non-NMDA receptors. For NMDA receptors expressed in *Xenopus* oocytes, treatment with concanavalin A has been shown to suppress desensitization (Geoffroy *et al.* 1989). However, in mouse hippocampal neurons concanavalin A had no efffect (Mayer and Vyklicky 1989). These apparently contradictory results might reflect differences in rat cortical and mouse hippocampal NMDA rceptor subtypes or differences in oocyte post-translational modification. However, recent evidence suggests that the inhibitory effect of concanavalin A on NMDA receptor expressed in oocytes is most likely artifactual (Mayer 1991). Despite the lack of effect of concanavalin A on hippocampal neurons, the reactivity of isolated NMDA receptor protein subunits with concanavalin A (see later) clearly demonstrates that this receptor complex is also significantly glycosylated (Chen *et al.* 1988, Cunningham and Michaelis 1990). In addition, a glutamate binding protein from which NMDA subunit-specific antibodies have been raised was purified by concanavalin A Sepharose chromatography (Michaelis *et al.* 1983).

Of the cortical KAIN and AMPA receptor subtypes, only the AMPA receptor has been shown to desensitize rapidly. In contrast, in dorsal root ganglion (DRG) cells, a KAIN and domoate current did desensitize on prolonged agonist stimulation or through a brief prior exposure of glutamate, quisqualate or AMPA. In all cases, exposure to concanavalin A slowly eliminated desensitization. As a result, in the DRG neurons all five agonists produced a sustained current of a similar amplitude at the same receptor (Mayer and Vyklicky 1989, Huettner 1990). However, perhaps reflecting differences in intrinsic receptor aggregation, the dimeric succinylated form of concanavalin A was equieffective with the non-derivatized monomer at reducing the desensitization of DRG neurons but was essentially inactive at hippocampal neurons (Huettner 1990, Mayer and Vyklicky 1989). Also in hippocampal cell culture, the pathophysiological importance of quisqualate receptor desensitization was graphically demonstrated by the exacerbation of quisqualate-induced neurotoxicity by application of WGA (Zorumski *et al.* 1990). The ability of concanavalin A to produce a non-desensitizing channel with the same conductance state suggests that the structure of AMPA and KAIN receptors in DRG neurons might differ simply in their degree of extracellular protein glycosylation. In hippocampus, however, receptors might also differ in the degree or nature of their self-association. Despite the lack of effect of concanavalin A on mouse cortical KAIN-activated channels, it is clear from their retention on WGA affinity columns that both cortical KAIN and quisqualate binding proteins are significantly glycosylated in the frog (Henley *et al.* 1989).

Although, in the foregoing, it is implied that plant lectins affect receptor function by a direct interaction with protein-linked sugar residues, an interaction with other membrane constituents is also possible. In this, lectins might produce an effect on receptor desensitization through an association with glycolipids that surround the receptor, such as gangliosides. Therefore what may differentiate receptor subtypes may not only be the glycosylation state of the receptor proteins but also the nature of the lipids that surround the protein.

Effect of ions

In addition to revealing the degree of glycosylation, chemically probing the extracellular protein skeleton has yielded additional insight into EAA receptor structure. Calcium and chloride ions significantly increase (fourfold) quisqualate-sensitive [^{3}H]glutamate binding (Nielsen *et al.* 1988), but affects [^{3}H]AMPA binding little. However, replacing calcium and chloride with a high concentration (100 mM) of the chaotropic thiocyanate ion changed quisqualate-sensitive [^{3}H]glutamate binding little but increases total [^{3}H]AMPA binding fourfold. In contrast to its effect on [^{3}H]AMPA binding, thiocyanate tended to decrease [^{3}H]KAIN binding. Such studies have suggested that the AMPA receptor exists as two interconvertible low- and high-affinity binding site conformations. Overall, thiocyanate's action to increase total specific binding might result from either increasing the density of both high- and low-affinity binding sites or by lowering the apparent K_d of the low-affinity AMPA binding site (Murphy *et al.* 1987, Honoré and Drejer 1988). By disorganizing their associated water sheath, chaotropic ions have been suggested to increase the water solubility of proteins (Hatefi and Hanstein 1969). In this case, the result of such an action might be to expose or stabilize a low-affinity [^{3}H]AMPA recognition site. The removal of an inhibitory modulatory protein appears unlikely. Interestingly, treatment of membranes with high-energy electrons also causes a similar and significant increase in [^{3}H]AMPA binding. This increase in ligand binding might result from a radiation-induced uncoupling of a modulatory protein (Honoré and Nielsen 1985). These results suggest that such a modulatory protein subunit might be coupled through a concealed disulphide bridge.

Although the NMDA receptor is affected little by the presence of chaotropic ions, in cultured hippocampal neurons replacing normal sodium with isomolar potassium resulted in a response that was reduced to 12% that of K^+-free, sodium 150 mM. These conditions reduced mean channel open time but not channel conductance or reversal potential (Ozawa *et al.* 1990). Neither KAIN nor AMPA responses were affected by a high potassium/low sodium solution.

Thiol and disulphide bridges

Probing the AMPA receptor for the presence of thiol and disulphide groups has added further complexity to this picture. Oxidizing synaptic membranes with 5,5′-dithio-bis-(2-nitrobenzoic acid) or irreversibly trapping thiol residues with *N*-ethylmaleimide did not significantly influence [^{3}H]AMPA binding (Terramani *et al.* 1988). However, similar treatment with mercuric chloride, *p*-chloromercuribenzoic acid or phenyl mercurisulphonic acid produced a significant increase in the number of high-affinity and a decrease in both the number and affinity of low affinity [^{3}H]AMPA sites (Terramani *et al.* 1988). This effect was reversed by the addition of dithiothreitol and inhibited by prior treatment with *N*-ethylmaleimide, suggesting that this interaction occurred at the same thiol residues. Treatment of membranes with dithiothreitol alone had no effect on [^{3}H]AMPA binding. The inability of saturating concentrations of agonist to protect [^{3}H]AMPA binding from organomercurial modification further suggests that these modified thiol groups are not incorporated in the agonist recognition site. In addition, the equal ability of both lipophilic and non-lipophilic organomercurials to derivatize resealed postsynaptic membranes suggests that these reactive thiol groups are likely to be located on the extracellular domain. Receptor cloning and sequencing has clearly confirmed these suggestions (see later). In contrast to their action at the [^{3}H]AMPA binding site, high-affinity [^{3}H]KAIN binding was essentially unaffected by the action of sulphydryl reagents, although dithiothreitol alone did reduce [^{3}H]KAIN binding significantly (Terramani *et al.* 1988).

Oxidation and treatment with the same thiophilic reagents depressed glutamate and KAIN-induced inward currents in rat hippocampal pyramidal neurons, reflecting a dissociating action of thiophilic reagents on ligand binding and channel activation. The depressant action of mercuric chloride and *p*-chloromercuribenzoic acid on the KAIN response was non-competitive and was reversed by treatment with cysteine or glutathione (Kiskin *et al.* 1986). A similar non-competitive action of mercuric ions on human brain KAIN receptors expressed in *Xenopus* oocytes has been reported (Umbach and Gundersen 1989). Similar but more readily reversible antagonism was observed with zinc. Surprisingly, although similarly antagonistic at higher concentrations, low levels of zinc actually potentiated the activation of AMPA and KAIN receptors expressed in oocytes (Rassendren *et al.* 1990). Perhaps supporting this observation, it has also been noted that the administration of the zinc chelators diethyldithiocarbonate and dithizone to Fischer-344 rats treated with KAIN both reduced the time to onset and increased the severity of resulting seizures (Mitchell *et al.* 1990). However, more in keeeping with its lack of effect on [^{3}H]KAIN and [^{3}H]AMPA binding, recent reports have concluded that dithiothreitol and 5,5-dithio-bis-(2-nitrobenzoic acid) do not inhibit KAIN and AMPA channel activation (Aizenmann *et al.* 1989).

Zinc has been shown to inhibit the activation of the NMDA receptor both at voltage-dependent and -independent sites (Mayer *et al.* 1989, Christine and Choi 1990). The voltage-dependent site is suggested to lie within the ion channel. However, from binding kinetics, the voltage-independent site does not appear to be associated with thiol chelation but perhaps reflects an interaction with extracellular histidine residues (Legendre and Westbrook 1990). This conclusion is supported by the observation that irreversible reaction of histidine residues with diethylpyrocarbonate appears to reduce the effectiveness of zinc inhibition of NMDA responses (Traynelis and Cull-Candy 1991). Interestingly, it has been suggested recently that zinc antagonizes the NMDA receptor through an indirect inhibition of glycine binding to one of its two recognition sites (see later) (Yeh *et al.* 1990).

From the action of reducing and oxidizing reagents, it is clear that, unlike either the KAIN or AMPA receptors, the NMDA receptor contains one or more critical disulphide bridges. Numerous *in vitro* studies have established that disrupting disulphide bridges through reduction with dithiothreitol greatly potentiates the activation and subsequent overall conductance of the NMDA receptor-associated ion channel (Aizenmann *et al.* 1989, Lazarewicz *et al.* 1989, Levy *et al.* 1990, Reynolds *et al.* 1990, Aizenmann *et al.* 1991). Although reducing rat cortical membranes did not modify the binding affinity of either glutamate or glycine (Reynolds *et al.* 1990), increases in [^{3}H]MK-801 binding produced by glutamate, glycine and spermidine were further enhanced by dithiothreitol's action. However, inhibition of [^{3}H]MK-801 binding by *cis*-(4-phosphonomethyl)piperidine-2-carboxylic acid (CGS 19755) and kynurenic acid were unaffected, as was the inhibition of NMDA-induced channel activation by 2-amino-5-phosphonopentanoic acid (AP5) (Aizenmann *et al.* 1989). Reoxidation with 5,5-dithio-bis-(2-nitrobenzoic acid) reversed the stimulatory action of dithiothreitol. These results suggest that a disulphide bridge is unlikely to be an integral part of the glutamate or glycine recognition site but may instead stabilize the NMDA receptor in its closed conformation, and thereby inhibit its transition to an open conformation. Providing further evidence that the NMDA receptor is not a close member but perhaps is only a distant relative of the ligand-gated super-family of receptors, the effects of dithiothreitol reduction on the nicotinic receptor are to abolish channel function completely (Aizenmann *et al.* 1989, Loring *et al.* 1989). Oxidation re-establishes full receptor function. This disruptive action on the nicotinic receptor has been ascribed to the disruption of a critical extracellular disulphide bridge which is deemed essential to maintaining the topography of the receptor binding site.

The importance of thiol groups in maintaining the tertiary structure of the NMDA receptor was demonstrated further by the report that 1 mM *N*-ethylmalemide reduces [^{3}H]3-(2-carboxypiperazin-4-yl)propyl-1-phosphonic acid ([^{3}H]CPP) binding by half without altering that of NMDA-sensitive glutamate binding (Ogita and Yoneda 1990a).

pH sensitivity of channel function

Both pH and chemical studies have suggested that several histidine groups are located on the external face of the channel protein and are involved in regulating KAIN and AMPA receptor channel gating. In isolated catfish cone horizontal cells, increases in extracellular hydrogen ion concentration (decreasing pH) reduced the magnitude of membrane current induced by application of AMPA and KAIN (Christensen and Hida 1990). This effect of pH was voltage-independent and did not modify agonist affinity, channel conductance or mean channel open time but did appear to affect channel gating. Chemical modification of histidine but not thiol residues also reduced agonist-induced current. Similar inhibitory effects of pH were observed for AMPA and KAIN-induced activation of rat cerebellar granule neurons (Traynelis and Cull-Candy 1991). Cloning and sequencing of KAIN and AMPA receptors (see later) have confirmed the existence of extracellular histidine residues. In these studies, between 8 and 12 histidine residues have been identified in the extracellular protein domains of the various receptor subunits.

Similar inhibition by extracellular protons of cerebellar neuronal NMDA responses has also been observed (Traynelis and Cull-Candy 1990, Traynelis and Cull-Candy 1991). As with inhibition of the AMPA receptor, no influence on the excitatory concentration (EC_{50}) of NMDA, aspartate or glycine was observed, suggesting no change in agonist affinity. However, a decrease in the frequency of NMDA channel opening was observed. A reduction in the relative proportion of longer-duration bursts was also observed. From these studies, Traynelis and Cull-Candy (1991) concluded that protons probably act at or near the extracellular face of the receptor to protonate an ionizable group within the receptor. Protonation therefore renders the protein non-functional by inhibiting a crucial conformational change of the receptor that leads to the open state. Traynelis and Cull-Candy speculated that protonation might occur within a portion of the NMDA receptor that is directly influenced by glycine binding rather than at the glycine binding site itself. Inhibition of channel opening in this way might change the conformation of the receptor protein such that glycine binds less effectively.

The NMDA receptor shows about half maximal activation at physiological pH and is significantly more sensitive to pH than are AMPA or KAIN receptors. At pH 7.4, neither KAIN nor AMPA responses were significantly affected. However as noted previously by Christensen and Hida (1990), AMPA and KAIN responses were inhibited under more acidic conditions, with an apparent IC_{50} of pH 6.3 and 5.7, respectively. Further illustrating the sensitivity of the NMDA receptor to pH, it has been established in several *in vitro* hypoxia models that mild acidosis (pH 6.5) gives virtually complete neuroprotection (Tombaugh and Sapolsky 1990), indicating that at this level of protonation the NMDA receptor is locked in a closed, inactivated or desensitized state. Although the inhibitory action of increased acidity on NMDA receptor activation would be compatible with the protonation of extracellular histidine and/or cysteine residues, chemical modification of histidine or thiol residues does not support this conclusion. In their recent study, Traynelis and Cull-Candy (1991) showed that, unlike AMPA and KAIN receptors, treatment of NMDA receptors with diethyl pyrocarbonate resulted in a potentiation rather than inhibition of agonist response. Furthermore, the pH sensitivity of these chemically modified receptors was unaltered, suggesting that protonation of histidine residues does not underlie the inhibitory action of decreasing pH. Although not directly implicated in the pH-dependent inhibition of NMDA responses, the potentiation of histidine-modified NMDA receptor responses does support a central role for histidine residues in the gating mechanism of the NMDA receptor.

Interaction with lipid bilayer

Although not affording much direct insight into the structure of the excitatory amino acid receptors, a close interaction of receptor protein with the lipid membrane has been suggested. In these studies the affinity of [^{3}H]AMPA was increased following the treatment of rat brain membranes with phospholipase A_2 and bacteria-derived phospholipase C (Massicotte *et al.* 1990, Massicotte and Baudry 1990). A more inositol-specific phospholipase C had no effect. Surprisingly, [^{3}H]KAIN binding was unaffected by all three treatments.

The NMDA receptor, although unresponsive to both phospholipase A_2 and inositol-specific phospholipase C, was modified by the less selective phospholipase C. Under these conditions, the affinity of [^{3}H]1-[1-(2-thienyl)cyclohexylpiperidine ([^{3}H]TCP) for its binding site was decreased in the activated receptor yet increased in the closed. In addition, although agonist and glycine binding were unaffected, glycine-stimulated glutamate binding was eliminated and the affinity of [^{3}H]CPP significantly reduced. However, the ability of AP5 to displace [^{3}H]glutamate binding was not affected. These results suggest that the lipid microenvironment plays a critical role in maintaining the structural integrity and conformational interrelationship of the NMDA receptor protein subunits. In addition, these data also suggest that the agonist, antagonist and glycine sites might be located on different receptor subunits. Using a membrane fluidizing agent it was also noted that the function of the NMDA–PCP receptor complex is impaired by increases in membrane fluidity (DePietro and Byrd 1990).

Location of binding sites

The combined results from the action of chaotropic ions and membrane lipid-modifying agents on the AMPA receptor suggest that the AMPA recognition site is located at a deeper and perhaps less accessible site in the receptor protein structure. However, from the lack of influence of similar treatments on [^{3}H]KAIN binding, its recognition site may be different and located much closer to the receptor protein surface. Although not providing evidence to locate the binding site on the receptor complex, data from electrophysiological studies of isolated stingray horizontal cells also support the existence of multiple but not identical binding sites for KAIN, AMPA and glutamate on the same receptor protein (O'Dell and Christensen 1989). In this study KAIN and glutamate are proposed to bind to different recognition sites, whereas AMPA and quisqualate bind to the same recognition site. Additional evidence that KAIN and AMPA receptors are actually located on the same protein complex has been reviewed in detail recently (Barnard and Henley 1990).

The position of the NMDA agonist and antagonist recognition sites on the receptor complex is at present also unknown. However, careful electrophysiological studies have shown that glutamate may remain bound to its recognition site for hundreds of milliseconds rather than dissociating rapidly. Such a result might suggest that the agonist recognition site may lie at a deeper rather than a shallow site within the protein complex (Lester *et al.* 1990). Similarly, it has been shown that the more conformationally rigid an antagonist molecule is made, its rate of association and dissociation from the NMDA receptor are both slowed significantly, dissociation being more severely affected (Benveniste *et al.* 1990b). These observations suggest that the reduction in antagonist conformational flexibility makes it more difficult for the molecule to mould itself to or accommodate the overlaying receptor skeleton, and hence reach and bind to its particular recognition site.

It was reported that trypsin-mediated acute dissociation of hippocampal neurons irreversibly removes the NMDA response with little effect on that of KAIN or quisqualate (Allen *et al.* 1988, Akaike *et al.* 1988), initially suggesting differences in the structure or relationship of the external protein domains of NMDA and non-NMDA receptors. This result implies that the NMDA receptor may contain a trypsin-accessible extracellular protein domain. More recent and perhaps more careful studies with trypsin and other proteases have shown that NMDA receptor integrity can be maintained during protease-mediated cell-dissociating procedures (Kay and Connor 1990, Gündel *et al.* 1990).

Intracellular domains and phosphorylation

Few studies are available that provide insight into the nature or structure of the intracellular domains of the excitatory amino acid receptors. Extrapolating from other ligand and voltage-gated ion channels, the function of such regions are likely to be influenced by phosphorylation (Huganir and Greengard 1990). Receptor cloning experiments on KAIN and AMPA binding proteins have demonstrated the existence of several PKC and Cam-kinase phosphorylation consensus sequences (see later). However, biochemical evidence supporting the function and thereby existence of such sites on NMDA and non-NMDA receptors is limited. Following quisqualate-stimulated phosphorylation of Na^+, Ca^{2+} and γ-aminobutanoic acid (GABA) (type A) gated ion channels expressed in *Xenopus* oocytes, channel gating was clearly influenced. In this system, the KAIN-induced response was essentially unaffected (Sigel and Bauer 1988). Given the probable unity of quisqualate and KAIN receptors, clear interpretation of this result is difficult. In contrast, using perch retinal horizontal cells, dopamine stimulation of a cAMP-dependent protein kinase significantly enhanced KAIN-induced currents (Liman *et al.* 1989). In a recent study it was established that the 49-kDa polypeptide KAIN binding protein isolated from chick cerebellum (see later) underwent phosphorylation by cAMP-dependent protein kinase (Ortega and Teichberg 1990). Up to two equivalents of phosphate were incorporated for each KAIN binding site. Surprisingly, KAIN inhibited phosphorylation in a concentration-dependent manner. Although 1 mM NMDA or quisqualic acid alone had no effect, KAIN inhibition of phosphorylation could be antagonized by the addition of 10 μM quisqualate. Clearly, these data suggest that the binding of KAIN to its extracellular recognition site brings about an intracellular conformational change which inhibits access of the kinase to its phosphorylation sites. The probable unity of quisqualate and KAIN binding protein is also supported by these data.

Arising from its ability to flux calcium, the NMDA receptor is recognized to be closely associated with protein kinase C and calmodulin-dependent kinase. Such a close association suggests that the NMDA receptor might also be intimately modulated by phosphorylation. However, only a few workers have so far examined this question. The effect of phosphorylation on the NMDA receptor has been studied in murine cell cultures (McDonald *et al.* 1989b). Adding to the complexity of known NMDA receptor modulatory influences, it was concluded that phosphorylation/dephosphorylation also modulates this receptor. It was further suggested that phosphorylation of the NMDA channel or some related protein appeared to be required to maintain the channel in a functional state (McDonald *et al.* 1988). Providing additional evidence for the existence of phosphorylation sites on the intracellular membrane of the NMDA receptor, it was noted by Leonard and Kelso (1989) that for NMDA receptors expressed in *Xenopus* oocytes stimulation of protein kinase C greatly enhanced the magnitude of NMDA-induced currents (Leonard and Kelso 1989).

SIZING EXCITATORY AMINO ACID RECEPTORS

Efforts to establish the molecular weight of lipid-bound excitatory amino acid receptors through indirect techniques have afforded some interesting insights into the nature and cooperativity of these receptor component proteins. However, these studies have also added to the overall complexity of the picture. Using high-energy electron inactivation techniques, Honoré and coworkers have shown that KAIN, in the absence of calcium, binds to two proteins of M_r 76 600 $\pm$ 5000 and 52 400 $\pm$ 5500. The high-molecular-weight protein, which corresponds to the high-affinity [^{3}H]KAIN binding site (K_D 3.5 nM) was lost with the addition of a high concentration of calcium (IC_{50} 2.5 mM) and only the low-affinity site (apparent K_D of 65 nM) remained (Honoré *et al.* 1986). This calcium-insensitive low-affinity binding site shows a similar molecular target size as both the [^{3}H]AMPA and

[^{3}H]6-cyano-7-nitroquinoxaline-2,3-dione ([^{3}H]CNQX) binding sites (Honoré and Nielsen 1985, Honoré *et al.* 1989a). The curvilinear Scatchard plot of these binding sites again indicated cooperativity or multiplicity of binding. Although, as expected, the total number of binding sites was reduced by radiation treatment, the affinity of KAIN for the low-affinity site was increased threefold by the radiation treatment. The affinity of quisqualate for this low-affinity KAIN site was similarly enhanced. In studies using [^{3}H]AMPA binding as a probe for the quisqualate receptor, an almost identical protein size (M_r 51 600 $\pm$ 3800) was established (Honoré and Nielsen 1985). [^{3}H]AMPA affinity was also enhanced by radiation inactivation. In this case the increase in ligand affinity was ascribed to the presence of a larger (128 000-Da) modulatory (inhibitory) protein. It is unclear, however, whether this 128 000-Da protein in fact represents two smaller receptor protein subunits.

Although on first inspection the existence of two dissimilar size KAIN binding proteins is indicated by these experiments, it should be emphasized that the radiation inactivation technique of assessing molecular weight relies on the assumption that the protein target size is directly proportional to molecular weight and that this relationship is the same over the range of proteins used in calibration. However, in any condition which changes this relationship, for example in the case above, the presence of a high concentration of calcium might afford slightly different apparent molecular masses for the same protein targets. A hypothetical example of this is illustrated diagrammatically in Fig. 1. An alternative explanation of a calcium-induced change in apparent molecular weight of a binding protein might be that a high concentration of calcium brings about a dissociation of a smaller modulatory

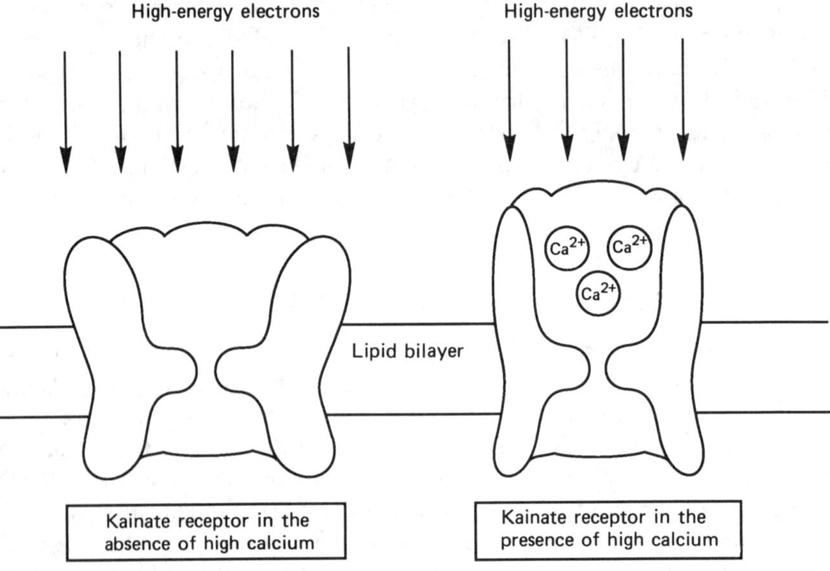

Fig. 1 — This figure, which is exaggerated for clarity, represents the hypothetical action of calcium ion binding to modify the cross-sectional area of a protein undergoing high-energy electron inactivation. The result of this calcium-induced conformational change is to suggest that the protein has undergone a reduction in apparent molecular weight.

subunit from the larger 76 400-Da protein. Calcium-activated protein proteolysis might also explain the observed change in molecular weight.

More recent electron inactivation studies with *Xenopus* central nervous system (CNS) tissue have suggested that the [^{3}H]KAIN binding site is located on an oligomeric protein of 240 kDa (Ambrosini *et al.* 1990). Interestingly, an identical molecular weight has been reported for the purified GABA A receptor (Mamalaki *et al.* 1989). Using [^{3}H]-glutamate binding to the cockroach nervous system, radiation inactivation afforded a binding protein molecular weight of 77 800 ± 7900 Da (Sepulveda and Sattelle 1989). However, in this case, KAIN did not displace [^{3}H]glutamate binding, suggesting significant differences between apparently similar-molecular-weight excitatory amino acid binding proteins.

Estimation of NMDA binding protein molecular weight by radiation inactivation techniques have been reported by several workers. In the first study, [^{3}H]glutamate, [^{3}H]TCP and [^{3}H]glycine bound to similar sized proteins of 121 000 ± 17 000, 118 000 ± 10 000 and 115 000 ± 10 000 Da, respectively (Honoré *et al.* 1989b). (In a separate study, [^{3}H]MK-801 was found to bind to a protein of 128 000 ± 9000 Da molecular weight (Wong and Nielsen 1989).) In their most simple interpretation, these results suggest that all three ligands bind to the same protein subunit — although from the curvilinearity detected in the binding of [^{3}H]glutamate, the possible presence of a high-molecular-weight modulatory protein was also suggested. Surprisingly, the competitive antagonist [^{3}H]CPP bound to a significantly larger protein of apparent molecular weight 209 000 ± 19 000 Da. Clearly, these results indicate that the binding of competitive antagonists is dependent on the cooperative interaction of additional NMDA receptor protein subunits.

RECEPTOR SOLUBILIZATION AND ISOLATION

Using both ionic and non-ionic detergents, excitatory amino acid receptors have been solubilized from vertebrate CNS tissue. Binding studies on KAIN receptors solubilized from *Xenopus* CNS tissue revealed a similar agonist affinity in both solubilized and non-solubilized tissue (Henley and Barnard 1989). These observations suggest that the KAIN receptor configuration is little disturbed by isolation from its lipid matrix. Using antibodies raised against KAIN receptors solubilized from frog brain, a number of homologous proteins of molecular weight 48 000 Da were isolated by immunoprecipitation of proteins isolated by antiserum-based affinity chromatography (Hampson *et al.* 1989). In addition to immunoreacting with numerous and closely related 48 000-Da proteins, two monoclonal antibodies, mAb KAR-D1 and mAb KAR-B1, also cross-reacted with a protein of 99 000 molecular weight. In addition, a 55 000-Da protein was also recognized by this antibody. On immunoblotting, reducing agents were capable of modulating the relative intensity of the 48 000 and 99 000-Da bands, suggesting a possible monomer/dimer relationship.

By cross-reacting frog-derived antibodies with rat brain tissue, proteins of similar molecular weight were identified. However, in this case, reduction had no effect on the intensity of the 99 000-Da band. Furthermore, in addition to a 55 000-Da band, mAb KAR-D1 also immunoreacted with proteins of 42 000 and 57 000 molecular weight. Using domoic acid-based affinity chromatography, a similar 48 000-Da protein was also isolated from frog brain (Hampson and Wenthold 1988). In addition, the active purified receptor complex also migrated on Sepharose gel filtration with a molecular weight of 570 000 Da. Scatchard analysis of binding to this purified receptor was best fit by a two-site (high/low affinity) model. From both domoate and (*RS*)-2-amino-3-(3-hydroxy-5-bromomethylisoxazol-4-yl)propanoic acid (ABPA) derivatized affinity columns, a major band (> 90%) of 42 000 Da was also isolated from *Xenopus* CNS tissue (Ambrosini *et al.* 1990). However, two bands of 48 000 and 55 000 Da were also consistently identified. These, and additional observations clearly support the existence of a unitary AMPA/KAIN receptor, at least in *Xenopus* CNS. Affinity purification of KAIN receptor protein from goldfish has afforded as the major product a 49 000-Da protein. A less intense band of 43 000 Da was also observed (Ziegra and Oswald 1989).

Quisqualate-sensitive glutamate binding activity has been solubilized from rat retina and pituitary (Yoneda and Ogita 1989a, Yoneda and Ogita 1989b). Ligand binding to solubilized and membrane-bound protein was similar and was significantly attenuated by cooling. AMPA binding sites have also been solubilized and partially purified from rat brain (Hunter *et al.* 1990). In both solubilized and membrane-bound protein, addition of thiocyanate both increased total binding and revealed high- and low-affinity [^{3}H]AMPA binding sites. Both solubilized and membrane-bound receptors recognized ligands with a similar rank order. However, with the interesting exception of the antagonist CNQX, agonists and antagonists bound with higher affinity to the solubilized protein than to membrane-bound. These AMPA binding sites migrated with a molecular weight of 425 000 Da on gel filtration chromatography. This high molecular weight, which approaches that of purified KAIN receptors (Hampson and Wenthold 1988) suggests that these proteins are either very heavily glycosylated or form (or naturally exist in membranes as) dimeric pairs under solubilizing conditions.

From the viewpoint of isolation and protein characterization, the NMDA receptor has received the most attention. The NMDA receptor channel complex has been solubilized from rat and porcine brain by numerous workers (Ambar *et al.* 1988, McKernan *et al.* 1989, McVittie and Sibley 1989, Ogita and Yoneda 1990b). In these studies the solubilized receptors bound [^{3}H]MK-801, [^{3}H]TCP and other non-competitive antagonists with a similar rank order but, with the exception of SKF 10047, with a slightly lower affinity to that recorded in membrane-bound protein — although in several of these studies it was recognized that high detergent concentrations inhibited the level of total binding. As with membrane-bound receptors, non-competitive binding was modulated by AP5 (Ambar *et al.* 1988), glutamate, glycine and divalent ions (McKernan *et al.* 1989), and polyamines (Ogita and Yoneda 1990b). However, both zinc and cadmium, which are recognized to bind to a voltage-independent site, were 100 times more potent in inhibiting MK-801 binding in solubilized receptor than in membrane-bound (McKernan *et al.* 1989). This inhibitory site may normally be masked by non-specific membrane-associated charges, or may simply be located in a part of the protein complex that is made more accessible by solubilization.

Although the NMDA receptor, solubilized by the procedure of Cunningham and Michaelis (1990), bound a range of ligands with an affinity comparable to that of the membrane-bound receptor, significant differences were noted in the modulation of

[^{3}H]CPP binding by magnesium, glycine and CNQX (Cunningham and Michaelis 1990). Rather than the slight (33%) inhibition of [^{3}H]CPP binding by 100 μM glycine observed in the membrane-bound receptor, a 195% enhancement of binding was seen in the solubilized receptor. However, a similar magnitude of inhibition of membrane-bound [^{3}H]CPP binding by 100 μM CNQX (44%) was transformed to an even greater inhibition (86%) in the solubilized form. This study also suggested that magnesium ions (other than those thought to block the ion channel in a voltage-dependent manner) may also be tightly associated with the NMDA receptor and increase binding to the competitive antagonist-associated site. A recent study has confirmed the existence of such a high-affinity magnesium binding site (Hori *et al.* 1990). Interestingly, in a separate study by Marvizon and Skolnick (1988), it was noted that in EDTA-washed membranes [^{3}H]glycine binding was markedly enhanced in a concentration-dependent manner by magnesium. CPP (10 μM) partially counteracted this stimulation, whereas L-glutamate had no effect. These results have been interpreted to suggest that this high-affinity magnesium site is more closely associated with the glycine and CPP recognition sites than with the glutamate site and that the agonist and antagonist sites might in fact be located on different but closely interacting receptor subunits (see below) (Cunningham and Michaelis 1990). More recent results from this group have suggested that the NMDA receptor is comprised of four different subunits of M_r 70 000, 58 000, 42 000 and 36 000 (Ly and Michaelis 1991, Michaelis 1991). Receptor stoichiometry has yet to be firmly established (see later). It is apparent that for NMDA recognition all four proteins are required to be associated (Michaelis 1991). Using sucrose density gradient centrifugation and size exclusion chromatography of the solubilized and purified receptor, Michaelis has shown that the strychnine-insensitive glycine binding site is located on the smaller 42-kDa protein (Michaelis 1991). In addition, it has been further noted that the 70-kDa protein appears to prefer to exist as a dimer. Prior dithiothreitol protein reduction and iodoacetamide alkylation did not interfere with this dimerization. Therefore this association does not appear to be covalent (disulphide-linked) in character (Michaelis 1991). Completing the observations of this group, the 36-kDa protein appears to be strongly associated with the other receptor subunits, whereas that of M_r 42 000 appears to be more loosely associated.

In a separate study, solubilization of glutamate binding sites from porcine brain afforded proteins which bound glutamate, L-aspartate, cysteate and homocysteate (all NMDA agonists) but not NMDA, KAIN or quisqualate (Chang *et al.* 1990). Ligand binding was removed by trypsin or proteinase k digestion or by fractionation on gel filtration chromatography. However, some ligand binding was detected upon recombinations of different fractions containing proteins of M_r 330 000 and 34 000. Protein cross-linking of the solubilized receptor with brief glutaraldehyde treatment also preserved glutamate binding even after gel filtration fractionation. In this experiment a single slightly higher-molecular-weight glutamate-sensitive protein band was eluted. These observations strongly suggest that for NMDA-sensitive glutamate binding at least two protein subunits are required for ligand recognition. In addition, the presence of phospholipid may play an important role in receptor function.

Both agonist and non-competitive antagonist-based affinity chromatography has been used to purify and characterize the NMDA receptor complex. Purifying sodium cholate-solubilized rat forebrain NMDA receptors through an amino-phencyclidine-agarose affinity column, followed by polyacrylamide electrophoresis in the presence of detergent and dithiothreitol, afforded four major protein bands of M_r 67 000, 57 000, 46 000 and 33 000 (Ikin *et al.* 1990). In this study [^{3}H]azido-PCP bound to purified receptor was irreversibly incorporated into each of these individual proteins upon UV irradiation. The 57- and 46-kDa bands were more strongly labelled than that at M_r 67 000. The 33-kDa band was labelled only after short incubation times.

Retention of solubilized NMDA receptors over an ibotenate-derived affinity column followed by elution with 2 mM magnesium afforded a 58–60-kDa protein which bound [^{3}H]CPP but not glutamate (Cunningham and Michaelis 1990). [^{3}H]CPP binding was also inhibited by 2-amino-7-phosphonoheptanoic acid (AP7), AP5 and CNQX. Of reference agonists, only quisqualate inhibited [^{3}H]CPP binding. A strong interaction of this 58–60-kDa protein with concanavalin A–biotin was also observed, suggesting the presence of significant protein glycosylation. Surprisingly, elution of residual protein from the ibotenate affinity column with ibotenate did not afford additional receptor protein. However, changing the eluent to 1 M KCl did wash out an additional protein which bound glutamate with a similar affinity to that observed in membrane-bound receptor. Binding of [^{3}H]glutamate was also inhibited by other agonists, including NMDA. Glutamate binding was weakly inhibited by phosphonate-based competitive antagonists. The surprising inability of ibotenate to elute receptor proteins was ascribed to the additional yet strong non-specific binding of receptor sugar residues to the agarose column matrix.

Looking further at protein identification by photoaffinity labelling, photolysing [^{3}H]azidophencyclidine ([^{3}H]azido-PCP) bound to the NCB-20 cell line followed by solubilization and gel electrophoresis (in the absence of dithiothreitol) afforded five radiolabelled bands of M_r at 90 000, 68 000, 49 000, 40 000 and 33 000 — a pattern similar to that seen in rat brain (Haring *et al.* 1990). Of these proteins, labelling to the 90 000 and 68 000-Da bands was inhibited by preincubation with MK-801 and AP5. Labelling to the 33-kDa band was partially inhibited by antagonist binding. Photolabelling guinea-pig membranes preincubated with [^{3}H]azido-MK-801 afforded a single protein band of M_r 120 000 upon gel electrophoresis (Sonders *et al.* 1990). Photolabelling of this protein was also inhibited by preincubation with MK-801 or PCP.

The source of tissue, the difference in photolabile ligand and procedure used (photolabelling solubilized rather than membrane-bound receptors) clearly complicates the interpretation of these apparrently contradictory results. However, the isolation of only smaller peptide units in the presence of the disulphide reducing agent dithiothreitol might suggest that the larger molecular bands observed in the studies of Haring *et al.* (1990) and Sonders *et al.* (1990) may actually represent one or more strongly associated or disulphide-linked receptor subunits. The fact that the larger protein bands were derived from photolabelling *membrane-bound* receptors might also suggest that an additional and larger membrane-associated protein was additionally exposed and incorporated upon photolabelling.

The function of the 33-kDa protein isolated from the purified NMDA receptor is unknown. However, its small size relative to other known ligand-gated ion channel receptors suggests that it, like the 43-kDa protein of the nicotinic receptor (Toyoshima and Unwin 1989), may be located intracellularly and may play a role in receptor regulation or membrane anchoring. A role in receptor aggregation is also possible.

Starting with a previously isolated 14-kDa fragment, antibodies have been raised which recognize and immunoextract two glutamate binding proteins of M_r 70 000 and 63 000 Da (Chen *et al.* 1988, Eaton *et al.* 1990). From concanavalin A labelling it appears likely that the 63-kDa protein is a non-glycosylated precursor of the larger 70-kDa protein. These antibodies, although ineffective at inhibiting the binding of [^{3}H]CPP to its binding protein (see above), did inhibit the binding of glutamate to solubilized NMDA receptors and to the 63- and 70-kDa proteins (Cunningham and Michaelis 1990). Recent data show that these antibodies also inhibit the binding of NMDA itself to its receptor (Michaelis 1991). Therefore these antibodies, which in immunogold staining label postsynaptic densities in brain regions enriched in glutamate neurotransmitter pathways, appear to be NMDA receptor specific (Eaton *et al.* 1990). Recently, a monoclonal antibody (B6) has been generated which has been suggested to modulate the NMDA receptor in a manner identical to that of glycine (Moskal *et al.* 1989). The structural implications of this observation with regard to the accessibility of the glycine recognition site have yet to be firmly established. Extending the use of antibodies to probe ligand binding site structure, antibodies have also been raised against glutamate and aspartate (Petrusz *et al.* 1990). Using spinal cord sections the immunocytochemical staining by the glutamate antibody was inhibited by quisqualate but not by KAIN or NMDA. Similarly, the staining by the aspartate antibody was blocked by KAIN and to lesser extent by quisqualate.

THE STOICHIOMETRY AND CONFORMATIONAL RELATIONSHIP OF RECEPTOR SUBUNITS

Using quantitative autoradiography, the densities of [^{3}H]glutamate and [^{3}H]TCP receptor binding in the same membrane preparation have been estimated (Maragos *et al.* 1988). From this study it was concluded that there existed four to five glutamate binding sites for each TCP recognition site. Using instead quantitative receptor binding techniques, Fagg and colleagues reported that the concentration of L-[^{3}H]glutamate, [^{3}H]CPP and [^{3}H]MK-801 binding sites were similar (Thedinga *et al.* 1988). However, twice as many strychnine-insensitive [^{3}H]glycine binding sites were detected, suggesting a glutamate:CPP:MK-801:glycine site stoichiometry of 1:1:1:2. Interpreting the suggestion of Monaghan *et al.* (1988) that the NMDA receptor may bind ligands in an agonist-preferring or antagonist-preferring state, Fagg postulated that perhaps the binding site concentrations of glutamate and CPP should be combined to give a true ligand stoichiometry of 2:2:1(agonist/antagonist: glycine:MK-801). Perhaps suggesting little cooperativity in ligand recognition, the Hill coefficients of binding to each site were close to unity. However, again using the glutamate stimulation of [^{3}H]MK-801 binding in rat cortex and hippocampus, a Hill coefficient of 2 was later observed (Javitt *et al.* 1990). These data suggest that binding of two glutamate molecules is required to open the associated NMDA ion channel fully.

Previous electrophysiology studies in patched clamped neurons or *Xenopus* oocyte-expressed NMDA receptors have been associated with Hill coefficients in excess of unity (Verdoorn and Dingledine 1988, Wong *et al.* 1988). A bimolecular action has been recorded for both NMDA agonists and antagonists in hippocampal slices (Williams *et al.* 1988). However, the complex conformational interrelationship of NMDA binding sites coupled with the inability to remove residual glycine completely has complicated the clear interpretation of many of these results. More recently, using fast application–concentration jump techniques in cultured hippocampal neurons, it has been confirmed that two molecules of both NMDA and glycine are required to bind to the NMDA receptor for channel activation to occur (Benveniste and Mayer 1991). This agonist stoichiometry agrees with results obtained from a previous equilibrium dose–response analysis using low concentrations of NMDA (Patneau and Mayer 1990). Extrapolating from stoichiometric and subunit molecular weight studies, plausible NMDA subunit compositions can be derived. Two of these possibilities are shown in Fig. 2. The interrelationship of NMDA receptor binding sites (and therefore perhaps receptor subunits) is complex. Using comparative [^{3}H]glutamate and [^{3}H]CPP autoradiography in the rat brain, Monaghan *et al.* (1988) noted that, relative to [^{3}H]glutamate, [^{3}H]CPP binding was low in the striatum and septum and high in the thalamus and inner cerebral cortex (Monaghan *et al.* 1988). In competition studies NMDA antagonists were relatively more potent than agonists in displacing [^{3}H]glutamate binding from the thalamus and cortical regions. The converse was true in the striatum and septum. These data were interpreted to suggest that the NMDA receptor might exist in either an agonist- or antagonist- preferring conformation or that these agonist- and antagonist preferring regions might represent distinct receptor subtypes. Perhaps supporting the existence of interconverting forms, Monaghan *et al.* (1988) also noted the addition of glycine increased the binding of [^{3}H]glutamate more in 'antagonist-preferring' than in the 'agonist-preferring' brain regions. In contrast, the binding of [^{3}H]CPP was inhibited by the addition of glycine.

The interaction of glutamate, glycine and their antagonists is becoming clearer. In general, a reciprocal relationship has been observed. That is, the glutamate antagonist CGS 19755 decreased the affinity of glycine agonists for their receptor (Hood *et al.* 1990). Conversely, the affinity of glycine antagonists is increased by the addition of CGS 19755. This interrelationship is, however, somewhat more complex in that not all

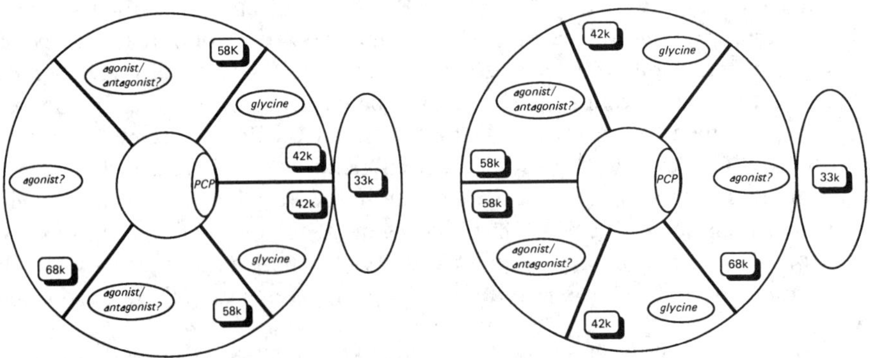

Fig. 2 — Hypothetical configurations of protein subunits of the NMDA receptor. From protein isolation and ligand binding site stoichiometry studies summarized in the text, this figure represents two hypothetical arrangements of protein subunits and their associated ligand binding sites that may comprise the NMDA receptor complex. As shown, these arrangements illustrate $\alpha_2\beta_2\gamma\delta$ subunit compositions.

competitive glutamate antagonists are equipotent in inhibiting glycine binding. Rather, the potency (but not efficacy) of inhibition was significantly greater for the C-5 than the C-7 phosphonoalkyl amino acid-based glutamate antagonists (Monahan *et al.* 1990). Interestingly, it is the IC_{50} of the C-5 antagonists for inhibition of glycine binding which more closely paralleled their potency in displacing [^{3}H]glutamate binding and inhibiting the binding of [^{3}H]MK-801. However, confirming the same antagonist recognition site, the addition of C-7 antagonists readily reversed the glycine inhibitory action of their C-5 homologues. Demonstrating further the reciprocal binding relationship of glycine and competitive antagonists it has been shown by Kaplita and Ferkany (1990) that glycine and D-serine also inhibit the binding of [^{3}H]CGS 19755, but not CPP (a C-7 analogue). Surprisingly, although [^{3}H]CPP binding is unaffected by glycine, it is stimulated (through an increase in affinity) by the glycine partial agonist HA-966 (Danysz *et al.* 1989). The heteroaromatic-based glycine antagonist 7-chlorokynurenic acid abolished [^{3}H]CPP binding. Completing the picture, HA-966 partially inhibited the binding of [^{3}H]glutamate to its receptor.

The absolute requirement for glycine for the operation of the NMDA channel was first recognized in NMDA receptors expressed in *Xenopus* oocytes (Kleckner and Dingledine 1988). Recording from either acutely isolated or cultured neurons, this absolute requirement for glycine has been confirmed. Looking further at the action of glycine at the receptor, it has been shown that the application of glycine potentiates both the peak and steady state responses of the NMDA receptor to the application of glutamate (Vyklicky *et al.* 1990, Shirasaki *et al.* 1990). At higher glycine concentrations this potentiation is also accompanied by reduced receptor desensitization In assigning a mechanism to glycine's action Vyklicky *et al.* (1990) suggested that a combination of regulation of transitions of the NMDA channel complex to the active state and a reduction in desensitization was likely. A similar conclusion was also reached by Lerma *et al.* (1990). Investigating further the kinetics of glycine's action, Benveniste *et al.* (1990a) found that, in a concentration-dependent manner, glycine both increased the rate of onset of desensitization and the rate of recovery. With increasing glycine concentration, it was noted that the rate of recovery from desensitization increased more rapidly than the rate of onset. Glycine analogues of lower affinity produced desensitization with faster kinetics. From this study Benveniste *et al.* (1990a) concluded that for the NMDA receptor, transitions to the open state do not occur unless glycine has first bound to a closed state of the receptor. Potentiation of NMDA responses by glycine results from the absolute requirement for glycine in promoting transitions from the closed to the open state. The paradoxical glycine-sensitive desensitization results from an agonist-triggered lowering of the affinity for glycine which, at low glycine concentrations, causes it to dissociate from the closed state of the receptor complex. This relationship, which is reciprocal, is an example of negative cooperativity between the binding of glutamate and glycine to the NMDA receptor. As noted previously, the model which best fits the observed data suggests that the NMDA receptor has four independent amino acid binding sites — two for glutamate and two for glycine (Fig. 2) (Benveniste *et al.* 1990a, Benveniste and Mayer 1991).

Although reduction in desensitization provides a rational explanation for the action of glycine, glycine-insensitive desensitization of NMDA responses has been reported in outside-out patch clamping in cultured mouse embryonic neurons (Sather *et al.* 1990). Leaving aside possible methodological explanations for such an effect, these observations may be rationalized by the existence of differences in cell body (from which patches are drawn) or dendritic localized NMDA receptors (Sather *et al.* 1990) or a cytoplasmic regulation of NMDA receptor properties by an as yet unknown mechanism (Mayer 1991). A further explanation may be found in the ontogenic development of the NMDA receptor and its component protein subunits. Using quantitative autoradiography, McDonald *et al.* (1990) have clearly shown that over a range of brain regions NMDA-sensitive [^{3}H]glutamate binding developed and peaked more rapidly than either [^{3}H]glycine or [^{3}H]TCP binding, which themselves developed at a similar slower rate. Interestingly, the rate of development of NMDA-sensitive [^{3}H]glutamate binding varied significantly over different brain regions. These data might suggest that the glycine-insensitive desensitization observed in embryonic neurons may result from over-expression of the glutamate subunit or incomplete development or coupling of the glycine receptor protein to the NMDA receptor complex. Furthermore, the study of McDonald *et al.* (1990) points to the likelihood that the NMDA receptor arises from the combination of different gene products since protein expression appears to be differentially controlled. The observation that the [^{3}H]glycine and [^{3}H]TCP binding sites develop at a similar rate also suggests that these binding sites may be located on the same protein subunits.

A varied or changing relationship of the strychnine-insensitive glycine binding site and the NMDA-sensitive glutamate receptor has been suggested by several other studies. Glycine-insensitive NMDA receptors were expressed in *Xenopus* oocytes injected with guinea-pig cerebellar mRNA (Sekiguchi *et al.* 1990). NMDA receptors derived from expression of cerebral mRNA were, however, normally sensitive to glycine. Perhaps relating the modulation of receptor expression to pathology, it has been shown from autoradiography in aged rats that in specific brain regions, notably the telencephalon, [^{3}H]glycine binding is severely decreased relative to [^{3}H]CPP binding sites (Kito *et al.* 1990). Comparing tissue from post-mortem Alzheimer's patients and age-matched controls, a significant reduction in the stimulation of [^{3}H]MK-801 and [^{3}H]TCP binding by glycine was detected (Procter *et al.* 1989a, 1989b, Steele *et al.* 1989). These data suggest that coupling of the glycine receptor to the NMDA complex is in some way impaired or that the structural composition of the receptor has changed. Further observations suggesting regional variation in glycine/NMDA receptor coupling have also been reported (Brugger *et al.* 1990, Danysz and Wroblewski 1989).

ASSESSING THE INTERNAL STRUCTURE OF EAA RECEPTORS

It seems reasonable to assume that the ion channel made up by component receptor protein subunits may have a physical shape (Fig. 3) similar to that established for the nicotinic acetylcholine receptor (Toyoshima and Unwin 1988, Toyoshima and Unwin 1990). Characterization of the anionic charges that might line the first vestibule of the non-NMDA receptor channels has in part been revealed by cloning of these receptors (see below). However, such data for the NMDA receptor can only be indirectly inferred.

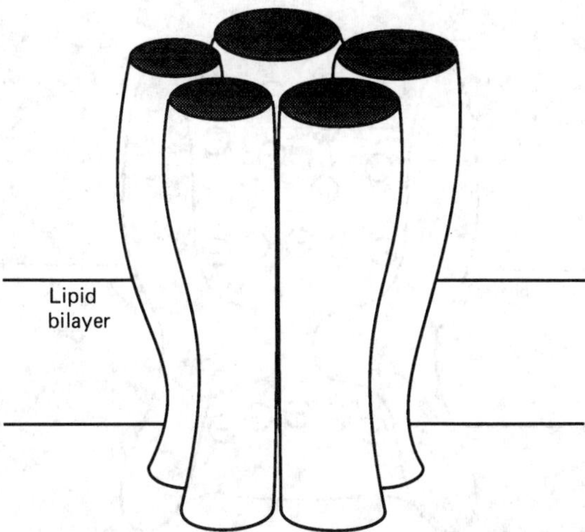

Fig. 3 — Extending observations from the pentameric nicotinic receptor channel complex, it is likely that AMPA/Kainate (KAIN), and probably NMDA receptor channel complexes, are composed of similar pentameric arrangements of protein subunits anchored in the cellular lipid membrane. The pore described by these proteins makes up the path through which permeant ions enter the cell.

The transmembrane spanning pore dimension for both NMDA and non-NMDA receptors has been investigated by comparing the permeabilities of different size cations (Vyklicky *et al.* 1988). From this study, it was noted that the relative monocationic permeability of the KAIN and nicotinic channel were very similar (the nicotinic channel has been modelled as a water-filled cation-selecting pore about 0.65 × 0.65 nm in size (Dwyer *et al.* 1980)). Using different monovalent cations, the NMDA channel pore was estimated to be about 10% smaller than that of the KAIN channel. The rat KAIN receptor expressed in *Xenopus* oocytes is blocked in a highly voltage-dependent manner by tetrabutylammonium ion (Randle 1990). Shorter-chain analogues were relatively ineffective and longer-chain analogues more effective (but toxic) in blocking conductance. The rapid rate of association and dissociation of tetrabutylammonium ion and higher homologues indicates that at least in oocytes the KAIN channel has a broad external opening. Tetrabutylammonium ion appears to traverse 80–90% of the membrane electrical field before reaching its blocking site. This site may simply represent a narrowing of the channel. Since tetrapropylammonium ion is weakly permeant in the KAIN channel, the membrane spanning pore is thought to be at least 0.8–1.0 nm in diameter (Randle *et al.* 1988).

Although suggested to have similar pore dimensions, KAIN and AMPA receptor channels strongly discriminate in favour of sodium ions rather than calcium, whereas the NMDA receptor is permeant to both (a calcium-permeant KAIN channel has been described recently by Iino *et al.* (1990). Rationalizing this ionic selectivity, it has been suggested that the outer vestibule (and inner surface) of the NMDA receptor is lined with numerous anionic sites (Fig. 4) (Ascher and Johnson 1990). This anionic environment stabilizes and hence 'concentrates' divalent ions more than monovalent.

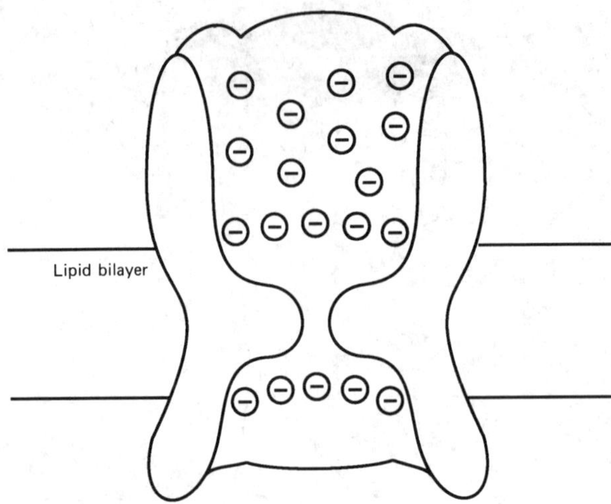

Fig. 4 — From the electrophysiology of the NMDA receptor, shown here in cross-section, it has been concluded that the outer vestibule and area immediately adjacent to the lipid membrane must contain a high concentration of negatively charged amino acid residues.

As a result, the NMDA receptor is more permeable to calcium ions than either KAIN or AMPA receptors. In addition to external anionic sites, from studies with the magnesium channel block at least two anionic sites are also thought to be located within the transmembrane electrical field (Ascher and Johnson 1990).

CLONING AND SEQUENCING OF EAA RECEPTORS

In the final weeks of 1989, three reports were published simultaneously that described for the first time the sequence determination of KAIN binding proteins (KBP) isolated from rat, frog and chick brain (Hollmann *et al.* 1989, Wada *et al.* 1989, Gregor *et al.* 1989, Wenthold *et al.* 1990). Protein from frog and chick were of similar molecular weight (48 and 49 kDa, respectively) and appeared to contain four possible membrane spanning domains in addition to other structural features including anionic, glycosylation and phosphorylation sites which supported a role for these proteins as part of a cation selective ligand-gated ion channel. However, regardless of these superficial similarities, these proteins also exhibited rather weak overall homology to the well-established ligand-gated ion channel super-family comprising the nicotinic, glycine and GABA receptors.

When expressed in COS-7 cells and *Xenopus* oocytes, the frog-derived KBP exhibited a similar pharmacology to that of biochemically purified KBP. However, this protein was electrophysiologically inactive. These observations suggest that although providing the necessary receptor binding domains a homo-oligomeric receptor comprising this 48 000-Da protein was insufficient for full receptor channel function. The chick-derived KBP, which was shown to be localized exclusively on Bergmann glial membranes, was not expressed in cell lines.

Isolated by an alternate cloning strategy, the rat-derived KBP reported by Hollmann *et al.* (1989) (GluR-K1) was found to be significantly larger, with an M_r of 99 800. Upon expression of the cDNA-derived mRNA in *Xenopus* oocytes, this homo-oligomeric protein complex formed an ion channel possessing pharmacological and electrophysiological properties similar to the mammalian KAIN receptor. Although similar in sequence in the membrane spanning regions, this rat-derived protein has a significantly larger amino terminal region than either that of frog or chick KBPs. In addition, due to a difference in the assignment of the first transmembrane spanning domain (TMD) by Hollmann *et al.* (1989) and other workers, differences in the character and length of the first intracellular loop were also apparent (Fig. 5). The pharmacology of the homomeric receptor has been recently studied in more detail (Dawson *et al.* 1990).

From the tools provided by these early successes, families of closely related genes encoding KAIN and AMPA receptors have since been isolated and expressed (Boulter *et al.* 1990, Keinänen *et al.* 1990). The first of these families contains three non-glycosylated proteins of M_r 99 800 (GluR1 ≡ GluR-K1), 96 400 (GluR2) and

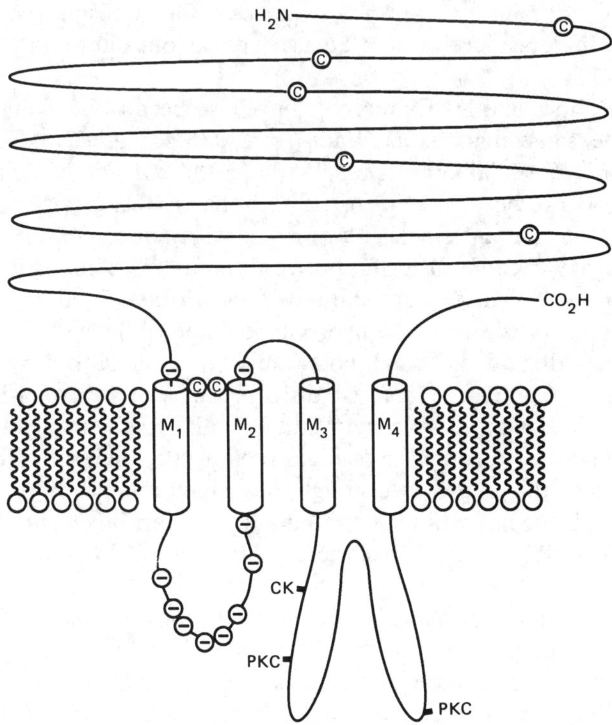

Fig. 5 — A simplified diagrammatic representation of a cloned and sequenced 90-kDa AMPA/Kainate (KAIN) protein subunit embedded in a lipid bilayer. Key structural features are: ©, cysteine residues; ⊖, glutamic acid or aspartic acid residues; PKC, protein kinase C consensus sequence; CK, calmodulin-dependent protein kinase II consensus sequence.

98 000 (GluR3). Each sequence contains numerous N-terminal glycosylation sites and five hydrophobic regions (four membrane spanning and one characteristic of a signal peptide). Each sequence shows significant overall identity with the others (69–74%). Greatest homology was found in the large intracellular loop spanning transmembrane domains III and IV.

A second family of AMPA-selective glutamate receptor proteins reported by Keinänen *et al.* contained four members: GluR-A, B, C and D (Keinänen *et al.* 1990). Each protein was of similar size (approximately 900 residues) and showed significant sequence homology with each other. Although determined to be AMPA-selective, GluR-A was identical to the KAIN-selective GluR1 of Hollmann *et al.* (1989). Membrane spanning regions were assigned in a manner similar to that of Wada *et al.* (1989) and Gregor *et al.* (1989). Resulting from these assignments of membrane spanning domains, those regions expected to be part of the inner and outer channel mouth were shown to have a preponderance of negatively charged residues (Fig. 5). Of particular interest is the identification of two glutamate residues flanking the TMD II. This second membrane spanning unit is, in other ligand-gated ion channels, the one most commonly associated with lining the channel pore. This arrangement of negative charges is somewhat similar to that seen in the nicotinic channel (Barnard *et al.* 1990) and is presumed to determine overall channel ion selectivity and conductance. In addition to extracellular glycosylation sites, the intracellular loop spanning TMB III and IV also contains consensus sequences for phosphorylation by both Ca^{2+}-calmodulin-dependent protein kinase type II (one site) and protein kinase C (two sites) (Fig. 5).

Additional clones of a KAIN receptor have also been isolated and TMD regions assigned in the same manner as Wada *et al.*, Gregor *et al.* and Keinänen *et al.* (Nakanishi *et al.* 1990, Sakimura *et al.* 1990). In these assignments transmembrane domain II is noted to be similar in structure to the GABA, glycine and nicotinic ion channels in that it contains small polar or neutral residues in every fourth position such that they are capable of forming a continuous ridge on the TMD II helical surface (Unwin 1989). This surface, along with four others in a pentameric complex (Fig. 3), is thought to comprise the hydrophilic lining of the channel pore. As in the nicotinic channel, this ridge of small polar amino acids is flanked by a ridge of large hydrophobic residues. This conserved distribution would be consistent with one suggested gating mechanism whereby channel closure results from a helical displacement in which larger hydrophobic residues replace the smaller hydrophilic residues from the lining of the channel. Although electrophysiologically inactive when expressed alone, the conservation of this amino acid arrangement in the TMD II domains of the KBPs suggests that these proteins are also components of an ion channel.

From the biochemical evidence reviewed previously, however, the function of thiol residues in controlling KAIN receptor structure and function is uncertain. The strongly conserved distribution of such residues revealed in the cloned receptors suggests that they play a critical role in maintaining the shape of the extracellular protein matrix. Furthermore, for the more commonly accepted hydropathic assignment of membrane spanning domains I and II, two highly conserved cysteine residues are found at the extracellular face of each of these linked helical domains (Nakanishi *et al.* 1990). From their probable juxtaposition in the assembled receptor, it seems likely that these thiol residues may, under normal conditions, be oxidized to form a disulphide bridge. Alternatively, one of these residues may instead be linked to a neighbouring protein subunit or, as demonstrated for the nicotinic receptor, to a second receptor complex (Toyoshima and Unwin 1990). Interestingly, in the frog and chick kainate binding protein sequences, two cysteines were also found in membrane spanning domains. However, in these proteins both cysteines were localized in TMD I; none were identified in TMD III.

Although the isolation and sequencing of these glutamate binding proteins in itself represents a major achievement, it is the homomeric and heteromeric expression of these proteins in both *Xenopus* oocytes and mammalian cells which has finally clarified the confusion surrounding the unity or dissimilarity of KAIN and AMPA receptors (Boulter *et al.* 1990, Keinänen *et al.* 1990). These studies have clearly shown that the pharmacological selectivity for KAIN or AMPA and the resulting electrophysiological characteristics of the stimulated channels is completely dependent on the homomeric or heteromeric protein composition of the individual channels.

Adding to the complexity of this situation, it has more recently been shown that for all four proteins GluR-A, B, C and D, alternate splicing of the separate gene products can occur (Sommer *et al.* 1990). The alternate splicing of the mRNA, which shows distinct expression patterns particularly in the CA1 and CA3 regions of the hippocampus, results in the variation of a small (9–11 residues) intracellular region bordering the fourth membrane spanning domain. These alternate intercellular protein sequences were designated as 'flip or flop'. The electrophysiological properties of both homomeric and heteromeric channels with either flip and flop modules expressed were different when activated by glutamate and AMPA from when activated by KAIN. In heteromeric receptor complexes of GluR-A(flip)/GluR-B-(flip), saturating concentrations of glutamate and KAIN elicited similar magnitude current. However, in GluR-A(flop)/GluR-B(flop), KAIN generated a much larger current than glutamate. In heteromeric complexes of GluR-A(flip)/GluR-B(flop) and GluR-A(flop)/GluR-B(flip), electrophysiological responses appear to be dominated by GluR-A.

Interestingly, despite these intracellular changes, agonist binding affinity was essentially unaffected. Therefore these rather modest changes to the TMD III/IV protein loop appear to modify channel function through controlling channel activation, gating, and desensitization rather than by modifying agonist affinity.

Although those clones isolated by Nakanishi *et al.* (1990) were shown to be equivalent to GluR-B(flip) and GluR-C(flip) of Keinänen *et al.* (1990), these workers also recognized that 100 amino acid segments of both frog and chick KAIN binding proteins (amino terminal residues 22–130) and three GluR subunits (395–510) had 20–30% sequence identity and 15% conservative substitution with the periplasmic glutamine binding protein (GlnH) of *Escherichia coli* (an essential component of *E. coli* glutamine permease — a glutamine uptake system) (Nakanishi *et al.* 1990). This regional homology, which for the KBPs represents the majority of their amino terminal domains, strongly suggests that this region is likely to constitute the glutamate binding site. Although the three-dimensional structure of the glutamine binding protein has not been elucidated, related bacterial binding proteins are known to possess very similar globular domains. Upon binding to a cleft formed by the two domains, it has been proposed that the ligand brings about the required conformational change. A similar relationship for the agonist binding domains of these KAIN and AMPA-selective receptors might then exist (Fig. 6) (Nakanishi *et al.* 1990). Extrapolating yet further, the apparent homology of the *E. coli* glutamine binding protein and these glutamate receptors might also give a clue as to from where the non-NMDA glutamate receptors were originally derived.

A novel fifth glutamate binding protein has been cloned and expressed in *Xenopus* oocytes (Bettler *et al.* 1990). Pharmacological stimulation of this protein expressed alone did not elicit membrane depolarization. Co-expression with one of the earlier clones gave only a weak electrophysiological response. This protein displays a reduced (40–41%) amino acid identity with other glutamate receptor clones although it is slightly larger (920 amino acid residues). Suggesting that this protein might represent a G-protein coupled receptor, hydropathy plots reveal that this novel clone protein might comprise seven transmembrane spanning domains. However, from homology with other glutamate receptor clones, its incorporation in an ion channel appears more probable. Interestingly, this new clone had a unique distribution in the adult central and peripheral nervous system and also appears to play a role in neuronal differentiation and synapse formation.

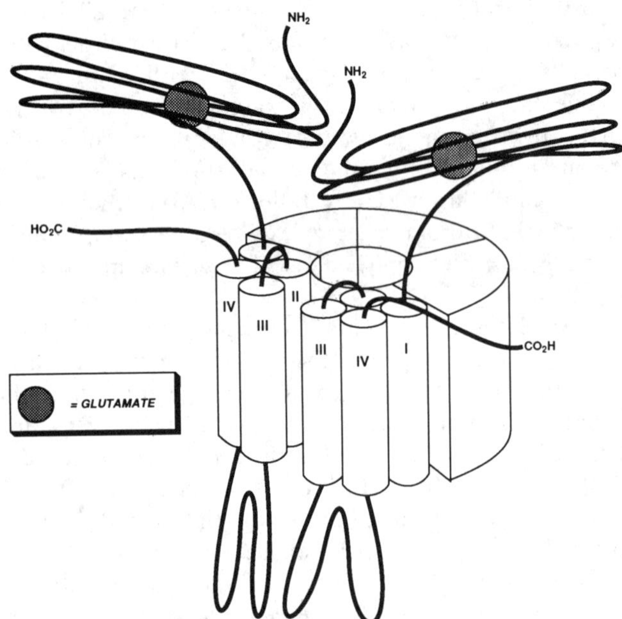

Fig. 6 — The amino terminal protein sequence of an AMPA/KAIN receptor subunit cloned and expressed by Nakanishi *et al.* (1990) contains an amino acid sequence similar to the glutamine binding domain of a known bacterial glutamine binding protein. Since other closely related bacterial amino acid binding proteins have been shown to bind ligand in a cleft formed by two globular domains, it is suggested that a similar binding cleft might comprise the glutamate binding domain of this AMPA/KAIN receptor subunit. Such a globular binding domain is represented in this figure.

A functional metabotropic glutamate receptor has been recently cloned, sequenced and expressed in *Xenopus* oocytes (Masu *et al.* 1991). Surprisingly, this receptor shows little sequence similarity to conventional G protein-coupled receptors and has a unique structure with large hydrophilic sequences at both ends of the seven membrane spanning domains. Although similar in size to the previously reported ionotropic AMPA/KAIN receptors, the overall sequence homology with these receptors is rather low, with 16.7% of identical amino acids and 36% of exact matches, and conservative substitutions extending over the whole amino terminal domain. In addition to a number of charged residues adjacent to the membrane spanning regions, several potential sites for glycosylation and phosphorylation were also identified (Fig. 7). Perhaps confirming a distant link to other G protein-coupled receptors, several unusual cytosolic sequences of proline/glutamine and glutamate/aspartate-rich regions were identified in the intracellular carboxy terminal protein. Similar amino acid sequences have been observed in both adrenergic and muscarinic receptors.

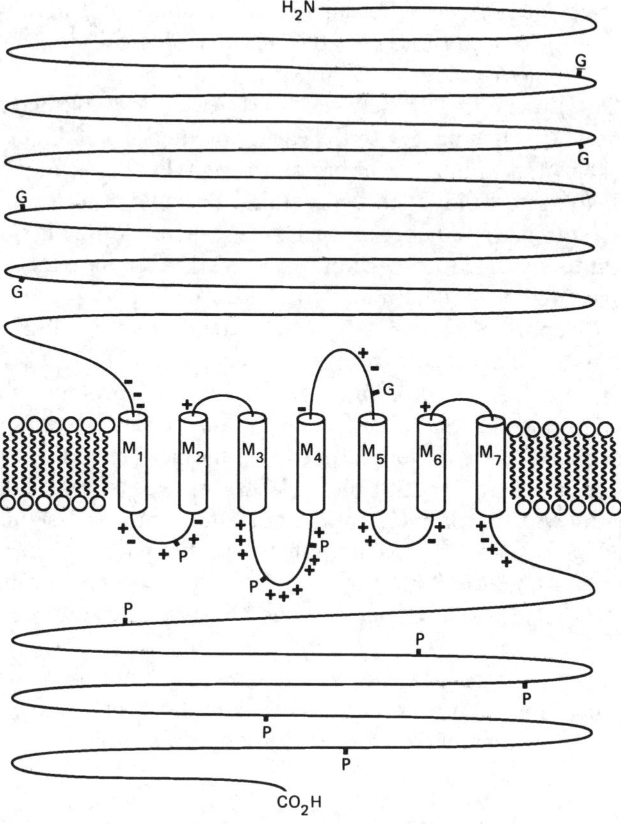

Fig. 7 — This figure, which is adapted from Masu *et al.* (1991), represents a glutamate metabotropic receptor protein sequence embedded in a lipid membrane. Key structural features of this protein are: —, acidic amino acid residues; +, basic amino acid residues; G, potential *N*-glycosylation sites; P, potential sites for *O*-phosphorylation.

CONCLUSION

The application of biochemical and molecular biological tools to the study of excitatory amino acid research has dramatically increased the pace with which new discoveries are being made. The power of these techniques to reveal the structure of receptor proteins and to permit the detailed study of receptors expressed in natural and unnatural forms will no doubt dramatically accelerate our understanding of the structure and function of excitatory amino acid receptors. The next few years promise to be a very exciting time in excitatory amino acid research.

NOTE ADDED IN PROOF

Following the completion of this manuscript, the cloning and expression of two dissimilar NMDA receptor proteins from rat brain has been reported (Moriyoshi *et al.* 1991, Kumar *et al.* 1991). In the first of these reports (Moriyoshi *et al.* 1991), a single protein of M_r 105 500 which comprised 938 amino acid residues was identified and cloned using a strategy similar to that of Hollmann *et al.* (1989). Expression of this protein, named NMDAR1, in *Xenopus* oocytes afforded a homomeric receptor displaying all of the electrophysiological and pharmacological properties characteristic of the native NMDA receptor. Significant structural similarity and sequence homology (22–26%) was observed between NMDAR1 and published AMPA/KAIN receptor sequences. In agreement with the earlier predictions of Ascher and Johnson (1990) (see earlier), the amino terminus of the NMDAR1 sequence contains 56 glutamate and aspartate residues thus providing the high concentration of anionic sites necessary to stabilize and hence 'concentrate' calcium ions (Fig. 4). Interestingly, *in situ* hydridization revealed a distribution of NMDAR1 message which is rather broader than that previously determined for NMDA receptor protein.

In the second report, Kumar *et al.* (1991) extend their previous studies on the solubilization and purification of the NMDA receptor (Michaelis 1991) with the resolution of the NMDA receptor into four component protein subunits and the cloning in *E. coli* of one of these subunits, the glutamate-binding protein (GBP). This protein has a deduced sequence length of 516 amino acid residues with a molecular weight of 57 020 Da. A hydropathy plot of GBP reveals at least four membrane-spanning domains with a relatively short amino terminus and significantly longer carboxy terminus. It has no sequence homology to other glutamate receptor proteins. Using Northern blotting, GBP message was found only in brain, where, unlike the message for NMDAR1, it was restricted to the hippocampus, cerebrum and cerebellum.

Clearly these two reports pose interesting questions and opportunities for those working in the field. No doubt now that these key discoveries have been reported, progress in this field, like that of the non-NMDA receptors, will now accelerate rapidly.

ACKNOWLEDGEMENTS

I would like to thank Dr Christopher Bigge, Dr Jorgen Drejer, Dr Mark Mayer, and Dr Elias Michaelis for reading early drafts of this review and making numerous helpful comments and suggestions. In particular, I would like to thank Dr Mark Mayer and Dr Elias Michaelis for allowing me to incorporate unpublished results from their laboratories into this manuscript.

REFERENCES

Aizenmann, E., Lipton, S. A. and Loring, R. H. (1989) *Neuron* **2** 1257–1263.

Aizenmann, E., Hartnett, K. A. and Reynolds, I. J. (1991) *Neuron* (in press).

Akaike, N., Kaneda, M, Hori, N. and Krishtal, O. A. (1988) *Neurosci. Lett.* **87**, 75–79.

Allen, C. N., Brady, R., Swann, J., Hori, N. and Carpenter, D. O. (1988) *Brain Res.* **458**, 147–150.

Ambar, I., Kloog, Y. and Sokolovsky, M. (1988) *J. Neurochem.* **51** 133–140.

Ambrosini, A., Henley, J. H. and Barnard, E. A. (1990) *Biochem. Soc. Trans.* **18** 401–402.

Ascher, P. and Johnson, J. (1990) *Prog. Cell. Res.* **1** 149–158.

Barnard, E. A. and Henley, J. M. (1990) *Trends Pharmacol. Sci.* **11** 501–507.

Barnard, E. A., Darlinson, M. G., Harvey, R., Marshall, J., Moss, S. J., Sattelle, D. B., Smart, T. G. and Vreugdenhill, E. (1990). In: Nishizuka, Y. *et al.* (eds) *The Biology and Medicine of Signal Transduction.* Raven Press, New York, pp. 20–29.

Baudry, M., Monaghan, D., Cotman, C. and Altar, C. A. (1990) *J. Neurochem.* **54** 1682–1688.

Benveniste, N. and Mayer, M. L. (1991) *Biophys. J.* **59** 560–573.

Benveniste, M., Clements, J., Vyklicky, L., Jr. and Mayer, M. L. (1990a) *J. Physiol.* **428** 333–357.

Benveniste, M., Mienville, J.-M., Sernagor, E. and Mayer, M. L. (1990b) *J. Neurophysiol.* **63** 1373–1384.

Bettler, B., Boulter, J., Hermans-Borgmeyer, I., O'Shea-Greenfield, A., Deneris, E. S., Moll, C., Borgmeyer, U., Hollmann, M. and Heinemann, S. (1990) *Neuron* **5** 583–595.

Boulter, J., Hollmann, M., O'Shea-Greenfield, A., Hartley, M., Deneris, E., Maron, C. and Heinemann, S. (1990) *Science* **249** 1033–1037.

Brugger, F., Pozza, M. F., Wicki, U., Nassenstein, D., Olpe, H.-R. and Fagg, G. E. (1990) *Experientia* **46** A76.

Chang, Y.-C., Hon, Y.-S. and Chow, W.-Y. (1990) *Neurochem. Int.* **16** 173–185.

Chen, J. W., Cunningham, M. D., Galton, N. and Michaelis, E. K. (1988) *J. Biol. Chem.* **263** 417–426.

Christensen, B. N. and Hida, E. (1990) *Neuron* **5** 471–478.

Christine, C. W. and Choi, D. W. (1990) *J. Neurosci.* **10** 108–116.

Cunningham, M. D. and Michaelis, E. K. (1990) *J. Biol. Chem.* **265** 7768–7778.

Danysz, W. and Wroblewski, J. T. (1989) *Neurosci. Res. Commun.* **5** 9–18.

Dansyz, W., Fadda, E., Wroblewski, J. T. and Costa, E. (1989) *Mol. Pharmacol.* **36** 912–916.

Dawson, T. L., Nicholas, R. A. and Dingledine, R. (1990) *Mol. Pharmacol.* **38** 779–784.

DePietro, F. R. and Byrd, J. C. (1990) *J. Mol. Neurosci.* **2** 45–52.

Dwyer, T. M., Adams, D. J. and Hille, B. (1980) *J. Gen. Physiol.* **75** 469–492.

Eaton, M. J., Chen, J.-W., Kumar, K. N., Cong, Y. and Michaelis, E. K. (1990) *J. Biol. Chem.* **265** 16195–16204.

Geoffroy, M., Lambolez, B., Prado de Carvalho, L., Rossier, J. and Stinnakre, J. (1989) *Eur. J. Pharmacol.* **166** 355.

Gregor, P., Mano, I., Maoz, I., McKeown, M. and Teichberg, V. (1989) *Nature* **342** 689–692.

Gündel, J., Steinhäuser, Ch. and Matthies, H. (1990) *Neurosci Lett.* **119** 249–252.

Hampson, D. R. and Wenthold, R. J. (1988) *J. Biol. Chem.* **263** 2500–2505.

Hampson, D. R., Wheaton, K. D., Deschesne, C. J. and Wenthold, R. J. (1989) *J. Biol. Chem.* **264** 13329–13345.

Haring, R., Zukin, R. S. and Zukin, S. R. (1990) *Neurosci. Lett.* **112** 92–98.

Hatefi, Y. and Hanstein, W. G. (1969) *Proc. Nat. Acad. Sci. USA* **62** 1129–1136.

Henley, J. M., Barnard, E. A. (1989) *J. Neurochem* **52** 31–37.

Henley, J. M., Ambrosini, A., Krogsgaard-Larsen, P. and Barnard, E. A. (1989) *New Biologist* **1** 153–158.

Hollmann, M., O'Shea-Greenfield, A., Rogers, S. W. and Heinemann, S. (1989) *Nature* **342** 643–648.

Honoré, T. and Drejer, J. (1988) *J. Neurochem.* **51** 457–461.

Honoré, T. and Nielsen, M. (1985) *Neurosci. Let.* **54** 27–32.

Honoré, T., Dejer, J. and Nielsen, M. (1986) *Neurosci. Lett.* **65** 47–52.

Honoré, T., Drejer, J., Nielsen, E. O. and Nielsen, M. (1989a) *Biochem. Pharmacol.* **38** 3207–3212.

Honoré, T., Drejer, J., Nielsen, E. O., Watkins, J. C., Olvermann, H. J. and Nielsen, M. (1989b) *Eur. J. Pharmacol. Mol. Pharmacol. Sec.* **172** 239–247.

Hood, W. F., Compton, R. P. and Monahan, J. B. (1990) *J. Neurochem.* **54** 1040–1046.

Hori, T., Yamamato, T., Hatta, K. and Moroji, T. (1990) *Neurosci. Lett.* **119** 9–11.

Huettner, J. E. (1990) *Neuron.* **5** 255–266.

Huganir, R. L. and Greengard, P. (1990) *Neuron* **5** 555–567.

Hunter, C., Wheaton, K. D. and Wenthold, R. J. (1990) *J. Neurochem.* **54** 118–125.

Iino, M., Ozawa, S. and Tsuzuki, K. (1990) *J. Physiol.* **424** 151–165.

Ikin, A. F., Kloog, Y. and Sokolovsky, M. (1990) *Biochemistry* **29** 2290–2295.

Javitt, D. C., Frusciante, M. J. and Zukin, S. R. (1990) *Mol. Pharmacol.* **37** 603–607.

Kaplita, P. V. and Ferkany, J. W. (1990) *Eur J. Pharmacol. Mol. Pharmacol. Sec.* **188** 175–179.

Kay, A. R. and Connor, J. A. (1990) *J. Neurosci. Methods* **33** 77–79.

Keinänen, K., Wisden, W., Sommer, B., Werner, P., Herb, A., Verdoorn, T. A., Sakmann, B. and Seeburg, P. H. (1990) *Science* **249** 556–560.

Kiskin, N. I., Krishtal, O. A., Tsyndrenko, A. Ya. and Akaike, N. (1986) *Neurosci. Lett.* **66** 305–310.

Kito, S., Miyoshi, R. and Nomoto, T. (1990) *J. Histochem. Cytochem.* **38** 1725–1731.

Kleckner, N. W. and Dingledine, R., (1988) *Science* **241** 835–837.

Kumar, K. N., Tilakaratne, N., Johnson, P. S., Allen, A. E. and Michaelis, E. K. (1991) *Nature* **354** 70–73.

Lazarewicz, J. W., Wroblewski, J. T., Palmer, M. E. and Costa, E. (1989) *Neurosci. Res. Commun.* **4** 91–97.

Legendre, P. and Westbrook, G. L. (1990) *J. Physiol.* **429**, 429–449.

Leonard, J. P. and Kelso, S. R. (1989) *Soc. Neurosci. Abstr.* **15** 535.

Lerma, J., Zukin, S. and Bennett, M. V. L. (1990) *Proc. Nat. Acad. Sci. USA* **87** 2354–2358.

Lester, R. A. J., Clements, J. D., Westbrook, G. L. and Jahr, C. E. (1990) *Nature* **346** 565–567.
Levy, D. I., Nikolaus, J. S. and Lipton, S. J. (1990) *Neurosci. Lett.* **110** 291–296.
Liman, E. R., Knapp, A. G. and Dowling, J. E. (1989) *Brain Res.* **481** 399–402.
Loring, R. H., Aizenmann, E., Lipton, S. A. and Zigmond, R. E. (1989) *J. Neurosci.* **9** 2423–2431.
Ly, A. M. and Michaelis, E. K. (1991) *Biochemistry* (in press).
Mamalaki, C. Barnard, E. A. and Stephenson, F. A. (1989) *J. Neurochem.* **52** 124–134.
Maragos, W. F., Penney, J. B. and Young, A. B. (1988) *J. Neurosci.* **8** 493–501.
Marvizon, J. C. G. and Skolnick, P. (1988) *Eur. J. Pharmacol.* **151** 157–158.
Massicotte, G. and Baudry, M. (1990) *Neurosci. Lett.* **118** 245–248.
Massicotte, G., Kessler, M., Lynch, G. and Baudry, M. (1990) *Mol. Pharmacol.* **37** 278–285.
Masu, M., Yasuto, T., Tsuchida, K., Shigemoto, R. and Nakanishi, S. (1991) *Nature* **349** 760–765.
Mayer, M. L. (1991) Personal communication.
Mayer, M. L. and Vyklicky, L. (1989) *Proc. Nat. Acad. Sci. USA* **86** 1411–1415.
Mayer, M. L., Vyklicky, L., Jr. and Westbrook, G. L. (1989) *J. Physiol.* **415** 329–350.
McDonald, J. F., Mody, I. and Salter, M. W. (1989a) In: Cavalheiro, E. A., Lehmann, J. and Turski, L. (eds) *Frontiers of Excitatory Amino Acid Research* Liss, New York, pp. 159–166.
McDonald, J. F., Mody, I. and Salter, M. W. (1989b) *J. Physiol.* **1** 17–34.
McDonald, J. F., Johnston, M. V. and Young, A. B. (1990) *Exp. Neurol.* **110** 237–247.
McKernan, R. M., Castro, S., Poat, J. A. and Wong, E. H. F. (1989) *J. Neurochem.* **52** 777–785.
McVittie, L. D. and Sibley, D. R. (1989) *Life Sci.* **44** 1793–1802.
Michaelis, E. K. (1991) Personal communication.
Michaelis, E. K., Michaelis, M. L. Stormann, T. M., Chittenden, W. L. and Grubbs, R. D. (1983) *J. Neurochem.* **40** 1742–1753.
Mitchell, C. L., Barnes, M. I. and Grimes, L. M. (1990) *Brain Res.* **506** 327–330.
Monaghan, D. T., Olverman, H. J., Nguyen, L., Watkins, J. C. and Cotman, C. W. (1988) *Proc. Nat. Acad. Sci. USA* **85** 9836–9840.
Monahan, J. P. Biesterfeldt, J. P., Hood, W. F., Compton, R. P., Cordi, A. A., Vazquez, M. I., Lanthorn, T. H. and Wood, P. L. (1990) *Mol. Pharmacol.* **37** 780–784.
Moriyoshi, K., Masu, M., Ishii, T., Shigemoto, R., Mizuno, N. and Nakanishi, S. (1991) *Nature* **354** 31–37.
Moskal, J. R., Stanton, P. K. and Haring, R. (1989) *Soc. Neurosci. Abstr.* **15** 202.
Murphy, D. E., Snowhill, E. W. and Williams, M. (1987) *Neurochem. Res.* **12** 775–782.
Nakanishi, N., Schneider, N. A. and Axel, R. (1990) *Neuron* **5** 569–581.
Nielsen, E. O., Cha, J.-H. J., Honoré, T., Penny, J. B. and Young, A. B. (1988) *Eur. J. Pharmacol.* **157** 197–203.
O'Dell, T. J. and Christensen, B. N. (1989) *J. Neurophysiol.* **61** 162–172.
Ogita, K. and Yoneda, Y. (1990a) *J. Neurochem.* **54** 699–702.
Ogita, K. and Yoneda, Y. (1990b) *J. Neurochem.* **55** 1515–1520.
Ortega, A. and Teichberg, V. I. (1990) *J. Biol. Chem.* **265** 21404–21406.
Ozawa, S., Iino, M. and Tsuzuki, K. (1990) *J. Neurophysiol.* **64** 1361.
Palmer, E., Monaghan, D. T. and Cotman, C. W. (1988) *Mol. Brain Res.* **4** 161–165.

Petrusz, P., Van Eyck, S. L., Weinberg, R. J. and Rustoni, A. (1990) *Brain Res.* **529** 339–344.

Procter, A. W., Stirling, J. M., Stratmann, G. C., Cross, A. J. and Bowen, D. M. (1989a) *Neurosci. Lett.* **101** 62–66.

Procter, A. W., Wong, E. H. F., Stratmann, G. C., Lowe, G. C. and Bowen, D. M. (1989b) *J. Neurochem.* **53** 698–704.

Randle, J. C. R. (1990) *Can. J. Physiol. Pharmacol.* **68** 1069–1078.

Randle, J. C. R., Vernier, P., Garrigues, A.-M. and Brault, E. (1988) *Mol. Cell Biochem.* **80** 121–132.

Rassendren, F.-A., Lory, P., Pin, J.-P. and Nargeot, J. (1990) *Neuron* **4** 733–740.

Reynolds, I. J., Rush, E. A. and Aizenmann, E. (1990) *Br. J. Pharmacol.* **101**, 178–182.

Sakimura, K. Bujo, H., Kushiya, E., Araki, K., Yamazaki, M., Meguro, H., Warashina, A., Numa, S. and Mishina, M. (1990) *FEBS Lett.* **272** 73–80.

Sather, W., Johnson, J. W., Henderson, G. and Ascher, P. (1990) *Neuron* **4** 725–731.

Sekiguchi, M., Okamoto, K. and Sakai, Y. (1990) *J. Neurosci.* **10** 2148–2155.

Sepulveda, M.-I. and Sattelle, D. B. (1989) *Neurosci. Lett.* **100** 210–214.

Shirasaki, T., Nakagawa, T., Wakamori, M., Tateishi, N., Fukuda, A., Murase, K. and Akaike, N. (1990) *Neurosci. Lett.* **108** 93–98.

Sigel, E. and Baur, R. (1988) *Proc. Nat. Acad. Sci USA* **85** 6192–6196.

Sommer, B., Keinänen, K., Verdoorn, T. A., Wisden, W., Burnashev, N., Herb, A., Köhler, M., Takagi, T., Sakmann, B. and Seeburg, P. H. (1990) *Science* **249** 1580–1585.

Sonders, M. S., Barmettler, P., Lee, J. A., Kitahara, Y. Keana, J. F. W. and Weber, E. (1990) *J. Biol. Chem.* **265** 6776–6781.

Steele, J. E., Palmer, A. M., Stratmann, G. C. and Bowen, D. M. (1989) *Brain Res.* **500** 369–373.

Terramani, T., Kessler, M., Lynch, G. and Baudry, M. (1988) *Mol. Phamacol.* **34**, 117–123.

Thedinga, K. H., Benedict, M. S. and Fagg, G. E. (1988) *Neurosci. Lett.* **104** 217–222.

Tombaugh, G. C. and Sapolsky, R. M. (1990) *Brain Res.* **506** 343–345.

Toyoshima, C. and Unwin, N. (1988) *Nature* **336** 247–250.

Toyoshima, C. and Unwin, N. (1990) *J. Cell. Biol.* **111** 2623–2635.

Traynelis, S. F. and Cull-Candy, S. G. (1990) *Nature* **345** 347–350.

Traynelis, S. F. and Cull-Candy, S. G. (1991) *J. Physiol.* **433** 727–763.

Umbach, J. A. and Gundersen, C. B. (1989) *Mol. Pharmacol.* **36** 582–588.

Unwin, N. (1989) *Neuron* **3** 665–676.

Verdoorn, T. A. and Dingledine, R. (1988) *Mol. Pharmacol.* **34** 298–307.

Vyklicky, L. Jr., Krusek, J. and Edwards, C. (1988) *Neurosci. Lett.* **89** 313–318.

Vyklicky, L., Jr., Benveniste, M. and Mayer, M. L. (1990) *J. Physiol.* **428** 313–331.

Wada, K. Dechesne, C. J., Shimasaki, S., King, R. G., Kusano, K. Buonanno, A., Hampson, D. R., Banner, C., Wenthold, R. J. and Nakatina, Y. (1989) *Nature* **342** 684–689.

Wenthold, R. J., Hampson, D. R., Wada, K., Hunter, C., Oberdorfer, M. D. and Dechesne, C. (1990) *J. Histochem. Cytochem.* **38** 1717– 1723.

Williams, T. L., Smith, D. A. S., Burton, N. R. and Stone, T. W. (1988) *Br. J. Pharmacol.* **95** 805–810.

Wong, E. H. F. and Nielsen, M. (1989) *Eur. J. Pharmacol. Mol. Pharmacol. Sec.* **172** 493–496.

Yeh, G.-C. Bonhaus, D. W. and McNamara, J. O. (1990) *Mol. Pharmacol.* **38** 14–19.

Yoneda, Y. and Ogita, K. (1989a) *J. Neurochem.* **52** 1501–1507.

Yoneda, Y. and Ogita, K. (1989b) *Neuropharmacology* **28** 611–616.

Ziegra, C. J. and Oswald, R. E. (1989) *Soc. Neurosci. Abstr.* **15** 532.

Zorumski, C. F., Thio, L. L., Clark, G. D. and Clifford, D. B. (1990) *Neuron* **5** 61–66.

4

Electrophysiological techniques for the study of excitatory amino acids in mammalian neurons in slice preparations

John D. C. Lambert and **Mogens Andreasen**
PharmaBiotec Research Centre, Institute of Physiology, University of Aarhus, DK-8000 Århus C, Denmark

ABSTRACT

Electrophysiological techniques provide the means to make a precise assessment of the physiological action of excitatory amino acids (EAAs). Following binding of the agonist to its receptor, the receptor-coupled ionophores open and thereby increase the ionic permeability of the membrane. This gives rise to changes in the electrical characteristics of the individual neuron and the tissue as a whole. The net conductance to the environmental cations is increased, and these move down their electrochemical gradients. The resulting ionic current causes depolarization of the membrane. The bulk ionic shifts additionally cause a change in the extracellular potential. All of these parameters can be measured with electrophysiological techniques and this is illustrated with experiments performed on the *in vitro* hippocampal slice. Interpretation of the results, and the extent to which the different parameters reflect proportional occupation of the EAA receptors, are also discussed.

INTRODUCTION

Excitatory amino acid (EAA) agonists act on a number of different receptors (for reviews see Collingridge and Lester 1989, and Mayer and Westbrook 1987). On the basis of mechanistic and pharmacological studies, these receptors have been divided into two main groups: that which is preferentially activated by *N*-methyl-D-aspartate (NMDA), the NMDA receptor, and that which is activated by other prototypical ligands such as 2-amino-3-(3-hydroxy-5-methylisoxazol-4-yl)propanoic acid (AMPA), quisqualic acid (QUIS) and kainate (KAIN). The latter, which may constitute a number of subreceptors (see chapter 2), is known collectively as the non-NMDA receptor. Both the NMDA and non-NMDA receptors are coupled to ion-permeable channels in the membrane (ionophores). The unitary conductance of non-NMDA-operated ionophores (4–8 pS) is small compared with that of NMDA ionophores (40–50 pS) (Ascher and Nowak 1988). Both ionophores are permeable to Na^+ and K^+, while the NMDA-operated ionophore is additionally permeable to Ca^{2+} (Fig. 1). QUIS additionally activates a so-called metabotropic receptor (see chapter 15), which causes an increase in polyphosphoinositide metabolism and release of Ca^{2+} from internal stores. A fourth type of EAA receptor, which is activated by 2-amino-4-phosphonobutanoate (AP4) (see chapter 14), has yet to be character-

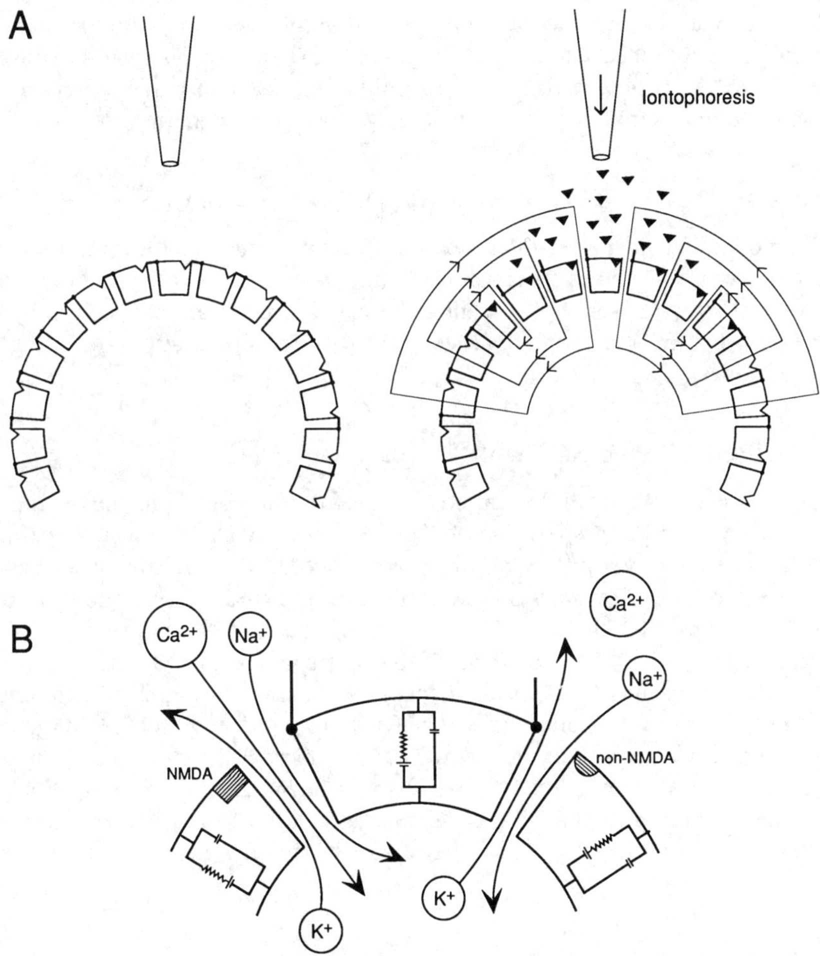

Fig. 1 — A: Schematic representation of the action of an excitatory amino acid (EAA) on the neuronal membrane. EAA molecules (▼) bind to receptors which are within range of a localized iontophoretic application of the agonist. This opens cation-specific ionophores through which a net inward current flows (current sink). Adjacent areas of membrane (current sources) are depolarized by the current flow as shown. B: Schematic diagram of a section of neuronal membrane showing an NMDA- and a non-NMDA-operated ionophore. The NMDA ionophore behaves as if it is physically larger than the non-NMDA ionophore. The former has a larger unitary conductance and, in addition to Na^+ and K^+, conducts Ca^{2+} ions. Equivalent circuit elements are shown in the membrane.

ized fully with respect to location and function (see Collingridge and Lester 1989). Although the latter two receptors do not directly activate ionophores on 'postsynaptic' neurons, it should be borne in mind that they make insidious contributions to EAA-evoked responses.

Effects mediated by operation of EAA receptor-coupled ionophores can be readily detected with electrophysiological techniques. The aim of this chapter is to provide an insight into various methods of quantifying responses to EAAs and to indicate how these response parameters reflect occupation of the EAA receptor by an agonist. To this end, experiments illustrating the action of EAAs on pyramidal cells in area CA1 of the rat hippocampal slice will be described. Although otherwise not discussed in this article, it should be emphasized that electrophysiological recordings also allow comparison of the effects of exogenously applied EAA agonists with synaptically mediated excitatory potentials (e.g., see Andreasen *et al.* 1989). Such direct, simultaneous comparisons are not usually available with other techniques.

ELECTROPHYSIOLOGICAL RECORDING TECHNIQUES

There is not sufficient space to describe the electrophysiological techniques in detail and the reader is referred to specialist texts on these topics. A very good book for the beginner is that by Purves (1981), while a more advanced treatment is given by Smith *et al.* (1985). Brown and Flaming (1986) deal with the preparation and use of glass microelectrodes.

THE HIPPOCAMPAL SLICE PREPARATION

The anatomical and morphological structure of the hippocampus makes this brain region amenable to the preparation of 300–500-μm thick slices in which the intrinsic neuronal circuitry is relatively well preserved. Indeed, the hippocampal slice is one of the most extensively used *in vitro* preparations for the study of synaptic transmission and drug actions (Kerkut and Wheal 1981, Dingledine 1984). The advantages of *in vitro* preparations over their *in vivo* counterparts include: quick and easy preparation for the experiment; following decapitation, no anaesthetic is present (not surprisingly, many anaesthetics have been shown to interfere with EAA receptors and ionophores); easy access to the neurons with microelectrodes; perfusion with a medium of defined composition to which pharmacologically active agents can be added; the possibility to change the ionic composition of the perfusion medium for mechanistic studies; a survival time of 10 hours or more, which is adequate for most experiments.

APPLICATION OF AGONIST

With direct bath perfusion of EAA agonists to the *in vitro* slice, there is a latent period before the agonist arrives at the impaled neuron, and it takes some time before equilibrium concentrations are achieved. Furthermore, long-duration exposures to EAAs are deleterious to the tissue. Although this method of application has the advantage that the precise concentration of agonist in the bathing medium is known, it is not very suitable for multiple additions of agonist (as, for example, is needed for the construction of a dose–response curve). For this purpose, it is better to use discrete applications of agonist in the vicinity of the recording site. Solutions containing agonists dissolved in perfusing fluid can be applied locally from blunt pipettes (MacDonald 1985). With large tip openings, the hydrostatic pressure of the column of fluid behind the tip is sufficient to allow droplet application. With finer tips it may be necessary to use positive pressure applied to the open barrel of the pipette. Even though the concentration of the agonist in the pipette is known precisely, it will be diluted as it spreads through the perfusing fluid and the tissue and therefore its concentration at the receptors will not be known.

For the consecutive application of agonists from a fine-tipped pipette (which is less intrusive to the tissue), iontophoresis is the method of choice. The technique has been described in detail elsewhere (Purves 1979, Kelly 1975, Salmoiraghi and Stefanis 1967, MacDonald 1985). Although iontophoresis is ideal for discrete, short-duration application of EAA agonists, the precise concentration of the agonist in the vicinity of the receptors is again unknown. The concentration of agonist decreases with distance from the point source of application, and receptors will be exposed to correspondingly different concentrations (see Fig. 1). The distribution profile will also be influenced by the presence or absence of an active uptake system for the agonist. Thus, agonists which are taken up will be confined to the vicinity of the application, while those which are not taken up will spread widely within the tissue. On the assumption that a given iontophoretic application always releases the same amount of agonist, and that this is linearly related to the ejection current (see Purves 1979), it is possible to construct dose–response curves for the agonist. The effect of a putative antagonist (applied in the perfusing medium at a known concentration) can then be investigated.

AGONIST-EVOKED RESPONSES IN NEURONAL TISSUES

The interaction between an EAA and its receptor causes opening of the ionophores, which leads to a net change in membrane permeability to the environmental ions. This is reflected as an electrically measurable parameter. For pharmacological analysis, it is advantageous to choose a parameter which bears a simple relationship to the interaction of the agonist with its receptor. This requires an understanding of the mechanisms involved in the generation of EAA-induced responses. These processes can be illustrated by considering the events which occur in a neuron following addition of an EAA agonist from a point source (Fig. 1). During the maintained presence of the agonist, the individual ionophores open and shut with a characteristic pattern. A population of ionophores will therefore behave as if a proportion (α) of these are conducting the whole time, while the remainder $(1 - \alpha)$ are permanantly shut. Since the EAA has opened new conductive channels through the membrane, the net *membrane conductance* (G_M) will increase. The resting membrane is 100–1000 times more permeable to K^+ than Na^+, with typical values of E_{Na^+} and E_{K^+} being $+50$ mV and -90 mV, respectively. The EAA-operated ionophores are essentially equipermeable to Na^+ and K^+. This leads to a net inward *current* as the ions move down their electrochemical gradients and a *depolarization* results as the membrane potential moves towards the new equilibrium potential (E_{EAA}), which is about 0 mV for all the EAA agonists. The size of the depolarization will depend on the number of conducting ionophores and the extent to which these shunt the resting membrane conductances. For the hypothetical example shown in Fig. 2, it is considered that the maximum conductance through the EAA-operated ionophores (g_{EAA}) is ten times that of the resting conductance (g_M) and, with activation of all ionophores, the potential at maximum depolarization will be -7 mV. When the depolarization crosses the threshold for action potential (AP) generation, *repetitive firing* results. Firing may fail with further depolarization ('depolarizing block'). The inward ionic current at the site of the conducting ionophores creates a sink into which current flows from other regions of the membrane (sources) which are not directly under the influence of the EAA. This depolarization decreases with distance from the active EAA ionophores, as determined by the length constant of the neuron. The ionic movements cause changes in the *ionic composition* of the bulk fluids. This, together with the local current circuits, gives rise to an *extracellular negativity* in the vicinity of the EAA application.

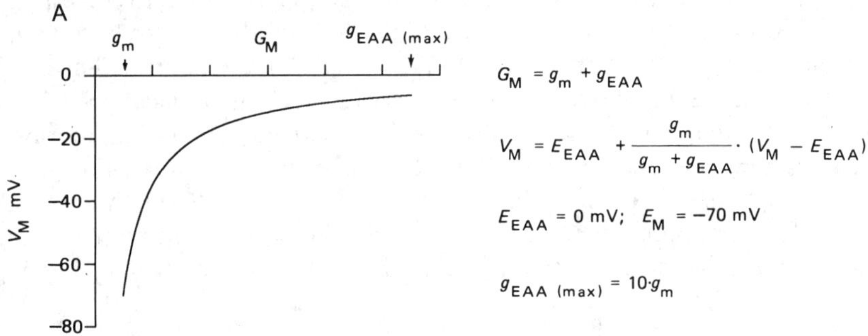

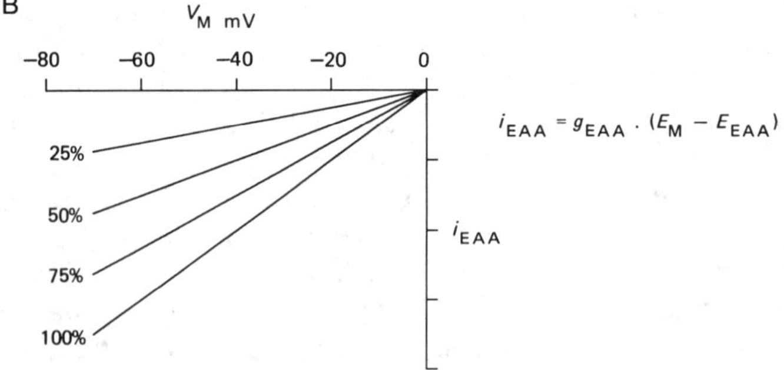

Fig. 2 — A: Relationship between membrane potential (V_M) and membrane conductance (G_M) where the resting membrane potential (E_M) has been set to -70 mV and the equilibrium potential for EAA action (E_{EAA}) to 0 mV. The plot has been derived from the relationships shown on the right (based on equations of Choi and Fischbach 1981). The maximum conductance of the EAA-operated ionophores (g_{EAA}(max)) has been arbitrarily set to ten times that of the resting membrane (g_M). The graph shows that as g_{EAA} increases, V_M asymptotes towards E_{EAA}. B: The current through the EAA-operated inophores (i_{EAA}) is a linear function of V_M. The percentage values represent the fractional occupation of the EAA receptors.

The parameters given in italics above can all be measured and used to quantify the action of an EAA. Measurements of each of these parameters, and the manner in which they are related to interaction of the agonist with the receptor, are discussed below. The effects of desensitization (i.e., occupied receptor, but non-conducting ionophore) are not considered.

QUANTIFICATION OF EXCITATORY AMINO ACID RESPONSES IN HIPPOCAMPAL SLICES

Extracellular negativity

The extracellular DC potential may be easily measured with a low-resistance, NaCl-filled microelectrode which is placed close to the site of iontophoretic application. EAA-evoked negative shifts in potential have been termed 'local potentials' (FP) by Flatman and Lambert (1979), who also suggested their utilization for pharmacological investigations. The maximum size of the FP, which is a function of the neuronal organization and distribution of the receptors, is typically -10 to -15 mV in the pyramidal cell layer of area CA1 (see Figs 3–5). The measurement of FPs has been adapted to a number of simple screening preparations on which the pharmacology of EAAs can be investigated. Slice preparations include the 'cortical wedge' where an insulating barrier (e.g. vaseline) is placed over the corpus callosum (Harrison and Simmonds 1985, Burton *et al.* 1987) and the hippocampal slice, where the alveus is separated and isolated electrically from area CA1 neurons (Blake *et al.* 1988). The potential difference (typically 1 mV or less) is measured between the two chambers using large, low-resistance electrodes (e.g. Ag/AgCl) and is thought to reflect the action of the agonist on the receptive membranes of neurons whose axons pass through the seal. However, in the hippocampal slice it has recently been noted that the alveus can be replaced by absorbent paper, which infers that all neurons in the vicinity of the grease barrier contribute to the potential recorded (Frenguelli *et al.* 1991). Thus, the potentials will be the counterparts of those recorded with extracellular microelectrodes, except that the latter are about ten times larger in amplitude.

Many factors contribute to the generation of the FPs. They are thought to arise in part from the EAA-induced inward current into the neurons, and in part from the local uptake of K^+ into neurons and glia (Somjen 1973, Dietzel *et al.* 1989). Depolarizing responses to EAAs are accompanied by a net movement of ions such that $[K^+]_o$ increases, while $[Na^+]_o$ and $[Cl^-]_o$ decrease (Lambert and Heinemann 1991). These movements give rise to an electrochemical potential between the region where the EAA is applied and the undisturbed areas. The movement of ions into the neurons is followed by water, which causes the neurons to swell. Under normal circumstances the extracellular resistance is very much smaller than the transmembrane resistance, which favours the flow of current in the extracellular medium rather than across the membrane. During maximal responses to EAAs, the transmembrane resistance breaks down, while the extracellular resistance increases because of the swelling of the neurons. In contrast to the resting situation, a major portion of the current will now flow across the membrane and through the intracellular fluid, thereby reducing the length constant of the neuron considerably. Thus, a multitude of events occur in the extracellular space following application of an EAA. This means that it is virtually impossible to predict the exact relationship between the size of the FP and occupation of the EAA receptors. In addition to being a quantifiable expression of the action of EEAs, the extracellular negativity has also important influences on the other indicators of EAA action, as discussed in the following sections.

Depolarization

The EAA-induced depolarization is the simplest parameter that can be measured with an intracellular recording electrode. An example of responses to a range of NMDA applications is shown in Fig. 3. Depolarization arises because the equilibrium potential for EAA action (E_{EAA}) is some 70 mV positive to the resting membrane potential (E_M). The magnitude of the current which flows through the conducting ionophores is linearly related to the electrochemical gradient ($E_M - E_{EAA}$) (Fig. 2). As the depolarization progresses, this inward driving force decreases. The recruitment of additional ionophores will therefore contribute progressively smaller increments to the total depolarization (Fig. 2). This gives rise to a parabolic relationship between g_{EAA} (which is directly related to the number of occupied receptors) and the depolarization.

In the equations given in Fig. 2, both V_M and E_{EAA} need further consideration. The potential recorded by the intracellular electrode is usually referred to a remote indifferent electrode. Thus, while the negative shift in extracellular potential will be 'seen' by the neuronal membrane, it will not be recorded. The actual transmembrane potential (TMP) is obtained by subtracting the locally recorded FP from the intracellular potential. In Fig. 3c it can be seen that the FP makes a proportionately greater contribution to the larger responses.

Since Na^+ and K^+ move down their electrochemical gradients, it is unlikely that E_{Na^+} and E_{K^+} will remain constant throughout the EAA application. However, these changes will tend to counteract each other such that there may be little overall change in E_{EAA}.

In summary, the EAA-induced depolarization is not a suitable parameter to reflect proportional occupation of the EAA receptors. Unless the intracellular electrode is referred to the immediate extracellular environment, part of the response will not be detected. Secondly, the depolarization will only discriminate adequately between responses where the overall receptor occupancy is small. Depolarization would not discriminate easily or reliably between a competitive and a non-competitive antagonist.

Firing

When the depolarization reaches E_{Th}, spontaneous APs arise. The firing pattern is determined by both the intrinsic properties of the neuronal membrane and the action of the EAAs. NMDA typically evokes burst firing, which arises at a lower threshold potential than that of the regular firing evoked by non-NMDA analogues. Firing complicates measurements of G_M and EAA-induced currents (see below). Fast Na^+-spiking can be readily abolished by the addition of tetrodotoxin (TTX) to the perfusing medium. TTX will also abolish activity in the presynaptic fibres and thereby block synaptic transmission. If this is undesirable, fast Na^+-spiking can be selectively blocked in the impaled neuron by using a recording electrode which contains a

(A)

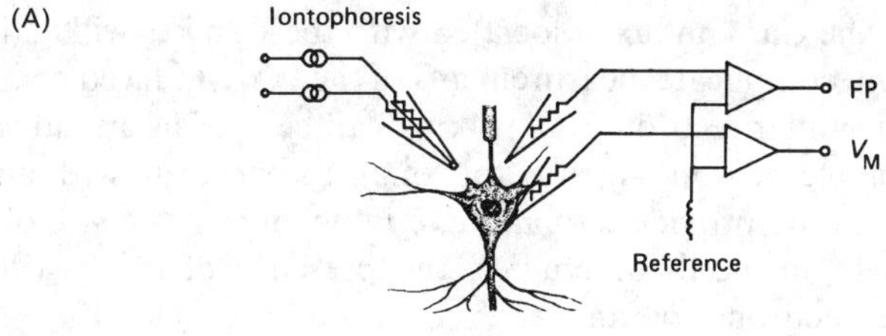

(B)

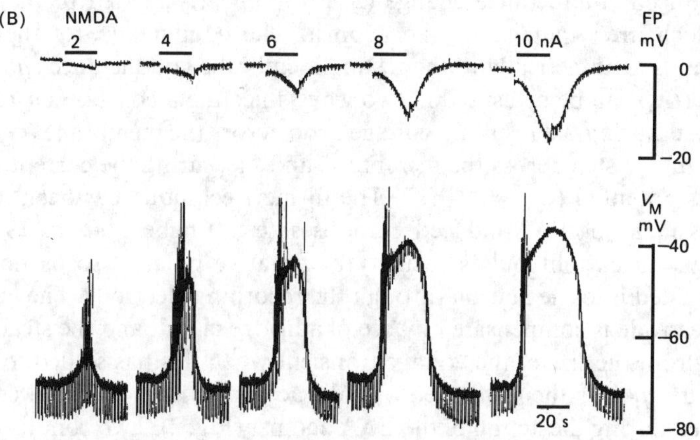

(C)

NMDA (nA)	ΔV_M (mV)	FP (mV)	TMP (mV)	FP:ΔV_M (%)
2	6	0.5	6.5	8%
4	20	1.5	21.5	8%
6	26	4	30	15%
8	31	9	40	29%
10	34	13	47	38%

Fig. 3 — The depolarizing action of NMDA on a hippocampal CA1 neuron. A: Schematic representation of electrode placement. Iontophoretic applications of NMDA were made in stratum pyramidale close to the site of an intracellular impalement from which the membrane potential (V_M) was recorded and, following withdrawal of the electrode, the extracellular field potential (FP). B: Responses to a series of NMDA applications. The deflections on the intracellular trace represent the voltage transients in response to injection of short-duration constant-current pulses. NMDA evoked a dose-dependent depolarization. The responses to NMDA evoked by ejecting currents of 2 and 4 nA were accompanied by an apparent decrease in G_M (the voltage transients during the response were larger), while the larger responses were all accompanied by an increase in G_M. Note that there are two distinct phases to the depolarization: a rapid rise to a potential of around − 45 mV, followed by a slower depolarization until the ejection was terminated. Following withdrawal of the electrode, NMDA was found to evoke FPs of up to − 13 mV in size. C: Table of results from the responses in B showing that the FP contributes a progressively larger proportion of the overall transmembrane potential (TMP) with increasing doses of NMDA.

quaternary local anaesthetic, such as QX 314 (Engberg *et al.* 1984). This in turn may reveal Ca^{2+}-spiking. The Ca^{2+} influx associated with these spikes will activate a K^+ conductance, which will complicate measurements of the EAA-induced changes in G_M.

In summary, extracellular recordings of spiking can be used as an indicator of the effect of EAAs on a single neuron which is too small to penetrate with a microelectrode. At best, this will only provide a qualitative indication of the size of the EAA-induced response. With intracellular studies, the presence of firing will influence measurements of other neuronal parameters.

EAA-induced conductance changes

The EAA-induced conductance change (g_{EAA}) is linearly related to the number of receptors which are occupied by the agonist. The conductance of the associated ionophores adds arithmetically to the resting membrane conductance (g_M). The total conductance (G_M) can be measured in two ways: injecting a constant current into the neuron and measuring the resulting voltage drop across the membrane ($G_M = I/\Delta V_M$); imposing a voltage step across the membrane and measuring the current required to maintain this potential ($G_M = \Delta I_M/V$). The former technique has been widely used for the assessment of EAA-induced responses (e.g. Engberg *et al.* 1979). Short-duration constant-current pulses (usually of negative polarity so as not to evoke firing) are injected into the neuron through the recording electrode. The voltage drop across the electrode is compensated by use of a bridge circuit and the steady-state G_M is calculated from the size of the voltage transient when this has settled to a constant (plateau) value. g_{EAA} is then obtained by subtracting g_M from the total conductance (G_M) measured during the action of the EAA agonist (Fig. 2). Two conditions must be satisfied in order that g_{EAA} determined in this way accurately reflects conduction through EAA-operated ionophores. First, the current–voltage (I/V) relationship of the membrane must be linear (i.e., g_M is independent of the potential across the membrane). Second, the operation of the EAA ionophores must be voltage-independent.

Membrane rectification

Most neuronal membranes are invested with voltage-dependent ionic channels which will be activated by the EAA-induced depolarization. If this conductances inactivate rapidly, they should not present a problem with long-duration applications of agonists. However, persistent, non-inactivating conductances will make a contribution to the net conductance change. The size of these conductances will have to be assessed and subtracted from the G_M pertaining during the application of the EAA. This can be done by depolarizing the neuron with current injection to the same levels of potential achieved during the EAA-induced responses. The extent to which the membrane can be depolarized by current injection is usually limited by the electrode, which is inclined to rectify during the passage of large currents. It is then not possible to compensate for the voltage drop across the electrode with the bridge circuit, and appreciable errors can result. Conduction through rectifying ionic channels can often be reduced using pharmacological techniques. For example, the delayed K^+ rectifier can be counteracted by injecting Cs^+ ions into the neuron through the recording electrode (Johnston *et al.* 1980). QX 314 (see above) will also reduce current flow through rectifying channels. Any channel-blocker which is used for this purpose should be shown to be without effect on the EAA-induced response.

Voltage-dependent EAA responses

While the I/V relationship of the non-NMDA-operated ionophore is linear, NMDA ionophores are gated by Mg^{2+} in a voltage-dependent manner (Mayer and Westbrook 1987). The net conductance of the NMDA ionophores decreases with hyperpolarization and their I/V relationship has a region of negative slope resistance. Conduction through the NMDA-operated ionophore will be decreased during passage of the anodal current pulse which is used to measure G_M. The voltage deflection will be larger; i.e., NMDA will apparently have caused a decrease in G_M (Lambert *et al.* 1981). Following removal of external Mg^{2+}, conduction through the NMDA ionophore is no longer voltage-dependent. This can be readily demonstrated in dissociated cultured neurons, which are in intimate contact with the perfusing fluid. In the slice preparation, however, it is difficult to achieve the very low levels of $[Mg^{2+}]_o$ required. First, trace amounts of Mg^{2+} are usually present in the salts used to prepare the perfusing medium. Second, the Mg^{2+} present in the tissue reservoirs has to be depleted to low levels. Intracellular Mg^{2+} is thought to be about 1 mM and, following the removal of Mg^{2+} from the extracellular fluid, Mg^{2+} will pass down its electrochemical gradient and leave the cells. This will mean that an appreciable concentration of Mg^{2+} will be present in the vicinity of the receptors for some considerable time. A few micromoles of Mg^{2+} are sufficient to confer appreciable voltage sensitivity on the NMDA ionophores (Mayer and Westbrook 1987).

In summary, g_{EAA} provides an accurate indication of the operation of EAA receptor-coupled ionophores. However, it is necessary to ascertain that rectification of neither the membrane nor the EAA ionophores themselves will influence the measurements.

EAA-induced ionic currents

The EAA-induced current declines as the membrane potential approaches E_{EAA} (Fig. 2). To measure the current, it is necessary to maintain a constant potential gradient across the membrane. For this purpose, the voltage-clamp technique is used. A current which is equal to that which would otherwise flow through the EAA-operated ionophores is injected into the neuron. Ideally, the neuron should be penetrated with two independent electrodes: one for measuring the potential and one for passing the current. In practice this is very difficult to achieve for neurons that cannot be visualized directly. Techniques have been developed for voltage-clamping with a single electrode (SEVC — Smith *et al.* 1985), whereby the electrode is switched rapidly between voltage-recording and current-passing configurations (Finkel and Redman 1985) (Fig. 4). Because of the compromise in using one electrode for both functions, it is very important to monitor the voltage across the electrode to ensure that this has settled before the potential is sampled.

Experiments illustrating voltage-clamping of responses to AMPA are shown in Fig. 4. Even with ideal clamping there are two possible sources of error. Firstly, only the membrane in the vicinity of the tip of the electrode will be held at the command potential (V_h). The efficacy of the clamp decreases with distance from the site of the

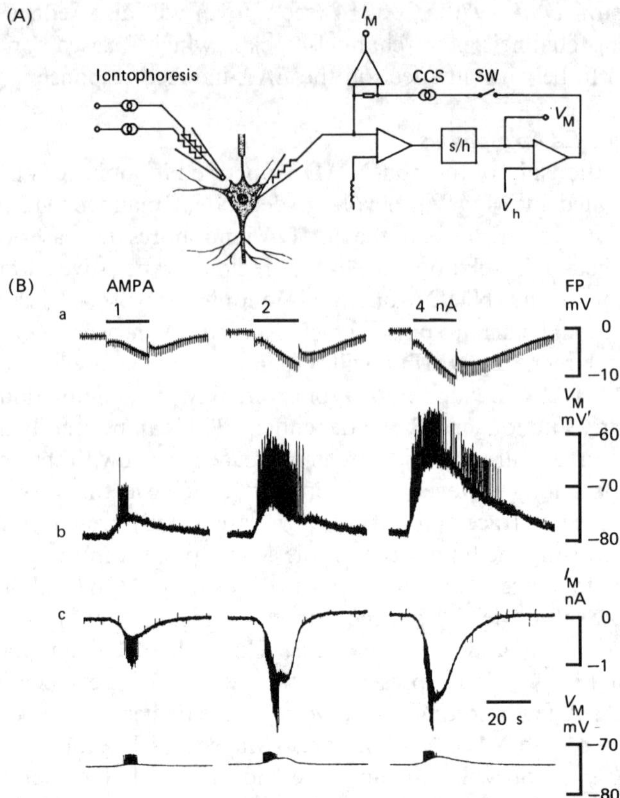

Fig. 4 — Responses of a hippocampal CA1 pyramidal cell to AMPA recorded with current- and voltage-clamp techniques. A: Schematic representation of the single-electrode voltage-clamp (SEVC) system. First, the intracellular depolarizing responses to three applications of AMPA were recorded (Bb) (see Fig. 3A). The recording system was then switched to the SEVC mode. The sample and hold (s/h) circuitry samples V_M. When this differs from the command potential (V_h) a controlling voltage is intermittently applied to a constant current source (ccs) by a rapid electronic switch (sw). The ccs injects current into the neuron (I_M), and the feedback circuit strives to keep $V_M = V_h (= -75$ mV in the lower record in Bc). During applications of AMPA, I_M represents the current which would have flowed through the ionophores had these been allowed to conduct freely. The electrode was then withdrawn to just outside the neuron and the FP responses recorded. Note that the V_M and I_M responses reached reasonably stable values before the end of the AMPA applications, while the FP was still rising. The deflections on the FP record represent responses to injection of short duration current pulses through the recording electrode. The size of the deflections is proportional to the resistance of the tissue.

impalement. A portion of the current which flows through the distally located ionophores (which are within range of the applied agonist) will be unclamped. Secondly, the actual transmembrane potential will change by an amount which is equivalent to the extracellular potential (see Fig. 3). In the experiment shown in Fig. 4, the membrane was apparently clamped adequately during application of AMPA. Even so, the response to AMPA was still accompanied by firing. Two factors are possibly involved in this: the firing arises at a remote, inadequately clamped membrane and attempts to invade the soma; the increase in extracellular negativity means that, although the membrane is apparently clamped at (or close to) V_h there is appreciable depolarization of the somal membrane.

In summary, measurement of the EAA-induced current will accurately reflect occupation of the EAA receptors. When the potential is clamped, recruitment of ionophores will add proportionate increments to the total current recorded. The following precautions would improve quantitative voltage clamp recordings: the membrane potential should be referred to an electrode placed just outside the neuron; the agonist should be applied close to the site of the recording, preferably with short-duration applications to avoid too much spread in the tissue.

Ionic activities

Extracellular ionic activities can be measured with appropriate ion-sensitive micro-

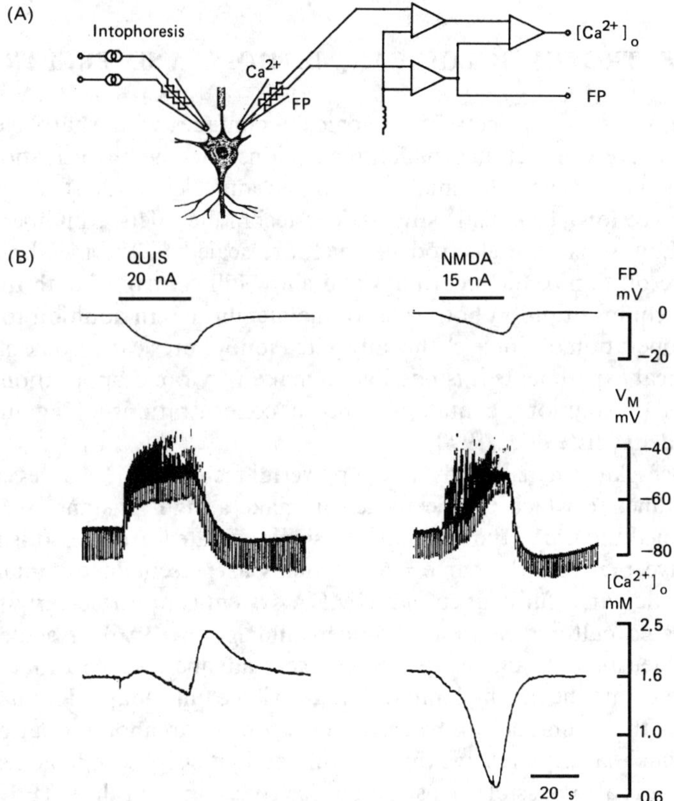

Fig. 5 — Responses to NMDA and QUIS recorded with a Ca^{2+}-selective electrode. A: Schematic diagram of the Ca^{2+}-sensitive electrode, from which the FP is also available from the reference barrel. In practice, the iontophoretic electrode is glued to the Ca^{2+}-sensitive electrode with a tip separation of about 50 μm. The calibrated Ca^{2+} signal is obtained by subtracting the FP from the potential recorded by the Ca^{2+} electrode. A simultaneous intracellular recording was made from a neuron in the near vicinity. B: The depolarizing response to QUIS is accompanied by a biphasic change in $[Ca^{2+}]_o$ with an overshoot in $[Ca^{2+}]_o$ following termination of the application. This type of response is typical of that to non-NMDA agonists such as KAIN and AMPA. The depolarizing response to NMDA is accompanied by a monophasic decrease in $[Ca^{2+}]_o$, which is also seen with mixed agonists.

electrodes (ISM). These can be glued to iontophoretic electrodes with a tip separation of 20–50 μm. The FP is also available from the reference barrel of the ISM (Fig. 5). While the ion signals themselves are not usually used to quantify the responses to the EAA analogues, measurement of the $[Ca^{2+}]_o$ signal is useful for determining the involvement of different EAA receptor subtypes. Operation of NMDA receptors alone causes a monophasic decrease in $[Ca^{2+}]_o$, while non-NMDA agonists, such as AMPA, QUIS and KAIN, cause characteristic multiphasic changes in $[Ca^{2+}]_o$ (Fig. 5). Mixed agonists such as L-glutamate, or mixtures obtained by conjoint applications of NMDA and a non-NMDA antagonist, cause monophasic falls in $[Ca^{2+}]_o$. It is also possible that operation of metabotropic receptors will be associated with characteristic changes in $[Ca^{2+}]_o$, thereby providing an electrophysiological parameter to reflect operation of this receptor subtype.

OTHER ELECTROPHYSIOLOGICAL METHODS AND PREPARATIONS

The foregoing summarizes electrophysiological techniques with 'sharp' electrodes as applied to relatively intact neuronal preparations for the investigation of EAA responses. The patch-clamp technique (Sakmann and Neher 1983) is, however, the method of choice for mechanistic studies. The technique offers a number of distinct advantages. Low resistance electrodes, which are sealed to the membrane, greatly increase the resolution of the recordings and allow full control of both the potential and the ionic composition on both sides of the membrane. In addition to whole-cell currents, the operation of single chemically gated ionophores can also be studied. For pharmacological experiments it is possible to make very rapid applications ('concentration-jump') of solutions containing known concentrations of agonists and/or antagonists (Benveniste *et al.* 1990).

While patch-clamping is clearly a very powerful technique, it is necessary to have 'clean' membranes to which the electrode can make a high-resistance seal. This has hitherto confined the use of the technique to single, isolated neurons. It is not certain that these have precisely the same EAA systems as present in the intact nervous system. In particular, a fully functional NMDA system is not necessarily present on either dissociated cultured neurons (Lambert and Jensen 1986) or acutely isolated hippocampal neurons (Akaike *et al.* 1988). Recent advances suggest that it is only a matter of time before the patch-clamp technique will be fully adapted for use on *in situ* neurons. Thus, the technique can be used on neurons in organotypic cultures (Llano *et al.* 1988), which largely preserve the structure and integrity seen in acutely isolated preparations. It is also possible to use patch electrodes on 'thin slices' (Edwards *et al.* 1989), where the neurons can be visualized with Nomarski optics. More recently still, the method has been adapted for use on intact slices (Randall *et al.* 1990), albeit that the neurons are not directly visualized. It is, however, necessary to use superficial neurons. On the one hand, this has the advantage that extracellular potentials will probably be kept to a minimum since the neuron is in close contact with the bathing medium. On the other hand the neuron will have been very close to the plane of the cut, which possibly means that dendritic processes and their synaptic inputs will have been damaged.

REFERENCES

Akaike, N., Kaneda, M., Hori, N. and Krishtal, O. A. (1988) *Neurosci. Lett.* **87** 75–79.

Andreasen, M., Lambert, J. D. C. and Jensen, M. S. (1989) *J. Physiol* **414** 317–336.

Ascher, P. and Nowak, L. (1988) *J. Physiol* **399** 227–245.

Benveniste, M., Mienville, J.-M., Sernagor, E. and Mayer, M. L. (1990), *J. Neurophysiol.* **63** 1373–1384.

Blake, J. F., Brown, M. W. and Collingridge, G. L. (1988) *Br. J. Pharmacol.* **95** 291–299.

Brown, K. T. and Flaming, D. G. (1986) *Advanced Micropipette Techniques for Cell Physiology. IBRO Handbook Series: Methods in the Neurosciences, Vol. 9.* Wiley, Chichester.

Burton, N. R., Smith, D. A. S. and Stone, T. W. (1987) *Comp. Biochem. Physiol.* **C88** 47–55.

Choi, D. W. and Fischbach, G. D. (1981) *J. Neurophysiol.* **45** 605–620.

Collingridge, G. L. and Lester, R. A. J. (1989) *Pharmacol. Rev.* **40** 143–210.

Dietzel, I., Heinemann, U. and Lux, H. D. (1989) *Glia* **2** 25–44.

Dingledine, R., (1984) *Brain Slices.* Plenum Press, New York.

Edwards, F. A., Konnerth, A., Sakmann, B. and Takahashi, T., (1989) *Pflugers Arch.* **414** 600–612.

Engberg, I., Flatman, J. A. and Lambert, J. D. C. (1979) *J. Physiol* **288** 227–261.

Engberg, I., Flatman, J. A. and Lambert, J. D. C. (1984) *Br. J. Pharmacol.* **81** 215–224.

Finkel, A. S. and Redman, S. J. (1985). In: Smith, T. G., Lecar, H., Redman, S. J. and Gage, P. W. (eds) *Voltage and Patch Clamping with Microelectrodes.* American Physiological Society, Bethesda, MD, pp. 95–120.

Flatman, J. A. and Lambert, J. D. C. (1979) *J. Neurosci. Methods* **1** 205–218.

Frenguelli, B. G., Blake, J. F., Brown, M. W. and Collingridge, G. L. (1991) *Br. J. Pharmacol.* **102** 355–362.

Harrison, N. L. and Simmonds, M. A. (1985) *Br. J. Pharmacol.* **84** 381–391.

Johnson, D., Hablitz, J. J. and Wilson, W. A. (1980) *Nature* **286** 391–393.

Kelly, J. S. (1975). In: Iversen, L. L., Iversen, S. D. and Snyder, S. H. (eds) *Handbook of Psychopharmacology, Vol. 2. Principles of Receptor Research.* Plenum Press, New York, pp. 29–67.

Kerkut, G. A. and Wheal, H. V. (1981) *Electrophysiology of Isolated Mammalian CNS Preparations.* Academic Press, London.

Lambert, J. D. C. and Heinemann, U. (1991) *Neuroscience* in press.

Lambert, J. D. C. and Jensen, M. S. (1986) *Neurosci. Lett.* **S26** S149.

Lambert, J. D. C., Flatman, J. A. and Engberg, I. (1981). In: Di Chiara, G. and Gessa, G, L. (eds) *Glutamate as a Neurotransmitter.* Raven Press, New York, pp. 205–216.

Llano, I., Marty, A., Johnson, J. W., Ascher, P. and Gähwiler, B. H. (1988) *Proc. Nat. Acad. Sci. USA* **85** 3221–3225.

MacDonald, J. F. (1985) In: Boulton, A. A. and Baker, G. B. (eds) *Neuromethods. Vol. 1. General Neurochemical Techniques.* Humana Press, Clifton, NJ, pp. 197–232.

Mayer, M. L. and Westbrook, G. L. (1987) *Prog. Neurobiol.* **28** 197–276.

Purves, R. D. (1979) *J. Neurosci. Methods* **1** 165–178.

Purves, R. D. (1981) *Microelectrode Methods for Intracellular Recording and Ionophoresis.* Academic Press, London.

Randall, A. D., Schofield, J. G. and Collingridge, G. L. (1990) *Biochem. Soc. Trans.* **18** 430–431.

Sakmann, B. and Neher, E. (1983) *Single-channel Recording.* Plenum Press, New York.

Salmoiraghi, G. and Stefanis, C. N. (1967) *Int. Rev. Neurobiol.* **10** 1–30.

Smith, T. G., Jr., Lecar, H., Redman, S. J. and Gage, P. W. (1985) *Voltage and Patch Clamping with Microelectrodes.* American Physiological Society, Bethesda, MD.

Somjen, G. G. (1973) *Prog. Neurobiol.* **1**, 199–237.

5

Receptor autoradiography of excitatory amino acid receptors: basic and clinical aspects

Sharin Y. Sakurai and **Anne B. Young***
Department of Neurology and Neuroscience Program, University of Michigan, Ann Arbor, MI, USA

ABSTRACT

Excitatory amino acids (EAAs) such as glutamate or aspartate are the predominant neurotransmitters mediating excitatory synaptic transmission at a majority of synapses in the mammalian central nervous system. EAAs interact with at least four receptor subtypes, which have been classified as follows: *N*-methyl-D-aspartate receptors, AMPA receptors, metabotropic receptors and kainate receptors. In this chapter, autoradiographic binding assays for each of the EAA receptor subtypes are described. The anatomical distribution of these receptors in rat brain and human brain tissue will be summarized. Furthermore, the role of EAAs and EAA receptor subtypes in human diseases such as Alzheimer's disease, Huntington's disease, lathyrism, domoic acid poisoning and olivopontocerebellar atrophy is discussed.

INTRODUCTION

The past twenty years have brought great advances in our understanding of excitatory amino acids (EAAs) and their role as neurotransmitters. The culmination of these endeavours is the widely accepted notion that excitatory acidic amino acids, such as glutamate or aspartate, are likely to be the predominant neurotransmitters mediating excitatory synaptic transmission in the mammalian central nervous system. EAAs exert their effects by interacting with several different receptors. At least four distinct glutamate receptor subtypes have been defined by electrophysiological and biochemical studies. Currently, these receptors are classified as follows: the *N*-methyl-D-aspartate (NMDA) receptor and the non-NMDA receptors (the 2-amino-3-(3-hydroxy-5-methylisoxazol-4-yl)propanoic acid (AMPA) receptor, the metabotropic

*Current address: Anne B. Young, M.D., Ph.D.,
Neurology Service,
Massachusetts General Hospital,
Boston, MA 02114, USA.

receptor and the kainate (KAIN) receptor (Dingledine *et al.* 1988, Monaghan *et al.* 1989, Young and Fagg 1990).

Due to the availability of selective antagonists, the NMDA receptor is the best-characterized glutamate receptor subtype (Watkins *et al.* 1990). NMDA receptor activation is involved in both physiological and pathological processes such as long-term potentiation, learning and memory, synaptic plasticity, hypoxic–ischaemic damage and chronic neurodegenerative disorders (Cotman and Iversen 1987, Rothman and Olney 1987, Watkins *et al.* 1990). The NMDA receptor is a ligand-gated ion channel that consists of several different modulatory sites. The ion channel associated with the NMDA receptor is gated by magnesium in a voltage-dependent manner (Nowak *et al.* 1984). At resting membrane potentials, magnesium blocks the flow of calcium and sodium ions through the channel; however, when the membrane is depolarized the magnesium blockade is relieved. The dissociative anaesthetics ketamine, phenycyclidine (abbreviation: PCP) and (+)-5-methyl-10,11-dihydro-5*H*-dibenzo[*a,d*]cyclohepten-5,10-imine maleate (MK-801) are non-competitive antagonists that bind to a site within the ion channel and prevent the passage of ions through the channel (Anis *et al.* 1983, Honey *et al.* 1985, Wong *et al.* 1986). In addition, the amino acid glycine, acting at a strychnine-insensitive site, enchances NMDA receptor activation and appears to be necessary for maximal channel activation (Johnson and Ascher 1987, Kleckner and Dingledine 1988). Finally, zinc ions and polyamines, acting at separate sites, also modulate NMDA receptor activity (Christine and Choi 1990, Peters *et al.* 1987, Ransom and Stec 1988, Reynolds and Miller 1988a, 1988b, Reynolds and Miller 1989, Westbrook and Mayer 1987).

Until recently, the non-NMDA receptors had been classified simply into two groups: KAIN receptors and quisqualate receptors. Biochemical and electrophysiological evidence, however, now suggests that the quisqualate receptors consist of two separate subtypes: the AMPA receptor and the metabotropic receptor (Watkins *et al.* 1990). Unlike the NMDA receptor, development of selective antagonists for the AMPA receptor has not been as successful. The AMPA receptor is a ligand-gated cation channel believed to mediate fast excitatory neurotransmission in mammalian brain. Recent studies have also suggested that the AMPA receptor may participate in neurodegenerative processes (Bridges *et al.* 1988, Morgan 1987) and in neurotoxicity associated with hypoxia–ischaemia (Sheardown *et al.* 1990). The metabotropic glutamate receptor is believed to be linked to the inositol phospholipid second messenger system (Sladeczek *et al.* 1988, Sugiyama *et al.* 1988). These receptors may be involved in developmental and synaptic plasticity (Dudek and Bear 1989, Dudek *et al.* 1989, Kano and Kano 1987) as well as neurotoxic processes (Garthwaite and Garthwaite 1989a, 1989b).

Currently there is considerable debate over whether a unique KAIN receptor exists. Electrophysiological studies suggest that kainate may exert its physiological actions at the AMPA receptor (Watkins *et al.* 1990). Evidence to support the existence of a separate unique KAIN receptor has come largely from binding studies using [^{3}H]KAIN (Berger and Ben-Ari 1983, Monaghan and Cotman 1982, Monaghan *et al.* 1986, Unnerstall and Wamsley 1983). These studies indicate that [^{3}H]KAIN binds to receptors with a unique regional distribution that is different from AMPA binding.

The NMDA receptor complex has been studied using several autoradiographic assays. The NMDA receptor can be labelled with the agonist [^{3}H]glutamate under conditions that optimize ligand binding to the NMDA site, specifically, while excluding binding to the other glutamate receptor subtypes by including AMPA, KAIN and metabotropic antagonists in the buffer. Alternatively, the competitive NMDA antagonists [^{3}H]($\pm$)-2-carboxypiperazine-4-yl-propyl-1-phosphonic acid (CPP) or [^{3}H]4-phosphonomethyl-2-piperidine carboxylic acid (CGS 19755) can also be used to label the NMDA site (Murphy *et al.* 1988, Murphy *et al.* 1987). Interestingly, the use of agonists and antagonists to label the NMDA binding site has revealed the possibility that subtypes or different forms of the NMDA receptor may exist. Comparing agonist and antagonist displacement of [^{3}H]glutamate versus [^{3}H]CPP in rat brain, Monaghan *et al.* (1988) report that agonists are more potent inhibitors of [^{3}H]glutamate binding than of [^{3}H]CPP binding whereas antagonists are more potent displacers of [^{3}H]CPP binding than of [^{3}H]glutamate binding. These findings suggest the presence of agonist-preferring and antagonist-preferring NMDA receptor sites in rat brain (Monaghan *et al.* 1988).

In addition to labelling the NMDA site with the agonist [^{3}H]glutamate, autoradiographic assays using [^{3}H]glycine to assess the glycine modulatory site are available (Bristow *et al.* 1986, McDonald *et al.* 1990). The NMDA receptor can also be studied using ligands such as [^{3}H]TCP and [^{3}H]MK-801 that bind with high affinity to the PCP site within the ion channel (Maragos *et al.* 1988, Sakurai *et al.* 1991). Binding of ligands to the PCP site is modulated by agents which regulate NMDA channel activity, such as glutamate and glycine, as well as by zinc, magnesium and the polyamines (Bonhaus and McNamara 1988, Javitt and Zukin 1989a, Javitt and Zukin 1989b, Ransom and Stec 1988, Reynolds *et al.* 1987). This regulation of binding to the PCP site by agents acting at the modulatory sites affords the opportunity to use the PCP site assays as functional assays for NMDA receptor complex (Young and Fagg 1990).

Previous autoradiographic assays utilized [^{3}H]glutamate under conditions that optimized binding to the quisqualate receptors; however, under such conditions it is likely that binding to both the metabotropic quisqualate receptor and the ionotropic AMPA receptor was being measured (Cha *et al.* 1988, Greenamyre *et al.* 1985). Now, [^{3}H]AMPA can be used to label the AMPA receptor directly (Monaghan *et al.* 1984, Nielsen *et al.* 1988, Young and Fagg 1990). Furthermore, we have developed an autoradiographic assay to measure binding to the metabotropic receptor using [^{3}H]glutamate in the presence of AMPA and NMDA which excludes binding to the other EAA receptors (Cha *et al.* 1990).

The development of quantitative autoradiographic binding assays for each of the EAA receptor subtypes has allowed the biochemical and pharmacological characterization of the glutamate receptor subtypes that correspond to the electrophysiologically defined receptors (Young and Fagg 1990). Furthermore, the anatomical localization of each of the EAA receptors can now be assessed. In addition, receptor autoradiographic studies can also be used to determine kinetic and equilibrium saturation information about the pharmacological properties of the EAA receptor subtypes. In this chapter the autoradiographic binding techniques for EAA receptor subtypes will be described. The distribution of these receptors in normal and disease tissue will also be summarized.

BINDING METHODS

Tissue preparation

Quantitative autoradiographic binding assays are carried out on frozen unfixed brain tissue sections. Animal or human brain tissue is removed and frozen rapidly under powdered dry ice. Before sectioning, the frozen blocks of tissue are equilibrated to cryostat temperature (−10°C to −20°C), after which 20–40 μm tissue sections are cut and thaw-mounted onto gelatin-coated slides. These tissue sections are stored at −20°C for 24–48 hours until used in the autoradiographic assays.

Several important issues that arise when applying receptor autoradiographic techniques to human post-mortem tissue will be reviewed here briefly (Young and Penney 1988). In order to assess the relation between clinical findings and the results of receptor binding studies, it is important to establish a record of the patient's clinical information prior to death. In addition to the routine clinical history, the following items from a patient's history should also be noted: medications and immediate clinical status prior to death, metabolic and nutritional status, as well as the presence of infections. Since biochemical changes in the brain are known to occur following death, the time between when a patient dies and when the body is placed at 4°C should be recorded. In addition the post-mortem delay — the time between when the patient dies and when the brain is frozen — should also be noted.

Glutamate receptor binding assays

Detailed descriptions of glutamate receptor autoradiography have been published previously (Greenamyre *et al.* 1983, 1984, 1985, Maragos *et al.* 1988, McDonald *et al.* 1990, Nielsen *et al.* 1988, Sakurai *et al.* 1991). The basic steps involved in receptor autoradiography are illustrated in Fig. 1. Briefly, the frozen tissue sections are pre-washed in buffer at 4°C for 30 minutes to remove any endogenous ligands and then dried under a stream of cool air. The buffer used in the pre-wash is the same buffer that is used in the remaining steps of the binding experiment. The tissue sections are incubated with radioligand and selective blocking agents under conditions that optimize binding to the desired EAA receptor. The sections are then rinsed and dried.

Sections are placed in an X-ray cassette and exposed to tritium-sensitive film such as Amersham Hyperfilm or LKB Ultrofilm ^{3}H. A complete set of radioactive standards that is calibrated against brain paste with known amounts of tritium is coexposed with the film. Following a two- to eight-week exposure period at 4°C films are developed in Kodak D19, fixed and air-dried. Analysis of radioligand binding densities from resultant autoradiograms is performed using microcomputer-assisted image densitometry. Optical densities from resultant autoradiograms are converted to binding densities using a polynomial regression curve derived from the calibrated radioactive standards. Ten to twenty readings per brain area from triplicate sections are analysed from the autoradiograms. Specific binding is calculated as the difference between total binding and non-specific binding.

Since tritium emits relatively low-energy β-particles, tissue absorbs some tritium emissions. In brain tissue, white matter absorbs β-particles more than grey matter (Kuhar and Unnerstall 1985, Unnerstall and Wamsley 1983). This issue is of importance when considering the binding of tritiated radioligands during brain development since changes in myelination may affect the degree of tissue quenching

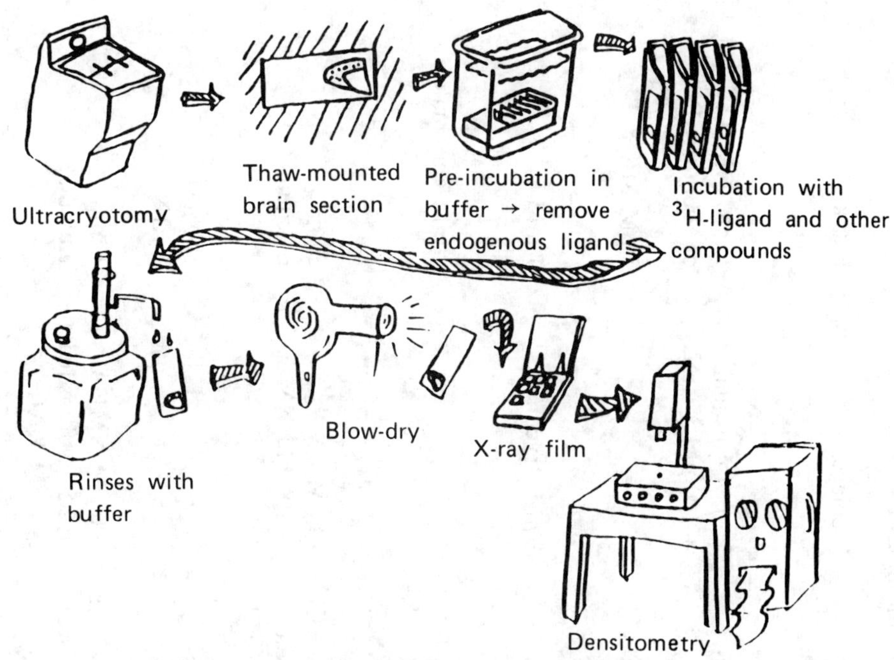

Fig. 1 — Receptor autoradiography of slide-mounted tissue sections. Frozen brain tissue is sectioned and mounted onto microscope slides. Tissue sections are pre-washed in buffer, incubated with radioligand, rinsed with buffer and blown dry. Sections are apposed to tritium-sensitive film with radioactive standards, developed and analysed with computer-assisted densitometry. See text for details.

and because brain regions differ in their myelin content. Comparisons of radioligand binding assays using [^{14}C]glutamate and [^{3}H]glutamate in brain tissue, however, have demonstrated that binding profiles are similar and that quenching due to myelination does not influence greatly binding densities in rat brain (Greenamyre *et al.* 1987, McDonald *et al.* 1990).

The autoradiographic binding assays for EAA receptors currently used in our laboratory are outlined in Table 1. For each autoradiographic assay, it is important to optimize the binding conditions in order to obtain a high ratio of specific to non-specific binding. The temperature, pH of the buffer, ionic conditions and rinse conditions should be assessed in order to optimize specific binding and minimize non-specific binding. Additionally kinetic studies must be performed in order to determine that binding studies are being performed at equilibrium conditions.

Ligands binding to the NMDA, KAIN, AMPA and metabotropic receptor have fast association rates and reach equilibrium binding in a short period of time; thus all assay incubation times are 45 minutes at a temperature optimum of 4°C, with the exception of [^{3}H]glycine, which has an even shorter incubation period of 35 minutes at 4°C (Cha *et al.* 1990, Greenamyre *et al.* 1983, 1984, McDonald *et al.* 1990).

Optimizing rinse conditions is crucial in order to generate autoradiograms with high specific binding. When association and dissociation rates are very fast, rapid rinses of less than 10 seconds are necessary to avoid underestimating the number of

Table 1 — Excitatory amino acid receptor binding assay conditions

Receptor	Ligand (concentration)	Temperature (incubation time)	Buffer	Displacers	Blank
NMDA	[^{3}H]Glutamate (65 nM)	4°C (45 min)	50 mM Tris-acetate pH 7.4	2.5 μM quisqualate + 100 μM KAIN	1 mM NMDA or 100 μM CPP
PCP	[^{3}H]MK-801 (5 nM)	25°C (120 min)	50 mM Tris-acetate pH 7.4		5 μM MK-801
Glycine	[^{3}H]Glycine (100 nM)	4°C (35 min)	50 mM Tris-citrate pH 7.4	1 μM strychnine	1 mM glycine
AMPA	[^{3}H]AMPA (37 nM)	4°C (45 min)	50 mM Tris-HCl + 2.5 mM $CaCl_2$ + 30 mM KSCN pH 7.2		1 mM glutamate
Metabotropic	[^{3}H]Glutamate (100 nM)	4°C (45 min)	50 mM Tris-HCl + 2.5 mM $CaCl_2$ + 30 mM KSCN pH 7.2	10 μM AMPA + 100 μM NMDA	2.5 μM quisqualate
KAIN	[^{3}H]KAIN (60 nM)	4°C (45 min)	50 mM Tris-acetate pH 7.2		100 μM KAIN
NNKQ	[^{3}H]Glutamate (100 nM)	4°C (45 min)	50 mM Tris-HCl + 2.5 mM $CaCl_2$ pH 7.2	100 μM NMDA + 2.5 μM quisqualate + 1 μM KAIN	1 mM glutamate

receptors. Our rinse conditions for all [^{3}H]glutamate assays, [^{3}H]glycine, [^{3}H]AMPA and [^{3}H]KAIN binding assays consist of three rapid rinses of cold buffer followed by one rinse with acetone/2.5% (v/v) glutaraldehyde at 4°C. The entire rinse procedure takes less than 10 seconds to complete. We have found that this rinse procedure minimizes the variation in binding from section to section and yields a high specific to non-specific binding ratio.

In contrast to the fast association and dissociation rates of ligands binding at the NMDA, KAIN, AMPA and metabotropic receptor, ligands binding to the PCP site display slower rates of association and dissociation. Binding to the PCP site by [^{3}H]MK-801 requires a longer incubation time of at least 2 hours at a temperature optimum of 23°C to reach equilibrium (Sakurai *et al.* 1991). Furthermore the optimal rinse time for [^{3}H]MK-801 binding is 80 minutes.

DISTRIBUTION OF EAA RECEPTORS

Animal studies

In adult rat brain, NMDA, AMPA and KAIN receptors have unique distributions (Fig. 2 and Table 2). NMDA receptor binding is densest in stratum moleculare of the dentate gyrus (SMDG), stratum oriens of CA1 (SO-CA1) and in the outer layers of the cerebral cortex. Less binding is observed in the inner layers of the cerebral cortex. NMDA receptor binding is intermediate in the striatum and is low in the brain stem. In the cerebellum, binding to the NMDA receptor is high in the granule cell layer, with very little binding in the molecular layer. In most brain areas the distribution of glycine and PCP binding sites is very similar to the distribution of NMDA binding sites (Maragos *et al.* 1988, McDonald *et al.* 1990, Sakurai *et al.* 1991). Autoradiographic studies, however, have revealed discrepancies in the ratio of these binding sites in certain brain areas. For example, in the granule cell layer of cerebellum there is a high density of NMDA and glycine binding sites, while there are relatively few PCP binding sites. It is possible that a low-affinity PCP binding site may be present in the cerebellum and is not detectable using standard autoradiographic techniques (Haring *et al.* 1987, 1988, Sakurai *et al.* 1991, Vignon *et al.* 1986). Alternatively, such differences may reflect the presence of NMDA receptor subtypes or differential regulation of the NMDA receptor in different brain regions (Young and Fagg 1990).

AMPA receptor binding is very dense in the SMDG, the SO-CA1 and CA3, outer layers of cortex and striatum. In the inner layers of cortex, AMPA binding is less dense. In the cerebellum AMPA binding is concentrated on the Purkinje cell dendrites in the molecular layer, with less binding in the granule cell layer.

Unlike AMPA or NMDA binding, KAIN binding is most dense in the inner layers of the cerebral cortex. In the hippocampal formation binding to the KAIN receptor is most prominent in the stratum lucidum of CA3 and SMDG, whereas there is virtually no binding in the CA1 region. In the striatum there is a moderate amount of KAIN binding. KAIN binding in the cerebellum is higher in the granule cell layer than in the molecular layer.

The regional distribution of the metabotropic receptor is distinct from AMPA, NMDA and KAIN binding sites. The highest amount of metabotropic receptor binding is found in the molecular layer of the cerebellum. There is a moderate amount

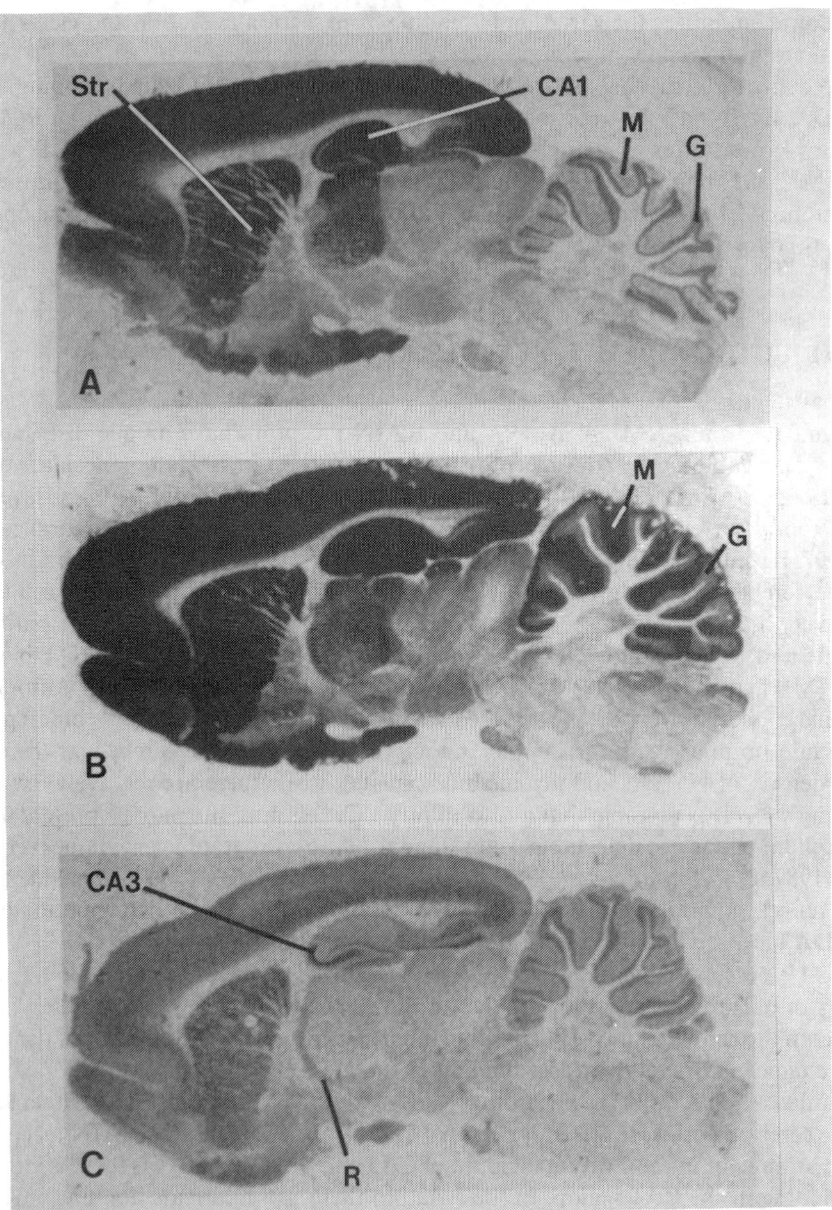

Fig. 2 — Representative autoradiograms of EAA binding sites in parasagittal sections of adult rat brain. A: NMDA-sensitive [^{3}H]glutamate binding sites were assayed in 200 nM [^{3}H]glutamate. B: [^{3}H]AMPA binding sites were measured with 40 nM [^{3}H]AMPA. C: [^{3}H]KAIN binding sites were assayed in 60 nM [^{3}H]KAIN. Abbreviations: CA1, stratum radiatum of CA1; CA3, stratum lucidum of CA3; G, granule cell layer of cerebellum; M, molecular layer of cerebellum; R, reticular nucleus of thalamus.

Table 2 — Regional distribution of [^{3}H]glutamate, [^{3}H]AMPA and [^{3}H]KAIN binding to NMDA, AMPA and KAIN receptors in rat brain

Regions	Protein (pmol/mg) ± SEM		
	[^{3}H]Glutamate	[^{3}H]AMPA	[^{3}H]KAIN
Cerebral cortex			
layers I–III	1.28 ± 0.06	2.42 ± 0.08	0.53 ± 0.07
Hippocampus			
SO-CA1	1.53 ± 0.09	3.52 ± 0.08	0.28 ± 0.05
CA3	0.86 ± 0.03	2.86 ± 0.05	2.72 ± 0.41
SMDG	2.11 ± 0.16	4.02 ± 0.12	0.72 ± 0.10
Striatum	0.67 ± 0.02	2.48 ± 0.09	1.00 ± 0.18
Cerebellum			
Granule cell layer	0.42 ± 0.09	0.73 ± 0.10	0.30 ± 0.01
Molecular layer	0.07 ± 0.04	1.72 ± 0.15	0.12 ± 0.01

Data represent mean ± SEM of 4 animals. [^{3}H]Glutamate binding to NMDA receptor was performed at 200 nM. [^{3}H]AMPA binding was measured at 40 nM and [^{3}H]KAIN binding was carried out at 60 nM. All assays were performed in triplicate in adjacent sections.

of binding in the cerebral cortex and denser binding in the striatum. In the hippocampal formation there is dense labelling of the SMDG and CA3, with less binding in the CA1 region.

Human studies

Using the same autoradiographic assays for the EAA receptors, it is also possible to examine glutamate receptors in human tissue. Specific [^{3}H]glutamate binding in human tissue has similar kinetic, pharmacological and regional properties to those described in rat (Greenamyre *et al.* 1985). With the development of specific EAA receptor assays we have examined different regions of human post-mortem brain tissue. As in rat brain, EAA receptors are distributed heterogeneously throughout the layers of the cortex and in the hippocampal formation (Fig. 3). EAA receptors are also found in the caudate nucleus and cerebellum (Fig. 4). Studies of EAA receptors in normal human and pathological human brain tissue using autoradiography provide detailed anatomical and pharmacological information which can increase our understanding of the receptor pathology of human diseases.

ROLE OF EAA RECEPTORS IN HUMAN DISEASES

Glutamate, the major EAA neurotransmitter in mammalian brain, is most likely the neurotransmitter used by intracortical association fibres and cortical fibres which project to the caudate and putamen, thalamus, brain stem nuclei and spinal cord (Fonnum 1984, McGeer *et al.* 1977, Young *et al.* 1981). Further evidence suggests that the cerebellar granule cells hippocampal commissural fibres and auditory, olfactory and somatosensory systems use glutamate as a neurotransmitter (Cotman and Iversen 1987, Fonnum 1984, Young *et al.* 1974). Since glutamate and glutamate-like analogues can be neurotoxic, abnormalities in glutamate neurotransmitter function may play a role in neurotoxicity and neurodegenerative disorders (Meldrum and Garthwaite 1990, Rothman and Olney 1987).

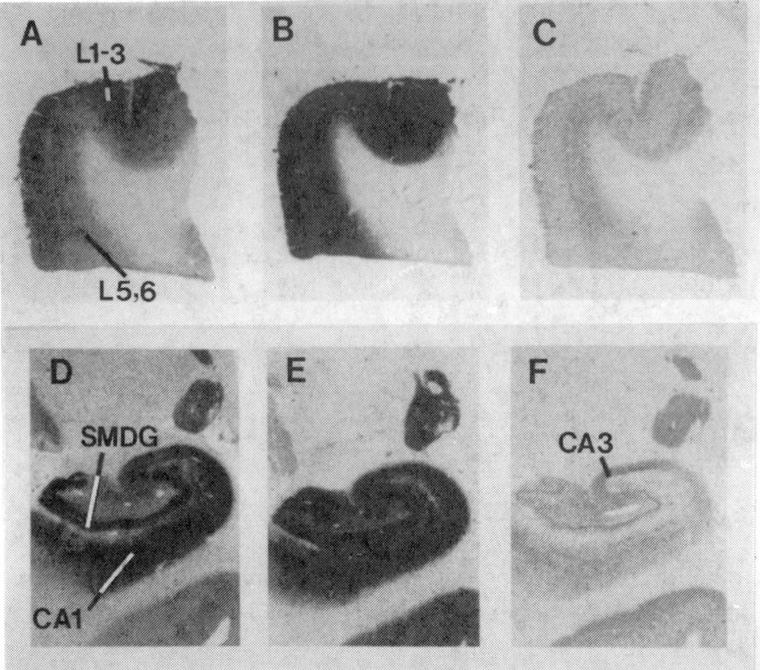

Fig. 3 — Representative autoradiograms of EAA binding sites in human middle temporal cortex (A, B, C) and human hippocampal formation (D, E, F). NMDA-sensitive [^{3}H]glutamate binding sites were assayed in 65 nM [^{3}H]glutamate (A,D). [^{3}H]AMPA binding sites were measured with 37 nM [^{3}H]AMPA (B, E). [^{3}H]KAIN binding sites were assayed in 60 nM [^{3}H]KAIN (C, F). Abbreviations: L1–3, layers 1–3 of middle temporal cortex; L5–6, layers 5 and 6 of middle temporal cortex; SMDG, stratum molecular of dentate gyrus; CA1, stratum radiatum; CA3, stratum lucidum.

Alzheimer's disease

Alzheimer's disease (AD) is a neurodegenerative disorder characterized by memory loss, learning deficits and personality changes (Greenamyre and Young 1990). Pathologically, AD is characterized by the presence of amyloid-containing neuritic plaques and neurofibrillary tangles (NFTs) in the cerebral cortex and hippocampus (Hyman *et al.* 1984, Khachaturian 1985). Furthermore, in AD the pyramidal neurons of cerebral cortex and hippocampus are markedly depleted and are predisposed to the formation of NFTs (Lewis *et al.* 1987, Mann *et al.* 1986, Pearson *et al.* 1985). Evidence suggests that both pre- and postsynaptic elements of EAA pathways may be involved in AD (Greenamyre and Young 1990). Neuritic plaques, NFTs and neurodegeneration characteristic of AD are observed in cortical and limbic areas known to be vulnerable to the neurotoxic effects of EAAs (Foster 1990). In fact it has been hypothesized that EAA toxicity may be involved in the pathogenesis of AD (Choi 1988, Greenamyre and Young 1990, Maragos *et al.* 1987).

EAAs appear to disrupt cytoskeletal elements, which may contribute to the neurodegenerative pathology in AD. It has been reported that paired helical filaments, a major component of NFTs, develop in neuronal cultures following exposure to glutamate and aspartate (DeBoni and Crapper-McLachlan 1985). The distribution of

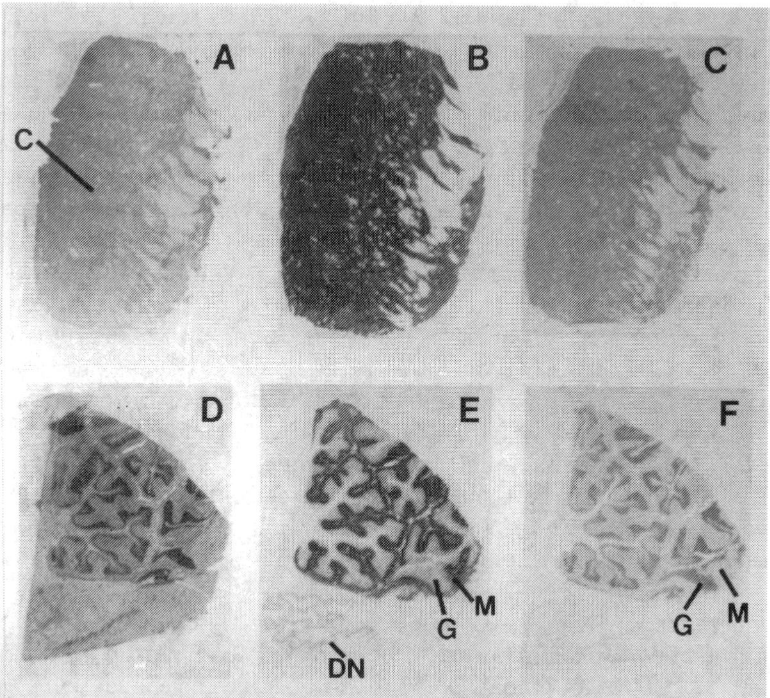

Fig. 4 — Representative autoradiograms of EAA binding sites in human caudate nucleus (A, B, C) and human cerebellum (D, E, F). NMDA-sensitive [^{3}H]glutamate binding sites were assayed in 65 nM [^{3}H]glutamate (A, D). [^{3}H]AMPA binding sites were measured with 37 nM [^{3}H]AMPA (B, E). [^{3}H]KAIN binding sites were assayed in 60 nM [^{3}H]KAIN (C, F). Abbreviations: C, caudate nucleus; DN, dentate nucleus; G, granule cell layer of cerebellum; M, molecular layer of cerebellum.

tau, a microtubule-associated protein associated with paired-helical filaments in neurons, is also modified following exposure to EAAs. NMDA and glutamate stimulated an increase in tau immunoreactivity in cell bodies and neurites, while NMDA antagonists blocked such a response (Bigot and Hunt 1990). Abnormal phosphorylation of neurofilaments in cell bodies and proximal neurites is observed in degenerating neurons following KAIN injections into spinal cord (Hugon and Vallat 1990). Furthermore, plaques and NFTs developed in rat cerebral cortex following excitotoxic lesions of the nucleus basalis magnocellularis (Arendash *et al.* 1987). These studies support the hypothesis that EAAs may be involved in the neurodegenerative changes observed in AD.

Numerous autoradiographic and homogenate binding studies reveal EAA receptor pathology in AD. Greenamyre and colleagues have reported significant reductions of NMDA- and quisqualate-sensitive [^{3}H]glutamate binding in the cerebral cortex of patients who died with AD (Greenamyre *et al.* 1985). Similarly, binding to the NMDA receptor was reduced in the hippocampi of AD patients (Greenamyre *et al.* 1985, Penney *et al.* 1990, Repressa *et al.* 1988). Two other groups, however, have reported no significant changes in NMDA receptor binding in the cortex or hippocampi (Cowburn *et al.* 1988, 1989, Geddes *et al.* 1986).

Binding to the PCP site, another site on the NMDA receptor complex, is also reduced in AD hippocampi (Maragos *et al.* 1987, Penney *et al.* 1990). Despite not having found reductions in NMDA receptor binding in AD hippocampi, Cotman and colleagues also report significant reductions in PCP sites (Cotman and Iversen 1987). In homogenate preparations of AD brain tissue, [^{3}H]TCP binding is decreased in frontal cortex but not hippocampus (Simpson *et al.* 1988); however, Mouradian and colleagues report no change in [^{3}H]MK-801 binding in temporal pole tissue (Mouradian *et al.* 1988).

Since binding to the PCP site is dependent upon receptor activation (Bonhaus *et al.* 1987, Bonhaus and McNamara 1988, Reynolds *et al.* 1987, Wong *et al.* 1987), an interesting question arises from the studies that demonstrate alterations in ligand binding to the PCP site in AD brain tissue. Is the change in binding due to a decrease in the number of binding sites or is there a change in the coupling of the sites on the NMDA receptor complex? Recently, Procter and colleagues demonstrated normal baseline PCP receptor binding but reduced glycine enhancement of [^{3}H]TCP and [^{3}H]MK-801 binding in homogenate preparations of frontal cortex from AD patients (Procter *et al.* 1989a, 1989b). This result would suggest the possibility that the allosteric coupling between sites on the NMDA receptor complex may be altered in AD.

Binding to the KAIN may also be altered in AD. [^{3}H]KAIN binding in the deep layers of frontal cortex was reported to be increased (Chalmers *et al.* 1990). In the hippocampus, [^{3}H]KAIN binding is reported to be unchanged (Jansen *et al.* 1990), decreased (Repressa *et al.* 1988) or there is a wider distribution of KAIN binding sites without a change in density (Geddes *et al.* 1985). No change in KAIN receptor binding was found in homogenate preparations of hippocampus or cortex (Cowburn *et al.* 1989).

Several factors could contribute to the discrepancies in the reports of EAA receptor binding in AD by different groups. Differences in assay protocols such as pre-wash conditions, type of buffer, and incubation and rinse procedures could produce different findings. The state of the human brain tissue also could contribute to differences in results. As mentioned above, factors such as post-mortem delay, tissue handling as well as the severity of the disease state contribute to receptor changes. Lastly, it is difficult to compare homogenate binding results to autoradiographic results since endogenous ligand concentrations differ between the two preparations and the anatomical localization of receptors is lost in membrane preparations. More studies are necessary in order to clarify these problems. Recently, Jansen and colleagues, using [^{3}H]glycine, [^{3}H]glutamate and [^{3}H]TCP autoradiography to label the three different sites on the NMDA receptor in serially adjacent brain tissue sections, reported significant reductions in binding to all three binding sites (Jansen *et al.* 1990). Since the binding studies were performed on serially adjacent tissue from the same AD brains, this report along with previous findings suggests that NMDA and PCP binding sites are decreased in AD.

Huntington's disease

Huntington's disese (HD) is an autosomal dominant neurodegenerative disorder characterized by onset, usually in mid-life, of progressive dementia, personality disorders, choreiform movements and death within 15–20 years of onset (Martin and

Gusella 1986). The HD gene is localized to the tip of chromosome 4; however, the gene has yet to be located and the biochemical defect remains unknown (Gusella *et al.* 1983). Pathologically, HD is characterized by extensive neuronal loss in the caudate nucleus and putamen, with relative sparing of neurons in the rest of the brain (Vonsattel *et al.* 1985). Neurochemically these changes are reflected by a decrease in striatal glutamate decarboxylase activity, and GABA and substance P immunoreactivity.

Although the pathogenesis of HD remains unknown, evidence from several studies has led to the hypothesis that glutamate or a glutamate-like substance acting at EAA receptors could contribute to neuronal degeneration observed in HD. In fact, the NMDA receptor and endogenous agonists acting at this receptor may play an important role in the pathogenesis of HD (DiFiglia 1990). In rats, intrastriatal injections of the endogenous NMDA agonist quinolinic acid reproduced some of the pathological and neurochemical alterations observed in HD more specifically than did kainic acid (Beal *et al.* 1986, Coyle and Schwarcz 1976, Davis and Roberts 1988, McGeer and McGeer, 1976). Interest has focused on the NMDA receptor since overactivation of this receptor has been shown to be excitotoxic to neurons in culture (Choi 1988). Recently, Beal and co-workers reported a decrease in the levels of kynurenic acid, an EAA antagonist, in HD putamen and suggested that a chronic deficiency of such an endogenous antagonist may lead to the development of an NMDA-mediated excitoxic neurodegenerative process (Beal *et al.* 1990). If one of the glutamate receptors such as the NMDA receptor is involved in the pathogenesis of HD, then NMDA antagonists may be clinically relevant. In fact, the non-competitive NMDA antagonist MK-801 has been shown to prevent the excitotoxic-induced behavioural alterations and pathological changes induced by intrastriatal quinolinic acid injections (Giordano *et al.* 1990).

Previous autoradiographic studies of EAA receptors in HD putamen indicated a decrease in total [^{3}H]glutamate receptors, with a preferential loss of the NMDA receptor subtype (Greenamyre *et al.* 1985, Young *et al.* 1988). Additionally, Albin and colleagues reported a decrease in NMDA and PCP binding in a patient with presymptomatic HD (Albin *et al.* 1990). More recent work in our laboratory, however, using autoradiographic binding assays to assess binding to specific receptor subtypes reveals that no single binding site is selectively decreased. Binding to NMDA, MK-801, glycine, KAIN and AMPA sites were all decreased by 50–60% in caudate nucleus. Binding to the metabotropic receptor and the non-NMDA, non-KAIN, non-quisqualate site was not altered in HD (Dure *et al.* 1991). These data would be consistent with the notion that a pre-existing metabolic abnormality renders some neurons more vulnerable to glutamate or quinolinic acid (Meldrum and Garthwaite 1990). In fact, neurons with compromised energy metabolism were more vulnerable to NMDA receptor-mediated damage (Novelli *et al.* 1988).

Lathyrism

Excitotoxicity also may be mediated by the non-NMDA class of EAA receptors. The subacute neurodegenerative syndrome lathyrism is an upper motor neuron disease linked to the excessive consumption of the chick-pea, *Lathyrus sativus*, during periods of famine in East Africa and southern Asia (Spencer *et al.* 1986). The toxin in this plant, identified as *β-N*-oxalylamino-L-alanine (L-BOAA = β-ODAP), produces seizures in neonatal rats, and tremors, spasticity and myoclonus in monkeys (Olney *et al.* 1976, Spencer *et al.* 1986). Furthermore, L-BOAA is excitotoxic in cultures of explants from spinal cord or frontal cortex as well as in cultures of cortical neurons (Nunn *et al.* 1987, Ross *et al.* 1987, Weiss *et al.* 1989). In electrophysiological studies L-BOAA appears to act as an agonist at the AMPA receptor (Bridges *et al.* 1989). Binding studies in synaptic membranes reveal that [^{3}H]AMPA binding is displaced potently by L-BOAA (Bridges *et al.* 1988, Ross *et al.* 1989).

Domoic acid poisoning

Another example of non-NMDA receptor-mediated neurotoxicity playing a role in human disease was recently documented during an outbreak of domoic acid poisoning in Canada in 1987 (Perl *et al.* 1990, Teitelbaum *et al.* 1990). The source of domoic acid, a potent KAIN receptor agonist, was found to be mussels which were contaminated with unusually high concentrations of domoic acid. Several victims developed seizures. Amnesia was also present in a small number of the patients. Neuropathological examination of patients who died revealed a pattern of neuronal loss in brain similar to the pattern of neuronal damage produced in rats by systemic injections of KAIN. Furthermore, the neuronal degeneration is similar to the regional distribution of high-affinity KAIN receptors. Electrophysiological studies support the suggestion that domoic acid exerts its excitatory actions at the KAIN receptor (Debonnel *et al.* 1989, Stewart *et al.* 1990). Recently, Sutherland and colleagues demonstrated that, in rat, intrahippocampal injections of domoic acid produced a pattern of hippocampal degeneration that is consistent with the hypothesis that domoic acid toxicity is mediated by interaction with KAIN receptors (Sutherland *et al.* 1990).

Olivopontocerebellar atrophy

Another brain region in which EAA receptors have been shown to be altered in disease states is the cerebellum. In the cerebellar cortex, glutamate is likely to be the neurotransmitter of mossy fibre afferents and granule cells, while aspartate may be the neurotransmitter of the climbing fibres (Sandoval and Cotman 1978, Stone 1979, Toggenberger *et al.* 1983, Wiklund *et al.* 1984, Young *et al.* 1974). Olivopontocerebellar atrophy (OPCA) is a neurological disorder in which cerebellar cortical neurons, particularly Purkinje and granule cells, degenerate, and mossy fibre and climbing fibre afferents are also lost (Berciano 1982, Harding 1984). Recently, Albin and colleagues reported a significant decrease in NMDA receptor binding in the granule cell layer of OPCA cerebellar cortex (Albin and Gilman 1990). Furthermore, binding to the AMPA receptor and metabotropic receptor in the molecular layer is significantly reduced in OPCA cerebellar cortex (Makowiec *et al.* 1990). These results, consistent with data from animal studies, suggest that NMDA binding sites are located on granule cells, while AMPA and metabotropic binding sites are localized on the Purkinje cell dendrites (Albin and Gilman 1990, Cha *et al.* 1990, Makowiec *et al.* 1990, Makowiec *et al.* 1991).

CONCLUSION

Quantitative autoradiography binding assays using selective ligands have permitted the characterization of anatomical and pharmacological properties of EAA receptor subtypes in brain and changes in these receptors in disease states. Elucidation of the pharmacological properties of EAA receptors will undoubtedly facilitate the development of therapeutic agents directed at specific EAA receptor subtypes. For example, drugs such as MK-801, a non-competitive NMDA antagonist, prevents neuronal damage following ischaemic brain injury (Gill *et al.* 1987, 1988).

In addition to autoradiographic binding, the application of molecular biology techniques to the study of EAA receptors is contributing to our understanding of EAA receptor subtypes. Recently, a family of AMPA receptor genes was cloned (Keinanen *et al.* 1990, Sommer *et al.* 1990). Using *in situ* hybridization it is now possible to characterize the cellular localization of the AMPA receptor. This information, however, is limited to the site of message synthesis and does not provide any information about processing of the message or the cellular distribution of the protein. Using immunocytochemical techniques, one can generate antibodies to cloned receptors and thereby characterize the cellular localization of receptor proteins. In order to characterize the pharmacology of receptors, however, radioligand binding methodologies must be relied upon. Autoradiographic binding techniques, combined with *in situ* hybridization methodologies and immunocytochemical assays, will enhance our understanding of the pharmacological, molecular and genetic properties of EAA receptor subtypes.

ACKNOWLEDGEMENTS

We appreciate the photographic assistance of R. H. Price and the artistic contribution of D. C. M. Chu on Fig. 1. This work was supported by USPHS grants NS 15655, NS 19613, and AG 08671.

REFERENCES

Albin, R. L. and Gilman, S. (1990) *Brain Res.* **522** 37–45.

Albin, R. L., Young, A. B., Penney, J. B., Handelin, B., Balfour, R., Anderson, K. D., Markel, D. S., Tourtellotte, W. W. and Reiner, A. (1990) *N. Engl. J. Med.* **322** 1293–1298.

Anis, N. A., Berry, S. C., Burton, N. R. and Lodge, D. (1983) *Br. J. Pharmacol.* **79** 565–575.

Arendash, G. W., Millard, W. J., Dunn, A. J. and Mayer, E. M. (1987) *Science* **238** 952–956.

Beal, M. F., Matson, W. R., Swartz, K. J., Gamache, P. H. and Bird, E. D. (1990) *J. Neurochem.* **55** 1327–1339.

Beal, M. G., Kowall, N. W., Ellison, D. W., Mazurek, M. F., Swartz, K. J. and Martin, J. B. (1986) *Nature* **321** 168–171.

Berciano, J. (1982) *J. Neurol. Sci.* **53** 253–272.

Berger, M. and Ben-Ari, Y. (1983) *Neurosci. Lett.* **39** 337–342.

Bigot, D. and Hunt, S. P. (1990) *Neurosci. Lett.* **111** 1275–280.

Bonhaus, D. W. and McNamara, J. O. (1988) *Mol. Pharmacol.* **34** 250–258.
Bonhaus, D. W., Burge, B. C. and McNamara, J. O. (1987) *Eur. J. Pharmacol.* **142** 489–490.
Bridges, R. J., Kadri, M. M., Monaghan, D. T., Nunn, P. B., Watkins, J. C. and Cotman, C. W. (1988) *Eur. J. Pharmacol.* **145** 357–359.
Bridges, R. J., Stevens, D. R., Kahle, J. S., Nunn, P. B., Watkins, J. C. and Cotman, C. W. (1989) *J. Neurosci.* **9** 2073–2079.
Bristow, D. R., Bowery, N. G. and Woodruff, G. N. (1986) *Eur. J. Pharmacol.* **126** 303–307.
Cha, J.-H. J., Greenamyre, J. T., Nielsen, E. O., Penney, J. B. and Young, A. B. (1988) *J. Neurochem.* **51** 469–478.
Cha, J. J., Makowiec, R. L., Penney, J. B. and Young, A. B. (1990) *Neurosci. Lett.* **113** 78–83.
Chalmers, D. T., Dewar, D., Graham, D. I., Brooks, D. N. and McCulloch, J. (1990) *Proc. Nat. Acad. Sci. USA* **87** 1352–1356.
Choi, D. W. (1988) *Neuron* **1** 623–634.
Christine, C. W. and Choi, D. W. (1990) *J. Neurosci.* **10** 108–116.
Cotman, C. W. and Iversen, L. L. (1987) *Trends Neurosci.* **10** 263–265.
Cowburn, R., Hardy, J., Roberts, P. and Briggs, R. (1988) *Brain Res.* **452** 403–407.
Cowburn, R. F., Hardy, J. A., Briggs, R. S. and Roberts, P. J. (1989) *J. Neurochem.* **52** 140–147.
Coyle, J. T. and Schwarcz, R. (1976) *Nature* **263** 244–246.
Davies, S. W. and Roberts, P. J. (1988) *Neuroscience* **26** 387–393.
DeBoni, U. and Crapper-McLachlan, D. R. (1985) *J. Neurol. Sci.* **68** 105–118.
Debonnel, G., Beauchesne, L. and DeMontigny, C. (1989) *Can. J. Physiol. Pharmacol.* **67** 29–33.
DiFiglia, M. (1990) *Trends Neurosci.* **13** 286–289.
Dingledine, R., Boland, L. M., Chamberlin, N. L., Kawasaki, K., Kleckner, N. W., Traynelis, S. F. and Verdoorn, T. A. (1988) *CRC Crit. Rev. Neurobiol.* **4** 1–96.
Dudek, S. M. and Bear, M. F. (1989) *Science* **246** 673–675.
Dudek, S. M., Bowen, W. D. and Bear, M. F. (1989) *Dev. Brain Res.* **47** 123–128.
Dure, L. S., Young, A. B. and Penney, J. B. (1991) *Ann. Neurol.* **30** 785–793.
Fonnum, F. (1984) *J. Neurochem.* **42** 1–11.
Foster, A. C. (1990) *Adv. Neurol.* **51** 97–102.
Garthwaite, G. and Garthwaite, J. (1989a) *Neurosci. Lett.* **97** 316–322.
Garthwaite, G. and Garthwaite, J. (1989b) *Neurosci. Lett.* **99** 113–118.
Geddes, J. W., Monaghan, D. T., Cotman, C. W., Lott, I. T., Kim, R. C. and Chang-Chui, H. (1985) *Science* **230** 1179–1181.
Geddes, J. W., Chang-Chui, H., Cooper, S. M., Lott, I. T. and Cotman, C. W. (1986) *Brain Res.* **399** 156–161.
Gill, R., Foster, A. C. and Woodruff, G. N. (1987) *J. Neurosci.* **7** 3343–3349.
Gill, R., Foster, A. C. and Woodruff, G. N. (1988) *Neuroscience* **25** 847–855.
Giordano, M., Ford, L. M., Brauckmann, J. L., Norman, A. B. and Sanberg, P. R. (1990) *Brain Res. Bull.* **24** 313–319.
Greenamyre, J. T. and Young, A. B. (1990) *Neurobiol. Aging* **10** 593–602.
Greenamyre, J. T., Young, A. B. and Penney, J. B. (1983) *Neurosci. Lett.* **37** 155–160.
Greenamyre, J. T., Young, A. B. and Penney, J. B. (1984) *J. Neurosci.* **4** 2133–2144.

Greenamyre, J. T., Olson, J. M. M., Penney, J. B. and Young, A. B. (1985a) *J. Pharmacol. Exp. Ther.* **233** 254–263.

Greenamyre, J. T., Penney, J. B., D'Amato, C. J. and Young, A. B. (1985b) *J. Neurochem.* **48** 543–551.

Greenamyre, J. T., Penney, J. B., Young, A. B., D'Amato, C. J., Hicks, S. P. and Shoulson, I. (1985c) *Science* **227** 1496–1499.

Greenamyre, J. T., Penney, J. B., Young, A. B., Hudson, C., Silverstein, F. S. and Johnston, M. V. (1987) *J. Neurosci.* **7** 1022–1030.

Gusella, J. F., Wexler, N. F., Conneally, P. M., Naylor, S. L., Anderson, M. A., Tanzi, R. E., Watkins, P. C., Ottina, K., Wallace, M. R., Sakaguchi, A. Y., Young, A. B., Shoulson, I., Bonilla, E. and Martin, J. B. (1983) *Nature* **306** 234–238.

Harding, A. E. (1984) *The Hereditary Ataxias and Related Disorders.* Churchill Livingstone, New York.

Haring, R., Kloog, Y., Kalir, A. and Sokolovsky, M. (1987) *Biochemistry* **26** 5854–5861.

Haring, R., Kloog, Y. and Sokolovsky, M. (1988). In: *Sigma and Phencyclidine-like Compounds as Molecular Probes in Biology.* NPP Books, Ann Arbor, MI, pp. 139–147.

Honey, C. R., Miljkovic, Z. and MacDonald, J. F. (1985) *Neurosci. Lett.* **61** 135–139.

Hugon, J. and Vallat, J. M. (1990) *Neurosci. Lett.* **119** 45–48.

Hyman, B. T., Van Hoesen, G. W., Damasio, A. R. and Barnes, C. L. (1984) *Science* **225** 1168–1170.

Jansen, K. L. R., Faull, R. L. M., Dragunow, M. and Synek, B. L. (1990) *Neuroscience* **39** 613–627.

Javitt, D. C. and Zukin, S. R. (1989a) *Mol. Pharmacol.* **35** 387–393.

Javitt, D. C. and Zukin, S. R. (1989b) *Proc. Nat. Acad. Sci. USA* **86** 740–744.

Johnson, J. W. and Ascher, P. (1987) *Nature* **325** 529–531.

Kano, M. and Kano, M. (1987) *Nature* **325** 276–279.

Keinanen, K., Widen, W., Sommer, B., Werner, P., Herb, A., Verdoorn, T. A., Sakmann, B. and Seeburg, P. H. (1990) *Science* **249** 556–560.

Khachaturian, Z. S. (1985) *Arch. Neurol.* **42** 1097–1105.

Kleckner, N. W. and Dingledine, R. (1988) *Science* **241** 835–837.

Kuhar, M. J. and Unnerstall, J. R. (1985) *Trends Neurosci.* **8** 49–53.

Lewis, D. A., Campbell, M. F., Terry, R. D. and Morrison, J. H. (1987) *J. Neurosci.* **7** 1799–1808.

Makowiec, R. L., Albin, R. L., Cha, J.-H. J., Young, A. B. and Gilman, S. (1990) *Brain Res.* **523** 309–312.

Makowiec, R. L., Cha, J.-H. J., Penney, J. B. and Young, A. B. (1991) *Neuroscience* (in press).

Mann, D. M. A., Yates, P. O. and Marcyniuk, B. (1986) *J. Neurol. Neurosurg. Psychiat.* **49** 310–312.

Maragos, W. F., Chu, D. C. M., Young, A. B., D'Amato, C. J. and Penney, J. B. (1987a) *Neurosci. Lett.* **74** 371–376.

Maragos, W. F., Greenamyre, J. T., Penney, J. B. and Young, A. B. (1987b) *Trends Neurosci.* **10** 65–68.

Maragos, W. F., Penney, J. B. and Young, A. B. (1988) *J. Neurosci.* **8** 493–501.

Martin, J. B. and Gusella, J. F. (1986) *N. Engl. J. Med.* **315** 1267–1276.

McDonald, J. W., Johnston, M. V. and Young, A. B. (1990a) *Exp. Neurol.* **110** 237–247.

McDonald, J. W., Penney, J. B., Johnston, M. V. and Young, A. B. (1990b) *Neuroscience* **35** 653–668.

McGeer, E. G. and McGeer, P. L. (1976) *Nature* **263** 517-519.

McGeer, P. L., McGeer, E, G., Scherer, U. and Singh, K. (1977) *Brain Res.* **128** 369–373.

Meldrum, G. and Garthwaite, J. (1990) *Trends Pharmacol. Sci.* **11** 379–387.

Monaghan, D. T. and Cotman, C. W. (1982) *Brain Res.* **252** 91–100.

Monaghan, D. T., Yao, D. and Cotman, C. W. (1984) *Brain Res.* **324** 160–164.

Monaghan, D. T., Nguyen, L. and Cotman, C. W. (1986) *Neurochem. Res.* **11** 1073–1082.

Monaghan, D. T., Olverman, H. J., Nguyen, L., Watkins, J. C. and Cotman, C. W. (1988) *Proc. Nat. Acad. Sci. USA* **85** 9836–9840.

Monaghan, D. T., Bridges, R. J. and Cotman, C. W. (1989) *Annu. Rev. Pharmacol. Toxicol.* **29** 305–402.

Morgan, D. T., Yao, D. and Cotman, C. W. (1984) *Brain Res.* **324** 160–164.

Morgan, I. G. (1987) *Neurosci. Lett.* **79** 267–271.

Mouradian, U. M., Contreras, P. C., Monaghan, J. B. and Chase, T. N. (1988) *Neurosci. Lett.* **93** 225–230.

Murphy, D. E., Schneider, J., Boehm, C., Lehmann, J. and Williams, M. (1987) *J. Pharmacol. Exp. Ther.* **240** 778–784.

Murphy, D. E., Hutchinson, A. J., Hurt, S. D., Williams, M. and Sills, M. A. (1988) *Br. J. Pharmacol.* **95** 932–938.

Nielsen, E. O., Cha, J. J., Honore, T., Penney, J. B. and Young, A. B. (1988) *Eur. J. Pharmacol.* **157** 197–203.

Novelli, A., Reilly, J. A., Lysko, P. G. and Henneberry, R. C. (1988) *Brain Res.* **451** 205–212.

Nowak, L., Bregestovski, P., Ascher, P., Herbert, A. and Prochiantz, A. (1984) *Nature* **307** 462–465.

Nunn, P. B., Seeling, M., Zagoren, J. C. and Spencer, P. S. (1987) *Brain Res.* **410** 375–379.

Olney, J. W., Misra, C. H. and Rhee, V. (1976) *Nature* **264** 659–661.

Pearson, R. C. A., Esiri, M. M., Hiorns, R. W., Wilcock, G. K. and Powell, T. P. S. (1985) *Proc. Nat. Acad. Sci. USA* **82** 4531–4534.

Penney, J. B., Maragos, W. F., Greenamyre, J. T., Debowey, D. L., Hollingsworth, Z. and Young, A. B. (1990) *J. Neurol. Neurosurg. Psychiat.* **53** 314–320.

Perl, T. M., Bedard, L., Kosatsky, T., Hockin, J. C., Todd, E. C. D. and Remis, R. S. (1990) *N. Engl. J. Med.* **322** 1775–1780.

Peters, S., Koh, J. and Choi, D. W. (1987) *Science* **236** 589–593.

Procter, A. W., Stirling, J. M., Stratman, G. C., Cross, A. J. and Bowen, D. M. (1989a) *Neurosci. Lett.* **101** 62–66.

Procter, A. W., Wong, E. H. F., Stratmann, G. C., Lowe, S. L. and Bowen, D. M. (1989b) *J. Neurochem.* **53** 698–704.

Ransom, R. W. and Stec, N. L. (1988) *J. Neurochem.* **51** 830–836.

Repressa, A., Duyckaerts, C., Tremblay, E., Hauw, J. J. and Ben-Ari, Y. (1988) *Brain Res.* **452** 403–407.

Reynolds, I. J. and Miller, R. J. (1988a) *Mol. Pharmacol.* **33** 581–584.

Reynolds, I. J. and Miller, R. J. (1988b) *Eur. J. Pharmacol.* **151** 103–112.

Reynolds, I. J. and Miller, R. J. (1989) *Mol. Pharmacol.* **36** 758–765.

Reynolds, I. J., Murphy, S. N. and Miller, R. J. (1987) *Proc. Nat. Acad. Sci. USA* **84** 7744–7748.

Ross, S. M., Seeling, M. and Spencer, P. S. (1987) *Brain Res.* **425** 120–127.

Ross, S. M., Roy, D. N. and Spencer, P. S. (1989) *J. Neurochem.* **53** 710–715.

Rothman, S. M. and Olney, J. W. (1987) *Trends Neurosci.* **10** 299–302.

Sakurai, S. Y., Cha, J.-H. J., Penney, J. B. and Young, A. B. (1991) *Neuroscience* **40** 533–543.

Sandoval, M. E. and Cotman, C. W. (1978) *Neuroscience* **3** 199–206.

Sheardown, M. J., Nielsen, E. O., Hansen, A. J., Jacobsen, P. and Honore, T. (1990) *Science* **247** 571–574.

Simpson, M. D. C., Royston, M. C., Deakin, J. F. W., Cross, A. J., Mann, D. M. A. and Slater, P. (1988) *Brain Res.* **462** 76–82.

Sladeczek, F., Recasens, M. and Bockaert, J. (1988) *Trends Neurosci.* **11** 545–549.

Sommer, B., Keinanen, K., Verdoorn, T. A., Wisden, W., Burnashev, N., Herb, A., Koehler, M., Takagi, T., Sakmann, B. and Seeburg, P. H. (1990) *Science* **249** 1580–1585.

Spencer, P. S., Ludolph, A., Dwived, M. P., Roy, D. N., Hugon, J. and Schaumburg, H. H. (1986) *Lancet* **ii** 1066–1067.

Stewart, G. R., Zorumski, C. F., Price, M. T. and Olney, J. W. (1990) *Exp. Neurol.* **110** 127–138.

Stone, T. W. (1979) *Br. J. Pharmacol.* **66** 291–296.

Sugiyama, H., Ito, I., Okada, D., Hirono, C., Ohmori, H., Shigemoto, T. and Furuya, S. (1988). In: Cavalheiro, E. A, Lehmann, F. and Turski, L. (eds) *Frontiers in Excitatory Amino Acid Research.* Liss, New York, pp. 21–28.

Sutherland, R. J., Hoesing, J. M. and Whishaw, I. Q. (1990) *Neurosci. Lett.* **120** 221–223.

Teitelbaum, J. S., Zatorre, R. J., Carpenter, S., Gendron, D., Evans, A. C., Gjedde, A. and Cashman, N. R. (1990) *N. Engl. J. Med.* **322** 1781–1787.

Toggenberger, G., Wiklund, L., Henke, H. and Cuenod, M. (1983) *J. Neurochem.* **41** 1606–1613.

Unnerstall, J. R. and Wamsley, J. K. (1983) *Eur. J. Pharmacol.* **86** 361–371.

Vignon, J., Privat, A., Chadieu, I., Thierry,, A., Kamenka, J. and Chicheportiche, R. (1986) *Brain Res.* **378** 133–141.

Vonsattel, J. P., Myers, R. H., Stevens, T. J., Ferrante, R. J., Bird, E. D. and Richardson, E. P. (1985) *J. Neuropathol. Exp. Neurol.* **44** 559–577.

Watkins, J. C., Krogsgaard-Larsen, P. and Honore, T. (1990) *Trends Pharmacol. Sci.* **11** 25–33.

Weiss, J. H., Koh, J.-Y. and Choi, D. W. (1989) *Brain Res.* **497** 64–71.

Westbrook, G. L. and Mayer, M. L. (1987) *Nature* **328** 640–643.

Wiklund, L., Toggenburger, G. and Cuenod, M. (1984) *Neuroscience* **13** 441–468.

Wong, E. H. F., Kemp, J. A., Priestly, T., Knight, A. R., Woodruff, G. N. and Iversen, L. L. (1986) *Proc. Nat. Acad. Sci. USA* **83** 7104–7108.

Wong, E. H. F., Knight, A. R. and Ransom, R. (1987) *Eur. J. Pharmacol.* **142** 487–488.

Young, A. B. and Fagg, G. E. (1990) *Trends Pharmacol. Sci.* **11** 126–132.

Young, A. B. and Penney, J. B. (1988). In: *Receptor Localization: Ligand Autoradiography, 13*. Liss, New York, pp. 177–189.

Young, A. B., Oster-Granite, M. L., Herndon, R. M. and Snyder, S. H. (1974) *Brain Res.* **73** 1–13.

Young, A. B., Bromberg, M. B. and Penney, J. B. (1981) *J. Neurosci.* **1** 241–249.

Young, A. B., Greenamyre, J. T., Hollingsworth, Z., Albin, R., D'Amato, C. J., Shoulson, I. and Penney, J. B. (1988) *Science* **241** 981–983.

6

NMDA receptors: heterogeneity and agonism

Donald J. Kyle, Raymond J. Patch, E. William Karbon and **John W. Ferkany**
Divisions of Computational Chemistry, Medicinal Chemistry and CNS Research,
Nova Pharmaceutical Corporation, Baltimore, MD, USA

ABSTRACT

The *N*-methyl-D-aspartate (NMDA) receptor was once thought to represent a simple ligand-gated ion channel. However, it is now apparent that at least a portion of NMDA recognition sites are present in the central nervous system as constituents of supramolecular complexes. Recent evidence suggests that in addition to the NMDA recognition site and ion channel, the complex also contains a co-agonist site activated by glycine, as well as modulatory sites for polyamines, Zn^{2+}, and possibly guanine nucleotides. Furthermore, it has been suggested that multiple agonist molecules may be required for complete channel activation, adding to the complexity of this receptor system. Whether heterogeneity exists among NMDA receptors is an important experimental and therapeutic issue which remains unresolved, although significant insight has been achieved. Results from autoradiographic, electrophysiological, biochemical, behavioural, functional and ontogenic studies have provided circumstantial evidence for multiple NMDA receptor populations.

In the present chapter, agonist and antagonist interactions within the NMDA receptor complex are discussed with particular emphasis on NMDA-sensitive [^{3}H]L-glutamate binding. Evidence is presented supporting the notion of multiple receptor populations or affinity states. In an initial attempt to define the structural requirements for agonist interactions at NMDA receptors, we introduce a simple and rapid computational program for modelling receptor topology and apply this to defining the molecular properties required for occupation of NMDA agonist recognition site(s).

INTRODUCTION

At the time of the initial classification (Watkins and Evans 1981) of excitatory amino acid (EAA) receptors into kainate (KAIN), quisqualate (now termed AMPA ((*R,S*)-2-amino-3-(3-hydroxy-5-methylisoxazol-4-yl)propanoic acid) and *N*-methyl-D-aspartate (NMDA) subtypes, it was generally accepted that transduction of EAA-mediated events occurred through a simple gating of ion channels. As evidenced by the present volume, the intervening decade has witnessed a rapid expansion in our knowledge of EAA neurotransmission and a concomitant appreciation for the complex role EAAs play in brain function. This intense scrutiny has been fuelled by a recognition of the ubiquitous nature of EAAs in brain, and an increased awareness of the physiological and pathophysiological effects of EAAs in the central nervous system.

Although the initial designation of the three ionotropic EAA receptors remains valid, it now appears that metabotropic recognition sites are present on neurons as well (see chapter 15). In contrast to the classical ionotropic subtypes, metabotropic recognition sites directly activate second messenger systems in target cells. The presence of a fifth type of EAA receptor preferentially recognized by the ω-phosphono amino acid, L(+)-2-amino-4-phosphonobutanoic acid (L-AP4) is also suspected (see chapter 14). The recent cloning and expression of seven distinct proteins, each of which can be functionally activated in isolation, or in various combinations, by KAIN and/or AMPA suggests that ionotropic receptors are likely to be heterogeneous at the molecular level (Hollmann *et al.* 1989, Bettler *et al.* 1990, Keinanen *et al.* 1990, Boulter *et al.* 1990).

NMDA RECEPTORS

Beyond these fundamental distinctions among EAA recognition sites, evidence supports the notion that at least a portion of NMDA receptors present in the brain are represented by complex molecular entities subject to multiple regulatory influences. The amino acid glycine (Gly), an inhibitory neurotransmitter in the spinal cord, functions through a strychnine-insensitive site in brain as a facilitatory or co-agonist substance in the gating of the NMDA receptor ion channel (Johnson and Ascher 1987, Snell *et al.* 1987, Foster and Kemp 1989). Both guanine nucleotides (Baron *et al.* 1989) and polyamines (see chapter 13), substances known to be present in high concentrations in brain, may play a regulatory role in NMDA receptor function. Recent evidence suggests that the divalent cation Zn^{2+} co-exists in neurons with L-glutamate (Glu) and is co-released with this EAA. Zinc, acting at a distinct site on the receptor, is thought to inhibit receptor activation (Peters *et al.* 1987, Weiss *et al.* 1989). Exogenous substances which have been suggested to interact with the NMDA molecular complex include some tricyclic antidepressants the psychotomimetic phencyclidine (PCP), and the popular cough-suppressant, dextromethorphan (Anis *et al.* 1983, Sills and Loo 1989, Ferkany *et al.* 1988).

In addition to its molecular complexity, some NMDA receptors may not function in the classical manner associated with pre- or postsynaptic information transduction. For example, investigations using the potent non-competitive antagonists (+)-5-methyl-10,11-dihydro-5*H*-dibenzo[*a*, *d*]cyclohepten-5,10-imine (MK-801) or *N*-[1-(2-thienyl)cyclohexyl]piperidine (TCP) have revealed that the action of these agents to block channel function is 'use-dependent', with antagonist effects increasing under depolarizing conditions (MacDonald and Nowak 1990, Fagg 1987). This has led to the suggestion that NMDA receptors are normally quiescent, with the ion channel blocked by the divalent cation Mg^{2+}. In normal function, synaptic excitation, such as that elicited by activation of KAIN or AMPA receptors, releases this Mg^{2+} block and allows NMDA receptor activation to induce inward neuronal currents. Conversely, by permitting receptor activation and channel opening, depolarizing stimuli provide the opportunity for non-competitive antagonists such as MK-801 to gain access to a binding site and to effect channel blockade. In a manner analogous to some other ion channels, NMDA receptors are 'voltage-dependent' and may be involved in a subtle modulation of chemical neurotransmission (Honey *et al.* 1985). This positive feedback and recruitment loop has been suggested to underlie physiological events as diverse as spreading field potentials, associated with some types of epileptogenesis, to forming the basis for learning and memory (Schwarcz and Ben-Ari 1986, Morris *et al.* 1986, Jones *et al.* 1989). A substantial amount of pre-clinical information indicates that EAA receptors in general, and NMDA receptors in particular, play a role in the aetiology of anxiety, psychosis, chronic neurodegenerative disorders and neuronal injury which occurs subsequent to hypoxic, ischaemic or traumatic insult to the brain (Cavalheiro *et al.* 1988).

NMDA RECEPTOR HETEROGENEITY

Empirically, it would seem that NMDA receptors should be heterogeneous since diversity within neurotransmitter receptor systems is more a rule than an exception. Surprisingly, however, direct demonstration of differences among NMDA recognition sites has proven difficult and evidence for receptor subtypes remains largely circumstantial.

The existence of regional distinctions in the central nervous system (CNS) in neuronal responsiveness to NMDA and quinolinic acid (QUIN), an endogeneous NMDA agonist, has been known for some time (Perkins and Stone 1983). Since the excitatory effects of both substances are antagonized by selective NMDA antagonists, it has been assumed that both compounds function through activation of an 'NMDA' receptor but that these receptors must be capable of recognizing different agonist configurations. Among NMDA agonists, there is a wide spectrum of excitatory and neurotoxic responses in brain. For instance, QUIN, which has a very low affinity for NMDA receptors *in vitro*, is equipotent with NMDA in producing axon-sparing lesions upon intracranial injection. Since NMDA has roughly a 100-fold greater affinity for NMDA receptors, it is difficult to understand why the compounds are similar as neurotoxic agents. This is particularly true since QUIN is present in brain, whereas NMDA is not, and it would seem reasonable to conclude that metabolic systems to inactivate QUIN but not NMDA would also be available in the CNS. Recently it has been found that the conformationally restricted NMDA analogue 2-amino-2-(3-hydroxy-5-methylisoxazol-4-yl)acetic acid (AMAA), a compound equipotent with NMDA *in vitro*, is an exceptionally potent neurotoxin with activity similar to that of KAIN (Madsen *et al.* 1990, Krogsgaard-Larsen *et al.* 1991). Thus, it is not immediately apparent that agonist potency at NMDA recognition sites is sufficient to explain the neurotoxic response to receptor activation. The existence of NMDA agonists which lack neurotoxic properties has been known for some time (Lehmann *et al.* 1985) and it remains possible that observed responses occur through different receptor populations.

More direct evidence for receptor heterogeneity has come from autoradiographic studies evaluating NMDA, TCP and AMPA receptors in cerebral cortex and putamen of individuals afflicted with Huntington's disease (Young *et al.* 1988). Whereas NMDA-sensitive [^{3}H]Glu binding was reduced by 93% in putamen, [^{3}H]TCP binding decreased only 67% in this same brain region. These data are inconsistent with the hypothesis that all TCP receptors represent an NMDA-sensitive ion channel since, if this were the case, both receptor populations would be expected to decline in unison during the course of the disease.

In a more extensive comparative evaluation of [^{3}H]TCP, [^{3}H]Gly and [^{3}H]Glu binding to rat brain a reasonable correlation of receptor densities and stoichiometry was found in most regions of the CNS (McDonald *et al.* 1990). However, in certain brain regions the stoichiometry of receptors was distinct and, for example, in stratum radiatum of the hippocampus the ratio of TCP/Gly/Glu receptors was found to be 1:3:4, respectively, suggesting a binding site density of NMDA > Gly > TCP. In contrast, in other brain regions, including the olfactory structures, basal ganglia, basal forebrain, thalamus and brain stem, the density of Gly receptors was consistently greater than Glu recognition sites and both agonists bound at much higher levels than TCP. The greatest disparity was observed in the cerebellar granule layer where Gly and Glu recognition sites were present in 60- and 50-fold excess, respectively, relative to those for TCP. Based upon these observations, it was argued that NMDA receptors were heterogeneous not only among brain regions but also within specific CNS substructures.

Using quantitative autoradiographic measurements, Monaghan *et al.* (1988) examined the regional distribution of agonist ([^{3}H]Glu) and antagonist ([^{3}H]3-(2-carboxypiperazin-4-yl)propanoic acid ([^{3}H]CPP)) binding to rat brain in the presence and absence of Gly. Not only were the regional distributions of agonist and antagonist binding distinct, but Gly was found to stimulate agonist binding while inhibiting the binding of antagonist ligand(s). Additionally, there were substantial regional differences in the efficacy of Gly to stimulate Glu binding, as well as differences in the potencies of NMDA agonists or antagonists to inhibit the binding of [^{3}H]Glu. As an example, binding in the striatum and cerebellar cortex was significantly more sensitive to inhibition by agonists (NMDA, QUIN, L-aspartic acid (Asp)), whereas binding to the thalamus and cortex was more sensitive to antagonist blockade. On the basis of these findings, Monaghan and colleagues proposed the existence of two distinct NMDA type receptors and/or two interconverting forms of a single receptor population.

Using target size analysis, Honoré *et al.* (1989) have provided molecular evidence for distinction between agonist and antagonist binding domains on NMDA receptors. High-energy irradiation of the receptor complex suggested that NMDA, TCP and Gly recognition sites resided on a similar protein of 115–121 kDa, whereas the binding site associated with CPP had an estimated mass of 209 kDa. Since a number of reports (Murphy *et al.* 1987, Lehmann *et al.* 1987) have documented that CPP functions as a competitive antagonist of NMDA, it was proposed that binding of some antagonists requires the presence of a second protein which could be tightly coupled to the NMDA/TCP/Gly receptor component. Such a hypothesis is attractive as it could provide an explanation for agonist- and antagonist-preferring subsets of NMDA receptors.

Contrasting with molecular target size data, however, are findings from ontogenic studies of NMDA, TCP and Gly recognition sites. Postnatal development of NMDA-sensitive [^{3}H]Glu binding occurs more rapidly and stabilizes at a higher density than recognition sites for either Gly or TCP, although sites for the latter ligands develop in parallel, suggesting possible co-localization to a single molecule (McDonald *et al.* 1990). These different ontogenic profiles and receptor densities are consistent with the notion that multiple forms of expression exist for NMDA receptors (McDonald and Johnston 1990) and argue against target size analysis, suggesting that a common molecular species contains binding domains for NMDA, TCP and Gly. Alternatively, however, the greater apparent density of NMDA receptors at adulthood, as well as the earlier appearance of NMDA binding sites, is consistent with recent observations (Javitt *et al.* 1990) that multiple agonist molecules are required for NMDA receptor activation. If this is the case, this might be reflected as an earlier appearance and greater final density of agonist recognition sites vis-à-vis TCP and Gly receptors.

In vivo evidence for NMDA receptor heterogeneity is more circumscribed, in part because of complications arising from differing patterns of biodistribution and pharmacokinetics of available compounds, as well as the inherent differences in test subjects and protocols employed by different laboratories. This notwithstanding, there is compelling evidence suggesting that significant departures in pharmacological and biological responses exist between competitive and non-competitive NMDA antagonists (Willetts *et al.* 1990). Whether this is due to heterogeneity among NMDA receptors or to multiple sites of action of non-competitive antagonists remains a focus of investigation.

Using drug discrimination methods (Willetts and Balster 1988, Koek *et al.* 1988), it has been convincingly demonstrated that MK-801 is more potent than PCP in substituting for PCP in animals trained to recognize the latter substance. Indeed, the initial discovery of a linkage between PCP and NMDA recognition sites almost derailed the drive to identify and develop competitive NMDA receptor blockers for therapeutic use since it was feared these compounds would be psychotomimetic in man, a notion supported by results from early studies (Koek *et al.* 1986, 1987). More extensive testing of this hypothesis using mammalian species has lessened this concern and it is now apparent that competitive and non-competitive antagonists elicit differential, although somewhat overlapping, discriminative cues in test subjects. For instance, competitive antagonists including CPP and 2-amino-4,5-(1,2-cyclohexyl)-7-phosphonoheptanoic acid (NPC 12626) only partially substitute for PCP in animals trained to recognize the latter substance (Ferkany *et al.* 1989). Conversely, PCP does not replace NPC 12626 in animals trained to recognize the competitive antagonist although recognition is complete when NPC 12626 is replaced by CPP (Willetts *et al.* 1989). Neither PCP nor MK-801 attenuate NMDA-mediated discriminative cues although these are potently antagonized by both NPC 12626 and CPP (Willetts and Balster 1989a). Unlike PCP, NPC 12626 and CPP effectively replace the discriminative cue elicited by the barbiturate pentobarbitone (Willetts and Balster 1989b). Work in our own laboratory has shown a reliable 'anxiolytic' response to NPC 12626 in rodent models whereas little or no effect was observed for MK-801 (Conti *et al.* 1990; however, see Leeson *et al.* 1990).

Other differences among competitive and non-competitive NMDA antagonists are also known. While systemic administration of both NPC 12626 (at doses which completely attenuate QUIN-induced destruction of the corpus striatum) and PCP produce hyperlocomotion in rodents, only the effects of the latter can be reversed by 6-hydroxydopamine lesion of the substantia nigra (French *et al.* 1991). Massive doses of NPC 12626 have no effect on dopaminergic neuronal firing in ventral tegmental area A_{10}, a brain region exquisitely sensitive to increased activity following administration of PCP (French *et al.* 1991). Similarly, while both (±)-*cis*-4-phosphonomethyl-2-piperidinecarboxylic acid (CGS 19755) and PCP are potent antagonists of harmaline-induced increases in cerebellar cGMP level, only PCP and MK-801 alter dopamine metabolism and release in the ventral tegmental area (Wood and Rao 1989). At pharmacologically effective doses, intravenous administration of MK-801 results in a profound hypotensive response in anaesthetized animals and a hypertensive response in conscious animals, whereas little effect on cardiovascular responses is seen with competitive antagonists (CPP, CGS 19755, NPC 12626) (I. A. Martin, personal communication). Whether these behavioural, electrophysiological and metabolic studies reflect the presence of multiple NMDA receptor subtypes (PCP and non-PCP linked) or the converse (PCP receptors not linked to NMDA receptors), or some other action of non-competitive antagonists such as blockade of potassium channels or cholinergic receptors (Albuquerque *et al.* 1980, Aronstam *et al.* 1980, ffrench-Mullen and Rogawski 1989, Galligan and North 1990, Wood and Rao 1989), remains to be determined. Nevertheless, the results clearly suggest that differential responses to competitive and non-competitive NMDA antagonists will be apparent at the level of the intact organism.

Differences in the excitatory and neurotoxic responses to agonists have been described above. Among competitive antagonists, subtle differences can also be found although these are again confounded by interpretive concerns. For example, the recently introduced compound (±)-*cis*-2-amino-4-methyl-5-phosphono-3-pentenoic acid (CGP 37849) and its ethyl ester homologue CGP 39551 have been reported to be potent, systemically active, long-lasting anticonvulsants in the maximal electroshock test (MES) procedure, but have little activity in attenuating pentylenetetrazole (PTZ)-induced convulsions (Schmutz *et al.* 1990). In contrast, CPP and CGS 19755 are essentially equipotent in blocking NMDA-, PTZ- and MES-induced convulsions, whereas NPC 12626 is more potent in inhibiting PTZ- and NMDA-induced activity than in blocking MES convulsions (J. W. Ferkany *et al.* unpublished). These differences are unlikely to be a consequence of biodistribution, given they are apparent following intracranial delivery of the compounds. Whether these are related to the differences in the receptor interactions described below remains to be determined.

RECEPTOR BINDING STUDIES

Early attempts to study NMDA receptors *in vitro* using [^{3}H]NMDA met with considerable frustration as the low affinity of this ligand made it unsuitable for standard protocols (Fagg and Baud 1985). Although 2-amino-5-phosphonopentanoic acid (AP5) was employed with some success (Olverman *et al.* 1984, Olverman *et al.* 1988a, 1988b, 1988c) specific binding was marginal for routine use. Likewise, whereas 2-amino-7-phosphonoheptanoic acid (AP7) was found to bind to membrane homogenates with a good signal, the pharmacological profile for binding was not consistent with the identification of the NMDA receptor subtype (Ferkany and Coyle 1983). Since the latter assays were performed in chloride-containing media, it seems likely that the Cl^--dependent Glu transport site or the quisqualate-sensitive transporter was labelled. Glu itself proved useful for identifying NMDA receptors using autoradiographic approaches (Monaghan *et al.* 1984, Greenamyre *et al.* 1984, 1985, Monaghan and Cotman 1985) or when assays were performed using purified synaptic plasma membranes (Fagg and Matus 1984), but these procedures were unsuitable for routine use or proved tedious for the detailed characterization of the receptor complex.

With the introduction of high-affinity antagonist ligands and the development of methods for labelling the NMDA receptor using [^{3}H]Glu in crudely purified membrane preparations, it became possible to study NMDA receptors *in vitro* using standard radioligand binding techniques. Indeed, when used in conjunction with [^{3}H]MK-801 and [^{3}H]Gly, examination of the individual recognition sites as well as the interactions which occur between Gly and NMDA receptors and the associated ion channel is feasible. Using these methods, differences are apparent in antagonist profiles as well as between the binding of agonist and antagonist ligands.

In our own laboratory we have devoted considerable effort to defining the characteristics of [^{3}H]CPP, [^{3}H]CGS 19755, [^{3}H]Glu and [^{3}H]Gly binding in the presence and absence of various agents. As discussed below, the use of each of these ligands in concert is crucial since each reveals subtly different characteristics of the NMDA receptor(s).

By virtue of its priority of introduction the interaction of CPP with NMDA recognition sites was examined initially. In accord with earlier studies (Murphy *et al.* 1987, Olverman *et al.* 1988c) it has been our experience that the pharmacological profile of the site labelled by CPP is wholly consistent with the identification of an NMDA recognition site. Curiously, both ourselves and others (Murphy *et al.* 1987) have consistently recorded shallow displacement curves for many inhibitors and routinely find pseudo Hill coefficients substantially less than unity. Attempts to discriminate two populations of CPP binding sites have been difficult since the affinity of the ligand (100–200 nM) does not allow for classical saturation analysis, and isotope dilution protocols must be used. It has been our experience upon repeated analysis that discrimination of multiple binding components is difficult using [^{3}H]CPP.

In practical terms, CGS 19755 has provided a somewhat more robust assay compared to CPP and is more amenable to routine use. As seen with CPP, however, inhibition curves using either agonist or antagonist displacers are frequently shallow, suggesting complex interactions. The moderate affinity of CGS 19755 for the NMDA receptor (100–200 nM) is associated with the same difficulties found with CPP.

The binding characteristics of CPP and CGS 19755, while similar with regard to inhibition by NMDA recognition site-directed agents, can be shown to be distinct under appropriate conditions, suggesting the compounds do not interact in an identical manner with the NMDA receptor complex. As reported previously, Gly, at concentrations identical to the K_D for the strychnine-insensitive Gly recognition site, can partially inhibit the binding of CGS 19755 (Fig. 1; Kaplita and Ferkany 1990). This inhibitory effect is reversed by 7-chlorokynurenic acid (7-Cl-KYN) and is mimicked by the Gly agonist D-serine. In contrast, in concentrations up to 1 mM neither Gly nor D-serine inhibit the binding of [^{3}H]CPP (Kaplita and Ferkany 1990).

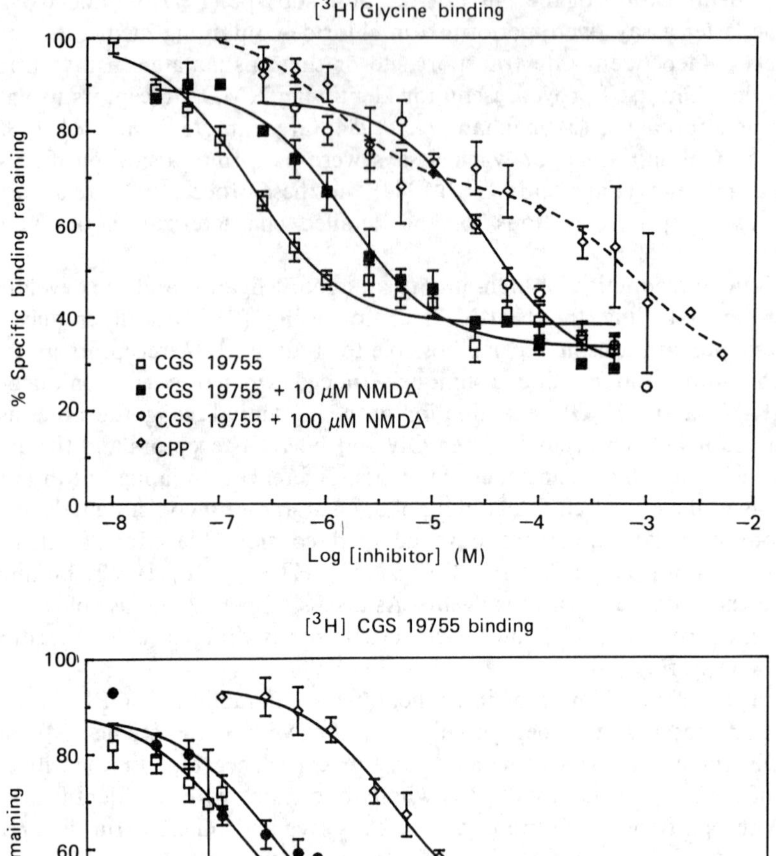

[³H] CGS 19755 binding

% Specific binding remaining

Glycine

D-Serine

Glycine + 10 μM 7-ClKYN

Log [inhibitor] (M)

Fig. 1 — Effect of CGS 19755 on specific [^{3}H]Gly binding, and of Gly on specific [^{3}H]CGS 19755 binding. The top panel shows the inhibition of [^{3}H]Gly binding to rat forebrain membranes by CGS 19755 in the absence (□) of NMDA or in the presence of 10 (■) or 100 μM (○) NMDA. Also shown is the effect of CPP (◇) (dashed line) on specific [^{3}H]Gly binding. The data demonstrate the partial inhibition of binding caused by 5-carbon atom-containing phosphono amino acids and the relative lack of activity of 7-carbon-containing homologues. The bottom panel demonstrates the inhibitory action of Gly (□) and the Gly agonist D-serine (●) on specific [^{3}H]CGS binding to NMDA recognition sites. Also shown is the inhibitory effect of Gly in the presence of 10 μM of the Gly antagonist 7-Cl-KYN (◇). In parallel experiments, Gly at concentrations up to 100 μM had no effect on the specific binding of [^{3}H]CPP. (Kaplita and Ferkany 1990.)

Several laboratories (Kaplita and Ferkany 1990, Hood *et al.* 1990, Monahan *et al.* 1990) have demonstrated that 5-carbon chain ω-phosphono carboxylic amino acids (AP5, CGS 19755, CGP 37849) partially inhibit the binding of [^{3}H]Gly to strychnine-insensitive sites with potencies identical to K_Is for the NMDA receptor and that this effect can be reversed by NMDA agonists (Monahan *et al.* 1990, Kaplita and Ferkany 1990, Hood *et al.* 1990, Fig. 1). Surprisingly, the higher homologues of these compounds (AP7, CPP, NPC 12626) have no effect on [^{3}H]Gly binding until micromolar concentrations are reached and, in fact, have been reported to reverse the inhibitory actions of 5-carbon atom-containing NMDA antagonists (Monahan *et al.* 1990a, Kaplita and Ferkany 1990). In one study, evidence has been provided suggesting that 5-carbon atom-containing phosphono amino acids alter the affinity of Gly binding; CGS 19755 has also been reported to increase the potency of the Gly antagonists HA-966 and 1-aminocyclobutane-1-carboxylate (ACBC) to inhibit [^{3}H]Gly binding (Hood *et al.* 1990). Interestingly, NMDA and Glu either have no effect on the binding of [^{3}H]Gly (Kaplita and Ferkany 1990) or only marginally stimulate binding to the receptor (Kessler *et al.* 1989).

The effects of Gly antagonists on [^{3}H]CPP binding have been reported, with variable results. Danysz *et al.* (1989) and Compton *et al.* (1990) demonstrated that HA-966 and ACBC stimulated the binding of [^{3}H]CPP, an action now known to reside in the *R*-enantiomer of the HA-966 (Pullan *et al.* 1990). The effect of HA-966 can be antagonized by 7-Cl-KYN and by Gly (Danysz *et al.* 1989). At high concentrations, both 6-cyano-7-nitroquinoxaline-2,3-dione (CNQX) a potent Gly antagonist (>10 μM) and 7-Cl-KYN ($>50\,\mu$M) inhibited the binding of both [^{3}H]CPP and [^{3}H]CGS 19755 (E. W. Karbon *et al.* unpublished). Both HA-966 and ACBC have been shown to inhibit the binding of [^{3}H]L-Glu (Compton *et al.* 1990).

Results from our own studies have confirmed the stimulatory effects of HA-966 and ACBC on [^{3}H]CPP binding (Fig. 2). Additionally, we have found that ACBC and HA-966 at concentrations $\leqslant 100$ μM have no effect on the binding of [^{3}H]CGS 19755, whereas the newly reported Gly antagonist indole-2-carboxylic acid (Huettner 1989) has no effect on the binding of either [^{3}H]CPP or [^{3}H]CGS 19755 (E. W. Karbon *et al.* unpublished). Thus, beyond differential influences of NMDA antagonists on the [^{3}H]Gly recognition site, it would appear that Gly agonists and antagonists differentially regulate the recognition site(s) for competitive NMDA antagonists.

In another approach examining interactions at the NMDA receptor complex, we have employed [^{3}H]Glu (10 nM) as a potent agonist label of the receptor. Assays were performed on freeze-fractured, extensively washed and, in some instances, detergent-treated (Triton-X-100) rat forebrain preparations. NMDA (10^{-3} M) was used to define non-specific binding and assays were performed in (0.05 M) Tris-acetate buffer to exclude the possibility of interference from Cl^--dependent transport sites (elemental analysis of lyophilized 20-ml buffer samples failed to detect Cl^- at a level of sensitivity of 0.05% w/w).

Unlike antagonist ligands, [^{3}H]Glu binds to an apparently homogeneous population of receptors (Fig. 3) and in the absence of any additions or Triton treatment the pseudo Hill coefficient (0.95 ± 0.02,; $n = 10$) was not different from unity. Addition of Gly (1 μM) had no effect on the affinity of [^{3}H]Glu ($K_D = 32.2 \pm 3.5$ nM, control; 35 ± 6.9 nM, Gly) but significantly increased the apparent receptor density (B_{max} =

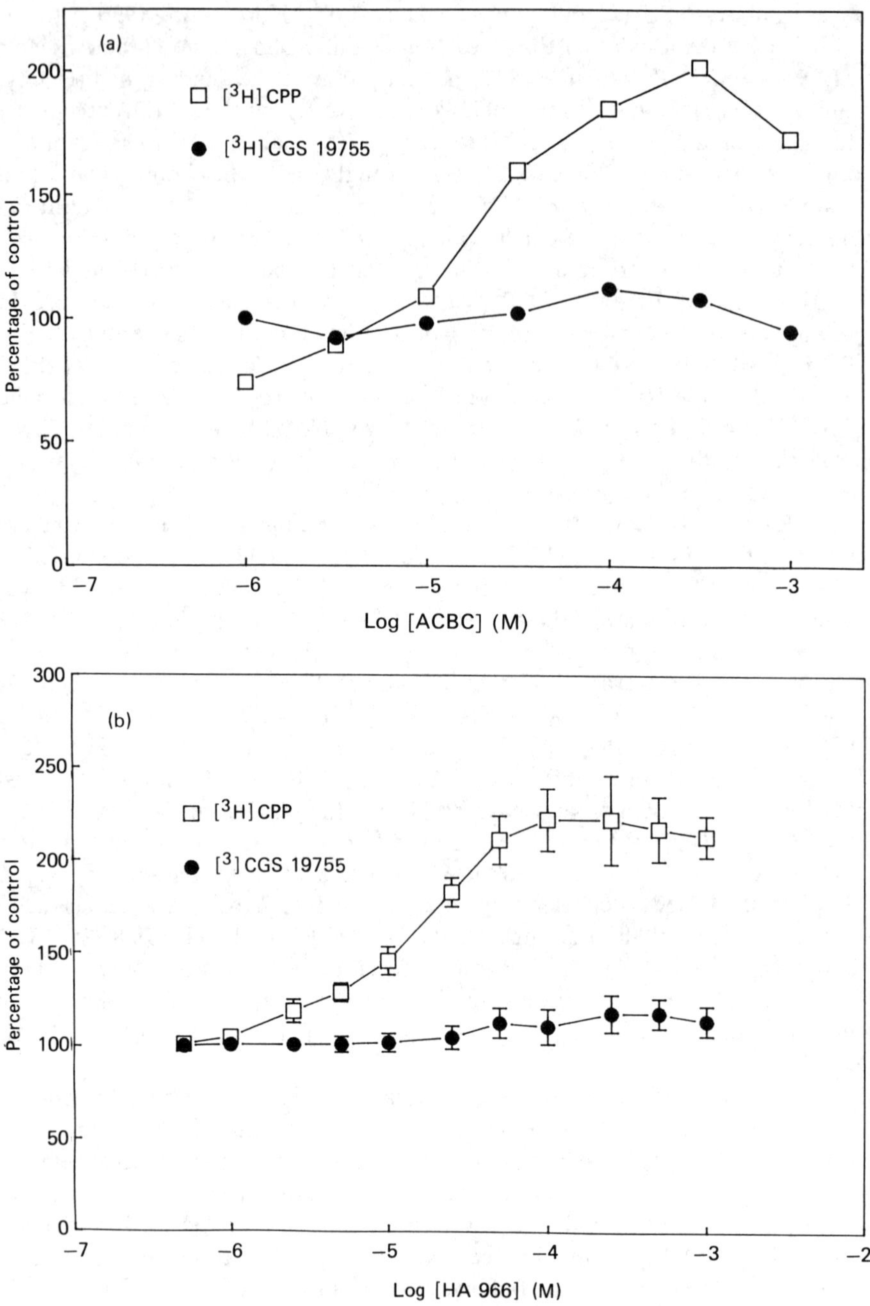

Fig. 2 — Effect of Gly antagonists on [^{3}H]CGS 19755 and [^{3}H]CPP binding to rat forebrain membranes. The data show the differential interactions of HA-966 (top panel) and ACBC (bottom panel) on the binding of 5-carbon- and 7-carbon-containing phosphono amino acid NMDA antagonists. Parallel studies have shown that indole-2-carboxylic acid has no effect on the binding of either ligand at concentrations up to 500 μM, whereas CNQX inhibits the binding of each with an IC_{50} of 50–100 μM. Likewise, at high concentrations (> 100 μM), 7-Cl-KYN is a weak inhibitor of the binding of competitive antagonists to the NMDA recognition site.

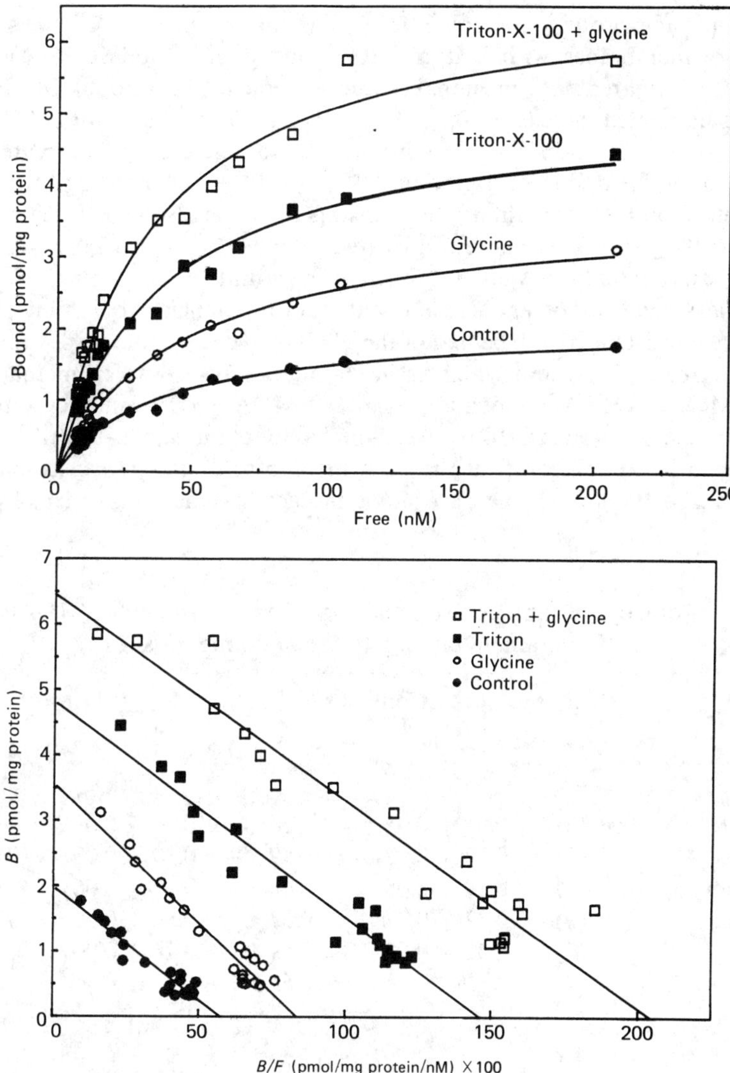

Fig. 3 — Effect of various conditions on NMDA-sensitive [^{3}H]Glu binding to rat forebrain membranes. Methods are described below; results are from a representative experiment. In contrast to the binding of competitive NMDA antagonists, [^{3}H]Glu binds to an apparent single population of receptors. When assays are performed in the presence of Gly (1 μM), the apparent B_{max} for [^{3}H]Glu increases with no change in the apparent K_d. This effect of Gly is observed in control and detergent-treated tissues.

Methods. Briefly, rat forebrain membranes were prepared according to the procedure described by Ferkany *et al.* (1989). On the day of assay, tissue was thawed to room temperature and resuspended in 20 vol. (w/v) of assay buffer (Tris acetate; 0.05 M; pH 7.1; 2°C) and incubated for 30 minutes (37°C) in the presence or absence of 0.05% Triton-X-100. Following the pre-incubation, the tissue was washed four times by centrifugation and resuspension in assay buffer. The final pellet was resuspended in buffer to a tissue concentration of 0.1 mg/ml and 1-ml portions were added to glass culture tubes containing 1 ml of buffer, [^{3}H]Glu (10 nM final) and various concentrations of unlabelled Glu (1–1000 nM). Non-specific binding was defined by 10^{-3} M NMDA; all incubations were performed in triplicate. The reaction was allowed to proceed for 30 minutes in an ice-water bath and was terminated by vacuum filtration using GF/C glass-fibre filters. Filters were washed with 2 × 3 ml of ice-cold buffer and processed for determination of radioactivity using standard methods. Typically under these conditions an 8 : 1 ratio of total to non-specific binding was obtained with total binding (control condition) equal to 10–12 000 dpm.

2.21 ± 0.31 pm/mg protein, control; 3.022 ± 0.59 pm/mg protein, Gly; $p \leqslant 0$ 0.05). Treatment of membranes with Triton-X-100 resulted in a modest non-significant increase in the apparent K_D, presumably due to removal of residual Glu from the assay, and a significant increase in B_{max} (3.74 ± 0.42 pm/mg protein; $p \leqslant 0.05$) relative to both basal and Gly-stimulated conditions. As was the case for untreated membranes, Gly stimulated [^{3}H]Glu binding ($B_{max} = 4.88 \pm 0.63$ pm/mg protein) while having no effect on receptor affinity. In all instances, pseudo Hill coefficients for Gly-stimulated (0.98 ± 0.10; $n = 6$), Triton-treated (0.95 ± 0.01; $n = 10$) or Triton plus Gly (0.92 ± 0.04) conditions were not different from unity.

Preliminary analysis of agonist and antagonist potencies to inhibit [^{3}H]Glu binding were consistent with labelling of the NMDA receptor complex. Rank orders for both control and Triton-treated tissue among agonists was *cis*-methanoglutamate (CMG) > AMAA > ibotenic acid ⩾ NMDA = *cis*-1-aminocyclopentane-1,3-dicarboxylate (*cis*-ACPD) > *trans*-methanoglutamate > 4-HPCA > QUIN; *trans*-ACPD, a putative agonist at the EAA metabotropic receptor, was inactive at concentrations ⩽100 μM (Table 1). Among antagonists, the rank order of potency

Table 1 — Potency of antagonists and agonists to inhibit NMDA-sensitive [^{3}H]L-glutamate binding to rat brain membranes

Compound	*n*	Control		Triton	
		IC_{50}	n_H	IC_{50}	n_H
Antagonists					
CGP 37849	(3)	0.05 ± 0.02	0.54	0.04 ± 0.01	0.56
CGS 19755	(7)	0.17 ± 0.02	0.59	0.19 ± 0.03	0.56
D(−)-CPP	(4)	0.19 ± 0.05	0.48	0.19 ± 0.06	0.51
L(+)-CPP	(3)	2.37 ± 0.77	0.57	0.86 ± 0.90	0.56
D(−)-AP7	(4)	2.52 ± 0.50	0.59	2.76 ± 0.90	0.56
L(+)-AP7	(2)	64.5	0.67	70.5	0.56
Agonists					
CMG	(4)	0.11 ± 0.02	0.61	0.10 ± 0.02	0.64
AMAA	(3)	0.79 ± 0.10	0.64	1.82 ± 0.70	0.77
Ibotenic acid	(3)	1.34 ± 0.19	0.92	1.02 ± 0.21	0.90
NMDA	(6)	2.15 ± 0.70	0.75	1.27 ± 0.17	0.70
cis-ACPD	(4)	3.94 ± 1.1	0.77	2.27 ± 0.60	0.71
trans-Methanoglutamate	(2)	8.8		6.9	0.70
4-HPCA[a]	(4)	32	—	—	—
QUIN[a]	(4)	> 100	—	> 100	—
trans-ACPD	(2)	⩾ 100	—	⩾ 100	—

Shown are the IC_{50}s (± SEM) and pseudo Hill coefficients for various compounds to inhibit NMDA (10^{-3} M)-sensitive [^{3}H]Glu (10 nM) binding to freeze-fractured preparations of rat forebrain. Details of the assay are given in the legend to Fig. 3. Values are expressed as micromolar concentration.
[a]From Madsen *et al.* (1990).

in either condition was CGP 37849 > CGS 19755 = D(−)-CPP > L(+)-CPP = D(−)-AP7 ≫ L(+)-AP7 (Table 1). With the exception of CMG and AMAA, pseudo Hill coefficients for agonist inhibition ranged from 0.70 to 0.92. In contrast, antagonist inhibition of binding consistently yielded pseudo Hill coefficients of 0.5–0.6 (Table 1). Biphasic inhibition of [^{3}H]Glu binding by antagonists was clearly apparent, while this was less evident for agonist inhibitors (Fig. 4).

More detailed iterative curve-fitting analysis indicated that CGP 37849 displaced [^{3}H]Glu with high affinity from a receptor population representing approximately 80% of the available sites (Table 2). In contrast, the distribution of sites displaced by D(−)-CPP was more evenly distributed, with 40% represented by a high-affinity component. Gly increased the percentage of sites recognized by D(−)-CPP as 'high afffinity', an effect that was mimicked by detergent treatment (Table 2). Using membranes previously exposed to Triton and in the presence of Gly, D(−)-CPP apparently inhibited binding to a single population of high-affinity receptors.

Results from these investigations demonstrate complex interactions between Gly, and agonist and antagonist interactions at the NMDA receptor complex. Importantly, use of potent antagonists such as CGP 37849 to displace agonist labelling of the NMDA recognition site provides evidence of two receptor populations or affinity states. We are currently extending this type of interactive analysis to agonist inhibition of NMDA-sensitive [^{3}H]Glu binding.

RECEPTOR MODELS

As reviewed above, any unitary model of the NMDA receptor complex must account for the diversity of findings summarized in Table 3.

At the macro level these would include differences in ontogenic profiles, recognition site densities and stoichiometries of Gly, TCP/MK-801 and NMDA receptors. From a pharmacological perspective consideration must be given to the differential excitatory and excitotoxic effects of NMDA agonists as well as regional distinctions in the efficacy of agonists to elicit depolarizing responses; growing evidence suggests this will most likely need to be extended to include competitive NMDA antagonists. Certainly, any model must account for the similarities, and profound pharmacological, physiological and behavioural differences, between competitive and non-competitive receptor antagonists.

At the level of the receptor, consideration must be applied to emerging evidence that two agonist molecules, in addition to Gly, are required for receptor activation. Coupled to results from target size analysis it would appear that multiple agonist/antagonist binding domains must be envisioned. In some manner it will be necessary for the topology of these to be similar since competitive antagonists appear to fully cross-react with other [^{3}H]-labelled antagonists or agonists in receptor binding studies. The model must also account for the differential modulation by C_5 and C_7 ω-phosphono amino acids of agonist and antagonist binding to the Gly recognition site. Conversely, the differential actions of Gly agonists and antagonists on agonist binding and/or C_5 and C_7 antagonist interactions with the NMDA recognition site(s) need consideration. Beyond these matters, a successful picture of NMDA receptor(s) must adequately explain the complex interactions seen in the presence of Zn^{2+},

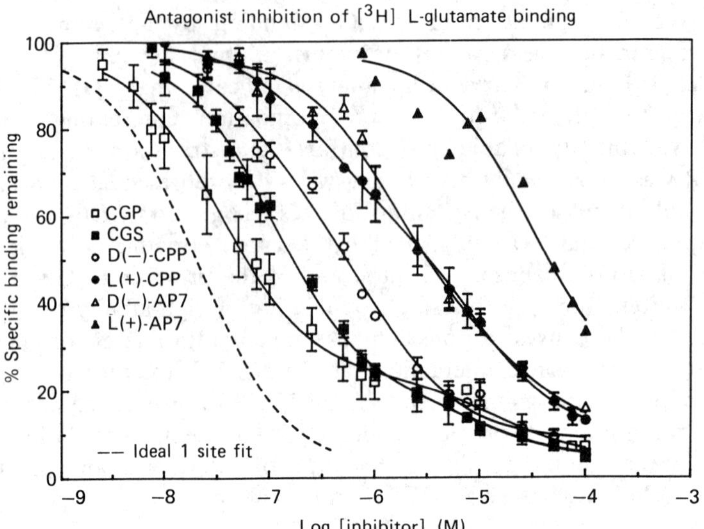

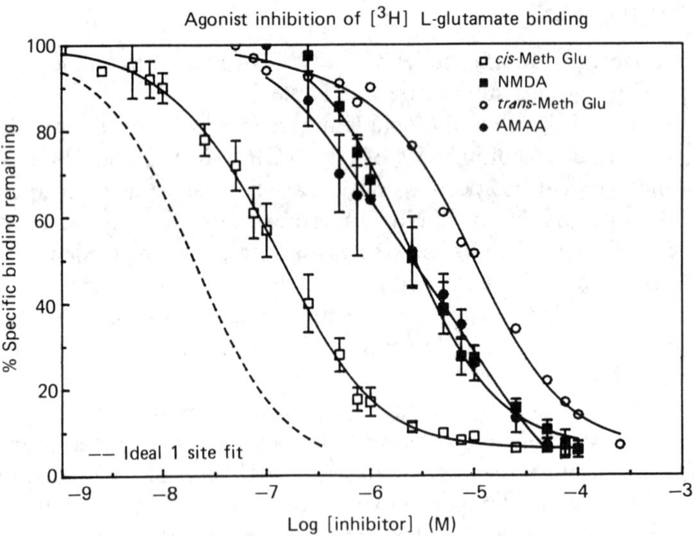

Fig. 4 — Inhibition of NMDA-sensitive [^{3}H]Glu (10 nM) binding by NMDA antagonists and agonists. Methods were identical to those described in the legend of Fig. 3; data are taken from control tissue with assays performed in the absence of Gly. The top panel shows the pattern of inhibition of binding seen in the presence of antagonists, whereas the bottom panel reflects inhibition by agonist ligands. In both panels, the dashed line represents a computer-generated binding curve for an ideal one-site receptor model. Most clearly seen in the top panel is the deviation from a monophasic pattern of inhibition for antagonist compounds. Where error bars are shown, data are the means ± SEM for 3–7 separate experimental determinations. Values for L(+)-AP7 and *trans*-methanoglutamate are from two experiments.

Table 2 — Two-site analysis of antagonist inhibition of NMDA-sensitive [^{3}H]L-glutamate binding to rat brain membranes

Compound	N	n_H	K_H (μM)	K_L (μM)	% K_H
CGP 37849					
Control	3	0.54	0.025	10.27	80
Triton	3	0.56	0.025	24.7	85
D(–)-CPP					
Control	4	0.57	0.067	0.82	40
Triton	4	0.58	0.114	2.80	77
Glycine	4	0.60	0.076	2.04	80
Triton + Gly	4	0.70	0.139	18.8	88

Values shown are the K_is (μM) for CGP 37849 and D-(–)-CPP to inhibit NMDA-sensitive [^{3}H]Glu binding to rat forebrain membranes in the presence and absence of Gly, or in tissue that had been pre-incubated with Triton-X-100 as described in the legend in Fig. 3.

Table 3 — Aspects of NMDA receptor function

Compound/ test	Ligand/receptor			
	Gly	Glu	CPP	CGS 19755
CGP 37849	Partial INH	Biphasic INH	INH	INH
CGS 19755	Partial INH	Biphasic INH	INH	INH
CPP	NE	Biphasic INH	INH	INH
Gly	INH	Stimulates	NE	NE/INH
7-Cl-KYN	INH	INH	NE/INH	Partial INH
ACBC	INH	INH	Stimulates	NE
HA-966	INH	Partial INH	Stimulates	NR
Glu	Stim/NE	INH	INH	INH
NMDA	Stim/NE	INH	INH	INH
Target size	115–130 kDa	115–130 kDa	208 kDa	?
CGS 19755	Causes Gly agonists to become less potent at Gly receptor			
CGS 19755	Causes Gly antagonists to become more potent at Gly receptor			
HA-966	Causes CPP to become more potent at NMDA receptor			

Shown are some of the known interactions which occur among compounds acting at the NMDA receptor complex. NE, no effect; Stim, stimulates; INH, inhibits.

guanine nucleotides and polyamines. It is likely that more regulatory factors will emerge with continued study. How these various factors affect channel modulation must also be considered.

The ultimate resolution of the topology of the NMDA receptor complex and the structural requirements of agonist and antagonist compounds will require the cloning and expression of the various components of the receptor. Nevertheless, using a combination of pharmacological, chemical and computational approaches, it is possible to elucidate key requirements for receptor recognition. In the second section of this chapter we introduce a simple and rapid method of computational analysis useful for this purpose. Because ligand/receptor interactions may be simpler for agonists, we have applied initial effort toward defining structural requirements for NMDA recognition site-directed agonist compounds.

MOLECULAR MODELLING

It is noteworthy that many of the compounds reported to be agonists of the NMDA-type receptor contain three chemically similar groups. Specifically these include (1) the carboxylic acid, (2) an amine nitrogen of an α-amino acid-like moiety, and (3) a second acidic group, most typically carboxylate. This functional group homology suggests that this receptor might have particular acceptor points for each of these three moieties. A more detailed hypothesis dictates that each of these acceptor points are spatially oriented in a well-defined manner with regions of steric tolerance or intolerance flanking each. In the absence of X-ray crystallographic data corresponding to the receptor, one approach toward understanding this receptor's topology lies in a ligand-based investigation, in this case using a combination of molecular dynamics and energy minimization.

Of several NMDA receptor agonists that have been reported over the past thirty years, CMG is among the most potent both in terms of receptor affinity and excitatory effects (Lanthorn *et al.* 1990). Furthermore, this molecule is conformationally rigid, with the exception of the C^{α}–C^{β} bond. This inherent rigidity unambiguously orients the three key functional groups into particular quadrants of Cartesian coordinate space with little fluctuation.

Given the potency of CMG, we made the assumptions that (1) the compound's functional group orientation likely represented an optimal conformation with which Glu binds to this EAA receptor subtype, and (2) the 'scaffolding' (in this case a cyclobutyl ring) holding these groups in place was sterically accommodated by the receptor. Using this molecule as a template, a series of structurally and pharmacologically diverse NMDA receptor agonists (Fig. 5) were matched, functional group to functional group, in a two-step procedure. This was accomplished by an initial application of harmonic constraints between the key atoms of each while calculating a brief 3-picosecond (ps) molecular dynamics trajectory. During the calculation the scale factor on the constraints was incrementally increased (as discussed later) to fit the molecules together approximately. This was followed by a more tightly constrained energy minimization to overlay the atoms more precisely in the three functional groups while optimizing the overall geometric features of the model. Subsequently the constraints were removed and the energy minimization continued to determine the actual stability of the 'forced conformation'. By analogy, one could consider building a pair of Dreiding models: one for the template, the other for the test compound. The harmonic constraints, which serve as 'rubber bands' that can be switched on and off, were 'wrapped about' the functional group atoms in a pairwise fashion between the molecules and switched off. Next the 'rubber bands' were switched on during an energy minimization procedure which forced the intermolecu-

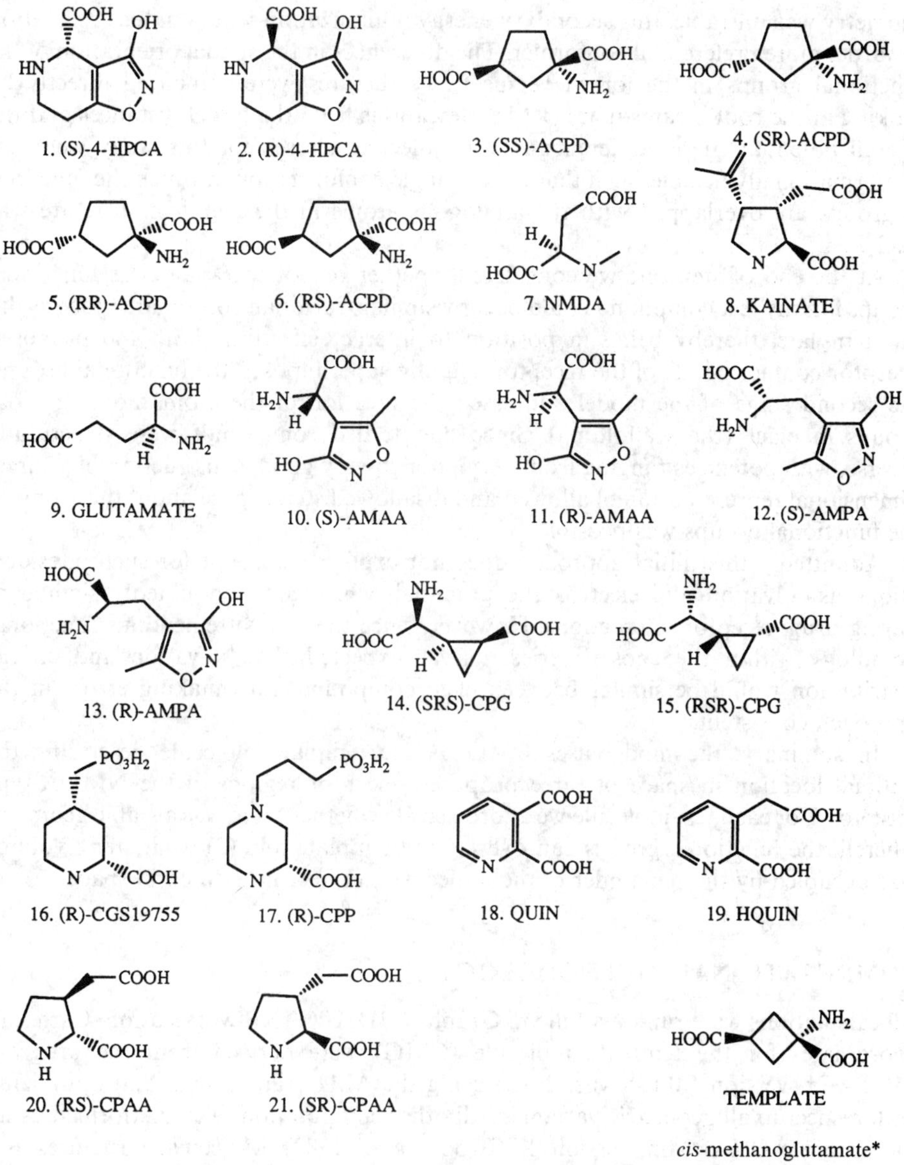

1. (S)-4-HPCA
2. (R)-4-HPCA
3. (SS)-ACPD
4. (SR)-ACPD
5. (RR)-ACPD
6. (RS)-ACPD
7. NMDA
8. KAINATE
9. GLUTAMATE
10. (S)-AMAA
11. (R)-AMAA
12. (S)-AMPA
13. (R)-AMPA
14. (SRS)-CPG
15. (RSR)-CPG
16. (R)-CGS19755
17. (R)-CPP
18. QUIN
19. HQUIN
20. (RS)-CPAA
21. (SR)-CPAA

TEMPLATE

cis-methanoglutamate*

Fig. 5 — Chemical structures of the molecules chosen for this modelling. The template molecule is CMG. *According to IUPAC nomenclature rules *cis*-methanoglutamate should be termed as *trans*-methanoglutamate. Similarly, *cis*-ACPD should be *trans*-ACPD.

lar atom pairs to overlay as the 'rubber bands' contracted to their equilibrium state. Simultaneously, the overall test molecule was being optimized with respect to energy.

At the end of this step the observed structure may not correspond to a reasonable conformational state since the 'rubber bands' might be distorting bond lengths, bond angles and so on. To determine the 'goodness' of the forced structure the 'rubber bands' were switched off and the energy minimization continued. If the forced geometry was unstable, this secondary energy minimization would induce relaxation toward a more preferential conformer. The differences in the spatial orientation of the functional groups in the forced versus free structures were compared directly by looking at the root mean square (RMS) deviations from the target distances and the overall potential energy differences. Test molecules chosen for this study were not conformationally flexible, such that only a single conformation in which the functional groups are overlapped with the analogous atoms in the template structure was likely.

At the end of step one we considered whether or not the analogous functional groups for any test compound could occupy similar Cartesian coordinate space as did our template, thereby being in position to interact effectively with the proposed acceptor contact points of the receptor. Having superimposed the functional groups, the second stage of the model addressed the location of the atoms holding those groups in place (the scaffolding). Since the studied compounds were structurally diverse, with potencies ranging from very poor to very good, construction of a three-dimensional representation of allowed and disallowed steric space about the atoms of the functional groups was possible.

Admittedly this initial approach does not explicitly account for such considerations as solvation and electrostatic potential, which are undoubtedly significant during drug–receptor interactions. However, since there was tremendous functional homology within the chosen series, it was expected that solvation and charge distribution would be similar between each compound, thus making errors in the approach consistent.

In summary, the model uses CMG as the template molecule describing the optimal location in space of three proposed acceptor regions of the NMDA-type receptor. For each test molecule we addressed (1) whether there was a stable geometry wherein the functional groups can overlay the template and (2) what steric volume was occupied by the remainder of the molecule and how it oriented in space.

COMPUTATIONAL METHODOLOGY

All calculations were run on a Silicon Graphics 4D120GTXB workstation. Cartesian coordinates for the template molecule (CMG) were derived from the program MOPAC†, version 5.0 (Stewart 1983) using the AM1 Hamiltonian and optimizing with respect to all geometric variables. All other computations were performed using the program CHARMm, version 21 (Brooks *et al.* 1983). Molecular structures and van der Waals (VDW) surfaces were viewed on the workstation with the program QUANTA, version 3.0 (Polygen Corporation 1990).

Like other programs which apply the molecular mechanics approach, CHARMm builds a non-bonded list using pairwise additivity of atoms. For this procedure to be succcessful, it was required first to eliminate the atom pairs in the non-bonded list between the template and test molecule such that there would not be any contribution to the overall potential energy caused by intermolecular forces. This was accomplished by translating the test molecule away from the template by a distance greater

† This program is available from the Quantum Chemistry Program Exchange, Indiana State University.

than the non-bond cut-off value. Since the non-bond cut-off was set to be 8.0 Å, the test molecule was translated 20.0 Å along the x-axis. While separated the non-bonded list was generated, then the test molecule was translated − 20.0 Å along the x-axis to its initial location in space. Subsequently, during evaluation of the overall energy expression, the non-bonded atoms list was not updated. A final aspect of this phase of the approach involved fixing the atoms of the pre-optimized template molecule in space. This ensured that only the geometry of the test molecule would change during energy calculations, not the geometry of the template. Likewise, all of the internal coordinate energies associated with the template were removed from consideration. Using this approach, the two molecules were overlaid without each 'feeling' the presence of the other energetically.

Since, at first, the template and test molecules were likely to be dissimilar (i.e. the test molecule may require bond rotations and overall translation to resemble the spatial disposition of the template more closely) a constrained molecular dynamics heating step was run. Template matching was acccomplished by imposing harmonic constraints between like atoms of the template and test molecule functional groups. The specific atoms used as template guides are illustrated in Fig. 6. During the dynamics step, the target distance was 0.1 Å for each atom, with an upper bound of 0.4 Å and a lower bound of 0.1 Å. The contribution of these internuclear distances was incorporated into the overall potential energy expression in the form of an additional potential energy term given by:

$$E_{\mathrm{NOE}} = \begin{matrix} K_1(R - R_o)^2, \text{ for } R < R_o \\ K_2(R - R_o)^2, \text{ for R} > \mathrm{R}_o \end{matrix}$$

where R and R_o are calculated and target internuclear distances, respectively, and K_1 and K_2 are force constants given by:

$$K_1 = \frac{S\left(k_B \frac{T}{2}\right)}{(\Delta R_{\mathrm{LOW}})^2}; \quad K_2 = \frac{S\left(k_B \frac{T}{2}\right)}{(\Delta R_{\mathrm{HIGH}})^2}$$

where k_B is the Boltzmann constant, T is the temperature and S is a scale factor. ΔR_{LOW} and ΔR_{HIGH} are deviations above and below the target distance R_o, respectively. During the 3-ps heating step of the molecular dynamics, the temperature was raised

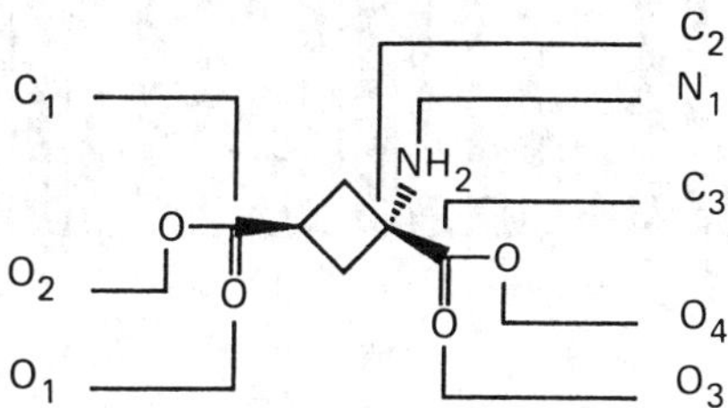

Fig. 6 — Definition of the atoms used as spatial template markers in CMG. This number scheme applies also to Table 4.

from 0 K to 300 K in steps of 20 K every 0.2 ps. Since the target distances were poorly satisfied in the starting structures, E_{NOE} was applied gradually by increasing the scale factor in a non-linear fashion such that it was 0.0 after 0.2 ps, 0.1 after 0.1 ps, and 1.0 after the full 3.0 ps. By slowly raising the scale factor on the constraints as the target distances became better satisfied, the test molecule was gently forced onto the template without geometric distortions. This is in contrast to our initial attempts wherein the constraints were imposed directly with a scale factor of 1.0 during energy minimization. In the latter cases test molecules were abruptly forced toward the template, often causing undesirable distortions.

After the molecular dynamics had 'eased' the test molecule onto the template, the target distances were tightened such that both upper and lower bounds were 0.1 Å, and 500 iterations of conjugate gradients energy minimization were performed. The RMS deviations for this forced structure from the target atoms in the template were calculated and plotted. Finally, the constraints were cleared and the energy minimization continued for another 500 iterations. Again, when complete, the deviations from the target atoms were tabulated. The overall RMS fluctuations for forced and free test molecules are graphically depicted in Fig. 7. For further comparison, the calculated overall potential energies for both the forced and free structures are likewise shown in Fig. 8. Atom-by-atom fluctuations from target distances were tabulated for each test molecule, forced and free, in Table 4. For each free test molecule, a mesh contour plot representing the VDW surface was computed. Fig. 9 illustrates the two perspectives of the template used throughout the remainder of this chapter in figures depicting these surfaces.

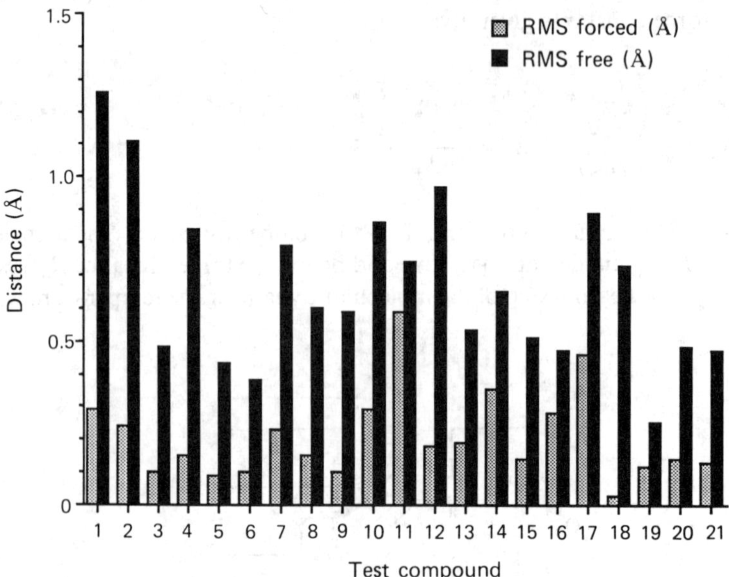

Fig. 7 — Summary of the RMS deviations from the template atoms for each of the test compounds. Grey corresponds to the structure with constraints applied and black corresponds to the unconstrained structure. Units are in Å.

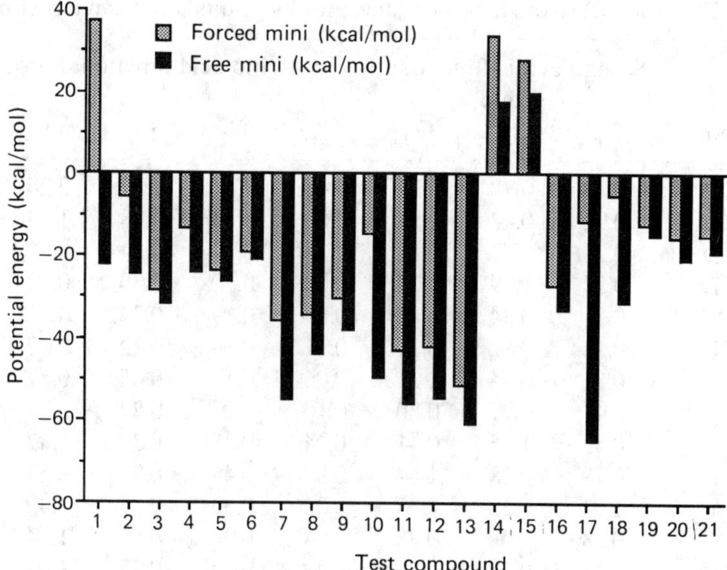

Fig. 8 — Summary of the overall potential energy from each of the test compounds. Grey corresponds to the structure with constraints applied and black corresponds to the unconstrained structure. Units are kcal/mol.

RESULTS AND DISCUSSION

The structures of Glu and NMDA superimposed in a 'best-fit' orientation with *cis*-methanoglutamate are depicted in Fig. 10. While it is evident that Glu can well adopt the template conformation, it is clear that NMDA is unable to attain a tight three-point fit. Stated quantitatively, though net RMS differences for Glu and NMDA are similar (Fig. 7), individual atom deviations relative to corresponding template atoms (specifically C_1 and O_2) are somewhat greater for NMDA (Table 4). In addition, the potential energy difference between free and constrained conformations of NMDA ($\Delta E = 19.31$ kcal/mol) is also greater than that of Glu ($\Delta E = 7.84$ kcal/mol).

QUIN (Fig. 11), which represents conformationally constrained NMDA, favours this three-point template fit to an even lesser extent (ΔRMS $= 0.70$ Å, $\Delta E =$ 26.26 kcal/mol). However, as can be seen in this figure, much closer proximity is attained by the rigid glutamate analogue homoquinolinic acid (HQUIN) (ΔRMS $=$ 0.13 Å) and with little expenditure of potential energy ($\Delta E = 2.42$ kcal/mol).

The effects of incorporating the pyridine nucleus into the backbone of NMDA and Glu can be seen by comparing the VDW surfaces shown in Fig. 11a,b with Fig. 10a,b. Most pronounced is the expansion of surface area in the upper regions of QUIN and HQUIN relative to NMDA and Glu. Given the decreased receptor affinities of HQUIN and QUIN relative to Glu and NMDA, respectively (Patneau and Mayer 1990), it might be concluded that lipophilic bulk in this region is not sterically tolerated by the NMDA agonist binding site. The combination of this disfavoured bulk and poor three-point fit would account for the observed rank order of NMDA receptor affinity: QUIN $<$ HQUIN $\leqslant$ NMDA $<$ Glu.

Table 4 — Deviation (Å) from target distances for individual atoms in test compounds

	Second acidic functional group			Amino acid functional groups				
Test compound	C_1[a]	O_1[b]	O_2	N_1	C_2	C_3	O_3	O_4
(*S*)-4-HPCA	0.73	0.88	2.27	0.48	0.57	0.82	2.24	0.27
(*R*)-4-HPCA	0.92	0.42	1.81	0.89	0.13	0.78	1.04	1.80
(*SS*)-ACPD	0.23	0.51	0.52	0.15	0.17	0.24	0.82	0.72
(*SR*)-ACPD	0.24	1.39	1.22	0.36	0.10	0.39	0.92	1.00
(*RR*)-ACPD	0.21	0.34	0.43	0.13	0.19	0.34	0.72	0.68
(*RS*)-ACPD	0.18	0.82	0.26	0.11	0.16	0.22	0.32	0.40
NMDA	0.81	0.85	1.21	0.06	0.5	0.65	0.82	0.91
KAIN	0.41	0.97	0.80	0.10	0.27	0.27	0.99	0.07
Glu	0.42	1.26	0.41	0.14	0.30	0.30	0.87	0.41
(*S*)-AMAA	0.78	0.28	1.54	0.85	0.48	0.33	0.67	1.12
(*R*)-AMAA	1.04	1.22	0.97	0.32	0.32	0.34	0.49	0.48
(*S*)-AMPA	0.38	1.08	0.88	0.94	0.09	0.87	1.14	1.56
(*R*)-AMPA	0.28	0.53	0.58	0.42	0.24	0.27	1.03	0.44
(*S,R,S*)-CPG	0.37	0.95	0.76	0.49	1.1	0.28	0.54	0.03
(*R,S,R*)-CPG	0.15	0.98	0.47	0.47	0.18	0.26	0.21	0.72
(*R*)-CPP	1.55	0.45	0.60	0.81	0.45	0.70	0.77	0.81
(*R*)-CGS 19755	0.90	0.52	0.35	0.27	0.41	0.19	0.39	0.34
QUIN	0.72	1.21	0.47	0.02	0.28	0.53	1.04	0.79
HQUIN	0.32	0.40	0.10	0.24	0.26	0.02	0.17	0.26
(*RS*)-CPAA	0.14	0.86	0.52	0.14	0.32	0.21	0.79	0.23
(*SR*)-CPAA	0.46	0.72	0.61	0.01	0.17	0.33	0.72	0.18

Deviation of the individually constrained atoms in the test compounds from the target atoms in the template molecule.
[a]CG 19755 and CPP have a phosphorus atom in the analogous position to the carbon.
[b]The non-amino acid-like carbonyl oxygen of 4-HPCA, AMAA, and AMPA is replaced by a nitrogen atom.

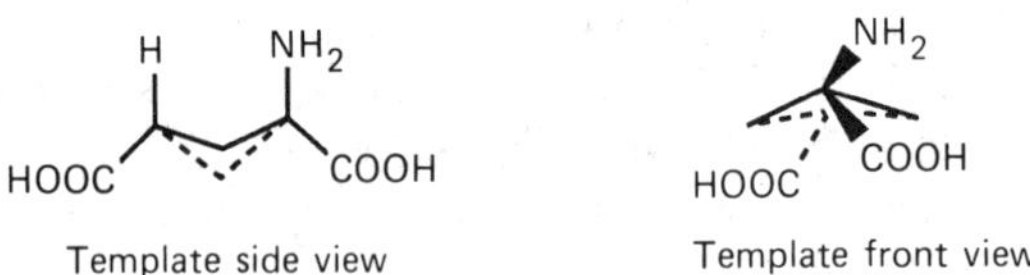

Fig. 9 — Two perspective views of the template used in all subsequent VDW surface representations.

This same region of steric bulk, which apparently has detrimental effects on NMDA agonist site affinity, may in fact be responsible for imparting neurotoxic properties to QUIN-related structures. Thus, neurotoxic activity might be due to the additive effects of good three-point functional group fit and steric occupation of this extended spatial volume. This region may be partially defined by the six-membered rings of QUIN and HQUIN as well as by the *N*-methyl group of NMDA. HQUIN's improved three-point functional group fit with respect to QUIN is also consistent with its observed higher potency in terms of eliciting excitatory effects (Stone 1984) and presumably also neurotoxicity.

Phthalic acid (PA), which is structurally related to QUIN with the exception that it contains a benzene rather than a pyridine nucleus, has also been shown to be excitotoxic, with potency essentially equal to QUIN (Lehmann *et al.* 1985). Somewhat paradoxically, homophthalic acid (HPA), which on the basis of the aforementioned arguments would be expected to be more potent then PA in this regard, was found to be much less excitatory than PA (Stone 1984), and may presumably be much less excitotoxic as well.

Two orientations of binding are available to HPA which allow for carboxylate group overlap with the template and benzene ring interaction with the 'excitotoxic domain'. In the first ('*syn*' orientation), the aromatic and homoaromatic carboxylates of HPA align with the α- and ω-acids of the template, respectively. In the second ('*anti*' orientation) the aromatic and homoaromatic carboxylates of HPA align in an opposite fashion with the ω- and α-acids of the template, respectively. In this orientation the benzenoid ring is translated away from the proposed 'excitotoxic pocket' (data not shown). Without the nitrogen functionality to direct HPA toward the binding site, both orientations are of equal probability, though only the '*syn*' mode of attachment would be expected to give rise to excitotoxicity. Because of its inherent symmetry, hypothetical '*syn*' and '*anti*' orientations of PA are not distinguishable. Both orientations would provide the same VDW surface and template fit as QUIN. Hence, it is possible to rationalize the relative excitatory effects of HPA and PA, as well as equipotency of PA and QUIN as neurotoxins.

Use of this model is not meant to imply that quinolinate-related neurotoxins exert such effects by interacting with the NMDA agonist site in a manner similar to CMG. Nonetheless, the model may prove useful in predicting the neurotoxic potential of related structures. Precise spatial boundaries of the 'excitotoxic pocket' have not yet been defined, though it is apparent that this region is subject to steric tolerances. Acridinic acid (not shown), whose VDW surface area extends beyond this region, is known to antagonize the excitatory effects of NMDA in rat cortex and may well antagonize NMDA's excitotoxic effects as well, though this remains to be established (Curry *et al.* 1986).

The eight stereoisomers comprising the series of 3,4-cyclopropylglutamate (CPG) analogues have recently been synthesized (Pellicciari *et al.* 1990, Monahan *et al.* 1990a). Of these, two isomers, namely (2*R*,3*S*,4*R*)-CPG and (2*S*,3*R*,4*S*)-CPG, exhibited high affinities for the NMDA receptor, being essentially equipotent with CMG. Given that the NMDA receptor and most neurotransmitter receptors in general exhibit a great degree of stereoselective preference for their ligands, it was somewhat surprising to identify an isomeric pair whose individual enantiomers bind to similar extents with such high affinity. Thus, we were particularly interested in examining these isomers in the present model.

With respect to functional group overlap, both (2*R*,3*S*,4*R*)-CPG and (2*S*,3*R*,4*S*)-CPG are well accommodated by the template (ΔRMS = 0.37 Å and 0.30 Å, respectively) (Fig. 12c). This is reflected in the potential energy terms, wherein the former isomer seems slightly better able to adopt the geometry dictated by the template (ΔE = 7.89 and 15.96 kcal/mol, respectively). Though the angles of approach to these

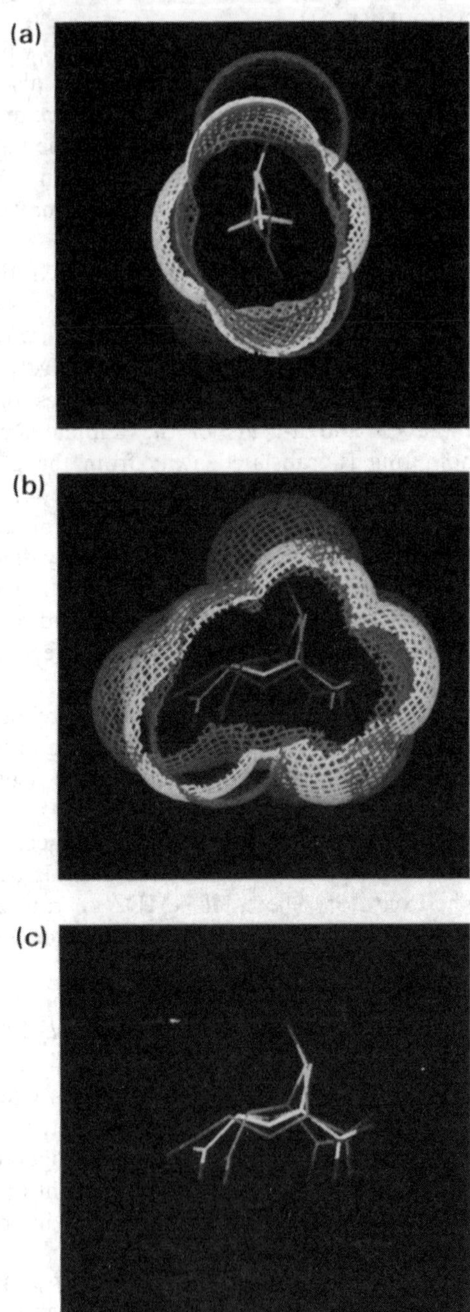

Fig. 10 — (**See colour section**.) CMG (white), Glu (blue), NMDA (pink). (a) Front view, VDW surface and Dreiding representation; (b) side view, VDW surface and Dreiding representation; (c) side view, Dreiding representation.

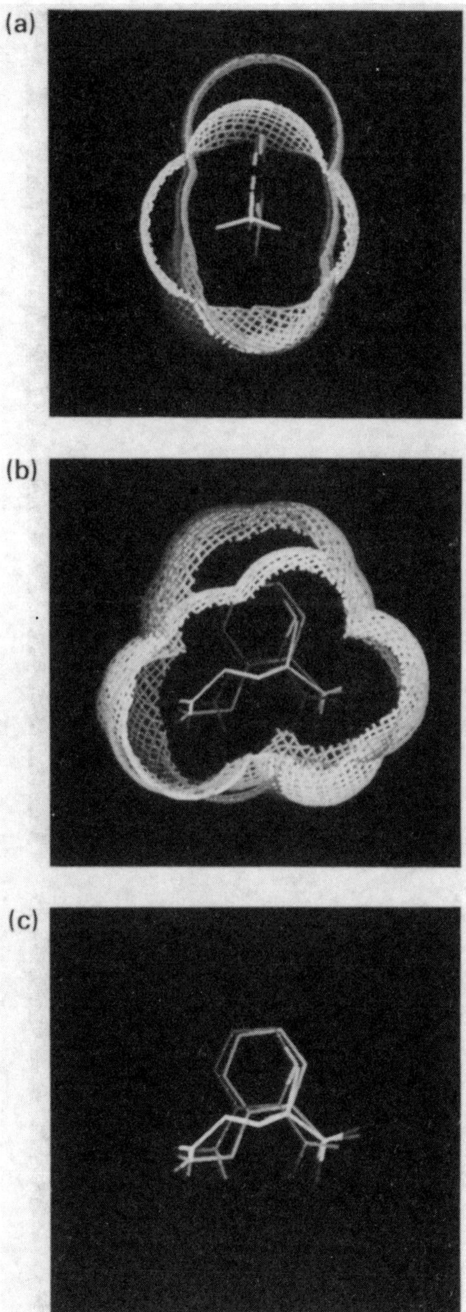

Fig. 11 — (**See colour section**.) CMG (white), HQUIN (green), QUIN (orange). (a) Front view, VDW surface and Dreiding representation; (b) side view, VDW surface and Dreiding representation; (c) side view, Dreiding representation.

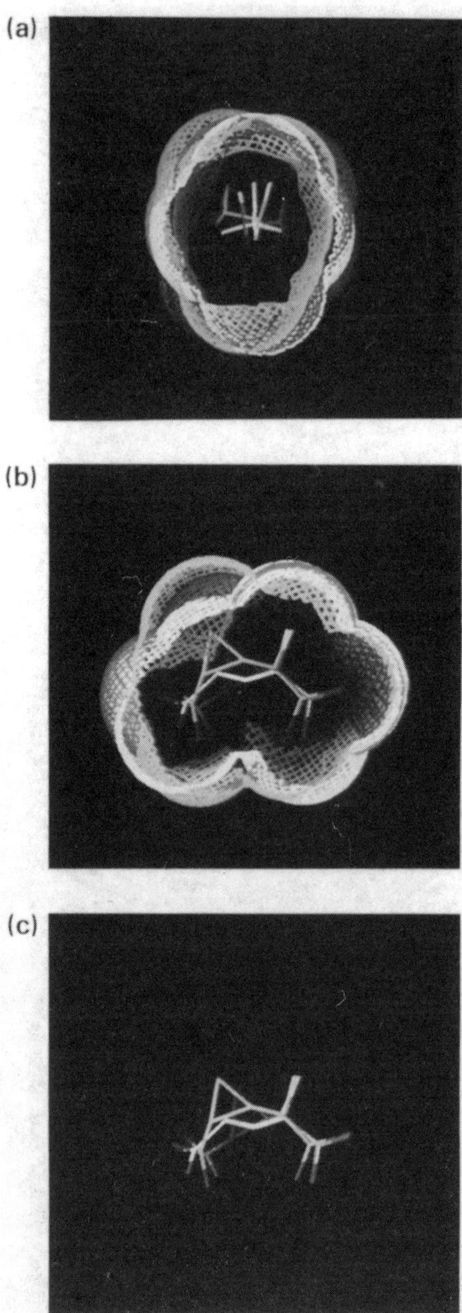

Fig. 12 — (**See colour section.**) CMG (white), (*R*,*S*,*R*)-CPG (green), (*S*,*R*,*S*)-CPG (orange). (a) Front view, VDW surface and Dreiding representation; (b) side view, VDW surface and Dreiding representation; (c) side view, Dreiding representation.

three points are inverted relative to each other, the frameworks connecting these functionalities lie within the spatial confines defined by the cyclobutane ring of CMG. Furthermore, both isomers' VDW surfaces show an excellent correlation with that of CMG (Fig. 12a,b). The methylene bridge of each cyclopropyl ring imparts only a slight extension of the surfaces relative to CMG. Such increases in steric bulk in these specific regions are apparently small enough so as not to provide deleterious interactions with the receptor. Thus, in light of these results it is not surprising that both enantiomers should exhibit similar high affinities for the NMDA agonist site.

The 3-hydroxyisoxazole AMAA, initially reported as a weak but selective NMDA agonist on spinal neurons (Krogsgaard-Larsen *et al.* 1980), has recently been shown to be a slightly more potent agonist than NMDA in receptor binding and electrophysiological studies using rat cortical slice preparations (Madsen *et al.* 1990). Additionally, in these same studies, AMAA was found to elicit much more potent neurotoxic effects in the rat than NMDA and QUIN (see chapter 2).

The (*R*)- and (*S*)-isomers of AMAA are shown in Fig. 13, oriented in such a way as to align their respective amino acid centres with the amino acid centre of CMG. Respective hydroxy and amino portions of the isosteric 3-hydroxyisoxazole rings are aligned with the ω-carboxylate oxygens of CMG. As shown in the figure, the most striking feature is the large extension of VDW surface areas relative to CMG which are imparted by the methyl group. Aside from these regions, the remainder of the VDW surfaces coincide quite closely with that defined by CMG. Based on the aforementioned potency of AMAA, it would appear that the extended surface region is well tolerated by the NMDA agonist site and may in fact represent a so-called 'agonist pocket' region. Similar to what we observed for the enantiomeric CPGs, chirality of the amino acid centre bears little consequence on the overall molecular volume since in this region of the molecule neither protrudes beyond the volume defined by the heart of the template. Likewise, the amino acid chirality appears to effect only subtle differences in the locations of these methyl groups relative to each other. Thus, surface areas corresponding to (*R*)-AMAA and (*S*)-AMAA reveal only slight differences.

However, it is apparent from the respective energy calculations that substantially more energy is involved in constraining (*S*)-AMAA into the orientation dictated by the template than (*R*)-AMAA. Whereas $\leqslant$ 15 kcal/mol is required to lock (*R*)-AMAA in this fit, the cost of similarly constraining (*S*)-AMAA is of the order of 35 kcal/mol. The overall better fit of (*R*)-AMAA is reflected in terms of ΔRMS (0.15 Å versus 0.57 Å for (*S*)-AMAA). Taken together, this would imply that greater NMDA agonist activity likely resides with (*R*)-AMAA (see chapter 2). Similar energy considerations reveal better fit at the expense of less energy for this isomer relative to both NMDA and QUIN, a factor that may in part explain the greater observed neurotoxicity of AMAA relative to these two agonists.

Resolution of (*RS*)-AMAA, though not yet accomplished, will help to provide important information allowing further evaluation of the scope and limitations of this computational model. In addition, a series of analogues wherein the methyl group is systematically replaced by other substituents would be of value for probing the boundaries and nature of this 'agonist pocket'. Replacement of the methyl group with hydrogen is expected to give rise to the corresponding isomeric pair ((*RS*)-desmethyl-AMAA), which on the basis of overall template fit should still retain NMDA agonist

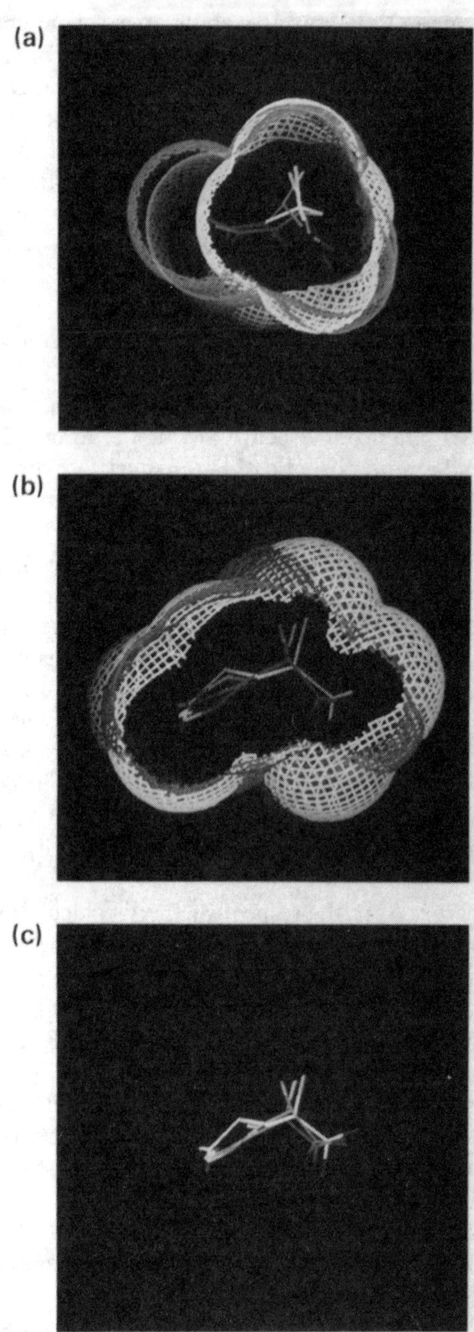

Fig. 13 — (**See colour section.**) CMG (white), (*R*)-AMAA (orange), (*S*)-AMAA (purple). (a) Front view, VDW surface and Dreiding representation; (b) side view, VDW surface and Dreiding representation; (c) side view, Dreiding representation.

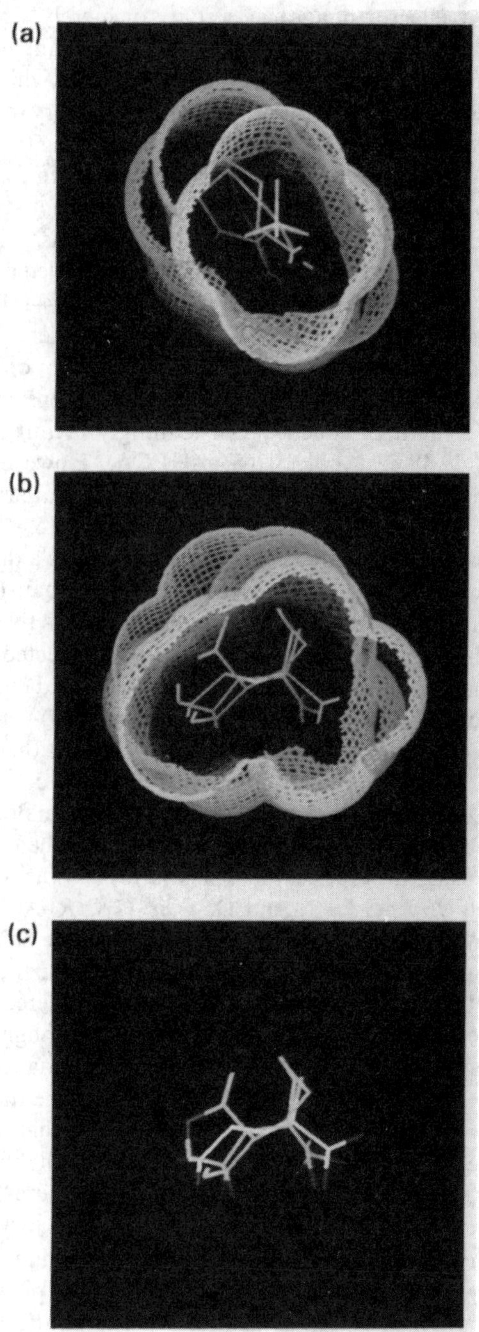

Fig. 14 — (**See colour section.**) CMG (white), (*R*)-4-HPCA (green), (*S*)-4-HPCA (yellow). (a) Front view, VDW surface and Dreiding representation; (b) side view, VDW surface and Dreiding representations; (c) side view, Dreiding representation.

properties. Whether this results in an increase or decrease in activity would be of particular significance, and would help to delineate the role of the 'agonist pocket' as either that of a positive effector of agonist activity or merely a bulk-tolerating site.

(*RS*)-3-Hydroxy-4,5,6,7-tetrahydroisoxazole[4,5-*c*]pyridine-4-carboxylic acid (4-HPCA) represents a bicyclic structural analogue of AMAA which has been reported as a weak NMDA agonist with receptor affinity and excitatory (depolarizing) potency similar to QUIN (Madsen *et al.* 1990). The effects of cyclizing the AMAA structure in the configuration of 4-HPCA are substantial, as can be seen in Fig. 14. Chirality of the amino acid centre is important as it establishes unique overall configurational arrangements and relative proximities of the ring systems. The methyl group of AMAA that defined the so-called 'agonist pocket' is now pulled away from this region — an effect that appears more pronounced in the case of (*S*)-4-HPCA. Taken by itself, this might suggest (*R*)-4-HPCA to have somewhat greater receptor affinity and depolarizing ability than its enantiomer. This is supported not only by a better overall VDW surface fit of this isomer with CMG but also by the energy difference between constrained and free structures which overwhelmingly favour (*R*)-4-HPCA ($\Delta E =$ 18.71 kcal/mol versus 59.48 kcal/mol for (*S*)-4-HPCA). However, relative to CMG, neither isomer provides a particularly close VDW surface match with the template. It is therefore not surprising that NMDA agonist effects of (*RS*)-4-HPCA are weak.

Curiously, (*S*)-4-HPCA appears partially to overlap with a distinct region defined by KAIN (see below), and is somewhat more effective than (*R*)-4-HPCA in this regard. Though we are unaware of reports confirming KAIN receptor activity of (*RS*)-4-HPCA in higher brain regions, such activity is expected to be weak (on the basis of poor functional group overlay indicated by RMS and energy differences) and may have gone unnoticed. On feline spinal neurons, (*RS*)-4-HPCA produced no significant excitatory effects, nor did it significantly reduce the excitatory effects of KAIN (Madsen *et al.* 1986).

A series of ACPD compounds, which, like the CMG analogues, represent rigid Glu analogues, has proven interesting in terms of receptor heterogeneity. In electrophysiological studies using rat hippocampal slices, three of the four isomers comprising this series, namely *trans*-(1*R*,3*S*)-ACPD, *trans*-(1*S*,3*R*)-ACPD and *cis*-(1*S*,3*S*)-ACPD, produced KAIN-like responses; *cis*-(1*R*,3*R*)-ACPD behaved as an NMDA agonist (Curry *et al.* 1988). However, in rat spinal cord preparations, *trans*-(1*R*,3*S*)-ACPD resembled QUIN in its excitatory effects and differed from NMDA. The remaining three isomers elicited NMDA-like responses, although (1*R*,3*R*)-ACPD was the most potent (Magnuson *et al.* 1988). As a mixture of enantiomers, *trans*-ACPD has been reported to be a somewhat selective agonist at the metabotropic receptor (Palmer *et al.* 1989), though recent data have shown that this activity resides exclusively in the 1*S*,3*R*-isomer (Irving *et al.* 1990). In light of these seemingly contradictory results, we were interested in examining these four isomers in our template model.

As can be seen in Fig. 15, only subtle differences in volume areas are apparent relative to CMG, the most prominent of which occur in the planar region defined by the cyclic ring systems. The surfaces of *cis*-(1*S*,3*S*)-ACPD and *trans*-(1*S*,3*R*)-ACPD extend slightly into the region previously termed the 'agonist pocket' and on this basis alone might be expected to exert more potent NMDA agonist effects. However, of the four ACPD isomers, *trans*-(1*S*,3*R*)-ACPD shows the greatest potential energy difference and degree of RMS deviation between constrained and unconstrained conforma-

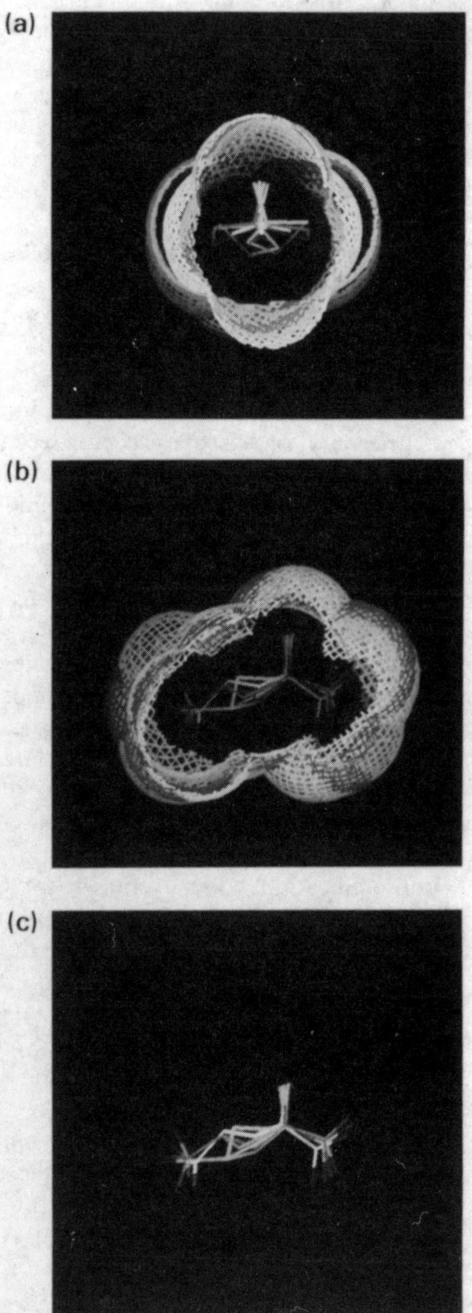

Fig. 15 — (**See colour section**.) CMG (white), (*RR*)-ACPD (pink), (*SS*)-ACPD (orange), (*RS*)-ACPD (yellow), (*SR*)-ACPD (green). (a) Front view, VDW surface and Dreiding representation; (b) side view, VDW surface and Dreiding representation; (c) side view, Dreiding representation.

tions (Fig. 7) and is expected to be the least likely to act as an NMDA agonist. This would tend to support the findings of Irving *et al.* (1990), mentioned above. Both *cis*-(1*R*,3*R*)-ACPD and *trans*-(1*R*,3*S*)-ACPD have similar surfaces which extend slightly beyond that of the template in a region directly opposite to the 'agonist pocket' region. In addition, RMS differences and energy calculations suggest that both of these isomers can adopt the template-preferring conformation in similar fashion without a significant expenditure of energy (Figs 7 and 8). Thus, it would be predicted from this model that *cis*-(1*R*,3*R*)-ACPD and *trans*-(1*R*,3*S*)-ACPD would behave similarly as NMDA agonists and may be slightly less potent than *cis*-(1*S*,3*S*)-ACPD in this regard.

These predictions are not in agreement with the findings of Magnuson *et al.* (1988) or Curry *et al.* (1988) as described above. With the exception of the metabotropic agonist *trans*-(1*S*,3*R*)-ACPD, the model cannot account for the observed relative differences in activities of these isomers, thereby suggesting a limitation in this computational approach. Clearly, other factors than those accounted for in this modelling system are responsible for conferring the unique physical properties to the individual ACPD isomers necessary for selective receptor recognition.

Though the present model is designed to provide insight into the topographical requirements of the NMDA agonist recognition site, it was nonetheless of interest to examine the nature of NMDA antagonist interactions with this site. The structures of CGS 19755 and CPP, two well-known NMDA antagonists, were chosen for this purpose, the former representing a semi-rigid analogue of AP5, the latter an analogue of AP7 (see chapter 8). The antagonist molecules are aligned with the template molecule in a fashion similar to agonists such that ω-phosphonate groups are constrained to fit the ω-carboxylate of CMG. These compounds are shown in Fig. 16 along with (*R*)-AMAA, which is included to highlight the so-called 'agonist pocket'. When oriented in this fashion, both CGS 19755 and CPP share a common extended volume which spans the entire 'backside' (with respect to the front view) of the template surface. This large 'antagonist pocket' lies in a spatial region that is quite distinct from the proposed 'agonist pocket'.

While RMS calculations indicate a good overall fit for CGS 19755 with the template, the fit of CPP is not as tight (ΔRMS = 0.19 Å and 0.43 Å, respectively). That this is a less than optimal orientation of CPP is clearly indicated in the potential energy difference between free and constrained conformations (ΔE = 53.35 kcal/mol). In contrast, little energy is required to constrain the functional groups of CGS 19755 into the orientation dictated by CMG (ΔE = 6.03 kcal/mol). Thus, it can be inferred that, while CGS 19755 can well adopt the template orientation of three-point fit, CPP will favour an alternative orientation.

It is tempting, on this basis, to propose two antagonist binding sites: one which may be similar to that defined by CGS 19755 in adopting the template fit; the other which is represented by an as yet undetermined, more favourable orientation of CPP. The anomalies described earlier in this chapter between AP5- and AP7-based antagonists would certainly tend to support such a proposal. A more in-depth study would be required to substantiate this hypothesis and we are extending this work to include a variety of NMDA receptor antagonists.

As stated at the outset of this chapter, Glu interacts with at least two other ionotropic receptor subtypes which are identified by their respective selective agonists,

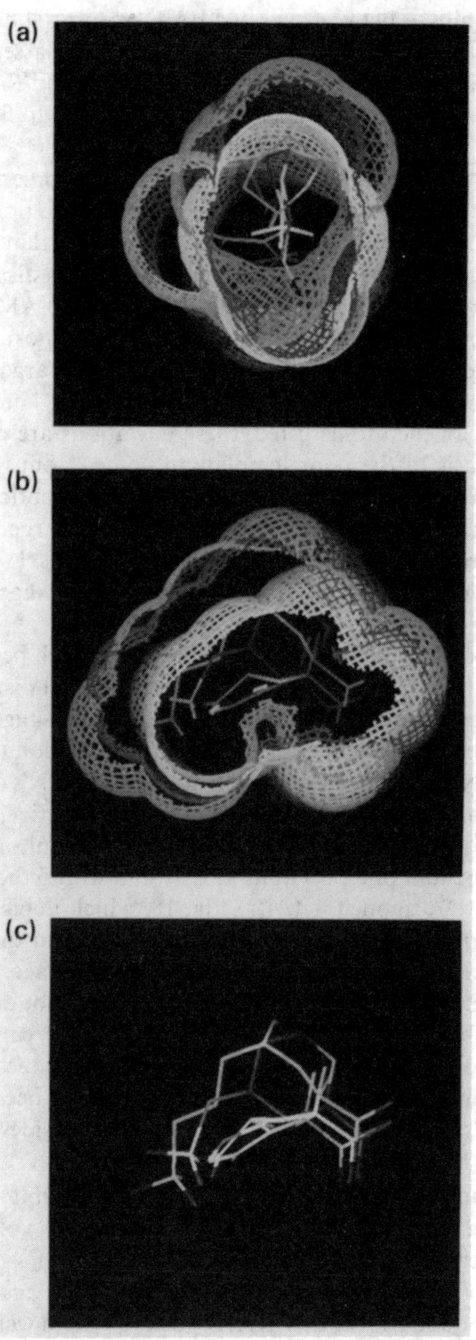

Fig. 16 — (**See colour section.**) CMG (white), (*R*)-CG 19755 (orange), (*R*)-CPP (yellow), (*R*)-AMAA (peach). (a) Front view, VDW surface and Dreiding representation; (b) side view, VDW surface and Dreiding representation; (c) side view, Dreiding representation.

AMPA and KAIN. However, the lack of selectivity of various agonist substances, for example quisqualic acid and ibotenic acid, coupled with the inabilities of several antagonists to discriminate between various EAA receptor subtypes, suggest certain degrees of common topography between these receptor classes. We were therefore interested in examining the structural features of AMPA and KAIN as they relate to each other as well as to NMDA agonists when constrained to fit this CMG template model.

The structure of (*S*)-AMPA, which is the active enantiomer (Krogsgaard-Larsen *et al.* 1982; Hansen *et al.* 1983), and KAIN superimposed on our template model, are shown in Fig. 17. From these perspectives, it is evident that these EAA agonists occupy unique volume regions with respect to CMG. It is also apparent (Fig. 17c) from the rather poor fit of acid and amino functions of (*S*)-AMPA on this template that this selective agonist likely adopts a much different conformational arrangement of functionalities when interacting with its binding site. Target distance deviations (Table 4) for (*S*)-AMPA show the worst correlation among compounds examined in this study. For many of the atoms listed, these deviations are of the order of 1 Å or more. Further emphasis of the poor template fit comes from the RMS calculations wherein the difference between constrained and free conformations (ΔRMS) is 0.79 Å.

The most striking feature of KAIN in this model is the large steric volume, unique from CMG, that is provided by the isopropylene group. This 'KAIN pocket' occupies space that is quite distinct from the proposed NMDA agonist and antagonist 'pockets' (Fig. 16).

In terms of functionality fit, it is clear that KAIN is much better able to attain the model conformation than (*S*)-AMPA, as evidenced by target distance deviations (Fig. 4). That this in fact represents a favourable orientation is suggested by the minimal amount of energy required to constrain KAIN in this fashion ($\Delta E = 9.17\,\text{kcal/mol}$). Aside from the 'KAIN pocket' region, the remainder of KAIN's surface volume shows very close correlation with that of CMG. From this perspective, it would appear that removal of the isopropylene moiety from KAIN would result in a compound where VDW surfaces match those of CMG quite closely and would therefore be expected to possess intrinsic NMDA agonist activity. Fig. 18, which depicts this structure (2*S*-carboxy-3*R*-pyrolidine acetic acid ((*SR*)-CPAA)) along with the enantiomeric (*RS*)-CPAA in our model system confirms this predicted surface fit. RMS deviations, potential energy differences and target distance deviations not only indicate a tight fit with the template but also suggest that both isomers can accommodate this orientation of functional groups equally well. The NMDA receptor agonist activity predicted for these molecules by this model has recently been demonstrated with *trans*-(±)-CPAA (Tsai *et al.* 1988). Whether this activity resides in both enantiomers, however, remains to be established.

Previously, we suggested that (*S*)-4-HPCA might exhibit weak KAIN agonist activity based on the partial extension of its VDW surface volume into the 'KAIN pocket' region. This can be seen in Fig. 19, which depicts both 4-HPCA isomers along with KAIN and CMG. Given the structures of other KAIN agonists (see chapter 12) it is evident that the boundaries of this 'KAIN pocket' extend well beyond that defined by KAIN's isopropylene group. If this is the case, then (*S*)-4-HPCA in actuality might only extend partially into the proposed 'KAIN pocket' and would therefore be expected to possess very weak KAIN-like activity.

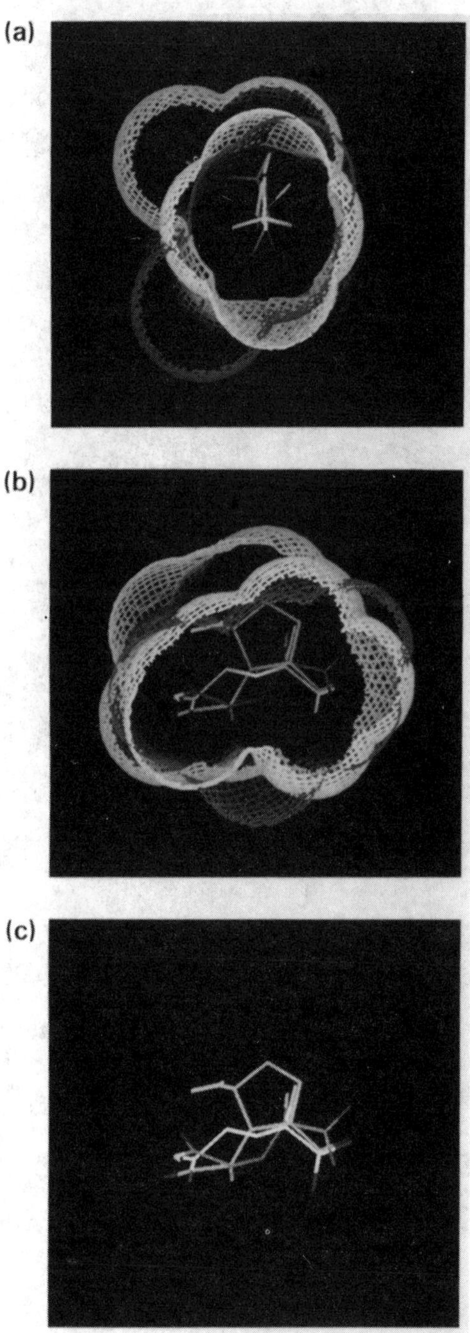

Fig. 17 — (**See colour section.**) CMG (white), KAIN (yellow), (*S*)-AMPA (orange). (a) Front view, VDW surface and Dreiding representation; (b) side view, VDW surface and Dreiding representation; (c) side view, Dreiding representation.

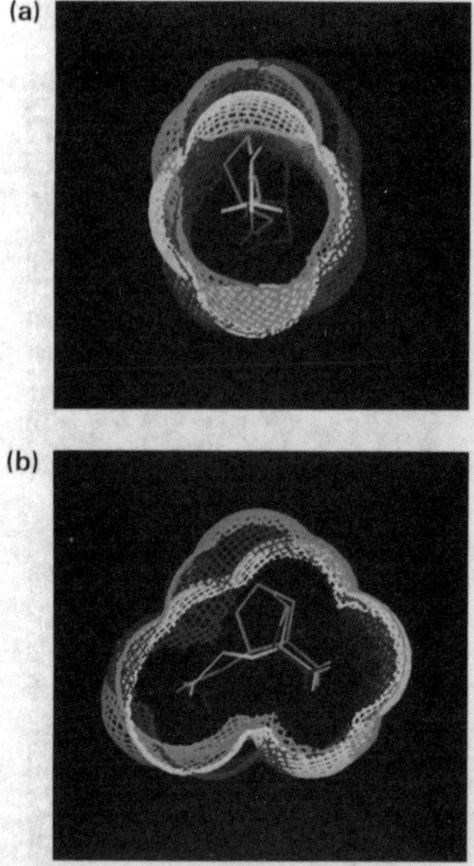

Fig. 18 — (**See colour section.**) CMG (white), (*RS*)-CPAA (green), (*SR*)-CPAA (blue). (a) Front view, VDW surface and Dreiding representation; (b) side view, VDW surface and Dreiding representation.

Based on these results, it is tempting to propose that L-Glu adopts very similar conformations when interacting with NMDA and KAIN agonist binding sites, these being quite distinct from that adopted for interaction with AMPA receptors. In addition, this 'KAIN pocket' might be viewed as a spatial region common to both NMDA and KAIN agonist sites. Relative to the NMDA site, this would represent a region of steric intolerance. Interaction of an otherwise suitable ligand with this region would render such a compound inactive as an NMDA agonist. In contrast, this region is not merely one of steric tolerance relative to the KAIN agonist site, but a region with which specific interaction is required for eliciting KAIN agonist responses. Clearly more detailed analyses are required to confirm and extend such a proposal.

SUMMARY

There exists substantial evidence supporting the notion of heterogeneity amongst NMDA-type excitatory amino acid receptors. Thus, several studies have now

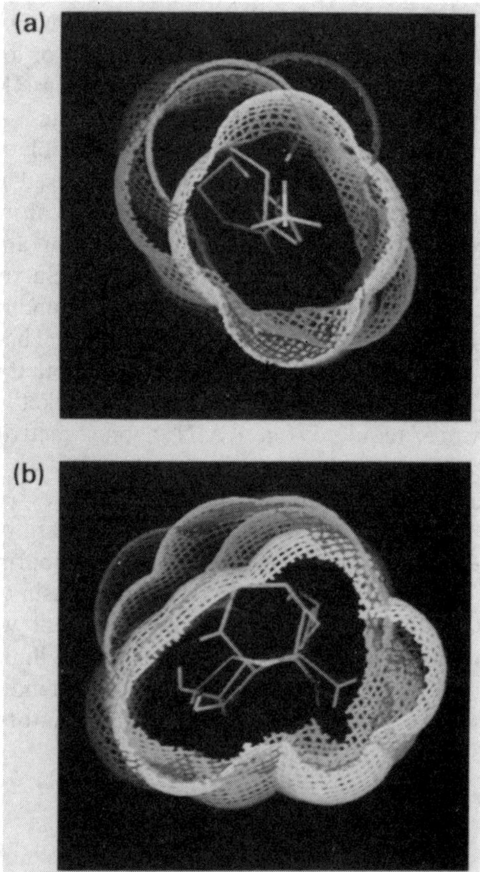

Fig. 19 — (**See colour section.**) CMG (white), KAIN (orange), (*S*)-4-HPCA (yellow), (*R*)-4-HPCA (green). (a) Front view, VDW surface and Dreiding representation; (b) side view, VDW surface and Dreiding representation.

reported differences in (1) agonist- and antagonist-labelled receptor densities and distribution in brain, (2) the molecular mass of agonist and antagonist recognition sites and (3) the stoichiometry and ontogenic profile of TCP, Gly and NMDA recognition sites.

The identification of NMDA agonists and antagonists having diverse pharmacological properties has provided additional support for NMDA receptor heterogeneity. *In vitro* labelling of the NMDA binding site, strychnine-insensitive Gly binding site, and ion channel has resulted in the discovery of functional interactions between various components of the NMDA receptor complex, and the recognition that NMDA agonists and antagonists differentially interact with NMDA as well as Gly recognition sites. Importantly, substantial differences exist in the neuroexcitatory and neurotoxic properties of NMDA agonists, and in the functional properties of competitive and non-competitive antagonists.

In order further to explore and define the structural requirements for agonist interactions with the NMDA receptor complex, we have developed a simple computational method. Because agonist/agonist interactions (e.g. NMDA-Glu) appear to be less complex than those associated with agonist-antagonist or antagonist/antagonist reactions, we have initially applied this method to modelling NMDA agonists using the potent and selective NMDA agonist CMG as a template.

As a result of such template matching, it has been possible to demonstrate both common and distinct features of the various molecules used in this study. These features, in turn, have allowed us to construct a hypothetical model of the NMDA agonist site in which an optimal functional group orientation and molecular volume are defined by an energy-minimized conformation of CMG. Surrounding this primary binding region in discrete spatial locations are secondary binding 'pockets', interactions with which result in distinct pharmacological activities. These secondary regions include an additional 'agonist pocket' which extends beyond that defined by CMG, an 'excitotoxic pocket', an antagonist region, and a 'KAIN pocket'. Steric occupation of this KAIN region, while required for KAIN agonist activity, would preclude interaction with the NMDA agonist site.

While this model may help to explain some of the seemingly anomalous pharmacological observations that have been made regarding NMDA receptor agonist activities, it has limitations. For example, observed pharmacological activities of three of the four ACPD isomers would not be predicted using this NMDA agonist site model. Nonetheless, the model seems to be consistent with the majority of compounds examined, though admittedly these constitute a limited series of NMDA agonists. Our efforts continue to focus on refining and extending the scope of this method to include NMDA antagonists as well as modulators at the other EAA receptor subtypes.

REFERENCES

Albuquerque, E. X., Tsai, M.-C., Aronstam, R. S., Eldefrawi, A. T. and Eldefrawi, M. E. (1980) *Mol. Pharmacol.* **18**, 167–178.

Anis, N. A., Berry, S. C., Barton, N. R. and Lodge, D. (1983) *Br. J. Pharmacol.* **79**, 565–575.

Aronstam, R. S., Eldefrawi, M. E., Eldefrawi, A. T., Albuquerque, E. X., Jim, K. F. and Triggle, D. J. (1980) *Mol. Pharmacol.* **18** 179–184.

Baron, B. M., Dudley, M. W., McCarty, D. R., Miller, F. P., Reynolds, I. J. and Schmidt, C. J. (1989) *J. Pharmacol. Exp. Ther.* **250** 162–169.

Bettler, B., Boulter, J., Hermans-Borgmeyer, I., O'Shea-Greenfield, A., Deneris, E. S., Moll, C., Borgmeyer, D., Hollmann, M. and Heinemann, S. (1990) *Neuron* **5** 583–595.

Boulter, J., Hollmann, M., O'Shea-Greenfield, A., Hartley, M., Deneris, E., Maron, C. and Heinemann, S. (1990) *Science* **249** 1033–1037.

Brooks, B. R., Bruccoleri, R. E., Olafson, B. D., States, D. J., Swaminathon, S. and Karplus, M. J. (1983) *J. Comp. Chem.* **4** 187.

Cavalheiro, E. A., Lehmann, J. and Turski, L. (eds) (1988) *Frontiers in Excitatory Amino Acid Research*, Alan R. Liss, New York.

Compton, R. P., Hood, W. F. and Monahan, J. B. (1990) *Eur. J. Pharmacol. Mol. Pharmacol. Sect.* **188** 63–70.

Conti, L. H., MacIver, C. R., Ferkany, J. W. and Abreau, M. E. (1990) *Psychopharmacol.* **102** 492–497.

Curry, K., Magnuson, D. S., McLennan, H. and Peet, M. J. (1986) *Neurosci. Lett.* **66** 101–105.

Curry, K., Peet, M. J., Magnuson, D. S. K. and McLennan, H. (1988) *J. Med. Chem.* **31** 864–867.

Danysz, W., Fadda, E., Wroblewski, J. T. and Costa, E. (1989) *Mol. Pharmacol.* **36** 912–916.

Fagg, G. (1987) *Neurosci. Lett.* **76** 721–727.

Fagg, G. E. and Baud, J. (1985) *Soc. Neurosci. Abs.* **11** 109.

Fagg, G. E. and Matus, A. (1984) *Proc. Nat. Acad. Sci. USA* **81** 6876–6880.

Ferkany, J. W. and Coyle, J. T. (1983) *Life Sci.* **33** 1295–1305.

Ferkany, J. W., Borosky, S. E., Clissold, D. and Pontecorvo, M. (1988) *Eur. J. Pharmacol.* **151** 151–154.

Ferkany, J. W., Kyle, D. J., Willetts, J., Rzeszotarski, W. J., Guzewska, M. E., Ellenberger, S. R., Jones, S. M., Sacaan, A. I., Shell, I. P., Borosky, S., Jones, B. E., Johnson, K. M., Balster, R. L., Burchett, K., Kawasaki, K., Hoch, D. B. and Dingledine, R. (1989) *J. Pharmacol. Exp. Ther.* **250** 100–109.

Foster, A. C. and Kemp, J. A. (1989) *Nature* **338** 377–378.

French, E. D., Ferkany, J., Abreu, M. and Levenson, S. (1991) *Neuropharmacology.*

ffrench-Mullen, J. M. H. and Rogawski, M. A. (1989) *J. Neurosci.* **9** 4051–4061.

Galligan, J. J. and North, R. A. (1990) *Neurosci. Lett.* **108** 105–109.

Greenamyre, J. T., Young, A. B. and Penney, J. B. (1984) *J. Neurosci.* **4** 2133–2144.

Greenamyre, J. T., Olson, J. M. M., Penney, J. B. and Young, A. B. (1985) *J. Pharmacol. Exp. Ther.* **233** 254–261.

Hansen, J. J., Lauridsen, J., Nielsen, E. and Krogsgaard-Larsen, P. (1983) *J. Med. Chem.* **26** 901–903.

Hollmann, M., O'Shea-Greenfield, A., Rogers, S. and Heinemann, S. (1989) *Nature* **342** 643–648.

Honey, C. R., Miljkovic, Z. and MacDonald, J. F. (1985) *Neurosci. Lett.* **61** 135–139.

Honoré, T., Drejer, J., Nielsen, E. O., Watkins, J. C., Olverman, H. J. and Nielsen, M. (1989) *Eur. J. Pharmacol. Mol. Pharmacol. Sect.* **172** 239–247.

Hood, W. F., Compton, R. P. and Monahan, J. B. (1990) *J. Neurochem.* **54** 1040–1046.

Huettner, J. E. (1989) *Science* **243** 1611–1613.

Irving, A. J., Schofield, J. G., Watkins, J. C., Sunter, D. C. and Collingridge, G. L. (1990) *Eur. J. Pharmacol.* **186** 363–365.

Javitt, P. C., Frusciante, M. J. and Zukin, S. R. (1990) *Mol. Pharmacol.* **37** 603–607.

Johnson, J. W. and Ascher, P. (1987) *Nature* **325** 529–531.

Jones, K. W., Schaeffer, C. L. and DeNoble, V. J. (1989) *Pharmacol. Biochem. Behav.* **34** 181–185.

Kaplita, P. V. and Ferkany, J. W. (1990) *Eur. J. Pharmacol. Mol. Pharmacol. Sect.* **188** 175–179.

Keinanen, K., Wisden, W., Sommer, B., Werner, P., Herb, A., Verdoorn, T. A., Sakmann, B. and Seeburg, P. H. (1990) *Science* **249** 556–560.

Kessler, M., Terramini, T., Lynch, G. and Baudry, M. (1989) *J. Neurochem.* **52** 1319–1328.

Koek, W., Woods, J. H. and Ornstein, P. (1986) *Life Sci.* **39** 973–980.

Koek, W., Woods, J. H., Mattson, M. V., Jacobson, A. B. and Mudar, P. (1987) *Neuropharmacology* **26** 1261–1265.

Koek, W., Woods, J. H. and Winger, G. D. (1988) *J. Pharmacol. Exp. Ther.* **245** 969–974.

Krogsgaard-Larsen, P., Honoré, T., Hansen, J. J., Curtis, D. R. and Lodge, D. (1980) *Nature* **284** 64–66.

Krogsgaard-Larsen, P., Hausen, J. J., Lauridsen, J., Peet, M. J., Leah, J. D. and Curtis, D. R. (1982) *Neurosci. Lett.* **31** 313–317.

Krogsgaard-Larsen, P., Ebert, B., Johansen, T. N., Ferkany, J. W. and Madsen, U. (1991) In: Meldrum, B., Moroni, F., Simon, R. P. and Woods, J. H. (eds) *Excitatory Amino Acids.* Raven Press, New York, pp. 433–441.

Lanthorn, T. H., Hood, W. F., Watson, G. B., Compton, R. P., Rader, R. K., Gaoni, Y. and Monahan, J. B. (1990) *Eur. J. Pharmacol.* **182** 397–404.

Leeson, P. D., James, K., Carling, R. W., Wong, E. H. F. and Baker, R. (1990) In: Meldrum, B. S. and Williams, M. (eds) *Current and Future Trends in Anticonvulsants, Anxiety and Stroke Therapy.* Wiley/Liss, New York, pp. 495–512.

Lehmann, J., Ferkany, J. W., Schaeffer, P. and Coyle, J. T. (1985) *J. Pharmacol. Exp. Ther.* **232** 873–882.

Lehmann, J., Schneider, J., McPherson, S., Murphy, D. E., Bernard, P., Tsai, C., Bennett, D. A., Pastor, G., Steel, D. J., Boehm, C., Cheney, D. L., Liebman, J. M., Williams, M. and Wood, P. L. (1987) *J. Pharmacol. Exp. Ther.* **240** 737–746.

MacDonald, J. F. and Nowak, L. M. (1990) *Trends Pharmacol. Sci.* **11** 167–171.

Madsen, U., Schaumburg, K., Brehm, L., Curtis, D. R. and Krogsgaard-Larsen, P. (1986) *Acta Chem. Scand.* **B40** 92–97.

Madsen, U., Ferkany, J. W., Jones, B. E., Ebert, B., Johansen, T. N., Holm, T. and Krogsgaard-Larsen, P. (1990) *Eur. J. Pharmacol., Mol. Pharmacol. Sect.* **189** 381–391.

Magnuson, D. S. K., Curry, K., Peet, M. J. and McLennan, H. (1988) *Exp. Brain Res.* **73** 541–545.

McDonald, J. W. and Johnston, M. V. (1990) *Brain Res. Rev.* **15** 41–70.

McDonald, J. W., Penney, J. B., Johnston, M. V. and Young, A. B. (1990) *Neuroscience* **35** 653–668.

Monaghan, D. T., Holefs, V. R., Toy, D. W. and Cotman, C. W. (1984) *Nature* **306** 176–178.

Monaghan, D. T. and Cotman, C. W. (1985) *J. Neurosci.* **5** 2909–2919.

Monaghan, D. T., Olverman, H. J., Nguyen, L., Watkins, J. C. and Cotman, C. W. (1988) *Proc. Nat. Acad. Sci. USA* **85** 9836–9840.

Monahan, J. B., Hood, W. F., Compton, R. P., Cordi, A. A., Snyder, J. P., Pellicciari, R. and Natalina, B. (1990a) *Neurosci. Lett.* **112** 328–332.

Monahan, J. B., Biesterfeldt, J. P., Heed, W. F., Compton, R. P., Cordi, A. A., Vazquez, M. I. Lanthorn, T. H. and Wood, P. L. (1990b) *Mol. Pharmacol.* **37** 780–784.

Morris, R. G. M., Anderson, E., Lynch, G. S. and Baudry, M. (1986) *Nature* **319** 774–776.

Murphy, D. E., Schneider, J., Boehm, C., Lehmann, J. and Williams, M. (1987) *J. Pharmacol. Exp. Ther.* **240** 778–784.

Olverman, H. J., Jones, A. W. and Watkins, J. C. (1984) *Nature* **307** 460–462.

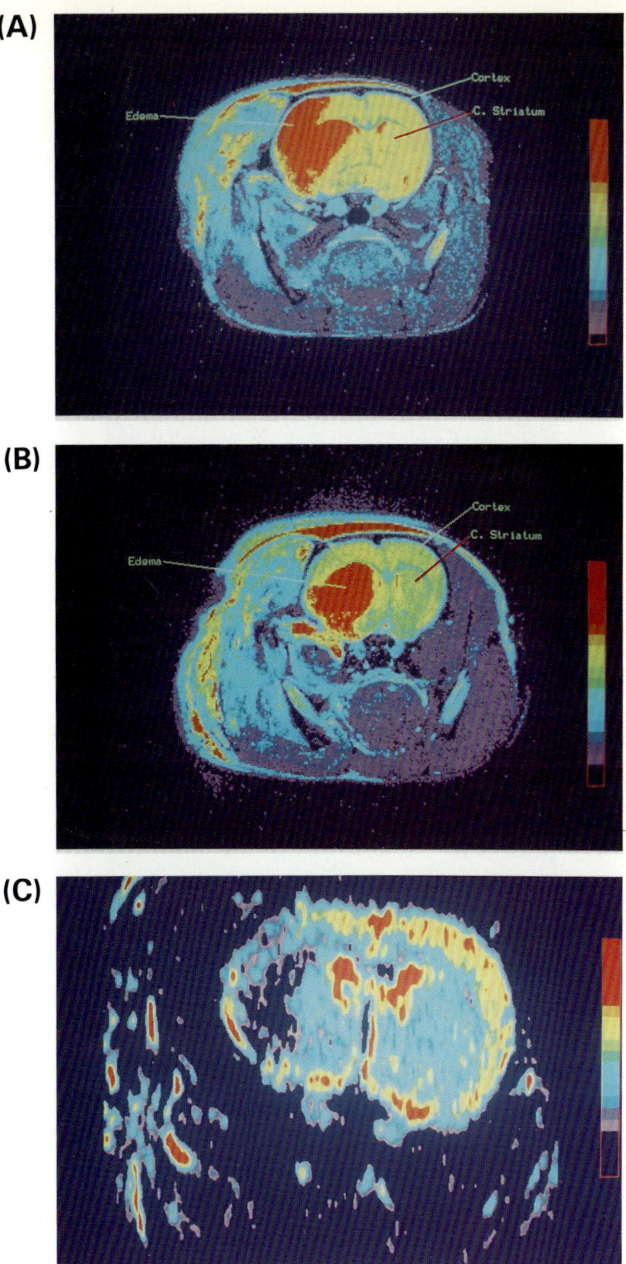

Fig: 1.4 — Magnetic resonance images (T_2 weighted, coronal planes) taken two days after middle cerebral artery occlusion in Fischer 344 rats which received either (A) no treatment or (B) 20 mg/kg i.v. CGP 40116 immediately post-occlusion. Images were taken using a Spectrospin BMT 47/30 (4.7T) instrument and colour-enhanced using a Bruker Aspect X32 workstation. Regions of oedema are colour-coded red; these regions were shown to be necrotic using histological techniques. Note the pronounced reduction in cortical oedema in the animal treated with CGP 40116. (C) Magnetic resonance subtraction image (before and after i.v. infusion of a gadolinium-based contrast agent) showing cerebral perfusion 1 hour following middle cerebral artery occlusion. Dark areas indicate regions of low vascular perfusion, and red high. Note the lack of perfusion in the striatum, and the moderate perfusion in cortical regions of the occluded side. (Courtesy of Dr P. Allegrini.)

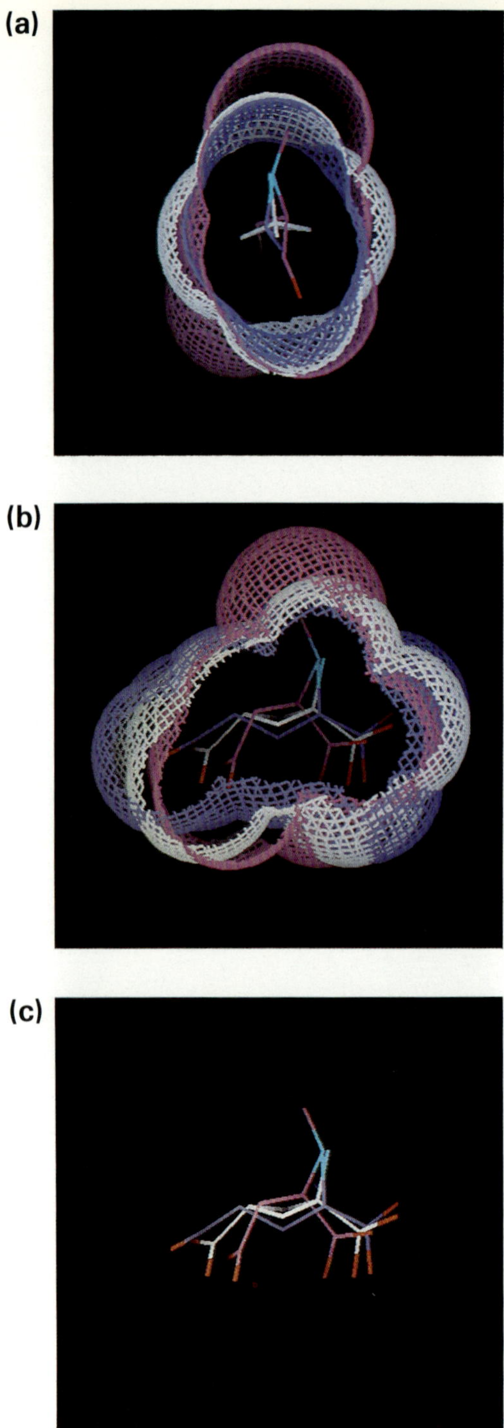

Fig: 6.10 — CMG (white), Glu (blue), NMDA (pink). (a) Front view, VDW surface and Dreiding representation; (b) side view, VDW surface and Dreiding representation; (c) side view, Dreiding representation.

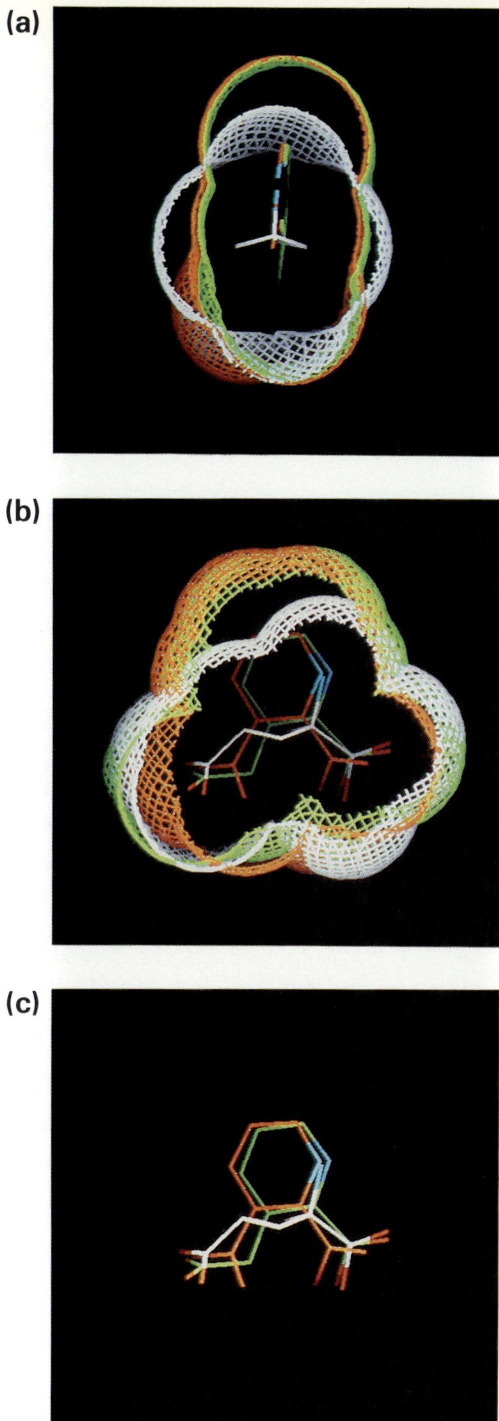

Fig: 6.11 — CMG (white), HQUIN (green), QUIN (orange). (a) Front view, VDW surface and Dreiding representation; (b) side view, surface and Dreiding representation; (c) side view, Dreiding representation.

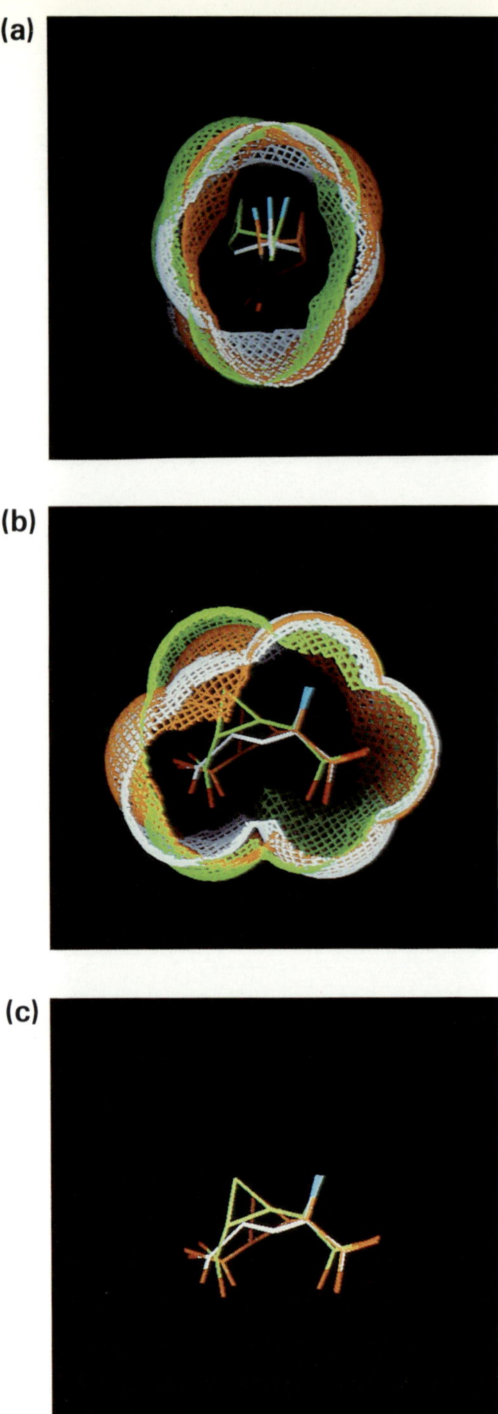

Fig: 6.12 — CMG (white), (R,S,R)-CPG (green), (S,R,S)-CPG (orange). (a) Front view, VDW surface and Dreiding representation; (b) side view, VDW surface and Dreiding representation; (c) side view, Dreiding representation.

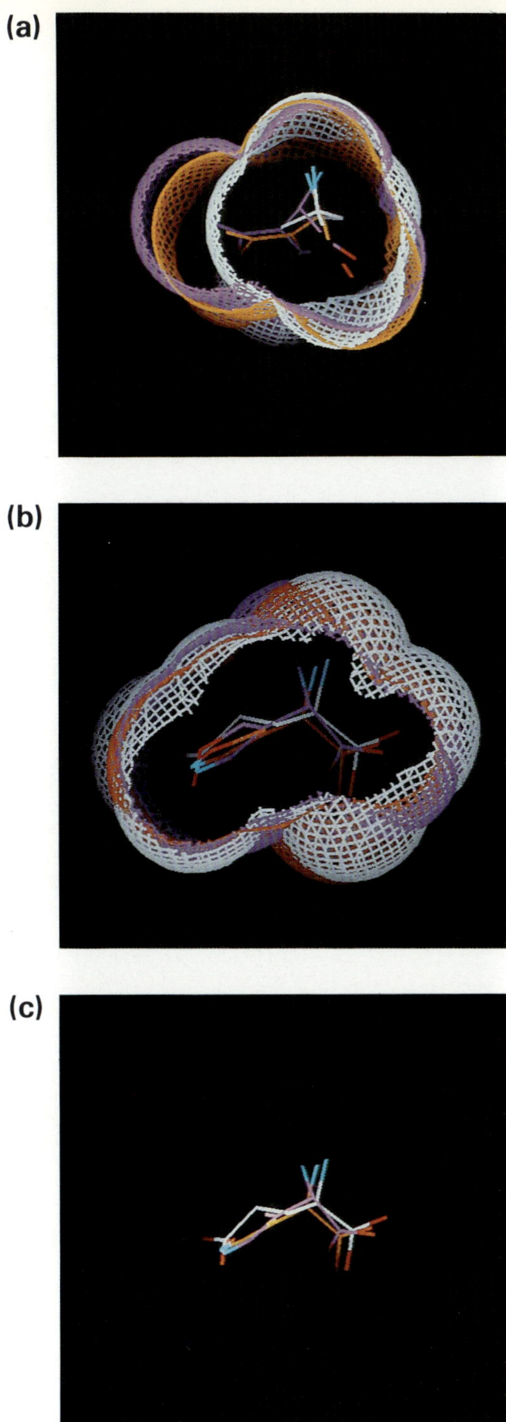

Fig: 6.13 — CMG (white), (*R*)-AMAA (orange), (*S*)-AMAA (purple). (a) Front view, VDW surface and Dreiding representation; (b) side view, VDW surface and Dreiding representation; (c) side view, Dreiding representation.

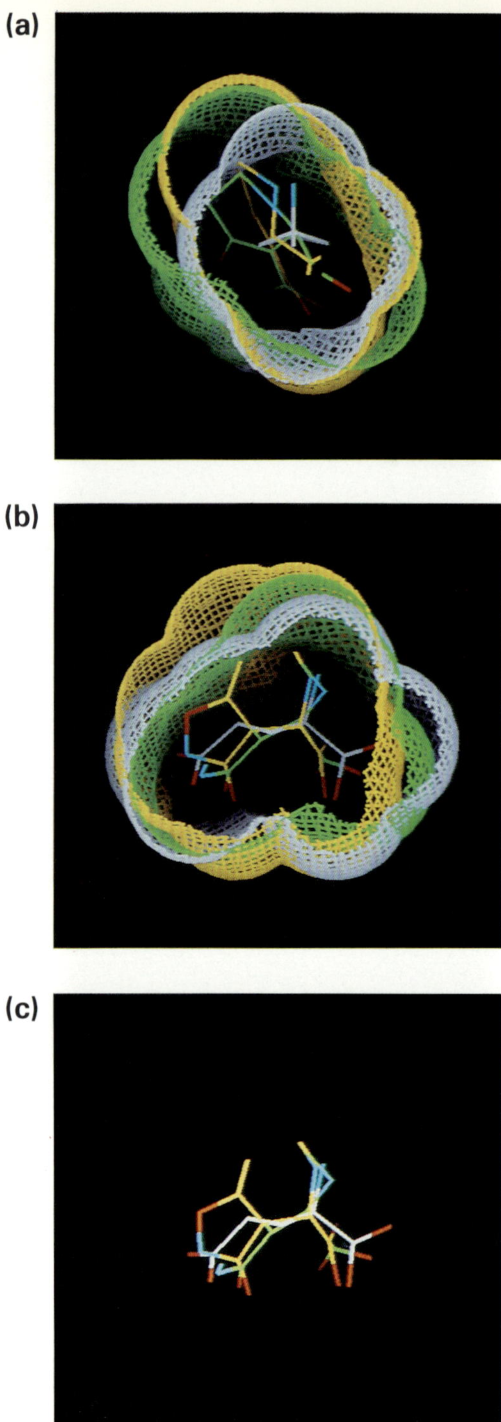

Fig: 6.14 — CMG (white), (*R*)-4-HPCA (green), (*S*)-4-HPCA (yellow). (a) Front view, VDW surface and Dreiding representation; (b) side view, VDW surface and Dreiding representation; (c) side view, Dreiding representation.

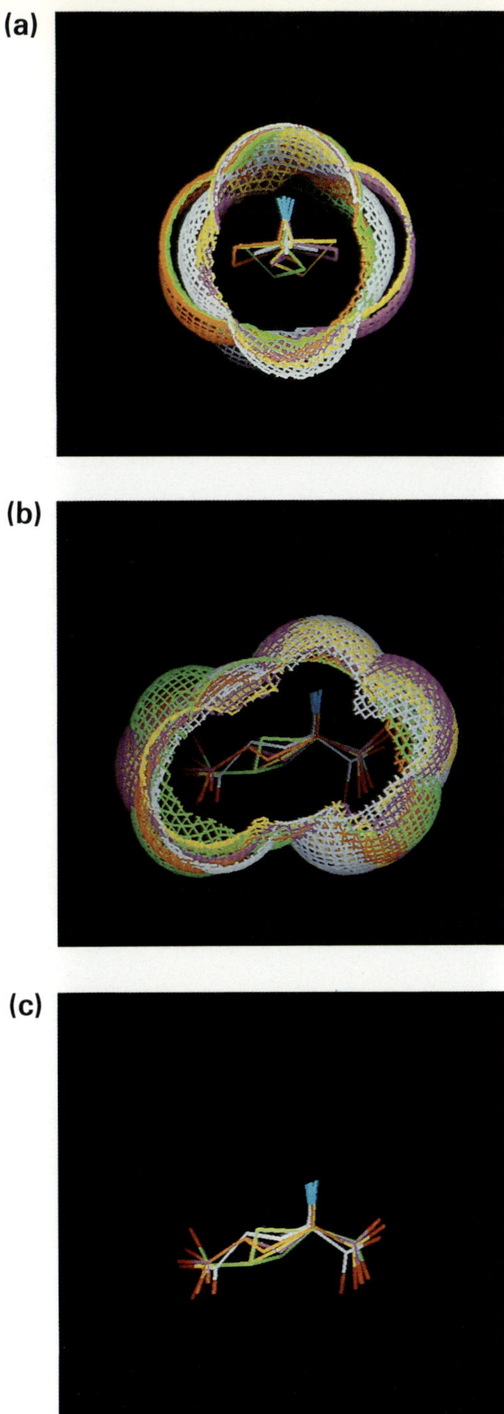

Fig: 6.15 — CMG (white), (*RR*)-ACPD (pink), (*SS*)-ACPD (orange), (*RS*)-ACPD (yellow), (*SR*)-ACPD (green). (a) Front view, VDW surface and Dreiding representation; (b) side view, VDW surface and Dreiding representation; (c) side view, Dreiding representation.

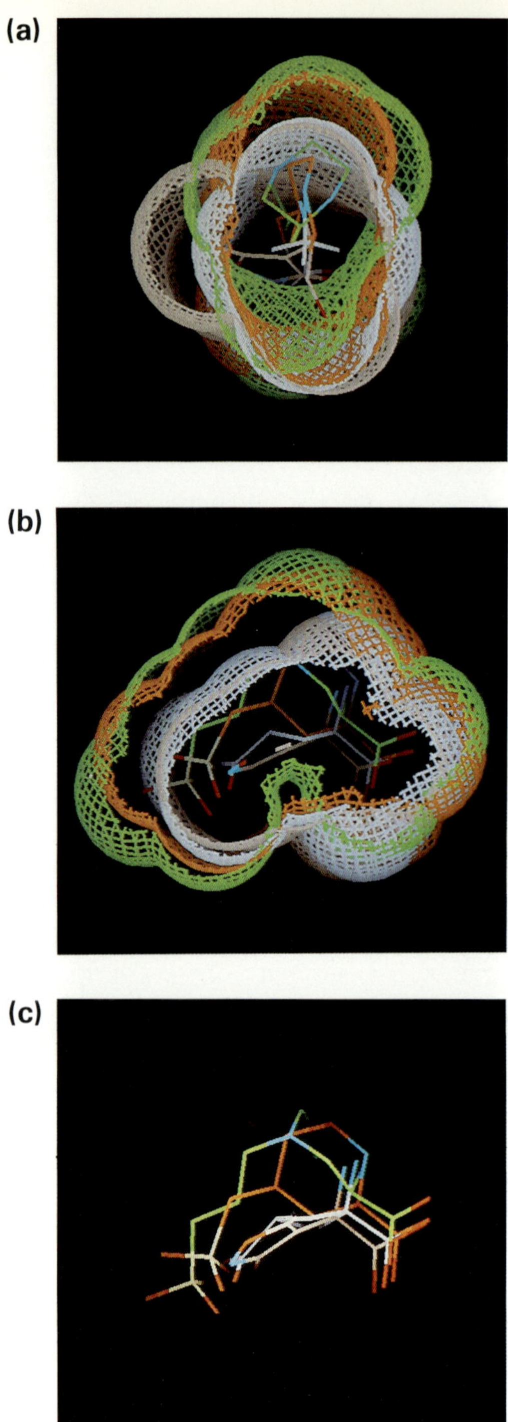

Fig: 6.16 — CMG (white), (*R*)-CGS 19755 (orange), (*R*)-CPP (yellow), (*R*)-AMAA (peach). (a) Front view, VDW surface and Dreiding representation; (b) side view, VDW surface and Dreiding representation; (c) side view, Dreiding representation.

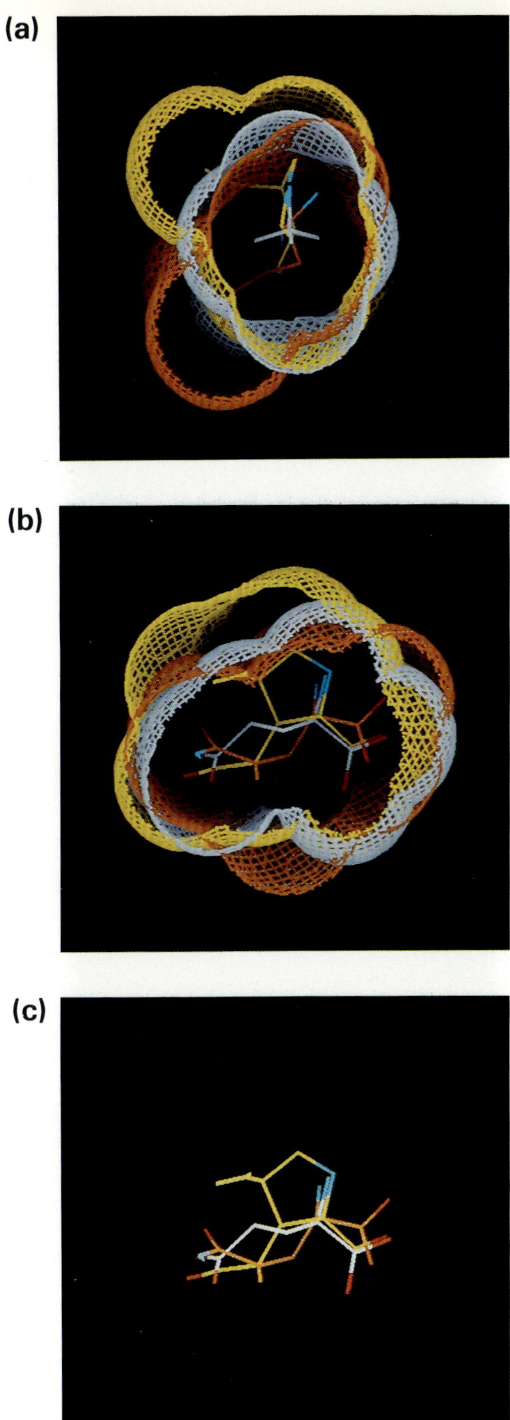

Fig: 6.17 — CMG (white), KAIN (yellow), (*S*)-AMPA (orange). (a) Front view, VDW surface and Dreiding representation; (b) side view, VDW surface and Dreiding representation; (c) side view, Dreiding representation.

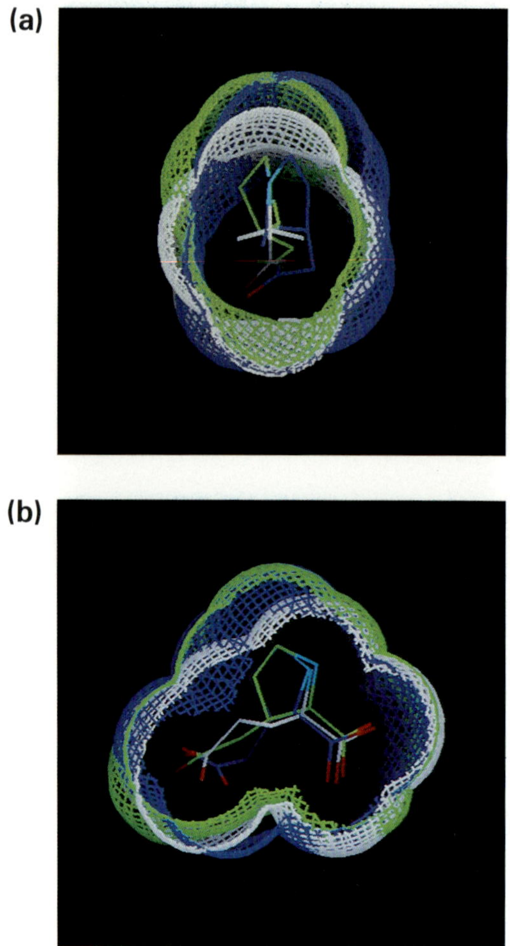

Fig: 6.18 — CMG (white), (*RS*)-CPAA (green), (*SR*)-CPAA (blue). (a) Front view, VDW surface and Dreiding representation; (b) side view, VDW surface and Dreiding representation.

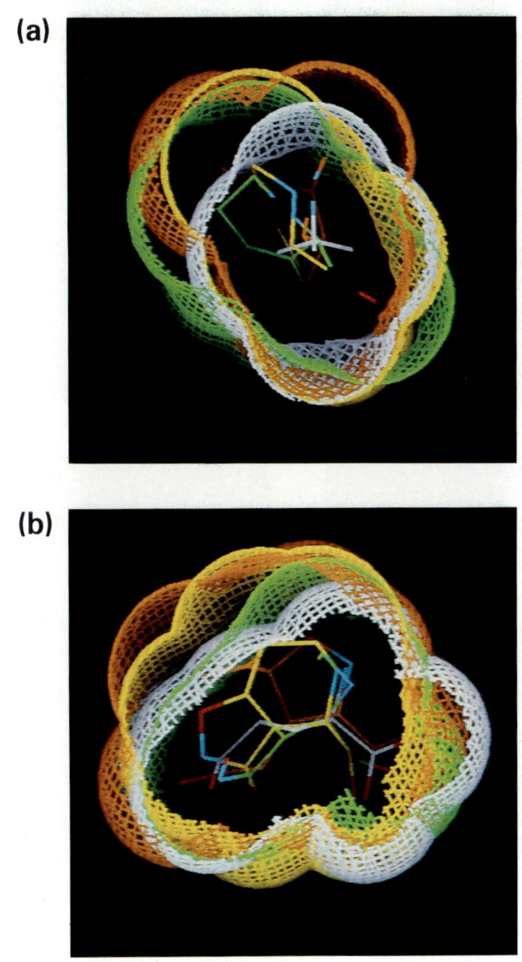

Fig: 6.19 — CMG (white), KAIN (orange), (*S*)-4-HPCA (yellow), (*R*)-4-HPCA (green). (a) Front view, VDW surface and Dreiding representation; (b) side view, VDW surface and Dreiding representation.

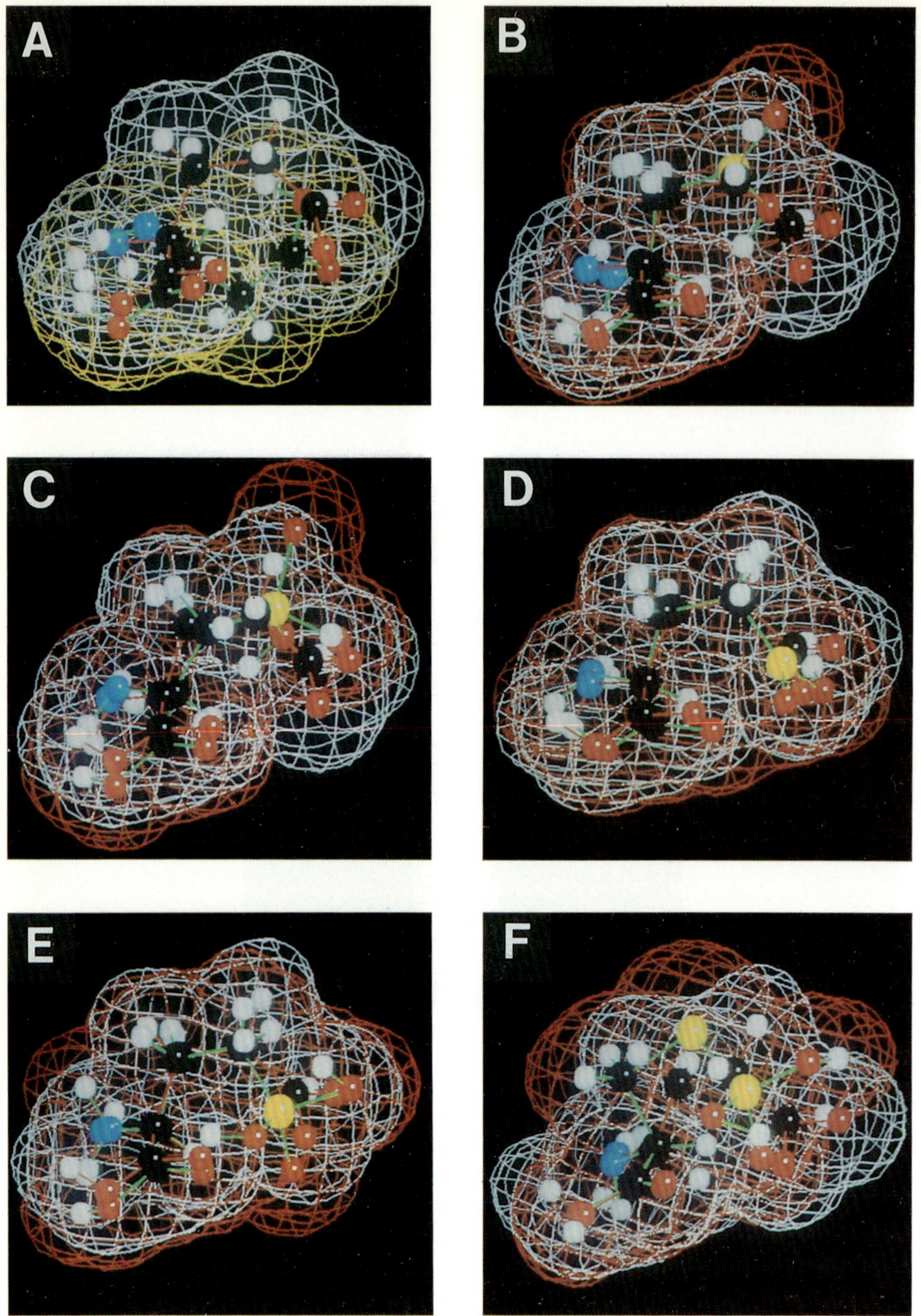

Fig: 7.4 — Volume maps of excitatory amino acids. The 3D contour maps were constructed and superimposed by computer-assisted use of the Chem-X software program as described in Griffiths *et al.* (1989). The plates (A – L) show the volume of D-Asp (yellow), D-Glu (orange) and the neuroexcitatory SAAs (L-enantiomers/red; D-enantiomers/yellow) superimposed over that of L-Glu (white). *Key* (with reference to inhibitors): A = D-Asp; B = L-CSA; C = L-CA; D = L-HCSA; E = L-HCA; F = L-SSC; G = D-Glu; H = D-CSA; I = D-CA; J = D-HCSA; K = D-HCA and L = D-SSC.

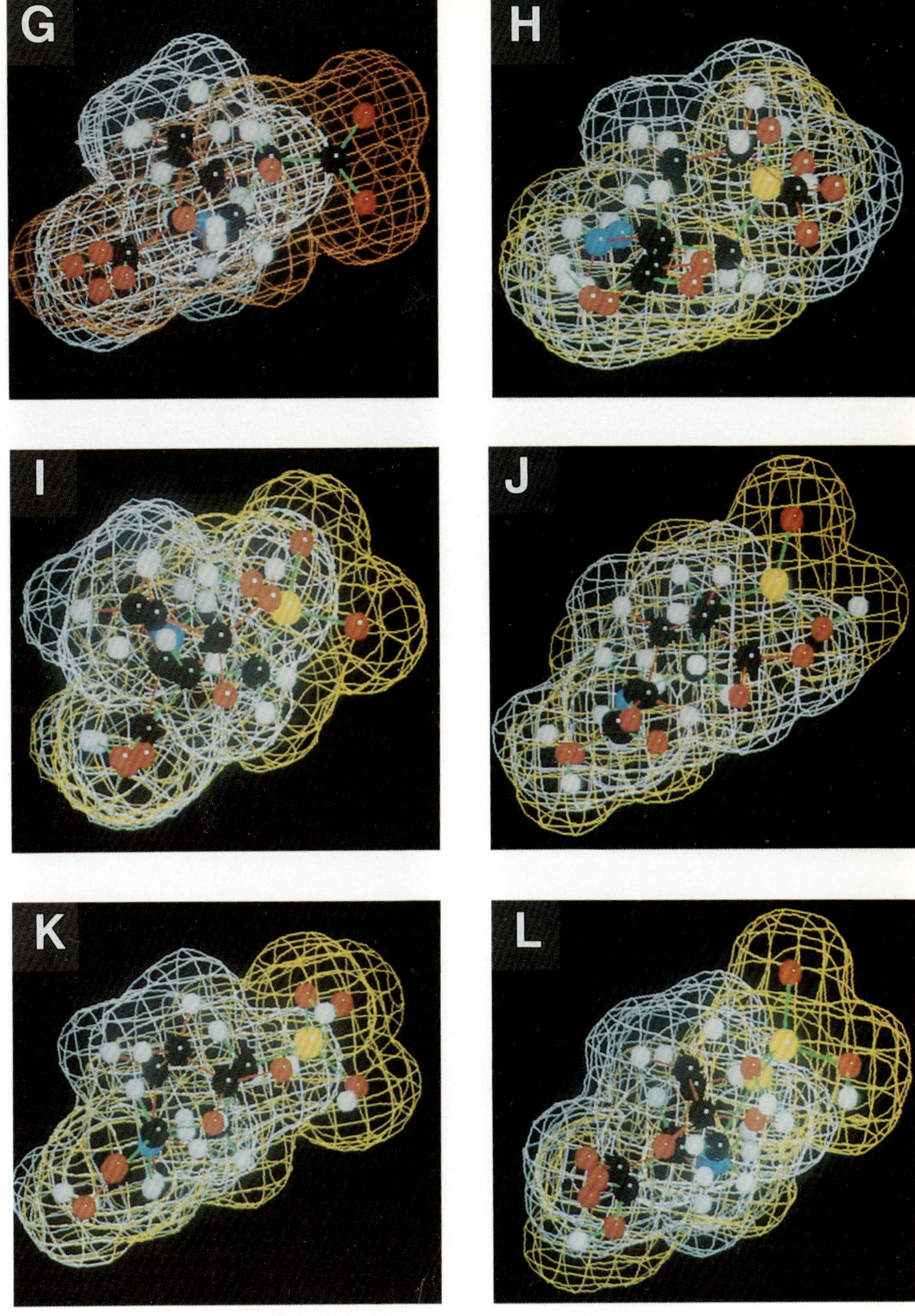
G
H
I
J
K
L

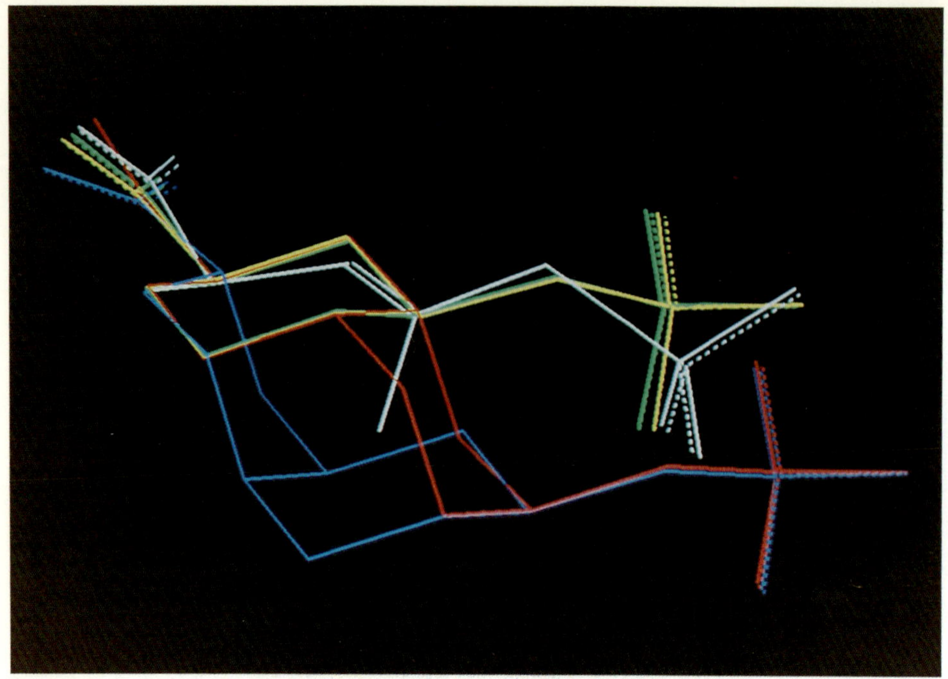

Fig: 8.5 — Overlap of 2*R*,4*S*-CGS 19755 (in green), 3*S*,4a*R*,6*S*,8a*R*-LY235959 (in blue), 3*R*,4a*S*,6*S*,8a*S*-LY288534 (in red), 2*R*-AP5 (in yellow), and 2*R*-CGP 40116 (in white). To provide some perspective in this two-dimensional drawing, the orientation of this overlap places the carboxylic acid and phosphonic acid groups behind the plane of the paper.

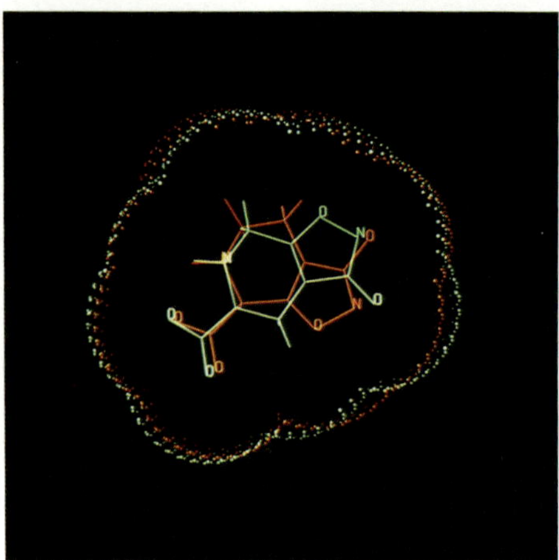

Fig: 10.6 — Structural similarities of 7-HPCA (red) and 5-HPCA (green) as shown by superposition of their solvent-accessible surfaces.

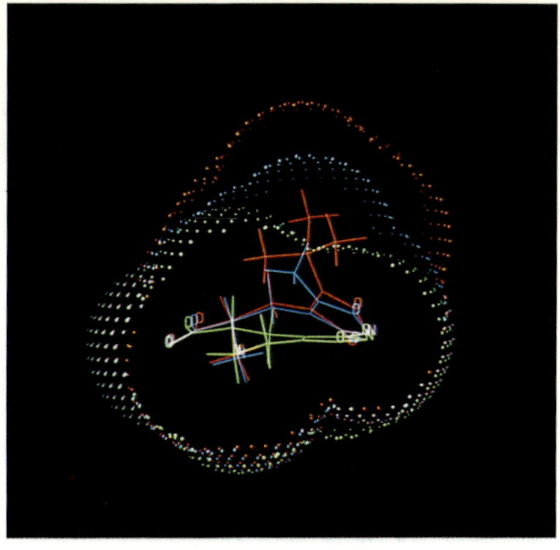

Fig: 10.9 — The solvent-accessible surfaces of ATPA (red), AMPA (blue) and 5-HPCA (green).

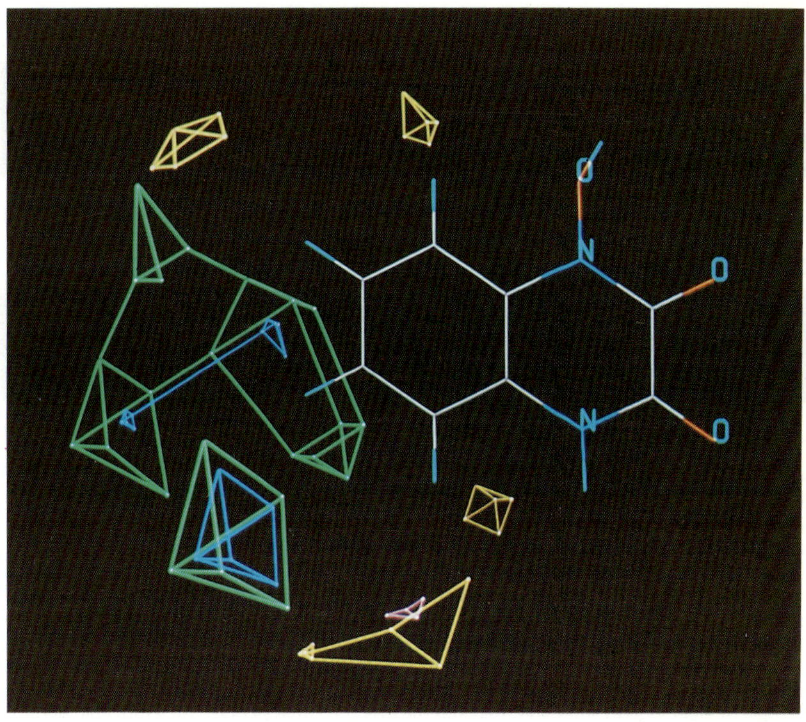

Fig: 11.13 — Steric 3D coefficient map.

7

Sulphur-containing excitatory amino acids

Roger Griffiths
Department of Biochemistry and Microbiology, University of St Andrews, North Street, St Andrews, Fife KY16 9AL, UK
Steven P. Butcher and **Henry J. Olverman**
Department of Pharmacology, University of Edinburgh, 1 George Square, Edinburgh, EH8 9JZ, UK

ABSTRACT

Naturally occurring acidic sulphur-containing amino acids (SAAs), namely L-cysteine sulphinate, L-cysteate, L-homocysteine sulphinate, L-homocysteate and *S*-sulpho-L-cysteine, structural analogues of L-glutamate and L-aspartate, may fulfil a transmitter or neuromodulatory role in the central nervous system by mediating transmission at certain excitatory synapses. In support of this proposal, a wealth of functional evidence is reviewed which demonstrates that (1) L- and D-enantiomers of SAAs are neuronal excitants eliciting responses in electrophysiological experiments that mimic those of the endogenous acidic amino acid transmitter L-glutamate and which can be attenuated by antagonists selective for the various EAA receptor subtypes, (2) SAAs are substrates for a plasma membrane transport system located in neurons and glia (3) certain SAAs are synthesized and may be heterogeneously distributed in brain tissue, (4) K^+-induced depolarization elicits a Ca^{2+}-dependent release of SAAs from various brain areas, (5) in membrane binding studies SAAs exhibit a range of affinities as inhibitors of radioligand binding to various populations of EAA sites and (6) SAAs evoke a Ca^{2+}-dependent and -independent release of excitatory and inhibitory transmitters from cultured neurons. The strength of this evidence is evaluated in the knowledge that the presence, cellular location and release of SAAs, in sufficient quantities required to elicit a physiological response, still remain to be unequivocally established.

INTRODUCTION

Naturally occurring amino acids other than L-glutamate (L-Glu) and L-aspartate (L-Asp) may play a role in excitatory transmission in the central nervous system (CNS). It has been shown in a number of studies that the acidic sulphur-containing amino acid (SAA) analogues of L-Asp (L-cysteine sulphinate (L-CSA), and L-cysteate (L-CA)) and of L-Glu (L-homocysteine sulphinate (L-HCSA) and L-homocysteate (L-HCA)) closely mimic the range of neurochemical actions of L-Glu, the major fast-acting neurotransmitter at excitatory synapses in the mammalian CNS. The thio ester *S*-sulpho-L-cysteine (L-SSC) also exhibits similar excitatory actions (see Fig. 1 for structures). Certain SAAs, particularly, L-CSA (Recasens *et al.* 1982, Griffiths 1990) and L-HCA (Cuenod *et al.* 1991, Lehmann *et al.* 1988), are regarded as strong candidates for neurotransmitter status. The aim of this chapter is to present the current perception of SAAs as transmitter candidates. Results of some recent functional studies from our respective laboratories will be presented.

BIOSYNTHESIS AND METABOLISM OF NEUROACTIVE SAAs

The biosynthetic route of neuroactive SAAs in mammalian brain originates from catabolism of the essential amino acid, L-methionine (Fig. 2). In a series of enzymatic reactions, L-methionine is demethylated to L-homocysteine, which lies at a metabolic branchpoint. In humans about half of the L-homocysteine formed is directed through trans-sulphuration to yield L-cysteine and other intermediates, the remainder being remethylated to L-methionine. Although L-methionine and L-cysteine are metabolized by a variety of reactions and pathways (not all present in brain) to at least two dozen intermediates and products, only pathways of immediate relevance to this review will be discussed. A number of reviews on mammalian sulphur amino acid metabolism are recommended for further reference (e.g. Huxtable 1986, 1989, Griffith 1987).

In brain, L-CSA is formed following oxidation of L-cysteine by the enzyme cysteine dioxygenase. This reaction is believed to be the major pathway of cysteine catabolism in mammals. L-CSA undergoes further rapid metabolism either by decarboxylation to hypotaurine or transamination to β-sulphinylpyruvate. The metabolic partitioning between transamination and decarboxylation shows considerable species variation. In the rat, the specific activity of CSA transaminase is much greater than that of CSA decarboxylase in cerebral cortex and striatum, suggesting that the pathway through β-sulphinylpyruvate may represent the major route of L-CSA metabolism at least in rat CNS. There is some doubt as to whether L-CA represents a true enzymatically formed metabolite of L-CSA. It has been suggested that L-CA is present in the diet as a result of non-enzymatic oxidation of L-cysteine and the actions of enteric bacteria. However, some evidence to support the enzymatic oxidation of L-CSA to L-CA by brain tissue is provided by the demonstration that [^{35}S]methionine is converted to [^{35}S]L-CA (Peck and Awapara 1967). β-Sulphinylpyruvate spontaneously desulphinates to sulphite and pyruvate. The sulphite subsequently interacts with β-mercaptopyruvate (formed by transamination of L-cysteine) to produce thiosulphate, which undergoes further reaction with metabolic intermediates in forming L-SSC. The biosynthetic route of L-HCSA and L-HCA from L-homocysteine remains obscure and can only be assumed by analogy to the metabolism of L-cysteine.

It has been recognized that SAA metabolism in brain may be ultimately involved in the regulation of excitability by excitatory and inhibitory amino acids (Huxtable 1986). These amino acids exist in a mutually close metabolic relationship in that

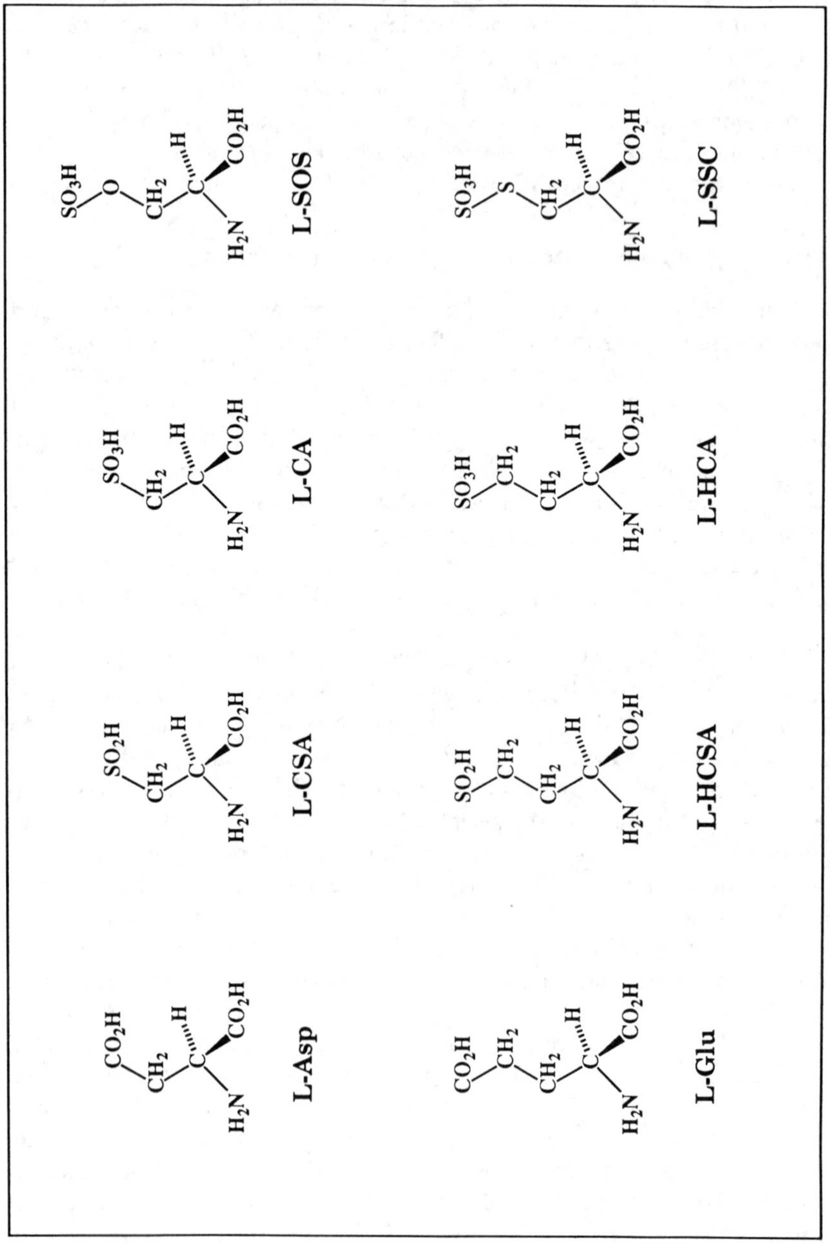

L-Asp **L-CSA** **L-CA** **L-SOS**

L-Glu **L-HCSA** **L-HCA** **L-SSC**

Fig. 1 — Structures of sulphur-containing analogues of L-glutamate and L-aspartate. Abbreviations: L-Asp, L-aspartic acid; L-CSA, L-cysteine sulphinic acid; L-CA, L-cysteic acid; L-Glu, L-glutamic acid; L-HCSA, L-homocysteine sulphinic acid; L-HCA, L-homocysteic acid; L-SSC, *S*-sulpho-L-cysteine; L-SOS, L-serine-*O*-sulphate.

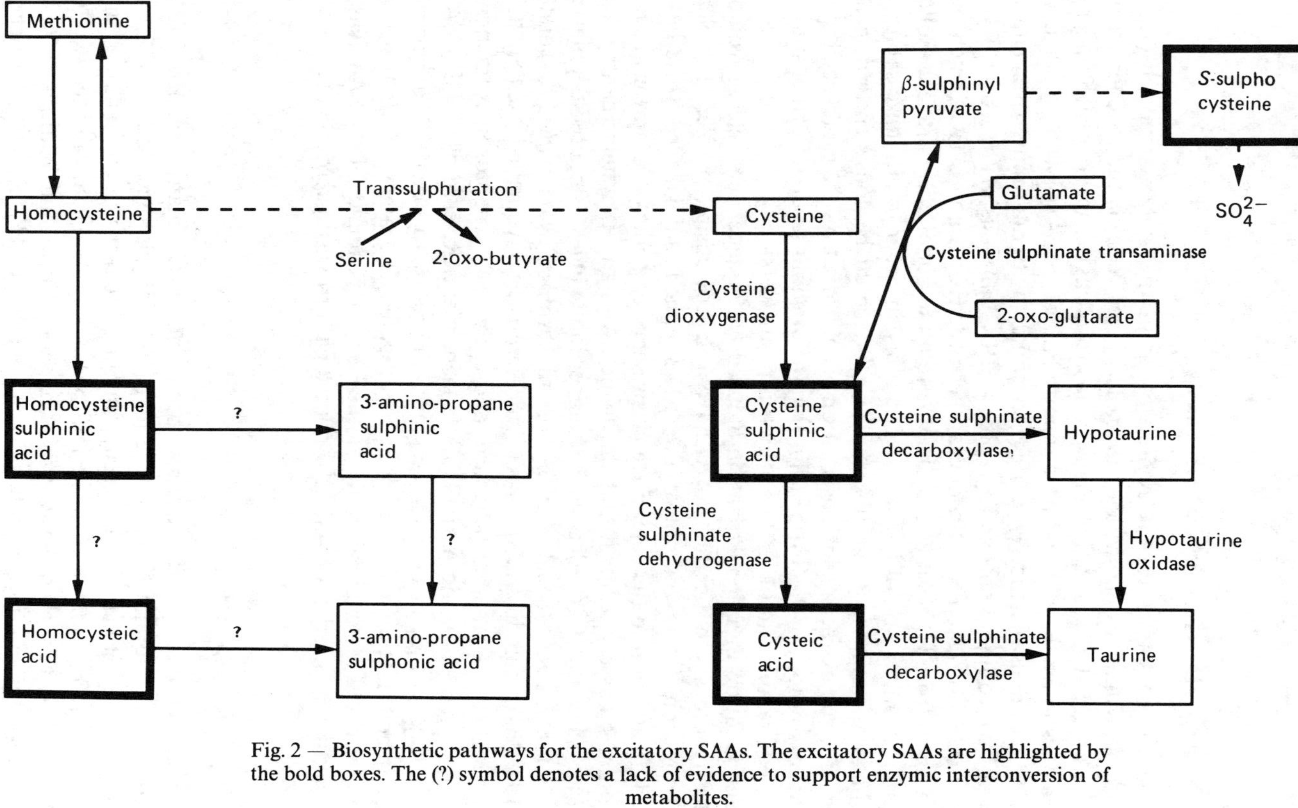

Fig. 2 — Biosynthetic pathways for the excitatory SAAs. The excitatory SAAs are highlighted by the bold boxes. The (?) symbol denotes a lack of evidence to support enzymic interconversion of metabolites.

decarboxylation of the excitatory SAAs yields those that exhibit inhibitory properties. Since the concentration ratios of all these intermediates can be markedly altered by slight modification of enzyme activity, it is conceivable that they may represent major determinants of brain excitability.

PRESENCE AND RELEASE OF SAAs

The presence of SAAs in various regions of rat brain has been reported (Do *et al.* 1986a, 1986b, Kilpatrick and Mozley 1986). However, recently the endogenous nature of L-HCSA and L-HCA, and thus their proposed role as excitatory neurotransmitters, has been questioned. First, it has been reported that L-HCA is localized in glia rather than nerve terminals (Cuenod *et al.* 1990). Secondly, with major improvements to the high-performance liquid chromatographic (HPLC) analysis of the excitatory SAAs Waller *et al.* (1991) claim that the appearance of L-HCSA and L-HCA in chromatograms derived from CNS tissue may be due to a methodological artefact. Following appropriate modification of the assay conditions, in order to minimize spontaneous oxidation of the sulphinate groups, only L-CSA could be consistently detected by analysis of tissue extracts. This observation is supported by the established metabolism and enzymology of the SAAs outlined in the previous section. Further studies are, nevertheless, required in order to clarify this controversy.

Demonstration of a depolarization-induced Ca^{2+}-dependent release would support the hypothesis that certain SAAs have a neurotransmitter role. The release of both endogenous SAAs and of exogeneously supplied radiolabelled SAAs has been reported. In addition to the demonstrated presence of acidic SAAs in various regions of rat brain it has been shown that the acidic SAAs and L-cysteine can be released from brain slices in a Ca^{2+}-dependent manner following K^+-induced depolarization (Do *et al.* 1986a, 1986b, Keller *et al.* 1989). Furthermore, Iwata *et al.* (1982) and Baba *et al.* (1983) have demonstrated the K^+- and veratridine-evoked release of [^{14}C]CSA from pre-loaded rat brain cortical slices and crude (P2) synaptosomal fractions, the release in both cases being partially Ca^{2+}-dependent. Release of [^{3}H]L-CSA from pre-loaded slices of various rat brain regions in response to either K^+- or veratridine-induced depolarization has also been investigated (Recasens *et al.* 1984). These workers showed a regional variability in the Ca^{2+}-dependence of L-CSA release and, in general, the depolarization-induced release pattern was heterogeneous.

PHARMACOLOGY OF SAAs

The excitatory actions of SAAs have been known for many years, L-CA being included with L-Glu and L-Asp in the first electrophysiological descriptions of the excitatory actions of acidic amino acids on single neurons in the mammalian CNS (Curtis *et al.* 1960). Soon after, the actions of L-CSA, DL-HCSA and the two optical isomers of HCA were reported (Curtis *et al.* 1961, Curtis and Watkins 1960, 1963). Using the lipophilic tetraphenylphosphonium cation ([^{3}H]TPP^+) as a biochemical correlate of electrophysiology to monitor membrane potential changes, it can also be shown that SAAs depolarize the plasma membrane of neuronal cells in culture (Table 1). This is seen as a reduction in the cellular content of [^{3}H]TPP^+ as the cell interior

Table 1 — Effect of EAAs and elevated K^+ on [^{3}H]TPP^+ accumulation by cultured neuronal cells

Treatment	Intracellular content of [^{3}H]TPP^+ (cpm × 10^{-3}/100 μg protein)	
	Cortex	Granule cells
K^+ (5 mM; control)	25.2 ± 2.6	41.4 ± 2.1
K^+ (25 mM)	15.1 ± 1.4**	29.9 ± 1.9**
K^+ (50 mM)	11.2 ± 0.4**	19.1 ± 1.3**
L-Glu (50 μM)	14.8 ± 1.9*	29.5 ± 3.8**
L-CSA (50 μM)	16.7 ± 1.4*	31.1 ± 1.5**
L-CA (50 μM)	19.3 ± 0.5*	37.3 ± 2.5
L-HCSA (50 μM)	19.0 ± 1.3*	34.1 ± 3.8*
L-HCA (50 μM)	19.4 ± 0.9*	37.0 ± 1.9*
L-SSC (50 μM)	17.9 ± 2.1*	34.6 ± 2.6*

Primary cultures of cortical neurons and cerebellar granule cells were incubated with [^{3}H]TPP^+ (to attain equilibrium) in the presence of test compounds at the concentrations indicated. Analysis of variance (ANOVA) was performed on overall data, then individual comparisons with control were made according to Dunnett (1964). Values taken from Dunlop *et al.* (1990, 1991b). Data represent the mean (±SEM) of two experiments in which assays were carried out in quadruplicate.
*$p < 0.05$, **$p < 0.01$ versus control.

becomes more positive (by depolarization) and hence acting as a barrier to diffusion of the cation.

The selectivity of the SAAs for the various EAA receptor subtypes is, by and large, difficult to analyse in electrophysiological experiments. Regional variations exist in the profile of receptor activity, which are most pronounced in the case of mixed agonists that interact with both *N*-methyl-D-aspartate (NMDA) and non-NMDA receptors. Electrophysiological studies in the cat and rat spinal cord and cerebellum show that L- and D-enantiomers of HCA and HCSA display a mixed NMDA/non-NMDA receptor activity (see Davies and Stanley 1988), the magnitude of the NMDA receptor-mediated response varying with the enantiomer and brain region. As excitants on the isolated hemisected spinal cord of immature rat the D- and L-isomers of HCSA and HCA, and L-SSC were considerably more potent when compared with L-Glu. Responses to D-CA, L-HCA, L-SSC and D-HCSA were inhibited by 80% or more by 2-amino-5-phosphonopentanoate (AP5) at concentrations sufficient to abolish responses to NMDA, whereas other SAAs were less sensitive or relatively resistant to inhibition (Mewett *et al.* 1983). The variations between brain areas presumably reflect differences in the distribution and ratio of NMDA and non-NMDA receptor subtypes.

Other reports based on electrophysiological experiments show that L-HCA, in particular, acts predominantly as an NMDA receptor agonist in striatum (Do *et al.* 1986b), cortex (Knopfel *et al.* 1987, Thomson 1989), hippocampus (Patneau and Mayer 1990), dorsolateral geniculate nucleus (Jones and Sillito 1989) and spinal cord (Mayer and Westbrook 1985). However, in the Purkinje cells of cerebellum L-HCA elicits excitation which is fully antagonized by the non-NMDA receptor antagonist 6-cyano-7-nitroquinoxaline-2,3-dione (CNQX) (Knopfel *et al.* 1989). Moreover, L-HCA stimulation of the inferior olive-climbing fibre system could also be abolished by CNQX but was unaffected in the presence of AP5 (Audinat *et al.* 1989). Where electrophysiology has been applied to the study of L-CSA (i.e. in striatum) it has been shown to act as a preferential non-NMDA receptor agonist (Herrling and Turski 1985).

The composite electrophysiological responses observed to SAAs are presumably a result of a balance between the proportion of different types of EAA receptors activated in the area of drug application and the relative affinities of the SAAs for those receptors. In addition, the concentrations at which SAA responses are elicited will be, in part, modulated by the rate at which they are transported by, for example, the sodium-dependent dicarboxylic amino acid uptake mechanism. In an attempt to obtain information about the interaction of SAAs at individual EAA receptors, the affinities of the compounds have been systematically examined in membrane binding studies using a variety of radioligands. These studies complement information gained from electrophysiological work and provide valuable data in some cases for determining the receptor selectivity of the SAAs. Indeed the SAAs, when tested in conjunction with a large number of other compounds, have had a key role in studies designed to determine the structural features required for agonist and antagonist binding to NMDA receptors and thus the topology of the transmitter recognition site (Olverman *et al.* 1988b, Olverman and Watkins 1989).

Radiolabelled L-CSA has been shown to bind saturably to whole brain membranes, binding being potently inhibited by L-Glu and L-CSA but much less effectively by DL-CA and DL-HCA (Recasens *et al* 1982, 1983). Most studies, however, have used ligands directed at defined populations of sites although the exact physiological relevance of some of the sites is not entirely clear. Thus tritiated 3-(2-carboxypiperazin-4-yl)propyl-1-phosphonate ($[^3H]$CPP), $[^3H]$D-AP5 and the NMDA-sensitive component of $[^3H]$L-Glu binding to brain membranes have been used to determine the affinity of SAAs for NMDA receptors (Olverman *et al.* 1984, 1988a, 1988b, Murphy *et al.* 1987, Monahan and Michel 1987, Pullan *et al.* 1987). Data are also available for tritiated 2-amino-3-(3-hydroxy-5-methylisoxazol-4-yl)propanoate ($[^3H]$AMPA), tritiated 2-amino-4-phosphonobutanoate ($[^3H]$AP4), $[^3H]$kainate and sodium-dependent and independent $[^3H]$L-Glu binding (Mewett *et al.* 1983, Murphy and Williams 1987, Pullan *et al.* 1987). The affinity of the D- and L-isomers of the SAAs for sites labelled in rat brain membranes by $[^3H]$D-AP5, $[^3H]$AMPA and $[^3H]$L-Glu (in the presence of 100 mM NaCl) are shown in Table 2.

For the NMDA receptor, SAA analogues of glutamate have higher affinity than their corresponding aspartate analogues and, with the exception of SSC, they show little enantiomeric preference. Apart from the D-Glu series of compounds, the SAAs have lower affinity than their ω-carboxylate analogues. The rank order for effectiveness of the ω-acidic group for L- and D- aspartate-length compounds is CO_2H, $SO_2H > SO_3H$, whereas glutamate-length compounds show an order for the L-series of $CO_2H > SO_2H$, SO_3H and for the D-series of SO_2H, $SO_3H > CO_2H$. L-SSC and L-SOS, analogues of L-HCA, where carbon in the inter-acidic chain has been replaced by either sulphur or oxygen, retain high or moderate activity, respectively. In fact,

Table 2 — K_i values for SAA inhibition of [^{3}H]D-AP5, [^{3}H]AMPA and sodium-dependent [^{3}H]L-Glu binding to rat brain membranes

Amino acid	K_i (μM)		
	[^{3}H]D-AP5	[^{3}H]AMPA	[^{3}H]L-Glu plus 100 mM NaCl
L-Asp	11	210	5.1
D-Asp	10	225	3.8
L-CSA	13	30	4.2
D-CSA	23	74	6.3
L-CA	120	23	2.1
D-CA	170	> 1 000	20
L-Glu	0.9	0.18	3.8
D-Glu	49	17	100
L-HCSA	10	4.9	41
D-HCSA	6.3	12	1 000
L-HCA	3.9	13	300
D-HCA	11	3.6	> 1 000
L-SSC	2.1	0.34	200
D-SSC	86	6.1	> 1 000
L-SOS	17	5.4	7.1

[^{3}H]D-AP5 and sodium-dependent [^{3}H]L-Glu binding are taken from Olverman *et al.* (1988b) and Mewett *et al.* (1983), respectively. For the [^{3}H]AMPA assay, crude synaptosomal membranes were resuspended in 30 mM Tris-HCl containing 2.5 mM $CaCl_2$, pH 7.1. Samples were incubated with 0.5 nM [^{3}H]AMPA for 60 minutes at 4°C, then bound and free ligand separated by rapid filtration through GF/B filters. Each value is the mean of three separate determinations (H. J. Sherriffs, H. J. Olverman and S. P. Butcher, unpublished data).

among short-chain agonists only L-Glu has higher affinity than L-SSC for NMDA receptors.

As with [^{3}H]D-AP5, glutamate-length SAAs have reasonably high affinity for [^{3}H]AMPA binding sites and are more active than their corresponding aspartate-length compounds; and neither Glu nor Asp analogues show a wide enantiomeric difference except perhaps SSC (see chapter 10). Indeed, Asp and its analogues show only moderate to low affinity for inhibition of [^{3}H]AMPA binding. Again, L-Glu and L-SSC show highest affinity. The results for [^{3}H]AMPA binding in Table 2 were obtained without thiocyanate ions being present during the incubations; however, a very similar pharmacological profile is seen for the SAAs when binding is carried out in the presence of 100 mM sodium thiocyanate. In contrast, most of the SAAs have low affinity for [^{3}H]kainate binding sites except for L-CA (22.9 μM) L-HCSA (32.6 μM) and L-SSC (7.7 μM), but even these compounds are only moderately active compared to L-Glu (0.343 μM) (Murphy and Williams 1987).

[^{3}H]L-Glu used as a ligand in the presence of 100 mM NaCl can be regarded as representing binding to the L-Glu transporter. The two optical isomers of aspartate and their SAA analogues have near-equal affinity to L-Glu for this site. Compounds of glutamate-length, apart from L-Glu itself and L-SOS, have only low affinity for the L-Glu transporter (Table 2). Thus the potent electrophysiological effects of these SAAs in comparison with L-Asp and L-Glu may be due in part to their lack of affinity for the transporter. Two compounds in particular, L-HCA and L-SSC, have high affinity for NMDA receptors, but low affinity for the transporter and are powerful neuronal excitants. From these results, however, it is not possible to distinguish between inhibition of uptake by SAAs and whether they are actually substrates for the transport mechanism. The substrate specificity of the L-Glu transporter with regard to the SAAs has been characterized in three cell types and the results are presented in the next section.

INACTIVATION OF SAAs

The presence of an efficient mechanism for the rapid termination of transmitter action following activation of appropriate receptors is regarded as an essential criterion of transmitter status. It is generally accepted that inactivation of amino acid transmitters occurs by their removal from the synaptic cleft via high-affinity carrier systems (Curtis and Johnston 1974). Relatively few studies have been undertaken in which SAA transport per se has been measured and characterized kinetically. It has been shown that a high-affinity uptake system ($K_m = 6\text{–}40\ \mu M$) is present: in synaptosomes, for L-CSA (Recasens *et al.* 1982, Iwata *et al.* 1982, Grieve *et al.* 1990, 1991) and L-CA (Wilson and Pastuszko 1986); and, in neuronal and glial primary cultures and cell lines for L-CSA (Abele *et al.* 1983); while, in brain slices, L-HCA is transported predominantly by a low-affinity ($K_m > 3$ mM) uptake system, although a minor contributory high-affinity system may be present (Cox *et al.* 1977). It is now well-established, however, that physiological inactivation of amino acid transmitter action is mediated by carrier systems located in the plasma membrane of both neurons and astrocytes (Drejer *et al.* 1982). The presence of distinct carriers for the excitatory SAAs in neurons and astrocytes has only recently been established (Grieve *et al.* 1991). These studies have been recently extended and show the uptake of L-CSA, L-CA, L-HCSA and L-HCA in astrocytes cultured from mouse prefrontal cortex, in neurons cultured from mouse cerebral cortex, in granule cells cultured from mouse cerebellum and also in isolated synaptosome fractions prepared from rat cortex. Each amino acid acts as a transport substrate for a distinct carrier located in both neurons and astrocytes.

Concentration dependence experiments show that for each SAA, uptake into neurons, astrocytes and synaptosomes is biphasic, composed of a saturable carrier-mediated system plus a non-saturable passive influx component. A representative plot showing L-HCSA uptake in astrocytes is shown in Fig. 3. The kinetic parameters, K_m and V_{max}, of SAA transport (corrected for passive influx) are shown in Table 3. Astrocytes, neurons and synaptosomes exhibit a high-affinity uptake for L-CSA and L-CA, with K_m values ranging from 14 μM to 100 μM, and a low-affinity uptake for L-HCSA and L-HCA, with K_m values ranging from 225 μM to 1550 μM.

The high affinity observed for neuronal (including synaptosomal) and astrocytic uptake of L-CSA and L-CA is consistent with the presence of a mechanism for rapid termination of their synaptic activity. The less potent depolarizing actions of L-CSA

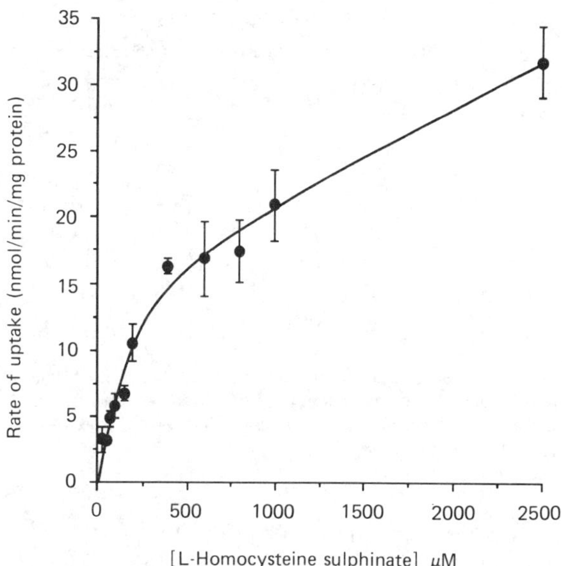

Fig. 3 — Transport of L-HCSA by cultured astrocytes. Primary cultures of mouse astrocytes derived from prefrontal cortex were incubated at 37°C in Hepes-buffered saline with varying concentrations (0–2500 μM) of L-HCSA. The rates of uptake presented in the figure were measured by HPLC. Uptake of L-HCSA (and all the other SAAs) was biphasic and could be most appropriately separated into two components comprising saturable carrier-mediated uptake and non-saturable passive influx:

$$V = V_{max}[1/(1 + K_m/S)] + kS$$

where V and V_{max} are the initial and maximum velocities, respectively, and K_m, the Michaelis constant for saturable uptake; k is the rate constant for the non-saturable influx (Neame and Richards 1972). Values for kinetic constants are given in Table 3.

and L-CA, compared to those of L-HCSA and L-HCA (Mewett *et al.* 1983), can, however, now be appreciated by the demonstrated presence of far more efficient re-uptake systems for the former compounds in both neurons and astrocytes.

The magnitude of the K_m values for the two L-Glu analogues L-HCSA and L-HCA indicate that they are transported solely by a low-affinity system in granule cells, cortical neurons, cortical astrocytes and synaptosomes (Table 3). This may be particularly significant when compared to the low levels of release of endogenous L-HCSA and L-HCA (Do *et al.* 1986a).

USE OF SAAs TO PROBE THE STRUCTURE OF THE GLUTAMATE TRANSPORTER

It has been recognized for some time that the dicarboxylic acid carrier synonymous with L-Glu transport is not specific for L-Glu, but is shared by L- and D-Asp, which are transported with essentially identical K_m and V_{max} values (Drejer *et al.* 1983). Furthermore, the results of inhibition kinetic studies using synaptosome fractions suggest the operation of a common carrier system for the high-affinity transport of L-Glu, D- and L-Asp, L-CSA and L-CA (Wilson and Pastuszko 1986, Erecinska and

Table 3 — Kinetic parameters for SAA uptake in synaptosomes and in primary cultures of neurons and astrocytes

Amino acid	K_m (μM)				V_{max} (nmol min^{-1} mg protein^{-1})			
	Gn	Cx	Syn	Ast	Gn	Cx	Syn	Ast
L-Glu	42[a]	43[a]	14[b]	67[a]	10.2[a]	5.9[a]	11.3[b]	14.9[a]
L-CSA	41	58	57	101	11.1	1.5	1.2	0.7
L-CA	41	14	23	88	25.2	5.4	3.6	27.3
L-HCSA	468	525	502	225	10.8	3.0	6.1	18.3
L-HCA	941	1 210	1 550	901	15.7	3.8	10.3	31.8

All uptake data were analysed by iterative curve-fitting to the equation in the legend of Fig. 3 by computer-assisted non-linear regression analysis using the ENZFITTER software program (Leatherbarrow 1987). Gn, cerebellar granule cells; Cx, cortical neurons; Syn, cortical synaptosomes; Ast, cortical astrocytes.
[a]Values taken from Drejer *et al.* (1982).
[b]Values taken from Debler and Lajtha (1987).

Troeger 1986). These studies were extended by Griffiths *et al.* (1989), who used the L- and D-enantiomers of CSA, CA, HCSA, HCA and SSC in an attempt to identify compounds which might be of use as selective markers for either the glia- or neuron-located L-Glu transporter. These results (Table 4) indicate that none of the acidic SAAs are useful as selective markers for a neuron- versus glia-located Glu transporter. However, comparison of Tables 2–4 indicates a strong correlation between the K_i

Table 4 — Inhibition constants for [^{3}H]D-Asp uptake in synaptosome fractions and cultured neurons and astrocytes in the presence of L- and D-enantiomers of SAAs

Inhibitor	K_{ei} (μM)		
	Gn	Ast	Syn
L-Asp	18[a]	82[a]	10
L-CSA	20	120	5
D-CSA	48	240	9
L-CA	12	47	8
D-CA	410	480	370
L-Glu	21[a]	75[a]	8
L-HCSA	960	2 730	608
D-HCSA	1 020	5 070	> 4 500
L-HCA	1 330	1 560	1 520
D-HCA	n.d.[b]	> 3 000	n.d.[c]
L-SSC	880	1 670	700
D-SSC	7 300	> 15 000	6 000

The inhibition constant (K_{ei}) values describe linear competitive inhibition in all cases. Kinetic characterization of uptake in the absence of SAAs gave K_m values for [^{3}H]D-Asp (used as a non-metabolizable analogue of L-Glu) of 21 μM, 6 μM and 84 μM, respectively, for cerebellar granule cells (Gn), cortical synaptosomes (Syn) and cortical astrocytes (Ast). Values taken from Griffiths *et al.* (1989).
[a]Values from Drejer *et al.* (1983).
[b]Not measurable at 5 mM.
[c]Not measurable at 10 mM.

values for inhibition of sodium-dependent [^{3}H]L-Glu binding, transport constants (K_m values) and inhibition constants (K_{ei} values) for L-CSA, L-CA and L-Glu, thus supporting previous claims that L-CSA and L-CA share a common transport system with L-Glu.

Complementary use of computer-assisted molecular modelling (Fig. 4) and *in vitro* kinetic studies (Tables 3 and 4) may be used to formulate an impression of the conformational requirements of the L-Glu carrier. The present study, in which three-dimensional volume maps have been included, provides an extension of the previous efforts (Schousboe *et al.* 1988) to characterize the uptake site. The results of the molecular modelling and kinetic studies, when taken together, indicate that the transporter (1) has high affinity for Asp and its ω-sulphinic and -sulphonic acid analogues of either L- or D-configuration, (2) recognizes L-Glu and its ω-sulphinic and -sulphonic acid homologues in the folded conformation, and (3) exhibits a stereospecificity for Glu and its analogues, in favour of the L-enantiomers. Thus, when used in combination with a more physiological (dynamic) system, molecular modelling allows some visual insight into the geometry of the uptake site and the requirements for high-affinity uptake.

SAA-EVOKED TRANSMITTER RELEASE

Only recently has there been a biochemical/pharmacological evaluation of the neurochemical actions of the excitatory SAAs. Dunlop *et al.* (1989a, 1989b, 1990, 1991a, 1991b), Minc-Golomb *et al.* (1989), Mount *et al.* (1990) and Weiss (1988) have shown that the SAAs evoke the release of established neurotransmitters in the absence of additional depolarizing agents. In detailed studies Dunlop *et al.* (1989a, 1989b, 1990, 1991b) applied a continuous superfusion system (Drejer *et al.* 1987) to study the SAA-evoked release of, respectively, [^{3}H]D-Asp from mouse cerebellar granule (glutamatergic) cells and [^{3}H]GABA from mouse cortical (GABAergic) neurons in primary culture (Fig. 5). All of the SAAs under study evoked a dose-dependent, saturable and Ca^{2+}-dependent release of [^{3}H]D-Asp and [^{3}H]GABA from, respectively, granule cells and cortical neurons. The stereospecificity of SAA-evoked release has also been investigated, and it has been shown that the D-enantiomers evoke considerably less release than equivalent concentrations of their L-enantiomers. A representative profile showing the SAA-evoked release of [^{3}H]D-Asp from granule cells is presented in Fig. 6.

The extent to which EAA receptor activation by the SAAs may be coupled to neurotransmitter release has also been studied. Dunlop *et al.* (1989a, 1990, 1991a, 1991b) demonstrated the EAA receptor-activated release of [^{3}H]D-Asp and [^{3}H]GABA from, respectively, cultured cerebellar granule cells and cortical neurons, while Mount *et al.* (1990) showed that the SAAs evoke an EAA receptor-mediated release of [^{3}H]dopamine from cultured mesencephalic cells. In both studies, the sensitivity of SAA-evoked release to selective EAA receptor antagonists suggests that NMDA and non-NMDA ionotropic receptors are involved.

The SAA-evoked release of [^{3}H]D-Asp from granule cells, although blocked by NMDA receptor antagonists, is insensitive to inclusion of 0.6 mM Mg^{2+} in the superfusion medium, whereas under similar ionic conditions the SAA-evoked release of [^{3}H]GABA from cortical neurons is significantly reduced (Dunlop *et al.* 1991b).

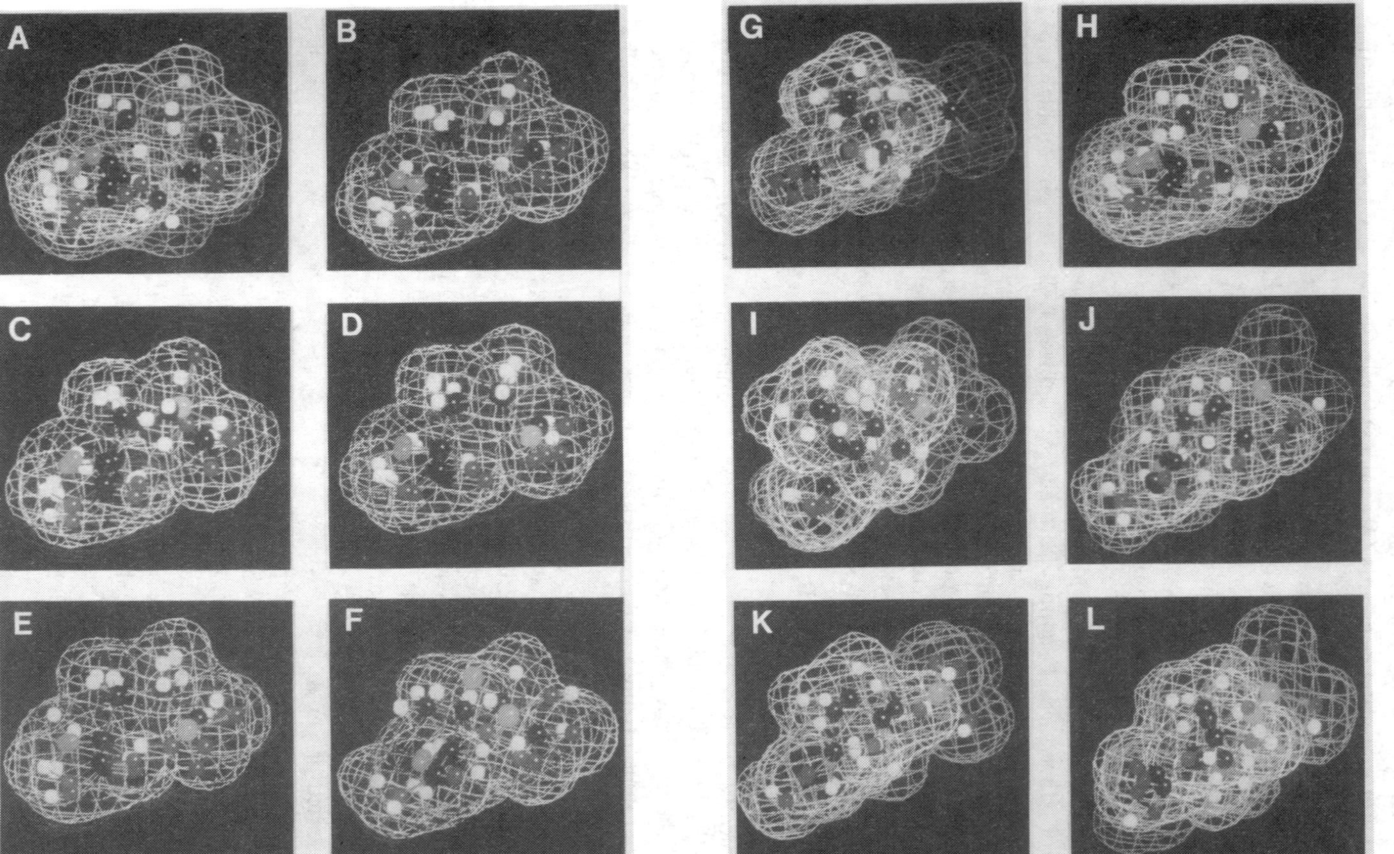

Fig. 4 — (**See colour section.**) Volume maps of excitatory amino acids. The 3D contour maps were constructed and superimposed by computer-assisted use of the Chem-X software program as described in Griffiths *et al.* (1989). The plates (A–L) show the volume maps of D-Asp (yellow), D-Glu (orange) and the neuroexcitatory SAAs (L-enantiomers/red; D-enantiomers/yellow) superimposed over that of L-Glu (white). *Key* (with reference to inhibitors): A = D-Asp; B = L-CSA; C = L-CA; D = L-HCSA; E = L-HCA; F = L-SSC; G = D-Glu; H = D-CSA; I = D-CA; J = D-HCSA; K = D-HCA and L = D-SSC.

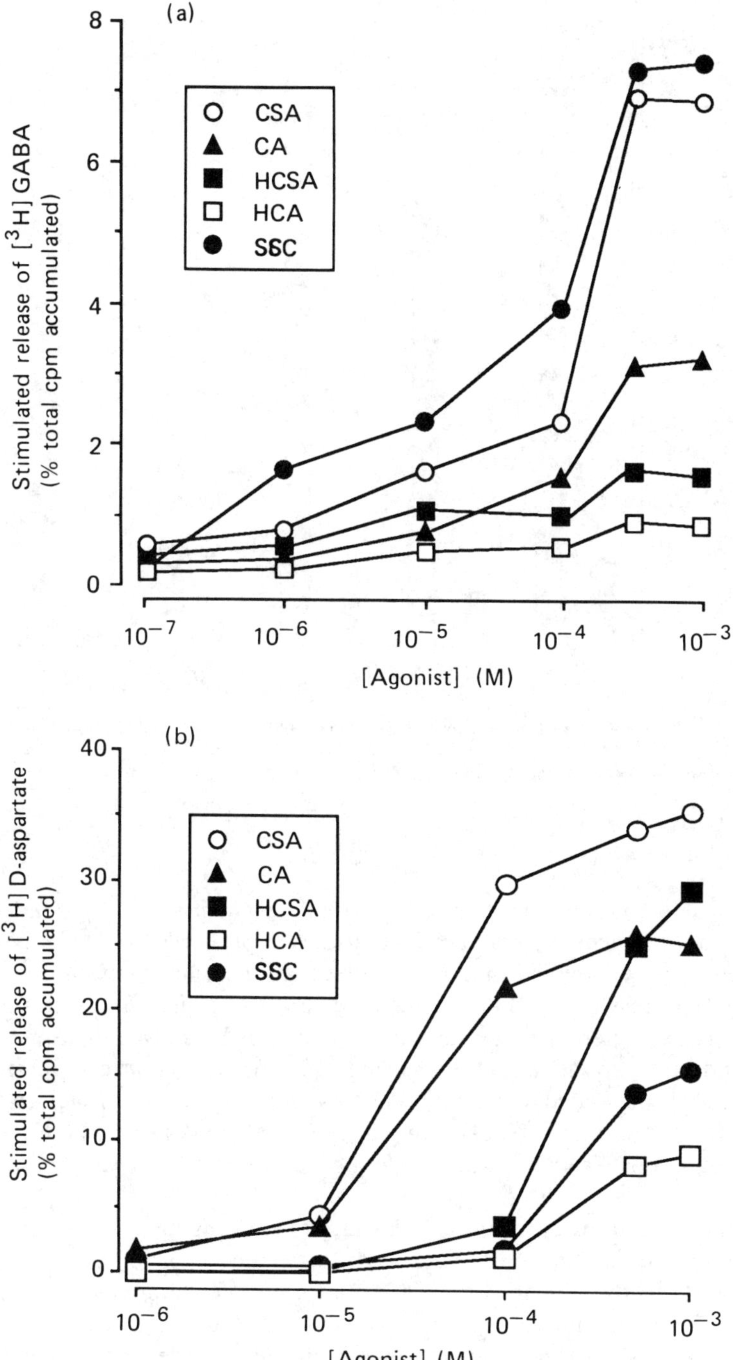

Fig. 5 — Dose-response profiles of SAA-evoked release of (a) [^{3}H]GABA from cerebral cortex neurons and (b) [^{3}H]D-aspartate from cerebellar granule cells. Cultured neurons were loaded with radioactive amino acid and stimulated with increasing concentrations of the various SAAs. Evoked release, corrected for basal (spontaneous) efflux, is expressed as a percentage of the total radioactivity accumulated by the cells during loading. Data represent mean values ($n = 4$) with an error no greater than $\pm$ 10%.

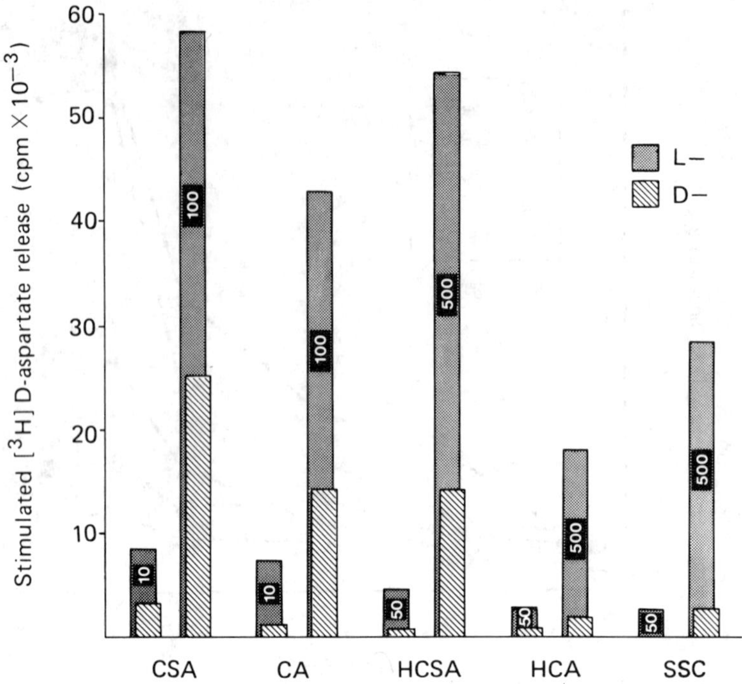

Fig. 6 — Stereospecificity of SAA-evoked [^{3}H]D-Asp release from cultured cerebellar granule cells. In separate experiments, granule cells were stimulated with either the L- or D- enantiomers of the various SAAs at concentrations (μM) indicated within the bars. (From Dunlop *et al.* 1990.)

The nature of these results is in agreement with other reports (Drejer *et al.* 1986, 1987) in which the pharmacological properties of these receptors were shown to be distinctly different in both types of neurons. Thus both cerebral cortical neurons and cerebellar granule cells possess Glu receptors, which mediate a Glu-induced transmitter release. However, unlike cerebral cortex neurons which can be shown to express AMPA, kainate and Mg^{2+}-sensitive NMDA receptors, cerebellar granule cells have receptors which can (1) be activated appreciably only by endogenous agents such as L-Glu, L-Asp and the SAAs, (2) be blocked by selective NMDA and non-NMDA receptor antagonists, and (3) exhibit a similar extent of release in the absence and presence of Mg^{2+}.

In primary cultures of cortical (GABAergic) neurons and of cerebellar granule (glutamatergic) cells, the SAAs evoke a distinct component of amino acid release which persists in the absence of extracellular Ca^{2+} (Dunlop *et al.* 1991a, 1991b). Under Ca^{2+}-free conditions, and using agonist concentrations two to three times their EC_{50} values, it is possible to amplify the Ca^{2+}-independent component of release in order to facilitate further investigation. This component of SAA-evoked [^{3}H]GABA release in cultured cortical neurons is (1) markedly attenuated (Table 5) by 15 μM SKF 89976-A, a non-transportable inhibitor of the GABA carrier (Larsson *et al.* 1988), and (2) abolished (Table 6) when either choline ions replaced sodium ions

Table 5 — Effect of the GABA transport inhibitor SKF 89976-A on the Ca^{2+}-indpendent EAA-evoked release of [^{3}H]GABA from cultured cortical neurons

EAA	Stimulated [^{3}H]GABA release (cpm × 10^{-3}) EAA alone	EAA + SKF 89976-A
L-Glu	9.04	1.88*
L-CSA	4.95	1.32*
L-HCSA	6.84	0.36*
L-HCA	6.83	0.85*
L-SSC	13.67	4.58*

Primary cultures of cortical neurons were loaded with [^{3}H]GABA and stimulated under Ca^{2+}-free conditions with various EAAs (500 μM) in the absence or presence of 15 μM SKF 89976-A. Stimulated [^{3}H]GABA release is expressed as total release (cpm) per stimulation minus basal (spontaneous) release. Data represent the mean of two experiments (using 4–6 cultures) where the stimulated release varied by < 10%. ANOVA was performed on overall data for each compound, while individual comparisons ($p < 0.001$ for all stimuli) with the control (basal release) were made according to Dunnett (1964). Values taken from Dunlop *et al.* (1991b).
*$p < 0.001$ versus stimulated release of EAA alone.

Table 6 — Effect of Na^+ replacement and EAA receptor antagonism on the Ca^{2+}-independent EAA-evoked release of [^{3}H]GABA from cultured cortical neurons

EAA	Stimulated [^{3}H]GABA release (cpm × 10^{-3})		
	+ sodium	+ choline	+ DNQX
L-Glu	10.98 ± 0.28	1.7 ± 0.30	n.d.*
L-CSA	6.79 ± 0.62	n.d.*	n.d.*
L-CA	3.66 ± 0.55	0.1 ± 0.04	n.d.*
L-HCSA	3.82 ± 0.50	n.d.*	n.d.*
L-HCA	5.74 ± 0.18	n.d.*	n.d.*
L-SSC	5.86 ± 0.51	0.3 ± 0.16	n.d.*

Primary cultures of cortical neurons were loaded with [^{3}H]GABA as described by Dunlop *et al.* (1989a). Cortical neurons were superfused in physiological medium containing either Na^+ (+ sodium) or one in which Na^+ was replaced with choline ions (+ choline). In separate experiments, using the same batch of cultures, cells were superfused in an Na^+-containing physiological medium and stimulated with various EAAs (500 μM) in the absence and presence of 200 μM 6,7-dinitroquinoxaline-2,3-dione (DNQX). Stimulated release is expressed as in Table 5. Data represent the mean (± sem) of at least four experiments (using 4–6 cultures). Values taken from Dunlop *et al.* (1991b). n.d., no detectable release.
*$p < 0.001$ versus EAA-evoked release in sodium-containing medium.

in the superfusion medium or in the presence of EAA receptor antagonists (Dunlop *et al.* 1991a, 1991b). These results provide evidence to show that the Ca^{2+}-independent component of release arises from a receptor-mediated, depolarization-induced reversal of the GABA carrier. Other studies by Pin and Bockaert (1989) and Harris and Miller (1989) have also shown that where EAAs are coupled to GABA release the mechanism involves reversal of the high-affinity GABA transporter. Pin and Bockaert

(1989) showed that in cultured striatal neurons NMDA, QUIS or kainate evoked a release of endogenous GABA that is wholly Ca^{2+}-independent and arises due to reversal of the GABA uptake. By comparison, Harris and Miller (1989) demonstrated that [^{3}H]GABA release from hippocampal neurons in primary culture comprises both a Ca^{2+}-dependent (NMDA-stimulated) component and a Ca^{2+}-independent (L-Glu- and kainate-stimulated) component; the sensitivity of the latter component to nipecotic acid being evidence to suggest that the GABA carrier is involved in the mechanism of release. It appears that the mode(s) of EAA-evoked GABA release may vary in different brain regions since the results of Dunlop *et al.* (1991a, 1991b) and Drejer *et al.* (1987) provide data to show that L-Glu- and SAA-evoked [^{3}H]GABA release from cultured cortical neurons comprises a major Ca^{2+}-dependent (presumably exocytotic) component of release, and an additional, minor but distinct Ca^{2+}-independent component of release which arises from non-exocytotic, receptor-mediated depolarization-induced reversal of the GABA carrier. The Ca^{2+}-independent component of GABA release is currently of unknown functional importance. It has been proposed that such a mechanism may be involved in the regulation of glutamatergic synapses directed to GABAergic neurons (Pin and Bockaert 1989). It may be speculated that this type of regulatory mechanism could be important in preventing the excess glutamatergic excitation characteristic of many neurodegenerative disorders.

By adopting a similar experimental approach, it has been possible to show that the Ca^{2+}-independent component of SAA- and L-Glu-evoked [^{3}H]D-Asp release from cultured granule (glutamatergic) cells also arises from carrier reversal, but in this case reversal of the L-Glu transporter (Dunlop and Griffiths 1990, Dunlop 1991). In the case of glutamatergic neurons, the Ca^{2+}-independent component of L-Glu release may be most relevant under pathological conditions, especially under conditions of cerebral ischaemia or anoxia. The ensuing energy depletion characteristic of cerebral ischaemia or anoxia would, for example, result in abolition of crucial ATP-driven systems located in the synaptic plasma membrane, such as the Na^{+}/K^{+}- and the Ca^{2+}-ATPases. In particular, inhibition of the Na^{+}/K^{+} pump would result in collapse of the Na^{+} electrochemical gradient across the plasma membrane, thus removing the driving force for accumulation of L-Glu via the Na^{+}/L-Glu transporter. As a consequence of these events the L-Glu transporter would proceed to work in the efflux direction, leading to massive release of L-Glu into the extracellular environment.

NEUROLOGICAL DYSFUNCTION AND SAAs

The cytotoxic (Olney *et al.* 1971, Kim *et al.* 1987, Pullan *et al.* 1987) and epileptogenic (Johnston 1973, Turski 1989, Jurson and Freed 1990) actions of various acidic SAAs are well-established, their underlying mode of action being consistent with excessive stimulation of both NMDA and non-NMDA receptors. Moreover, certain SAAs have been implicated as causative agents in a number of neuropathologies, such as the neurodegenerative disorder sulphite oxidase deficiency (Olney *et al.* 1975) and homocysteine-induced epilepsy (Blennow *et al.* 1979, Hammond *et al.* 1980, Dewhurst *et al.* 1983). Several lines of evidence support a role for brain amino acid transmitters in mediating homocysteine-induced seizure activity (Pestana *et al.* 1971, Tunnicliff and Ngo 1977, Allen *et al.* 1986); while disturbed SAA metabolism resulting in an imbalance of L-cysteine/inorganic sulphate ratios may contribute towards the aetiology of other neurodegenerative disorders, such as motor neuron disease, Parkinson's disease and Alzheimer's disease (Heafield *et al.* 1990). At elevated tissue levels ($>200\ \mu$M) free L-cysteine is neurotoxic, particularly in the immature animal (Olney *et al.* 1971, 1990). The observed neurotoxicity of L-cysteine, which reproduces the neurodegenerative processes of L-Glu-induced brain damage, appears to be mediated primarily through NMDA receptors (Olney *et al.* 1990). An alternative mechanism of L-cysteine neurotoxicity may be a result of its ability to form an adduct with pyridoxal 5′-phosphate (Buell and Hansen 1960, Mackay 1962), thus depleting the cell of an essential co-enzyme that participates in many biosynthetic reactions involving excitatory and inhibitory amino acids.

CONCLUSIONS

The L-enantiomers of the SAA analogues of L-Asp and L-Glu would appear to fulfil the criteria for a neurotransmitter or neuromodulatory role in the mammalian CNS. Thus the necessary metabolic enzymes are present; moreover the SAAs are powerful neuronal excitants displaying a spectrum of effects on EAA receptors comparable with that of L-Glu but exerting a receptor-selective action in some neuronal pathways, e.g. L-HCA. Consistent with electrophysiological data SAAs show a range of affinities as inhibitors of binding to various populations of EAA sites. Although they have lower affinity than L-Glu for almost all sites examined, this would not preclude a neurotransmitter role for an agonist at a fast-transmitter receptor. In addition, the SAAs (1) are substrates for the acidic amino acid transporter in neurons and glia, (2) are heterogeneously distributed in brain, (3) can be released in a Ca^{2+}-dependent manner and in turn (4) can evoke Ca^{2+}-dependent and -independent release of excitatory and inhibitory transmitters. An important role for L-SSC in the neurodegenerative disorder sulphite oxidase deficiency would seem possible if it could be shown to be present in sufficient quantities in brain tissue, since it is a potent agonist, presumably a result of its high affinity for EAA receptors, but is transported very inefficiently by the membrane uptake mechanism in neurons and glia.

This evidence should, however, be placed in perspective, since on closer inspection certain aspects of their proposed transmitter role are not as convincing. For example, the potent excitatory actions of L-HCSA and L-HCA may, paradoxically, be a reflection of their low affinity as substrates for the plasma membrane transporter and it is thus unclear what mechanism would be primarily responsible for rapid termination of their excitatory actions. Since the receptor population in a particular neuronal pathway presumably defines the type of excitatory response, the selective NMDA receptor actions of L-HCA would not argue in favour of its role as a transmitter. A role for such an agonist serving as a co-transmitter or modulator acting preferentially on a subpopulation of transmitter receptors is, however, conceivable. In conclusion, the most crucial questions, nevertheless, relate to whether or not any of the SAAs (1) are present in brain tissue, (2) are located in neurons or glia and (3) can be released in sufficient quantities required to elicit a physiological response. Although measurements of gross tissue levels are low it is possible that high concentrations of SAAs exist in some synapses. Until satisfactory answers are obtained, none of the SAAs can join the list of CNS neurotransmitters and modulators.

ACKNOWLEDGEMENTS

We thank Miss Audrey Kerr (Edinburgh) for typing the manuscript and Mr F. Simmonite (St Andrews) for colour photography. Some of the experimental work described was supported by the Scottish Hospital Endowments Research Trust, The Mental Health Foundation, The Wellcome Trust and the CEC (BRIDGE programme contract no. CT-90-0183).

REFERENCES

Abele, A., Borg, J. and Mark, J. (1983) *Neurochemistry* **8** 889–902.
Allen, I. C., Grieve, A. and Griffiths, R. (1986) *J. Neurochem.* **46** 1582–1592.
Audinat, E., Knopfel, T. and Gahwiler, B. H. (1989) *Soc. Neurosci. Abstr.* **15** 612.
Baba, A., Okumura, S., Mizuo, H. and Iwata, H. (1983) *J. Neurochem.* **40** 280–284.
Blennow, G., Folbergrova, J., Nilsson, B. and Siesjo, B. (1979) *Brain Res.* **179** 129–146.
Buell, M. V. and Hansen, R. E. (1960) *J. Am. Chem. Soc.* **82** 6042–6049.
Cox, D. W. G., Headley, M. H. and Watkins, J. C. (1977) *J. Neurochem.* **29** 579–588.
Cuenod, M., Do, K. Q., Grandes, P., Morino, P. and Streit, P. (1990) *J. Histochem. Cytochem.* **38** 1713–1715.
Cuenod, M., Do, K. Q. and Streit, P. (1991) *Trends Pharmacol. Sci.* **11** 477–478.
Curtis, D. R. and Watkins, J. C. (1960) *J. Neurochem.* **6** 117–141.
Curtis, D. R. and Watkins, J. C. (1963) *J. Physiol.* **166** 1–14.
Curtis, D. R. and Johnston, G. A. R. (1974) *Ergeb. Physiol.* **69** 97–188.
Curtis, D. R., Phillis, J. W. and Watkins, J. C. (1960) *J. Physiol.* **150** 656–682.
Curtis, D. R., Phillis, J. W. and Watkins, J. C. (1961) *Br. J. Pharmacol.* **16** 262–283.
Davies, J. and Stanley, M. D. A. (1988) In: Lodge, D. (ed.) *Excitatory Amino Acids in Health and Disease.* Wiley, Chichester, pp. 47–62.
Debler, E. A. and Lajtha, A. (1987) *J. Neurochem.* **48** 1851–1856.
Dewhurst, I. C., Hagan, J. J., Morris, R. G. M. and Griffiths, R. (1983) *J. Neurochem.* **40** 752–757.
Do, K. Q., Mattenberger, M., Streit, P. and Cuenod, M. (1986a) *J. Neurochem.* **46** 779–786.
Do, K. Q., Herrling, P. L., Streit, P., Turski, W. A. and Cuenod, M. (1986b) *J. Neurosci.* **6** 2226–2234.
Drejer, J., Larsson, O. M. and Schousboe, A. (1982) *Exp. Brain Res.* **47** 259–269.
Drejer, J., Larsson, O. M. and Schousboe, A. (1983) *Neurochem. Res.* **8** 231–243.
Drejer, J., Honoré, T., Meier, E. and Schousboe, A. (1986) *Life Sci.* **38** 2077–2085.
Drejer, J., Honore, T. and Schousboe, A. (1987) *J. Neurosci.* **7** 2910–2916.
Dunlop, J. (1991) *Characterisation of the Neurochemical Actions of Excitatory Sulphur Amino Acids.* PhD thesis, University of St Andrews, UK.
Dunlop, J. and Griffiths, R. (1990) *Biochem. Soc. Trans.* **18** 378–381.

Dunlop, J., Grieve, A., Schousboe, A. and Griffiths, R. (1989a) *J. Neurochem.* **52** 1648–1651.

Dunlop, J., Mason, H., Grieve, A. and Griffiths, R. (1989b) *J. Neural Transm.* **78** 195–208.

Dunlop, J., Grieve, A., Schousboe, A. and Griffiths, R. (1990) *Neurochem. Int.* **16** 119–132.

Dunlop, J., Grieve, A., Schousboe, A. and Griffiths, R. (1991a) *Biochem. Soc. Trans.* **19** 3S.

Dunlop, J., Grieve, A., Schousboe, A. and Griffiths, R. (1991b) *J. Neurochem.* **57** 1388–1397.

Dunnett, C. W. (1964) *Biometrics* **20** 482–491.

Erecinska, M. and Troeger, M. B. (1986) *FEBS Lett.* **199** 95–99.

Grieve, A., Cameron, D. and Griffiths, R. (1990) *Biochem. Soc. Trans.* **18** 426–427.

Grieve, A., Dunlop., J., Schousboe, A. and Griffiths, R. (1991) *Biochem. Soc. Trans.* **19** 5S.

Griffith, O. W. (1987). In: Jakoby, W. B. and Griffith, O. W. (eds) *Methods in Enzymology*, Vol. 413. Academic Press, New York, pp. 366–376.

Griffiths, R. (1990) *Progr. Neurobiol.* **35** 313–323.

Griffiths, R., Grieve, A., Dunlop, J., Damgaard, I., Fosmark, H. and Schousboe, A. (1989) *Neurochem. Res.* **14** 339–343.

Hammond, E. J., Hurd, R. W., Wilder, B. J. and Thompson, F. J. (1980) *Electroenceph. Clin. Neurophysiol.* **49** 184–186.

Harris, K. M. and Miller, R. J. (1989) *Brain Res.* **482** 23–33.

Heafield, M. T., Fearn, S., Steventon, G. B., Waring, R. H., Williams, A. C. and Sturman, S. G. (1990) *Neurosci. Lett.* **110** 216–220.

Herrling, P. L. and Turski, W. A. (1985) *Experientia* **41** 820.

Huxtable, R. J. (1986) *Biochemistry of Sulfur*. Plenum Press, New York.

Huxtable, R. J. (1989) *Progr. Neurobiol.* **32** 471–533.

Iwata, H., Yamagami, S., Mizuo, H. and Baba, A. (1982) *J. Neurochem.* **38** 1268–1274.

Johnston, G. A. R. (1973) *Biochem. Pharmacol.* **22** 137–140.

Jones, H. E. and Sillito, A. M. (1989) *J. Physiol.* **417** 97P.

Jurson, P. A. and Freed, W. J. (1990) *Pharmacol. Biochem. Behav.* **36** 177–181.

Keller, H. J., Do, K. Q., Zollinger, M., Winterhalter, K. H. and Cuenod, M. (1989) *J. Neurochem.* **52** 1801–1806.

Kilpatrick, I. C. and Mozley, L. S. (1986) *Neurosci. Lett.* **72** 189–193.

Kim, J. P., Koh, J. and Choi, D. W. (1987) *Brain Res.* **437** 103–110.

Knopfel, T., Zeise, M. L., Cuenod, M. and Zieglgansberger, W. (1987) *Neurosci. Lett.* **81** 188–192.

Knopfel, T., Audinat, E., Staub, C. and Gahwiler, B. H. (1989) *Eur. J. Neurosci.* (Suppl.) **2** 108.

Larsson, O. M., Falch, E., Krogsgaard-Larsen, P. and Schousboe, A. (1988) *J. Neurochem.* **50** 818–823.

Leatherbarrow, R. J. (1987) *Enzfitter: a non-linear regression data analysis program for the IBM-PC.* Elsevier Science Publishers, The Netherlands.

Lehmann, J., Tsai, C. and Wood, P. L. (1988) *J. Neurochem.* **51** 1765–1770.

Mackay, D. (1962) *Arch. Biochem. Biophys.* **99** 93–100.

Mayer, M. L. and Westbroook, G. L. (1985) *J. Physiol.* **361** 65–90.

Mewett, K. N., Oakes, D. J., Olverman, H. J., Smith, D. A. S. and Watkins, J. C. (1983). In: Mandel, P. and De Feudis, F. V. (eds) *CNS Receptors: From Molecular Pharmacology to Behaviour*. Raven Press, New York, pp. 163–174.

Minc-Golomb, D., Eimerl, S. and Schramm, M. (1989) *Brain Res.* **490** 205–211.

Monahan, J. B. and Michel, J. (1987) *J. Neurochem.* **48** 1699–1708.

Mount, H., Quirion, R., Chaudieu, I. and Boksa, P. (1990) *J. Neurochem.* **55** 268–275.

Murphy, D. E. and Williams, M. (1987). In: Hicks, T. P., Lodge, D. and McLennan, H. (eds) *Excitatory Amino Acid Transmission*, Alan R. Liss, New York, pp. 63–66.

Murphy, D. E., Schneider, J., Boehm, C., Lehmann, J. and Williams, M. (1987) *J. Pharmacol. Exp. Ther.* **240** 778–784.

Neame, K. D. and Richards, T. G. (1972) *Elementary Kinetics of Membrane Carrier Transport*. Blackwell, Oxford.

Olney, J. W., Ho, O. L. and Rhee, V. (1971) *Exp. Brain Res.* **14** 61–76.

Olney, J. W., Misra, C. H. and De Gubaref, M. S. (1975) *J. Neuropathol. Exp. Neurol.* **34** 167–177.

Olney, J. W., Zorumski, C., Price, M. T. and Labruyere, J. (1990) *Science* **248** 596–599.

Olverman, H. J. and Watkins, J. C. (1989). In: Watkins, J. C. and Collingridge, G. L. (eds) *The NMDA Receptor*. IRL Press, Oxford, pp. 19–36.

Olverman, H. J., Jones, A. W. and Watkins, J. C. (1984) *Nature* **307** 460–462.

Olverman, H. J., Jones, A. W., Mewett, K. N. and Watkins, J. C. (1988a) *Neuroscience* **26** 17–31.

Olverman, H. J., Jones, A. W. and Watkins, J. C. (1988b) *Neuroscience* **26** 1–15.

Patneau, D. K. and Mayer, M. L. (1990) *J. Neurosci.* **10** 2385–2399.

Peck, E. J., Jr and Awapara, J. (1967) *Biochem. Biophys. Acta* **141** 499–506.

Pestana, A., Sandoval, I. V. and Sols, A. (1971) *Arch. Biochem. Biophys.* **146** 373–379.

Pin, J. P. and Bockaert, J. (1989) *J. Neurosci.* **9** 648–656.

Pullan, L. M., Olney, J. W., Price, M. T. Compton, R. P,, Hood, W. F., Michel, J. and Monahan, J. B. (1987) *J. Neurochem.* **49** 1301–1307.

Recasens, M., Varga, V., Nanopoulos, D., Saadoun, F., Vincendon, G. and Bendavides, J. (1982) *Brain Res.* **239** 153–173.

Recasens, M., Saadoun, F., Varga, V., Defeudis, F. V., Mandel, P., Lynch, G. and Vincendon, G. (1983) *Neurochem. Int.* **5** 89–94.

Recasens, M., Fagni, L., Baudry, M., Maitre, M. and Lynch, G. (1984) *Neurochem. Int.* **6** 325–332.

Schousboe, A., Drejer, J. and Hertz, L. (1988). In: Kvamme, E. (ed.) *Glutamine and Glutamate in Mammals*, Vol. 2. CRC Press, Boca Raton, FL, pp. 21–38.

Thomson, A. M. (1989) *Trends Neurosci.* **12** 349–353.

Tunnicliff, G. and Ngo, T. T. (1977) *Can. J. Biochem.* **55** 1013–1018.

Turski, W. A. (1989) *Brain Res.* **479** 371–373.

Waller, S. J., Kilpatrick, I. C., Chan, M. W. and Evans, R. H. (1991) *J. Neurosci. Methods* **36** 167–176.

Weiss, S. (1988) *J. Neurochem.* **51** 435–441.

Wilson, D. F. and Pastuszko, A. (1986) *J. Neurochem.* **47** 1091–1097.

8

Competitive NMDA receptor antagonists

Paul L. Ornstein and **Valentine J. Klimkowski**
Lilly Research Laboratories, Lilly Corporate Center, Indianapolis, IN 46285, USA

ABSTRACT

In this chapter we describe the structure–activity relationships (SARs) of a number of competitive NMDA antagonists (compounds that bind at the glutamate recognition site of the NMDA receptor complex) that have been synthesized by various groups. These SARs serve to define the steric tolerances and spatial orientation for optimal receptor binding of the various ligands. The theme of conformational restraint through incorporation of rings or unsaturation into the acidic amino acid backbone is common throughout all of these SARs. We also highlight some significant aspects of the *in vivo* activity of a competitive NMDA antagonist, e.g. duration of action, protection from excitatory amino acid-induced striatal neuronal degeneration (as a measure of their potential as neuroprotective agents) and oral availability in anticonvulsant assays. Finally, we discuss some aspects of molecular modelling of receptor–ligand interactions based on the potent activity of a number of new competitive NMDA antagonists.

INTRODUCTION

The synthesis of antagonists that act at the glutamate recognition site of the *N*-methyl-D-aspartate (NMDA) receptor complex (competitive NMDA antagonists) is an important goal for medicinal chemists (Johnson 1989). NMDA receptors are involved in a number of physiological functions (Collingridge and Lester 1989), and may even play a key role in synaptic plasticity (Collingridge and Singer 1990). Excessive release of glutamate and/or selective vulnerability of NMDA receptors may be part of the pathophysiology of some acute and chronic neurodegenerative disorders (Meldrum and Garthwaite 1990; see chapter 2), including cerebral ischaemia (Albers *et al.* 1989), Alzheimer's disease (Greenamyre and Young 1989), AIDS-induced dementia (Heyes *et al.* 1989), and amyotrophic lateral sclerosis (Plaitakis 1990). Compounds that block the action of glutamate at its recognition site may prove to be useful therapeutic agents, and may offer a more advantageous side-effect profile than, for example, phencyclidine (PCP)-like non-competitive NMDA antagonists (Willets *et al.* 1990; see chapter 9). PCP-like compounds were limited in their clinical utility because of certain behavioural effects (e.g., abuse potential, hallucinations and central nervous system (CNS) stimulation), and recent studies have shown that not all of their behavioural pharmacology is shared by competitive NMDA antagonists (Balster *et al.* 1990, France *et al.* 1989, Ornstein *et al.* 1987). Thus, the synthesis of novel NMDA antagonists has been initiated to explore the viability of these agents for the treatment of a number of disorders and to understand more about the spatial and steric requirements to block activation of the NMDA receptor.

A number of competitive NMDA antagonists have been synthesized by ourselves and others, and these are shown in Figs 1 and 2. Clearly, the seminal work in this area was the discovery by Watkins (Evans *et al.* 1982) that the phosphono amino acids 2*R*-AP5 (2*R*-2-amino-5-phosphonopentanoic acid, also known as D-AP5) and 2*R*-AP7 (2*R*-2-amino-7-phosphonoheptanoic acid, also known as D-AP7) were potent and selective NMDA antagonists. The structural diversity of these novel compounds, which expand upon Watkin's discovery, provides useful insight into the struture–activity relationships (SARs) of NMDA antagonists. A common theme that pervades these new antagonists is the reduction of conformational mobility through the incorporation of ring systems and/or unsaturation. Figs 3 and 4 show the syntheses of a number of these novel NMDA antagonists. These examples are only a cursory glance at the chemisty involved in the preparation of these novel acidic amino acids, and the reader is encouraged to seek out the primary literature for more detailed procedures.

For a compound to possess antagonist activity at NMDA receptors, the basic structural requirement is an α-amino acid substituted at some distance by a carboxylic acid or isosteric group (acidic amino acid, see Figs 1 and 2), exemplified by 2*R*-AP5. The proximal acid is usually a carboxylic acid, incorporated into these molecules as an α-amino acid. The amino group can be primary or secondary, and can be part of a heterocyclic ring. It also appears that the nitrogen must bear at least one proton (e.g., substitution of CGS 19755 (2*RS*,4*SR*-4-(phosphonomethyl)piperidine-2-carboxylic acid) with a methyl group on the ring nitrogen (Hutchinson *et al.* 1989) completely abolishes the activity). Antagonist activity has been observed for compounds whose distal acid group is carboxylic or phosphonic acid or a tetrazole. The spacing between the two acidic groups is critical for antagonist activity: four or six atoms for the carboxylic and phosphonic acid compounds; and four, five or six atoms for the tetrazole-containing compounds. In certain cases, heteroatom substitution by nitrogen (as in 2*R*-CPP (2*R*-4-(3-phosphonoprop-1-yl)piperazine-2-carboxylic acid), 2*R*-CPP-ene (2*R*-4-(3-phosphonoprop-2-*E*-en-1-yl)piperazine-2-carboxylic acid), and LY275059 (2*RS*-4-(3-(1(2)*H*-tetrazol-5-yl)prop-1-yl)piperazine-2-carboxylic acid) or oxygen (e.g., MDL 100453 (2*R*-2-amino-4-oxo-5-phosphonopentanoic acid)) in or on the carbon framework is tolerated. It is in general true that only one enantiomer of each racemic NMDA antagonist that has been resolved is active, and that this isomer has the *R*-absolute stereochemistry at the α-amino acid centre. We have found, however, that the active enantiomers of LY274614 (3*SR*,4a*RS*,6*SR*,8a*RS*-6-(phosphonomethyl)-1,2,3,4,4a,5,6,7,8,8a-decahydroisoquinoline-3-carboxylic acid) and LY233536 (3*SR*,4a*RS*,6*SR*,8a*RS*-6-((1(2)*H*-tetrazol-5-yl)methyl)-1,2,3,4,4a,-5,6,7,8,8a-decahydroisoquinoline-3-carboxylic acid) (see Fig. 2) possess the *S*-absolute

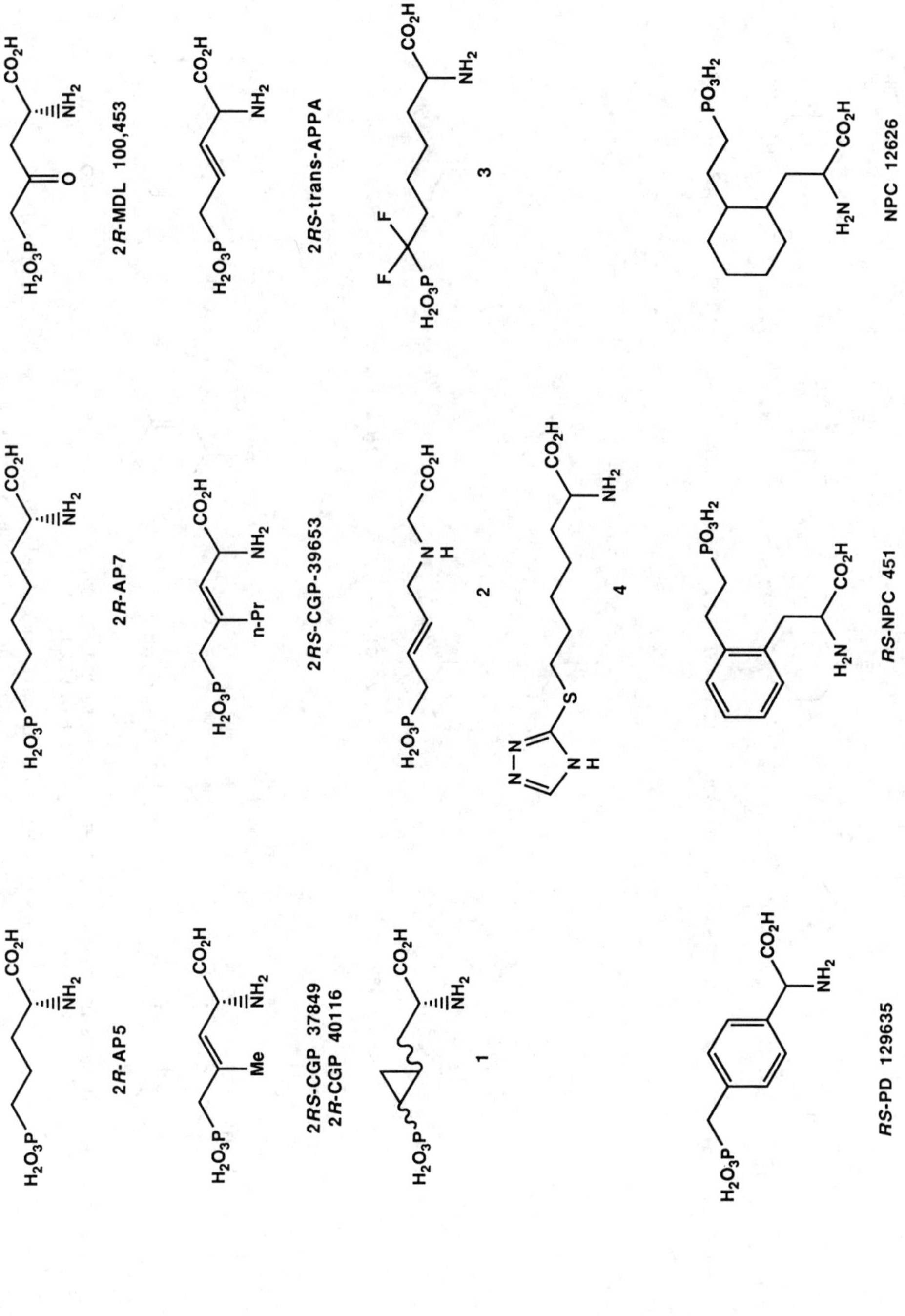

2*R*-AP5

2*R*-AP7

2*R*-MDL 100,453

2*RS*-CGP 37849
2*R*-CGP 40116

2*RS*-CGP-39653

2*RS*-trans-APPA

1

2

3

4

RS-PD 129635

RS-NPC 451

NPC 12626

Fig. 1 — Competitive NMDA antagonists: acyclic, phenylglycine and phenylalanine amino acids.

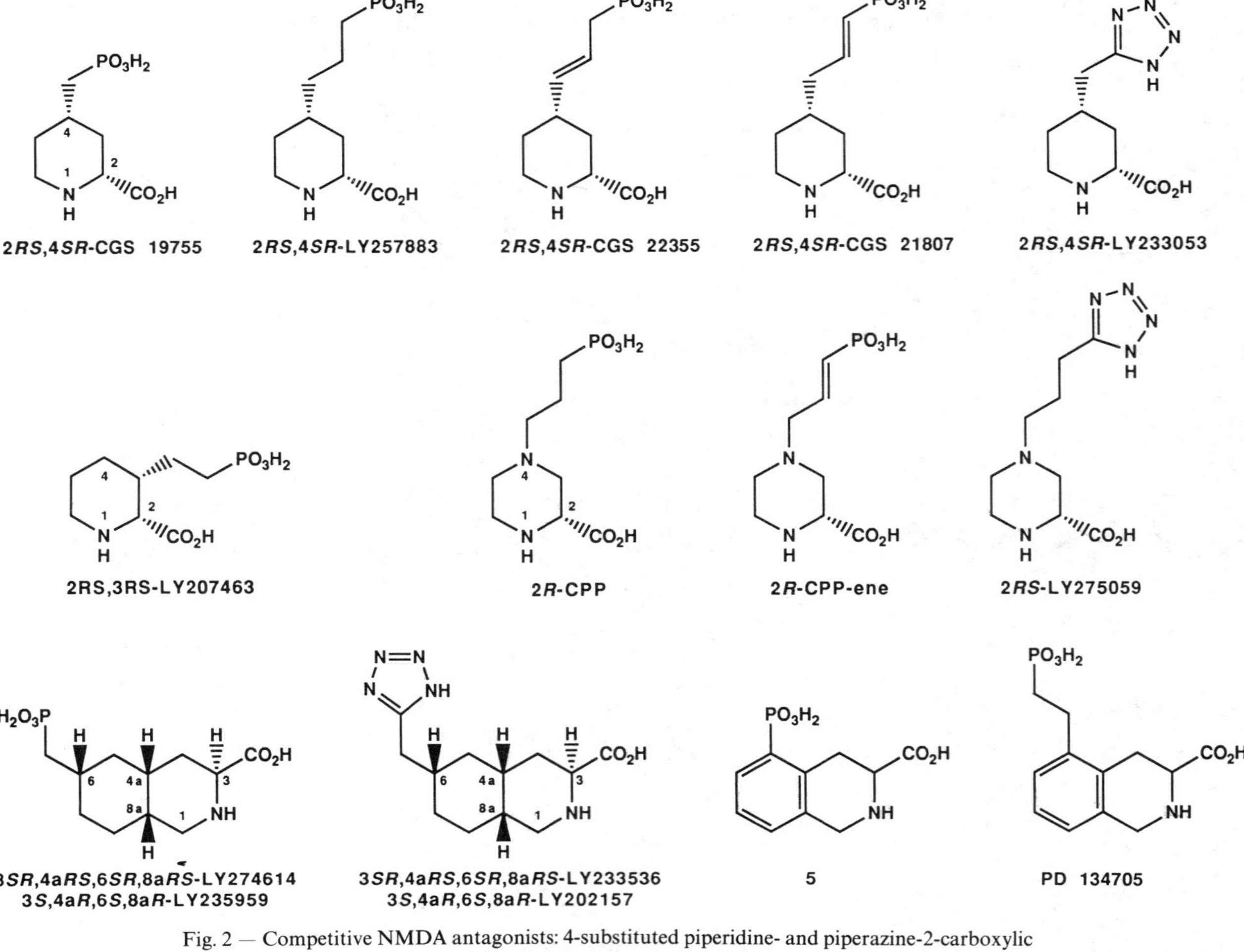

Fig. 2 — Competitive NMDA antagonists: 4-substituted piperidine- and piperazine-2-carboxylic acids and 5- and 6-substituted isoquinoline-3-carboxylic acids.

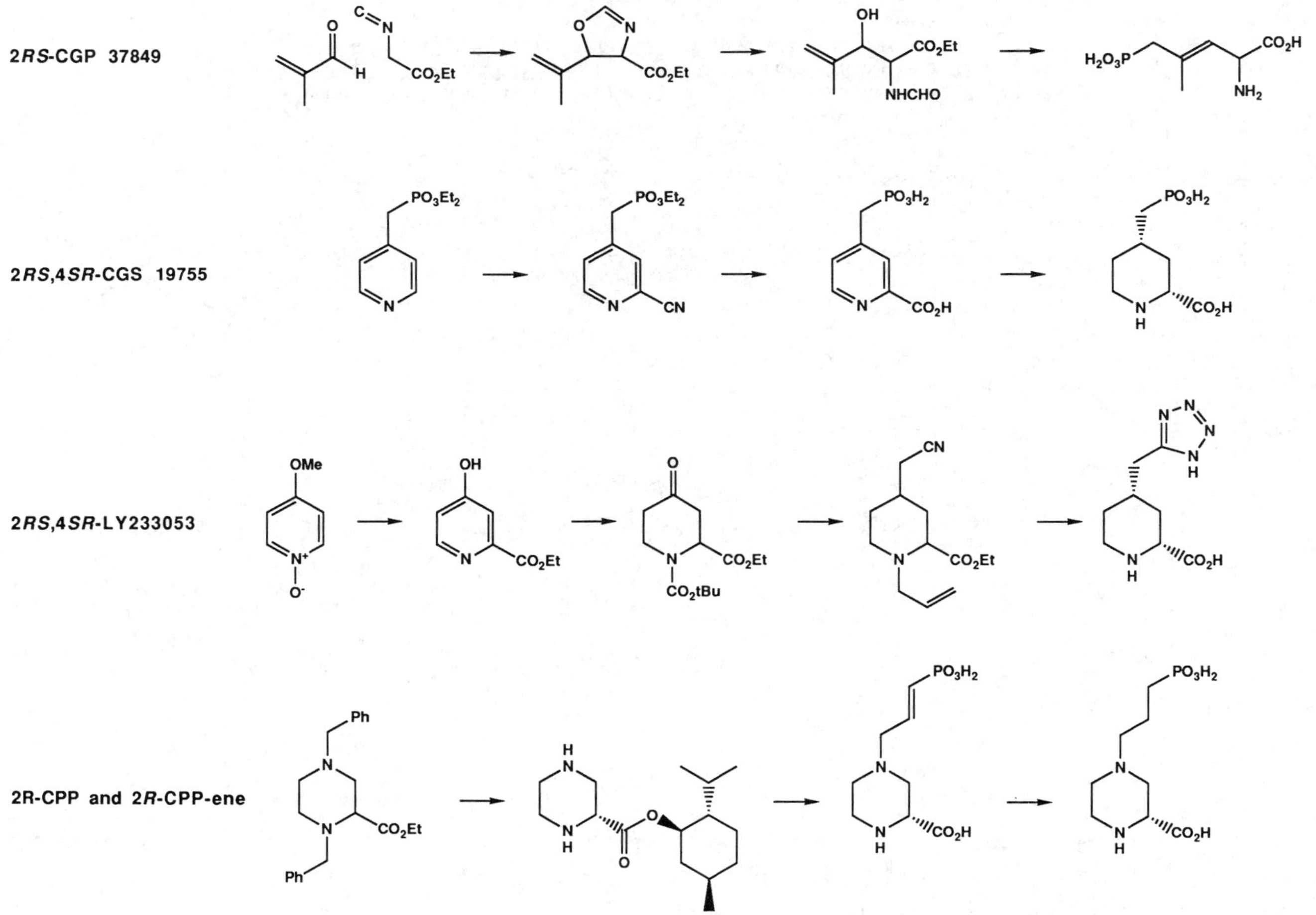

2*RS*-CGP 37849

2*RS*,4*SR*-CGS 19755

2*RS*,4*SR*-LY233053

2R-CPP and 2*R*-CPP-ene

Fig. 3 — Syntheses of selected competitive NMDA antagonists: synthesis of 2*RS*-CGP 37849 (Angst *et al.* 1987), 2*RS*,4*SR*-CGS 19755 (Ornstein *et al.* 1989), 2*RS*,4*SR*-LY233053 (Ornstein *et al.* 1991a), and 2*R*-CPP and 2*R*-CPP-ene (Aebischer *et al.* 1989).

2RS-PD 129635

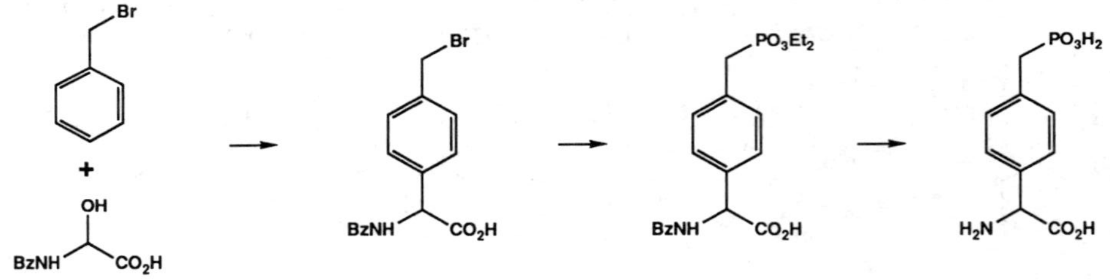

3*SR*,4a*RS*,6*SR*,8a*RS*-LY274614 and
3*SR*,4a*RS*,6*SR*,8a*RS*-LY233536

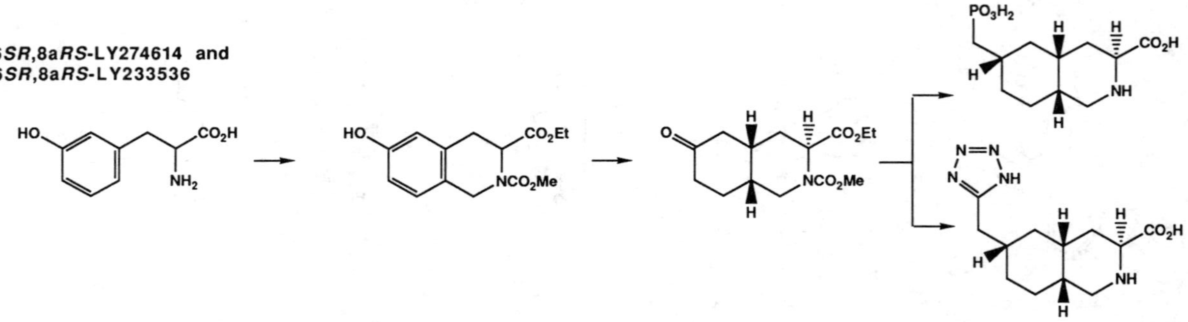

2*R*-MDL 100,453

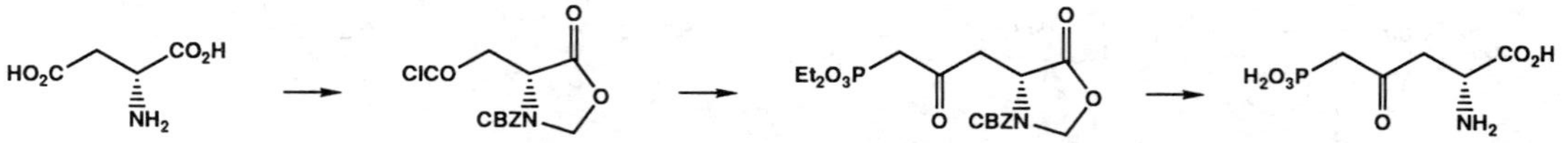

Fig. 4 — Syntheses of selected competitive NMDA antagonists: synthesis of 2*RS*-PD 129635 (Bigge *et al.* 1989), 3*SR*,4a*RS*,6*SR*,8a*RS*-LY274614 and 3*SR*,4a*RS*,6*SR*,8a*RS*-LY233536 (Ornstein *et al.* 1991b), and 2*R*-MDL 100453 (Whitten *et al.* 1989).

stereochemistry at the α-amino acid centre. The full implications of this finding will be discussed later in this chapter as we examine molecular modelling of NMDA receptor antagonists.

Our approach and the approach of others to the discovery of novel competitive NMDA antagonists has involved incorporation of the requisite acidic amino acid into a conformationally restrained framework. Making these molecules more rigid may serve to enhance activity both *in vitro* and *in vivo* by reducing conformational mobility. If this process is done in such a way that the molecule is held in an orientation that is more conducive for binding, then it is likely to provide an antagonist with greater potency, and thus a better understanding of the spatial and steric requirements for activity at NMDA receptors. We (Ornstein *et al.* 1991a) and others (Chenard *et al.* 1990) have also explored the use of bioisosteric replacements for the distal acid group other than phosphonate. The tetrazole moiety ($pK_a = 4.93$, Schaff and Hess 1979), for example, was found to be a useful surrogate for the phosphonic acid ($pK_{a1} = 2.73$, $pK_{a2} = 7.89$, Bigge *et al.* 1989) group in excitatory amino acids (Ornstein *et al.* 1991a). Because the tetrazole is planar and aromatic, and charge is diffused over the entire ring, it may interact with the receptor in a way that is distinct from other acidic moieties.

IN VITRO SAR ANALYSIS

It is worthwhile to delineate further some of the structural features that have been incorporated into many of these novel NMDA antagonists and the implications of these changes on the activity of these compounds. A significant limitation of AP5 and AP7 was their lack of potency following systemic administration (Meldrum *et al.* 1983). Therefore, modifications of the acyclic amino acid structure were sought that would lead to compounds that were more potent *in vitro* than these inaugural antagonists, while also showing improved activity *in vivo*. Table 1 provides NMDA receptor affinity data (from both [^{3}H]CPP and [^{3}H]CGS 19755 binding) for a number of the compounds shown in Figs 1 and 2. Because *in vivo* data for many of these compounds have not been reported, we have relied heavily on receptor affinity for our SAR analysis.

Acyclic amino acids

One series of AP5 analogues incorporates unsaturation to reduce conformational flexibility, and is exemplified by 2*RS*-CGP 37849 (2*RS*-*E*-2-amino-4-methyl-5-phosphonopent-3-enoic acid) (Fagg *et al.* 1990). Substitution by a small alkyl group at C_4, such as the methyl group of 2*RS*-CGP 37849 or the propyl group of 2*RS*-CGP 39653 (2*RS*-*E*-2-amino-4-propyl-5-phosphonopent-3-enoic acid, Sills *et al.* 1991), is well tolerated in these unsaturated amino acids and appears to be essential for potent activity. *Trans*-APPA (2*RS*-*E*-2-amino-5-phosphonopent-3-enoic acid), which bears only a proton at C_4, is about four-fold less potent in binding than CGP 37849 (Fagg *et al.* 1990). Alkyl substitution at C_4 appears to enforce a conformation (because of $A_{1,3}$ strain) where the phosphonic acid and carboxylic acid moieties are on the same face of the carbon backbone — a conformation that appears to be optimal for receptor

Table 1 — NMDA receptor binding for selected antagonists

Compound	[³H]CGS 19755[a] IC_{50} (nM)	[³H]CPP binding[b] IC_{50} (nM)
2*R*-AP5	350 ± 15[a]	183 ± 48[b]
2*RS*-AP7	900 ± 200[a]	1060 ± 120[b]
2*RS*-CGP 37849		35 ± 20[c]
2R-CGP 401116		19 ± 7[c]
2*RS*-CGP 39653		5[d]
2*R*-MDL 100453		109 ± 12[e]
2		1200[f]
3		27 000[g]
2*RS*-PD 129635		1010[h]
NPC 451		60 900[h]
NPC 12626		150 ± 85[i]
2*RS*,4*SR*-CGS 19755	54 ± 13[j]	95 ± 28[k]
2*RS*,4*SR*-LY257883		120 ± 19[k]
2*RS*,4*SR*-CGS 22355		14 ± 1[l]
2*RS*,4*SR*-CGS 21807		72 ± 2[l]
2*RS*,4*SR*-LY233053	107 ± 7[j]	
2*R*-CPP		140 ± 20[m]
2*R*-CPP-ene		40 ± 10[m]
3*SR*,4a*RS*,6*SR*,8a*RS*-LY274614	54 ± 14[n]	
3*S*,*4aR*,6*S*,8a*R*-LY235959	25 ± 11[o]	
3*SR*,4a*RS*,6*SR*,8a*RS*-LY233536	815 ± 205[n]	
3*S*,4a*R*,6*S*,8a*R*-LY202157	415 ± 122[o]	
PD 134705		120[p]

[a]Murphy *et al.* (1987). [b]Murphy *et al.* (1988). [c]Fagg *et al.* (1990). [d]Sills *et al.* (1991). [e]Whitten *et al.* (1990). [f]Bigge *et al.* (1989c). [g]Bigge *et al.* (1989d). [h]Bigge *et al.* (1989a). [i]Ferkany *et al.* (1989). [j]Ornstein *et al.* (1991a). [k]Ornstein *et al.* (1989). [l]Hutchinson *et al.* (1989). [m]K_is. Aebischer *et al.* (1989). [n]Ornstein *et al.* (1991b). [o]Ornstein (unpublished results). [p]Humbelt *et al.* (1990).

recognition (see below and Hutchinson *et al.* 1989). A ketone group at C_4 of 2*R*-AP5 gave 2*R*-MDL 100453 (Whitten *et al.* 1990). This substituent may also provide a similar conformational constraint through either keto–enol tautomerism or a favourable intramolecular hydrogen bond, or some combination of both factors. The incorporation of a cyclopropane group into the backbone of AP5 and AP7, as in a series of methano-AP5 derivatives exemplified by **1** (2*R*-2-amino-3-(1*RS*,2*RS*-2-phosphonocycloprop-1-yl)propanoic acid, Dappen *et al.* 1991), was deleterious to activity. It is probable that the changes in spatial orientation brought about by this moiety are not conducive to optimal binding. When a phosphonoalkenyl group was bonded via the nitrogen as in **2** (*N*-(4-phosphonobut-2-*E*-en-1-yl)glycine, Bigge *et al.* 1989d), only weak NMDA antagonist activity was observed. Again, this structural change does not appear to provide the proper orientation for potent receptor binding.

Conversion of AP7 to its 7,7-difluoro-analogue **3** (2*RS*-2-amino-7,7-difluoro-7-phosphonoheptanoic acid, Bigge *et al.* 1989c) completely abolished affinity for the receptor. It was postulated that the addition of the two fluorine atoms adjacent to the phosphonic acid moiety would render it fully ionized at physiological pH. It is not clear whether the lack of activity is due to steric or electronic perturbations of this amino acid. A series of thio-1,2,4-triazole derivatives of AP7 have been prepared, e.g. **4** (2-amino-8-(1*H*-1,2,4-triazol-5-yl)-8-thiaoctanoic acid), but none of them showed any NMDA antagonist activity (Chenard *et al.* 1990). Of the potent acyclic amino acids that have been resolved (AP5, AP7, CGP 37849 and MDL 100453), all of them have the *R*-absolute stereochemistry at the α-amino acid centre (the 2*R*-isomer of CGP 37849 is CGP 40116, Fagg *et al.* 1990).

Cyclic amino acids

A number of phosphonoalkyl-substituted phenylglycine and phenylalanine amino acids have been prepared, for example PD 129635 (2*RS*-4-phosphonomethylphenylglycine, Bigge *et al.* 1989a) and NPC 451 (2*RS*-(2-phosphono-eth-1-yl)phenylalanine, Rzeszatarski *et al.* 1987), respectively. None of these compounds are as potent as AP5 and AP7. This may indicate that the aromatic ring does not allow for the optimal spatial orientation for ligand binding, or that in a preferred binding orientation there is a deleterious steric or electronic interaction. When the aromatic ring is saturated, as in NPC 12626 (2*RS*-2-amino-3-(2*RS*-(2-phosphono-eth-1-yl)cyclohex-1*RS*-1-yl)-propanoic acid, Ferkany *et al.* 1989), there appears to be some enhancement of activity, possibly indicating that the saturated molecule can provide a more favourable orientation for binding. The low potency of this compound, however, may be ascribed to its being a mixture of diastereomers. It is therefore possible that one stereoisomer of NPC 12626 is particularly potent, but this remains to be elucidated.

Significantly enhanced activity has been observed for the piperidine and piperazine-2-carboxylic acids (see Fig 2 and Table 2), such as CGS 19755, CPP and LY233053 (2*RS*,4*SR*-4-((1(2)*H*-tetrazol-5-yl)methyl)piperidine-2-carboxylic acid). The SAR of the 4-phosphonoalkyl-piperidine- (Hutchinson *et al.* 1989, Ornstein *et al.* 1989) and piperazine-2-carboxylic acids (Bigge *et al.* 1989b, Hays *et al.* 1990) is interesting. The inter-acid spacing requirements parallels that of the acyclic amino acids AP5 to AP8, in that the compounds having four and six atoms separating the two acid groups gives the best activity, while the compounds with five and seven atoms separating the two acid groups are inactive. Optimal activity both *in vitro* and *in vivo* in the piperidine series is observed for the phosphonomethyl compound CGS 19755 (Lehmann *et al.* 1988), whereas the optimal activity in the piperazine series is observed with the phosphonopropyl compound CPP (Lehmann *et al.* 1987). In the piperidine series, attachment of the phosphonoalkyl side-chain in the 4-position is preferred, with the 3-phosphonoalkyl compounds considerably less active, e.g. LY207463 (2*RS*,3*RS*-3-(2-phosphonoeth-1-yl)piperidine-2-carboxylic acid; Ornstein *et al.* 1989); and in many cases the 2,4-*trans* compounds are nearly as active as the 2,4-*cis*-compounds (Hutchinson *et al.* 1989). Alkenyl substitution is well tolerated in the side-chain as, for example, with CPP-ene (Aebischer *et al.* 1989), 2*RS*,4*SR*-CGS 22355 (2*RS*,4*SR*-4-(3-phosphonoprop-1-*E*-en-1-yl)piperidine-2-carboxylic acid) and 2*RS*,4*SR*-CGS 21807 (2*RS*,4*SR*-4-(3-phosphonoprop-2-*E*-en-1-yl)piperidine-2-car-

Table 2 — *In vitro* and *in vivo* activity of 4-substituted piperidine- and piperazine-2-carboxylic acids

Compound[a]	*n*	[^{3}H]CPP binding[b] [^{3}H]CGS 19755[c] or IC_{50} (nM)	NMDA-induced lethality[d] in mice MED[e] (mg/kg, i.p.)
Piperidine phosphonates (A = CH)[f]			
CGS 19755	1	95 ± 28	1.25
6	2	6600 ± 1279	> 160
LY257883	3	120 ± 19	40
7	4	> 10 000	> 160
Piperazine phosphonates (A = N)[g]			
8	1	320 ± 90	
9	2	9800 ± 840	
CPP	3	79 ± 7	2.5[f]
10	4	48 000 ± 7400	
Piperidine tetrazoles (A = CH)[h]			
LY233053 (2,4-*cis*)	1	107 ± 7	5
11 (2,4-*trans*)	1	777 ± 204	80
12 (2,4-*cis*)	2	2280 ± 380	20
13 (2,4-*cis*)	3	5830 ± 910	10
14 (2,4-*cis*)	4	> 10 000	> 160
Piperazine tetrazoles (A = N)[i]			
15	1	1670 ± 620	> 80
16	2	781 ± 62	160
LY275059	3	3150 ± 1210	40
17	4	> 10 000	> 80

[a]All compounds are racemic. [b]Murphy *et al.* (1987). [c]Murphy *et al.* (1988). [d]Leander *et al.* (1988). Animals were given the test compound 30 minutes prior to a dose of 200 mg/kg of NMDA. [e]MED = minimum effective dose. This is the lowest dose where at least three of the five mice tested survived. [f]Ornstein *et al.* (1989). [g]Hays *et al.* (1990). [h]Ornstein *et al.* (1991a). [i]Ornstein *et al.* (1991b).

boxylic acid) (Hutchinson *et al.* 1989). Substitution with methyl groups on the phosphonate-bearing carbon of CGS 19755 lowers the activity, as does incorporation of a methyl group on C_3 or C_6 of the piperidine ring; a methyl group at C_5, however, is well tolerated (Hutchinson *et al.* 1989). Conversion of the CPP phosphonate-bearing side-chain to an arylalkyl group is also deleterious to activity (Hays *et al.* 1990). Some notable differences, however, are observed for the tetrazolylalkyl-substituted compounds. In the piperidine (Ornstein *et al.* 1991a) and piperazine (Ornstein *et al.* 1991b) series, the compounds having four, five or six atoms separating the two acidic groups are active, while the seven-atom-spaced compounds are inactive; and the 2,4-*cis* stereochemistry is preferred over the 2,4-*trans* stereochemistry. The tetrazolylmethyl compound LY233053 (Schoepp *et al.* 1990b) has the highest potency of the piperidine series both *in vivo* and *in vitro.* And even though they have highly limited *in vitro* potency, the *in vivo* activity of the ethyl and propyl compounds **12** (2*RS*,4*SR*-4-(2-(1(2)*H*-tetrazol-5-yl)eth-1-yl)piperidine-2-carboxylic acid) and **13** (2*RS*,4*SR*-4-(3-(1(2)*H*-tetrazol-5-yl)prop-1-yl)piperidine-2-carboxylic acid), respectively, is still quite good. This may reflect the ability of the less polar tetrazole group to provide better CNS availiabity. In the piperazine series, optimal activity *in vitro* is observed for the tetrazolylethyl compound **15** (2*RS*,4*SR*-4-(2-(1(2)*H*-tetrazol-5-yl)eth-1-yl)piperazine-2-carboxylic acid), but *in vivo* the tetrazolylpropyl compound LY275059 is clearly the best. We believe that the inter-acid spacing differences observed for the tetrazoles relative to the phosphonates may reflect the tetrazole's ability to spread charge over the entire ring, thereby allowing for a different orientation of these molecules within the recognition site. Both CPP and CPP-ene have been resolved, and the predominant NMDA antagonist activity resides with the 2*R*-isomer in both cases (Aebischer *et al.* 1989).

In the SAR of the decahydroisoquinoline series (Ornstein *et al.* 1991b), we have prepared a number of compounds that help to determine the optimal stereochemical arrangement for activity in this nucleus, as well as the optimal state of saturation in the cyclohexyl ring portion of these molecules (see Fig. 2 and Table 3). With four asymmetric centres in the molecule, the possibility exists to get eight diastereomerically different enantiomeric pairs. However, a stereocontrolled synthesis of these molecules allowed us to prepare many of the different diastereomers selectively. The most potent NMDA antagonist activity was observed for the diastereomer with relative stereochemistry corresponding to LY274614 and LY233536. The phosphonate and tetrazole acid isosteres provided compounds that showed better *in vitro* and *in vivo* activity than the corresponding carboxylic acids (e.g, LY274614 and LY233536 versus LY285719 (3*SR*,4a*RS*,6*SR*,8a*RS*-6-(carboxymethyl)-1,2,3,4,4a,5,6,7,8,8a-decahydroisoquinoline-3-carboxylic acid)). The C_6 epimers of these compounds, LY266845, LY247738 and LY285719 (3*SR*,4a*RS*,6*RS*,8a*RS*-6-(phosphomomethyl-, -6-(1(2)H-tetrazol-5-yl)methyl)- and -6-(carboxymethyl)-1,2,3,4,4a,5,6,7,8,8a-decahydroisoquinoline-3-carboxylic acid, respectively) were less active, as were the C_3 epimers LY215827 and LY262294 (3*SR*,4a*SR*,6*RS*,8a*SR*-6-(phosphonomethyl)- and -6-(1(2)*H*-tetrazol-5-yl)methyl)-1,2,3,4,4a,5,6,7,8,8a-decahydroisoquinoline-3-carboxylic acid, respectively). Moving the carboxylic acid group from C_3 to C_1, as in LY274827 and LY209795 (1*SR*,4a*SR*,6*RS*,8a*SR*-6-(phosphonomethyl)- and -6-(1(2)*H*-tetrazol-5-yl)methyl)-1,2,3,4,4a,5,6,7,8,8a-decahydroisoquinoline-1-carboxylic acid, respectively) showed a marked loss of activity, and the *trans*-ring isomers

Table 3 — *In vitro* and *in vivo* activity of 6-substituted decahydroisoquinoline-3-carboxylic acids

Compound[a]	X	[^{3}H]CGS19755[b] binding IC_{50} (nM)	NMDA-induced lethality[c] in mice MED[d] (mg/kg, i.p.)	Structure
LY274614	PO_3H_2	55 ± 14	1.25	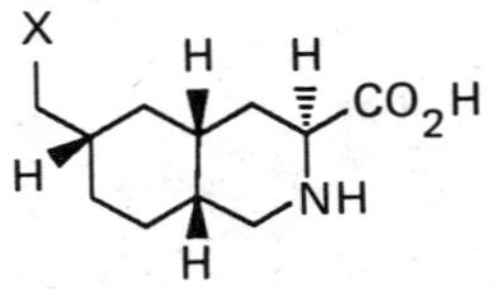
LY233536	Tetrazole	856 ± 136	2.5	
LY285720	CO_2H	4298 ± 589	10	
LY266845	PO_3H_2	815 ± 205	2.5	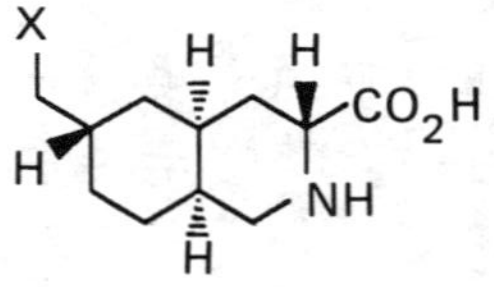
LY247738	Tetrazole	3380 ± 286	40	
LY285719	CO_2H	77 920 ± 7570	> 160	
LY215827	PO_3H_2	3557 ± 366	80	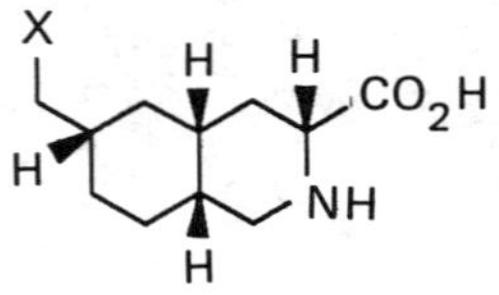
LY262294	Tetrazole	> 10 000	> 160	
LY266506	PO_3H_2	> 10 000	> 160	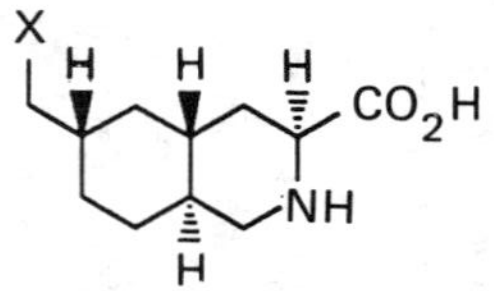
LY266505	Tetrazole	> 10 000	> 160	

Continued

LY274827 LY209795	PO_3H_2 Tetrazole	4870 ± 1000 2480 ± 260	> 160 160	
LY247251 LY202794	PO_3H_2 Tetrazole	2150 ± 122 11 470 ± 3531	40 > 160	
LY246308 LY246811	PO_3H_2 Tetrazole	13 810 ± 2170 26 400 ± 1946	> 160 > 160	

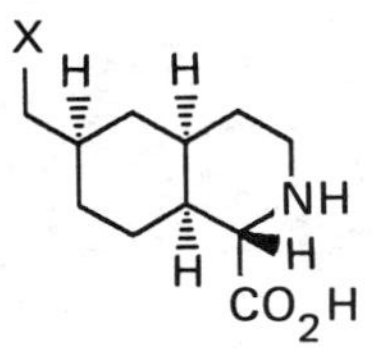

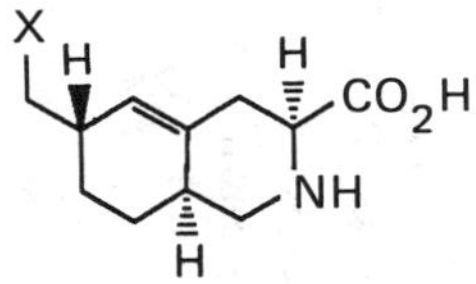

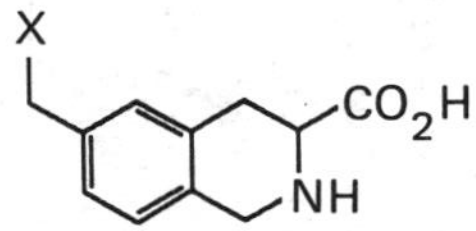

[a]Ornstein *et al.* (1991b). All compounds are racemic. [b]Murphy *et al.* (1988). [c]Leander *et al.* (1988). Animals were given the test compound 30 minutes prior to a dose of 200 mg/kg of NMDA. [d]MED = minimum effective dose. This is the lowest dose where at least three of the five mice tested survived.

LY266506 and LY266505 (3*SR*,4a*RS*,6*SR*,8a*SR*-6-(phosphonomethyl)- and -6-(1(2)*H*-tetrazol-5-yl)methyl)-1,2,3,4,4a,5,6,7,8,8a-decahydroisoquinoline-3-carboxylic acid, respectively) were inactive. The best NMDA antagonist activity in this series was observed for the fully saturated (decahydroisoquinoline) nucleus (e.g., LY274614 and LY233536); the octahydro compounds LY247251 and LY202794 (3*SR*,6*SR*,8a*SR*-6-(phosphonomethyl)- and -6-(1(2)*H*-tetrazol-5-yl(methyl)-1,2,3,4,6,7,8,8a-octahydroisoquinoline-3-carboxylic acid, respectively) were less active, and the tetrahydroisoquinolines LY246308 and LY2466811 (3*SR*-6-(phosphonomethyl)- and -6-(1(2)*H*-tetrazol-5-yl)methyl-1,2,3,4-tetrahydroisoquinoline-3-carboxylic acid, respectively) were inactive. This latter result is an interesting contrast to the 5-phosphono-substituted tetrahydroisoquinoline-3-carboxylic acids **5** (3*RS*-5-phosphono-1,2,3,4-tetrahydroisoquinoline-3-carboxylic acid; Cordi and Vazquez 1990) and PD 134705 (3*RS*-5-phosphonoethyl-1,2,3,4-tetrahydroisoquinoline-3-carboxylic acid; Humbelt *et al.* 1990), which both show potent affinity for the NMDA receptor.

IN VIVO SAR COMPARISONS

Most of the phosphonate-substituted NMDA receptor antagonists have been found to have a very long duration of activity *in vivo*. However, we have observed that tetrazole-substituted amino acids have a relatively short duration of action *in vivo*. For example, a single dose of CGS 19755 (20 mg/kg, i.p.) will protect neonatal rats from convulsions (Schoepp *et al.* 1990a) due to NMDA (20 mg/kg, i.p.) for 12 hours, while a single dose of LY233053 (100 mg/kg), the tetrazole counterpart, has only a 4-hour duration of action in the same assay (Leander *et al.* 1989).

A number of NMDA antagonists that have been prepared have been shown to be active following oral administration and include CPP and CPP-ene (Chapman *et al.* 1990), CGP 37849 (Fagg *et al.* 1990), LY274614 (Schoepp *et al.* 1991) and LY233053 (Schoepp *et al.* 1990) Most of this data has been generated in various models of anticonvulsant activity (CPP and CPP-ene, versus sound-induced seizures in DBA/2 mice; CGP 37849, versus maximal electroshock-induced seizures in mice; LY274614, versus NMDA-induced convulsions in neonatal rats and NMDA-induced lethality in mice; and LY233053, versus NMDA-induced convulsions in neonatal rats).

NMDA, when injected directly into the striatum of adult rats (at a dose of 300 nmol), produces a very characteristic loss of striatal neurons, as measured by decreases in choline acetyl transferase activity (Schoepp *et al.* 1989). We have used this assay to characterize the potential of a systemically administered NMDA antagonist to block neuronal degeneration induced by an excitatory amino acid agonist, and thereby gauge the potential for these agents to serve as neuroprotectants. A single dose of either LY274614 (20 mg/kg, i.p.) (Ornstein *et al.* 1991b) or CGS 19755 (40 mg/kg, i.p.) (Schoepp *et al.* 1989) given 30 minutes prior to a dose of 300 nmol of NMDA provided complete protection from agonist-induced striatal neuronal degeneration. In the same assay, LY233053 and LY233536 (Ornstein *et al.* 1991b) also completely blocked the effects of 300 nmol of NMDA, but to accommodate their shorter duration of action they were given in a series of doses equally spaced over a 12-hour period beginning 30 minutes prior to dosing with NMDA (LY233053, 4 × 100 mg/kg; LY233536, 6 × 100 mg/kg).

MOLECULAR MODELLING

We have resolved LY274614 and LY233536, and found that the NMDA antagonist activity resides in the enantiomer having the *S*-absolute stereochemistry at the α-amino acid centre (3*S*,4a*R*,6*S*,8a*R*-LY235959 and 3*S*,4a*R*,6*S*,8a*R*-LY202157, respectively), as shown in Fig. 1. To our knowledge, this is the first example of a potent NMDA antagonist having the *S*-configuration at the α-amino acid carbon. The C_6 epimer LY266845 has also been resolved and, interestingly, the active enantiomer 3*R*,4a*S*,6*S*,8a*S*-LY288534 has been shown to possess the 3*R*-absolute stereochemistry (as shown in Table 3). These remarkable findings led us to embark on a molecular modelling study of the possible receptor–ligand interactions for NMDA antagonists. Due to the limitations of space, our discussion will be focused primarily on the antagonists CGS 19755, 2*R*-CGP 40116, 2*R*-AP5, 3*S*,4a*R*,6*S*,8a*R*-LY235959 and 3*R*,4a*S*,6*S*,8a*S*-LY288534. While the absolute stereochemistry of CGS 19755 has not been determined, the compound has been resolved and the activity resides in the minus isomer (Lehmann *et al.* 1988). We have assumed that the active enantiomer is the 2*R*,4*S*-isomer, based on analogy to the piperazine amino acids 2*R*-CPP and 2*R*-CPP-ene.

Using molecular mechanics methods, conformational analysis by Hutchinson *et al.* (1989) has determined that for 2*R*,4*S*-CGS 19755 the 4,5-*gauche(−)* and *gauche(+)* conformations of the phosphonomethyl group are aproximately 1 kcal/mol different in energy. Of these, the *gauche(−)* form is the ground state. Using QUANTA/CHARMm software, we confirmed these findings (all molecular modelling was performed using version 3.0, Polygen Corporation, Waltham, MA. Details of the molecular modelling, as well as results comparing a wide range of analogues, will be published elsewhere). As Hutchinson earlier postulated, there is evidence that the *gauche(−)* isomer may be the bioactive form. Using this conformation of 2*R*,4*S*-CGS 19755 as a starting template, we derived low-energy structures for various acyclic and cyclic NMDA antagonists. Atom-to-atom overlaps, between the corresponding amino acid backbone atoms, reveal the close strutural similarity between the various molecules. This is illustrated for 2*R*,4*S*-CGS 19755, 2*R*-AP5, 2*R*-CGP 40116, 3*S*,4a*R*,6*S*,8a*R*-LY235959 and 3*R*,4a*S*,6*S*,8a*S*-LY288534 in Fig. 5. Ignoring 2*R*-CGP 40116 for the time being, it can be observed that the phosphonate group of 2*R*,4*S*-CGS 19755 and 2*R*-AP5 very closely occupy the same region of space. Likewise, the phosphonate groups of 3*S*,4a*R*,6*S*,8a*R*-LY235959 and 3*R*,4a*S*,6*S*,8a*S*-LY288534 overlap almost completely. Based upon the illustrated overlap, the distance between the phosphonate centres of the 4-atom and 6-atom analogues is 2.2 Å. Considering the apparent volume and the polarizability provided by a phosphonate group to a receptor site, this difference in the absolute positions may still be reasonable for the binding of both. However, we have determined that this distance can be reduced to approximately 1 Å if the phosphonomethyl group for each is rotated by 25° toward each other. The increase in energy accompanying this conformational change is less than 3 kcal/mol in all four cases. A consequence of this modification is that the phosphonate group of 2*R*-CGP 40116 now resides in the same region of space occupied by the other four analogues. If the restriction that the amino acid backbone atoms must overlay exactly is relaxed, it is clear that the phosphonate groups of each analogue can be made to overlay with even less structural deviation from their energy minimum.

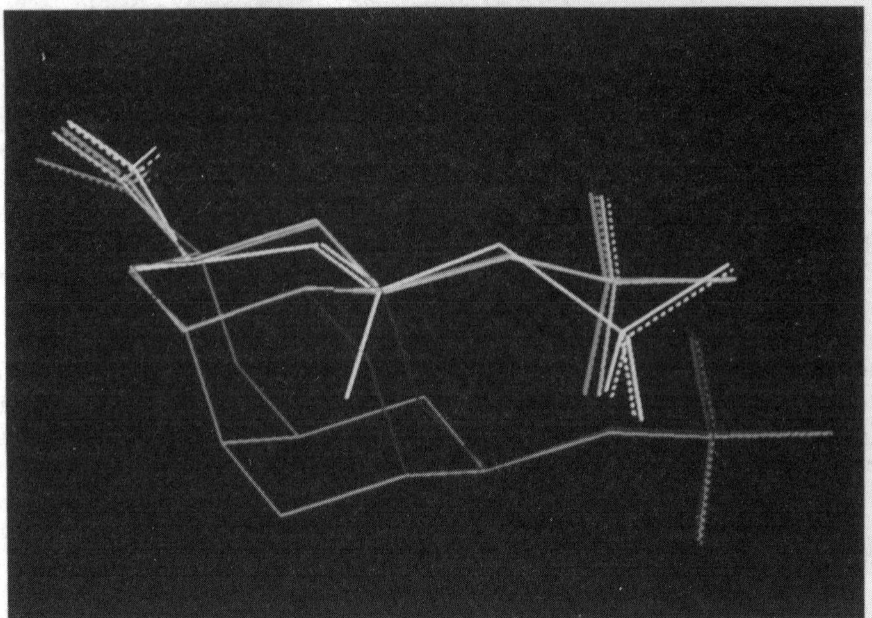

Fig. 5 — (**See colour section.**) Overlap of 2*R*,4*S*-CGS 19755 (in green), 3*S*,4a*R*,6*S*,8a*R*-LY235959 (in blue), 3*R*,4a*S*,6*S*,8a*S*-LY288534 (in red), 2*R*-AP5 (in yellow), and 2*R*-CGP 40116 (in white). To provide some perspective in this two-dimensional drawing, the orientation of this overlap places the carboxylic acid and phosphonic acid groups behind the plane of the paper.

CONCLUSION

Much has been elucidated since the discovery by Watkins that AP5 and AP7 were potent and selective NMDA receptor antagonists. Many new NMDA antagonists have been described which are more potent both *in vitro* and *in vivo* than AP5 and AP7 and which show excellent *in vivo* activity following parenteral and oral administration. The structural diversity of these novel competitive antagonists provides new insight into the steric and spatial requirements for activity at the glutamate recognition site. Considerable evidence supports the speculation that competitive NMDA antagonists will be useful in the treatment of major central nervous system disorders. Clinical evaluation in man is on the horizon, and will ultimately determine the therapeutic utility of these compounds.

REFERENCES

Aebischer, B., Frey, P., Haerter, H.-P., Herrling, P. L., Mueller, W., Olverman, H. J. and Watkins, J. C. (1989) *Helv. Chim. Acta* **72** 1043–1051.

Angst, C., Brundish, D. E., Dingwall, J. G. and Fagg, G. E. (1987) *Eur. Pat. Appl.* 0 233 154.

Bigge, C. F., Drummond, J. T., Johnson, G., Malone, T., Probert, A. W., Jr, Marcoux, F. W., Coughenour, L. L. and Brahce, L. J. (1989a) *J. Med. Chem.* **32** 1580–1590.

Bigge, C. F., Hays, S. J., Novak, P. M., Drummond, J. T., Johnson, G. and Bobovski, T. P. (1989b) *Tetrahedron Lett.* **30** 5193–5196.

Bigge, C. F., Drummond, J. T. and Johnson, G. (1989c) *Tetrahedron Lett.* **30** 7013–7016.

Bigge, C. F., Johnson, G., Marcoux, F. W., Probert, A. W., Jr, Coughenour, L. L. and Brahce, L. J. (1989d) *Soc. Neurosci. Abstr.* **15** 1167.

Chenard, B. L., Lipinski, C. A., Dominy, B. W., Mena, E. E., Ronau, R. T., Butterfield, G. C., Marinovic, L. C., Pagnozzi, M., Butler, T. W. and Tsang, T. (1990) *J. Med. Chem.* **33** 1077–1083.

Collingridge, G. L. and Lester, R. A. J. (1989) *Pharmacol. Rev.* **40** 143–210.

Collingridge, G. L. and Singer, W. (1990) *Trends Pharmacol. Sci.* **11** 290–296.

Cordi, A. A. and Vazquez, M. L. (1990) *Eur. Pat. Appl.* 0 364 996.

Dappen, M. S., Pellicciari, R., Natalini, B., Monahan, J. B., Chiorri, C. and Cordi, A. A. (1991) *J. Med. Chem.* **34** 161–168.

Evans, R. H., Francis, A. A., Jones, A. W., Smith, D. A. S. and Watkins, J. C. (1982) *Br. J. Pharmacol.* **75** 65–75.

Fagg, G. E., Olpe, H.-R., Pozza, M. F., Baud, J., Steinmann, M., Schmutz, M., Portet, C., Baumann, P., Thedinga, K., Bittiger, H., Allgeier, H., Heckendorn, R., Angst, C., Brundish, D. and Dingwall, J. G. (1990) *Br. J. Pharmacol.* **99** 791–797.

Ferkany, J. W., Kyle, D. J., Willets, J., Rzeszotarski, W. J., Guzewska, M. E., Ellenberger, S. R., Jones, S. M., Sacaan, A. I., Snell, L. D., Borosky, S., Jones, B. E., Johnson, K. M., Balster, R. L., Burchett, K., Kawasaki, K., Hoch, D. B. and Dingledine, R. (1989) *J. Pharmacol. Exp. Ther.* **250** 100–109.

France, C. P., Woods, J. H. and Ornstein, P. L. (1989) *Eur. J. Pharmacol.* **159** 133–139.

Hays, S. J., Bigge, C. F., Novak, P. M., Brummond, J. T., Bobovski, T. P., Rice, M. J., Johnson, G., Brahce, L. J. and Coughenour, L. L. (1990) *J. Med. Chem.* **33** 2916–2924.

Heyes, M. P., Rubinow, D., Lane, C. and Markey, S. P. (1989) *Ann. Neurol.* **26** 275–277.

Humbelt, C., Johnson, G., Malone, T. and Ortwine, D. F. (1990) Proceedings of the XIth International Symposium on Medicinal Chemistry, Jerusalem, Israel, September 2–7, p. 26.

Hutchinson, A. J., Williams, M., Angst, C., de Jesus, R., Blanchard, L., Jackson, R. H., Wilusz, E. J., Murphy, D. E., Bernard, P. S., Schneider, J., Campbell, T., Guida, W. and Sills, M. A. (1989) *J. Med. Chem.* **32** 2171–2178.

Johnson, G. (1989) *Ann. Rep. Med. Chem.* **24** 41–50.

Leander, J. D., Lawson, R. R., Ornstein, P. L. and Zimmerman, D. M. (1988) *Brain Res.* **448** 115–120.

Leander, J. D., Ornstein, P. L., Schoepp, D. D. and Salhoff, C. R. (1989) *Soc. Neurosci. Abstr.* **15** 1166.

Lehmann, J., Schneider, J., McPherson, S., Murphy, D. E., Bernard, P., Tsai, C., Bennet, D. A., Pastor, G., Steel, D. J., Boehm, C., Cheney, L., Liebman, J. M., Williams, M. and Wood, P. L. (1987) *J. Pharmacol. Exp. Ther.* **240** 737–746.

Lehmann, J., Hutchinson, A. J., McPherson, S. E., Mondadori, C., Schmutz, M., Sinton, C. M., Tsai, C., Murphy, D. E., Steel, D. J., Williams, M., Cheney, D. L. and Wood, P. L. (1988) *J. Pharmacol. Exp. Ther.* **246** 65–75.

Meldrum, B. S. and Garthwaite, J. (1990) *Trends Pharmacol. Sci.* **11** 379–387.

Meldrum, B. S., Croucher, M. J., Czuczwar, S. J., Collins, J. F., Curry, K., Joseph, M. and Stone, T. W. (1983) *Neuroscience* **9** 925–930.

Murphy, D. E., Schneider, J., Boehm, C., Lehman, J. and Williams, M. (1987) *J. Pharmacol. Exp. Ther.* **240** 778–784.

Murphy, D. E., Hutchinson, A. J., Hurt, S. D., Williams M. and Sills, M. A. (1988) *Br. J. Pharmacol.* **95** 932–938.

Ornstein, P. L. (1990) *US Pat.* 4 902 695.

Ornstein, P., Zimmerman, D. M., Hynes, M. D., III and Leander, J. D. (1987) In: Hicks, T. P., Lodge, D. and McLennan, H. (eds) *Neurology and Neurobiology, Vol. 24. Excitatory Amino Acid Transmission.* New York, pp. 123–126.

Ornstein, P. L., Schaus, J. M., Chambers, J. W., Huser, D. L., Leander, J. D., Wong, D. T., Paschal, J. W., Jones, N. D. and Deeter, J. B. (1989) *J. Med. Chem.* **32** 827–833.

Ornstein, P. L., Schoepp, D. D., Arnold, M. B., Leander, J. D., Lodge, D., Paschal, J. W. and Elzey, T. (1991a) *J. Med. Chem.* **34** 90–97.

Ornstein, P. L., Arnold, M. B., Augenstein, N. K., Schoepp, D. D., Leander, J. D. and Lodge, D. (1991b) In: Meldrum, B., Moroni, F., Simon, R. and Woods, J. (eds) *Excitatory Amino Acids 1990 (Fidia Research Foundation Symposium Series).* Raven Press, New York, pp. 539–545.

Plaitakis, A. (1990) *Ann. Neurol.* **28** 3–8.

Rzeszatarski, W. J., Hudkins, R. L. and Guzewska, M. E. (1987) *US Pat.* 4 657 899.

Schaff, T. K. and Hess, H.-J. (1979) *J. Med. Chem.* **22** 1340–1346.

Schoepp, D. D., Salhoff, C. R., Hillman, C. C. and Ornstein, P. L. (1989) *J. Neural Transm.* **78** 183–193.

Schoepp, D. D., Gamble, A. Y., Salhoff, C. R., Johnson, B. G. and Ornstein, P. L. (1990a) *Eur. J. Pharmacol.* **182** 421–427.

Schoepp, D. D., Ornstein, P. L., Leander, J. D., Lodge, D., Salhoff, C. R., Zeman, S. and Zimmerman, D. M. (1990b) *J. Pharmacol. Exp. Ther.* **255** 1301–1308.

Schoepp, D. D., Ornstein, P. L., Salhoff, C. R. and Leander, J. D. (1991) *J. Neural Transm.* **85**, 131–143.

Sills, M. A., Fagg, G., Pozza, M., Angst, C., Brundish, D. E., Hurt, S. D., Wilusz, E. J. and Williams, M. (1991) *Eur. J. Pharmacol.* **192** 19–24.

Whitten, J. P., Baron, B. M., Muench, D., Miller, F., White, H. S. and McDonald, I. A. (1990) *J. Med. Chem.* **33** 2963–2970.

Willets, J., Balster, R. L. and Leander, J. D. (1990) *Trends Pharmacol. Sci.* **11** 423–428.

9

Ligands for NMDA receptor modulatory sites

David Lodge
Department of Veterinary Basic Sciences, Royal Veterinary College, London NW1 0TU, UK

ABSTRACT

The *N*-methyl-D-aspartate (NMDA) subtype of glutamate receptor is one of the most studied ligand-gated ion channels in the mammalian central nervous system. The receptor–channel complex is modulated in a voltage-dependent manner by normal extracellular concentrations of magnesium ions. This, coupled with calcium permeability, provides the complex with a unique physiological function. Also within the channel are binding sites for organic molecules such as phencyclidine — this site has strong conformational and isomeric requirements. Some of the known pharmacological effects of phencyclidine and related compounds are mediated via the block of the NMDA receptor-coupled channel. The function of the channel is normally facilitated by normal extracellular concentrations of glycine, although further modulation by synaptically released glycine remains an open question. Antagonists and agonists for this site are now well documented. Facilitatory and inhibitory actions on NMDA receptor function have also been described for several polyamines, which may include some arthropod toxins. Structure–activity relationships for these established modulatory drugs, as well as some more speculative ones, are discussed.

INTRODUCTION

The development and acceptance of the notion of subclasses of receptors for glutamate, aspartate and other putative excitatory amino acid (EAA) transmitter candidates has depended on the development of agonists and antagonists selective for such subclasses. Amongst these have been several groups of substances which selectively affect responses to *N*-methyl-D-aspartate (NMDA) but in a manner that suggests they do not act directly at the recognition site for this EAA. Indeed the descriptions of the selective NMDA antagonistic action of magnesium (Evans *et al.* 1977) and HA-966 (Evans *et al.* 1978) were some of the earliest pieces of evidence suggesting that this EAA activated a subclass of glutamate receptor now known as the NMDA receptor. Furthermore the demonstrations that both magnesium and the selective NMDA antagonists related to phencyclidine (Lodge *et al.* 1988) act as channel blockers (see below) strengthened the concept of an NMDA receptor–ionophore complex distinct from those activated by 2-amino-3-(3-hydroxy-5-methylisoxazol-4-yl)propanoate (AMPA) and kainate. This chapter will deal with both inorganic and organic substances which modulate the actions of NMDA but which are thought not to act directly at the NMDA recognition site. In view of the extensive literature in this area only a selection of key primary references and review articles will be cited.

MAGNESIUM BLOCK OF THE NMDA RECEPTOR-COUPLED ION CHANNEL

The seminal observation that magnesium and related divalent cations are selective but not competitive NMDA antagonists was reported in more detail by Ault *et al.* (1980), using the frog spinal cord preparation. As stated above, this finding, together with development of competitive NMDA antagonists (see chapter 8), led to the subdivision of EAA receptors. However, it was not until the patch- and voltage-clamp studies of Nowak *et al.* (1984) and Mayer *et al.* (1984) that the mode of action of magnesium was elucidated. It is now clear that magnesium blocks the NMDA channel in a voltage-dependent manner, so that at membrane potentials between − 100 and − 90 mV NMDA receptor activation leads to little current flow (see MacDonald and Nowak 1990 for a recent review). Current flow increases (negative slope conductance) as the membrane potential falls to − 40 mV. Below − 40 mV, magnesium exerts little effect on channel conductance and the full effect of the large single channel openings, about 50 pS for 10 ms, becomes apparent. This action of magnesium occurs at micromolar concentrations — well below those that occur extracellularly. Although physiologically magnesium is the most important cation with respect to this phenomenon, the characteristic flickery voltage-dependent channel block is also shared by cobalt, nickel, manganese and, to a lesser extent, zinc (see below). The site of this action is thought to be deep within the channel, where there may be multiple binding sites for magnesium (see MacDonald and Nowak 1990).

The voltage-dependent effects of magnesium has extremely important physiological consequences. Within a range of normal resting membrane potentials, say − 80 to − 60 nV, current flow through the activated NMDA receptor-coupled channel will increase almost linearly with depolarization. Even at relatively hyperpolarized levels, however, NMDA receptor activation will lead to depolarizing current flow; the current–voltage curve only approaches the zero current axis at − 100 mV. The important consequences of the voltage-dependent block are:

(1) that prolonged NMDA receptor activation will have the effect of producing a regenerative current, a form of positive feedback; and
(2) that depolarization of a neuron by other means will result in greater response to NMDA receptor activation.

Since NMDA receptor-coupled ion channels are permeant to calcium, this action of magnesium is important for both rhythmogenicity (Headley and Grillner 1990) including epileptogenesis (Dingledine *et al.* 1990), for synaptic development and plasticity (Collingridge and Singer 1990) and for neurotoxicity (Meldrum and Garthwaite 1990). Changes in extracellular magnesium levels are likely to have profound effects on NMDA receptor-mediated events in addition to many other processes affecting neuronal excitability.

Although the interaction of many drugs with the NMDA receptor–channel complex is affected by magnesium, there is no hard evidence that any organic ligand acts at this site (see below).

ZINC AS AN NMDA ANTAGONIST

Other cations, such as zinc and cadmium, are also able to block the actions of NMDA but, unlike that described above for magnesium, the block is less voltage-dependent and does not require channel opening (see Mayer *et al.* 1989). Hence the binding site for these cations is likely to be near the extracellular surface of the channel. Zinc has a K_D of about 10 μM for this effect against NMDA whereas concentrations above 30 μM produce a magnesium-like flickery channel block which is not seen with cadmium (Mayer *et al.* 1989, Legendre and Westbrook 1990). Although they are selective NMDA antagonists in isolated cell systems, as might be expected from their known actions on enzyme and other sulphydryl-containing systems, these two cations have other effects in brain slices that prevent a demonstration of selectivity between the various EAA agonists (Lam *et al.* 1991). It is, however, possible that some of the NMDA antagonist effects of magnesium are mediated via this 'zinc site'. IC_{50} values for magnesium as an NMDA antagonist in brain slice and spinal cord preparations are between 0.5 and 1.0 mM, i.e. considerably greater than those that cause the flickery voltage-dependent block of NMDA channels in patch-clamp studies.

Beside these effects on NMDA actions, zinc is reported to block $GABA_A$ responses with similar potency (Mayer and Vyklicky 1989). A recent study of CA3 hippocampal neurons, however, suggested that $GABA_A$ and NMDA receptor-mediated events were relatively unchanged in 300 μM zinc but that $GABA_B$-mediated inhibitions were blocked (Xie and Smart 1991).

Interest in the zinc site as a possible target for modulating NMDA receptor function was awakened by the discovery that certain tricyclics, such as desipramine and promazine (see Fig. 4), had similar kinetics to zinc in MK-801 binding assays, albeit at concentrations way above those that therapeutically influence monoamines (Reynolds and Miller 1988). In electrophysiological studies, however, the block is somewhat more voltage-dependent and use-dependent than that of zinc (Sernagor *et al.* 1989). When tested on cortical slices *in vitro*, no selective NMDA antagonism was seen with a series of such compounds, but rather there was a non-selective reduction of responses to all EAA agonists tested, with little sign of recovery on return to drug-free medium (Jones *et al.* 1989). Development of more selective and more potent compounds than desipramine is likely and is awaited with interest.

PHENCYCLIDINE SITE

The discovery that ketamine and phencyclidine (PCP) selectively blocked both polysynaptic reflexes and the excitation by NMDA of central neurons *in vivo* (Lodge and Anis 1982) was followed by an extensive structure–activity study (see Lodge *et al.*

1988, Lodge and Johnson 1990) based largely on those compounds which displaced binding at a high-affinity PCP site on rat brain membranes (Zukin and Zukin 1979) and which produced similar behavioural effects to PCP in animals and man (Shannon 1983; see Willetts *et al.* 1990). In addition to the arylcyclohexylamines such as ketamine, PCP and their thienyl derivatives, tiletamine and *N*-(1-(2-thienyl)cyclohexyl)piperidine (TCP), other dissociative anaesthetics, such as dexoxadrol and etoxadrol, also proved to be selective NMDA antagonists. Several benzomorphans, known as sigma opiates, such as *N*-allylnormetazocine (SKF 10047), cyclazocine and pentazocine, and bridged benz(*f*)isoquinolines and morphinans such as dextrorphan and dextromethorphan were all shown to be NMDA antagonists (Lodge *et al.* 1988). Despite the structural diversity of compounds mimicking PCP (see Fig. 1), small changes to PCP itself markedly affect the binding affinity. Thus, 3-amino and 3-nitro substitutions produce compounds, respectively, 2.5 times more and 100 times less potent than PCP itself (Nadler *et al.* 1990). Stereoselectivity within these drugs was variable in relation to NMDA antagonism. Thus, the (+)-isomers of ketamine, 3-methyl-PCP, dextrorphan and *N*-allyl-normetazocine are more potent than the (−)-isomers, whereas the reverse is the case with α- and β-cyclazocine and pentazocine. Of the compounds acting like PCP, the most potent yet reported is dizocilpine, better known as MK-801 (Wong *et al.* 1986), although (−)-β-cyclazocine is almost as potent but shorter-acting (Church *et al.* 1991). The potency of all these compounds correlated closely with their ability to displace PCP binding and to evoke PCP-like discriminative effects in behavioural studies in animals (see Martin and Lodge 1989). Indeed the absolute values required to reduce responses of spinal neurons *in vivo* to NMDA following systemic administration are similar to those used in behavioural studies. It therefore seems likely that block of synaptic activity mediated via NMDA receptors underlies some of the behavioural effects common to this group of substances.

The subsequent elucidation of the mode of action of these PCP-like compounds resulted from multidisciplinary approaches from several laboratories throughout the world. First, non-parallel shifts in NMDA dose–response curves were obtained *in vitro* (Lodge and Johnston 1985, Snell and Johnson 1985). Secondly, the actions of PCP-like drugs, particularly of MK-801, were shown to be use-dependent, i.e. antagonism developed with exposure to the agonist (see Kemp *et al.* 1987). Thirdly, the channel block was shown to be voltage-dependent (and use-dependent) in a manner not dissimilar to that for magnesium but with no flickering and a somewhat flatter current–voltage curve (see MacDonald and Nowak 1990). Fourthly, although the regional distribution of binding in the brain for PCP closely correlated with that for NMDA (Maragos *et al.* 1988), binding to the NMDA recognition site was not displaced by PCP-like compounds (Fagg 1987, Foster and Wong 1987). Fifthly, NMDA agonists potentiated whereas antagonists prevented binding to the PCP site but agonists increased and antagonists decreased dissociation of bound PCP ligands (Loo *et al.* 1986, Kloog *et al.* 1988). The conclusion from the above is that the site of action of PCP is within the NMDA receptor channel and that access to and from this site is dependent on channel opening by NMDA receptor agonists. Indeed, the varying levels of endogenous agonists between preparations may have accounted for some of the differences in PCP binding in early non-equilibrium studies. Nevertheless the debate continues as to whether this PCP site is deep within the channel or near the

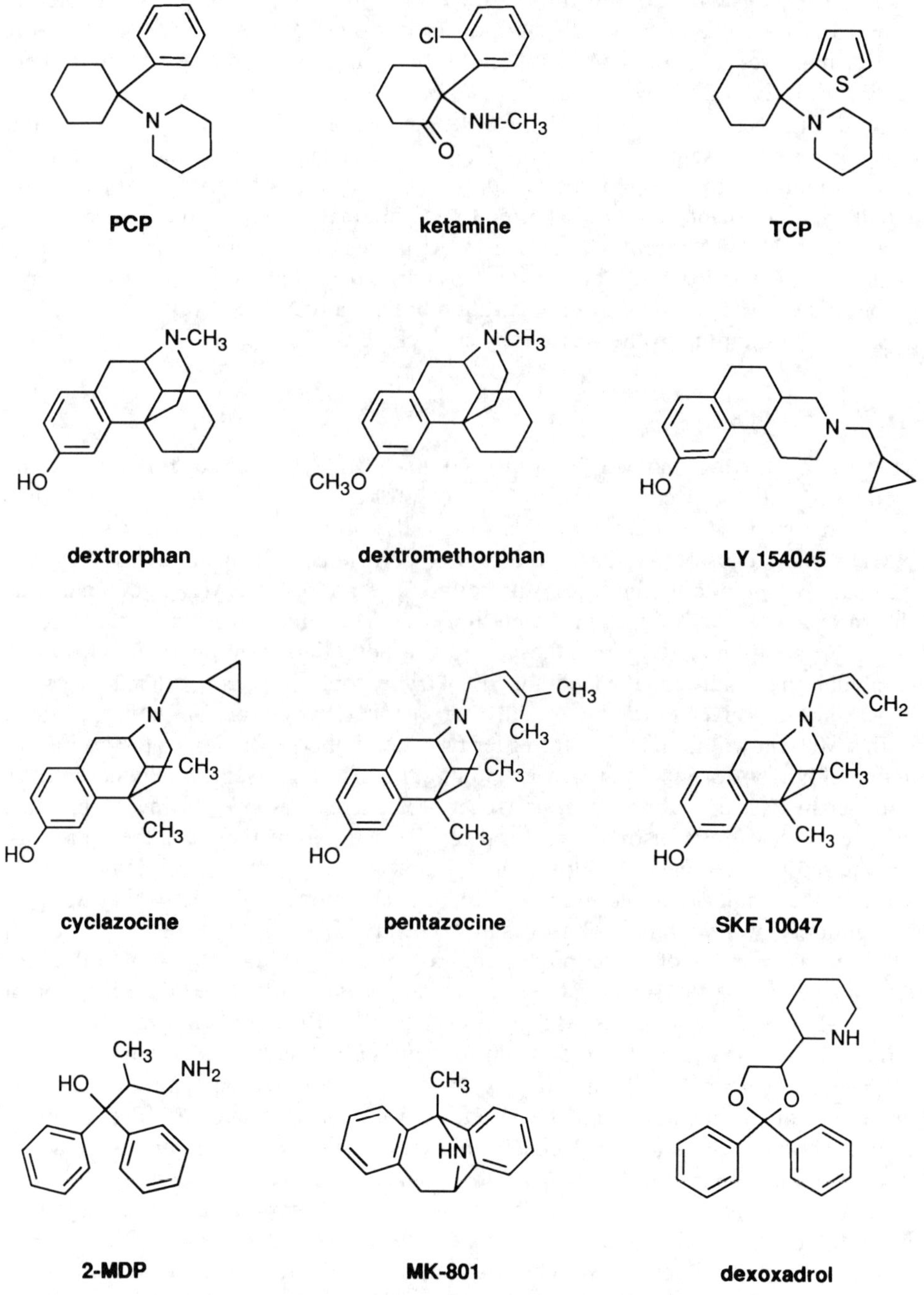

PCP **ketamine** **TCP**

dextrorphan **dextromethorphan** **LY 154045**

cyclazocine **pentazocine** **SKF 10047**

2-MDP **MK-801** **dexoxadrol**

Fig. 1 — Compounds active as NMDA antagonists. All have similar profile of action, displace TCP or MK-801 binding, and share behavioural properties and therapeutic potential as anticonvulsants and neuroprotectants.

vestibule and whether it interacts with a magnesium binding site (see MacDonald and Nowak 1990).

Following systemic administration, most PCP-like compounds cross the blood–brain barrier easily and in general, with the exception of MK-801, have considerably shorter half-lives than the competitive NMDA antagonists. This clearly gives them therapeutic advantages over competitive agents. On the other hand, PCP has other actions in the central nervous system (CNS) and several of the benzomorphans and morphinans act at naloxone-sensitive opiate sites. Such side-effects may detract from or contribute to the desired therapeutic actions. For example, the kappa opiate activity of cyclazocine might be expected to potentiate the neuroprotective effects that accrue from NMDA receptor antagonism. At present the potential for psychotomimetic activity and for acute but reversible neuronal vacuolation in electron micrographs of neocortex following administration of PCP and MK-801 (Olney *et al.* 1989) appear to be standing in the way of any such clinical trials.

GLYCINE SITE

The detective work that led Johnson and Ascher (1987) to the discovery of the modulatory site for glycine on the NMDA receptor complex is worthy of report. They observed that in their culture of neuronal cells the amplitude of the response to NMDA was inversely related to the rate of perfusion with medium and this was particularly true of older cultures. Furthermore, responses to NMDA got smaller as the patch was moved higher in the culture dish. They inferred from this that some factor, probably emanating from the culture, was potentiating the action of NMDA. A combination of analysing the medium and of trying some of the constituents led to the conclusion that glycine was the modulatory factor. Low micromolar concentrations of this well-known inhibitory transmitter provided full potentiation of the excitatory action of NMDA or glutamate! By examining the effects of glycine analogues in their patch-clamp system, Johnson and Ascher were able to show, however, that this glycine site had quite distinct structural requirements from those for the inhibitory glycine receptor coupled to chloride ion channels. For example, D-alanine and D-serine rather than β-alanine mimicked this novel action of glycine, which was not antagonized by strychnine (Johnson and Ascher 1987).

The precise mode of action of glycine is as yet unclear. Mayer and his colleagues (e.g. Vyklicky *et al.* 1990a) provide evidence that glycine reduces the desensitization of the NMDA receptor whereas Ascher (Sather *et al.* 1990) found that in outside-out patches there was still evidence of desensitization in high micromolar concentrations of glycine. This issue will only be resolved with more experimentation but it is apparent from most studies that glycine site activation is needed for NMDA receptors to be functional. Omission of glycine from the extracellular medium almost prevents NMDA-induced currents (Kleckner and Dingledine 1988) and binding of TCP or MK-801 (Snell *et al.* 1987, Reynolds *et al.* 1987). Furthermore, glycine antagonists block these functional effects of NMDA (Thomson 1990, Fletcher *et al.* 1990).

The previously described strychnine-insensitive binding (see Bristow *et al.* 1986), which has almost identical distribution to that for NMDA and PCP ligands and identical pharmacology, has proved to be the biochemical substrate for this electrophysiological effect (see Fletcher *et al.* 1990). Combination of binding and electrophysiological studies have increased our knowledge of the structure–activity requirements for this site. Only small substitutions are allowable on the parent glycine compound; methyl (alanine), hydroxymethyl (serine) and cyclopropyl on the α-carbon are all good agonists (Kleckner and Dingledine 1989, Lodge and Johnson 1990); whereas larger substituents here reduce potency or result in partial agonists or antagonists (see below and Fig. 2).

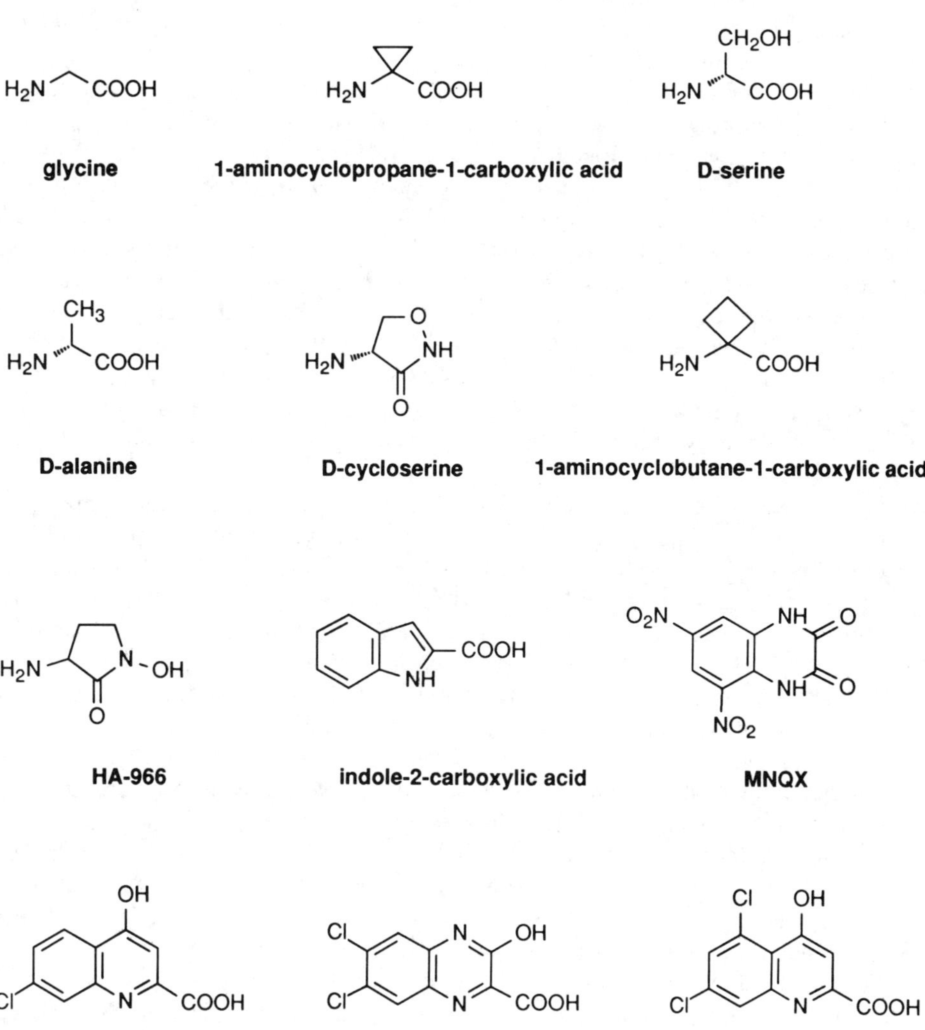

Fig. 2 — Compounds active at the glycine site on the NMDA receptor complex. Structures from glycine through to D-cycloserine are agonists in decreasing order of potency. 1-Aminocyclobutane-1-carboxylate and HA-966 are weak antagonists with partial agonist properties. Indole-2-carboxylate to 5,7-dichlorokynurenate are antagonists in increasing order of potency, although 6,7-dichloro-3-hydroxyquinoxaline-2-carboxylate is also a non-NMDA antagonist.

On more intact tissue, we and others (Fletcher and Lodge 1988, Kemp *et al.* 1988, Birch *et al.* 1988) were unable to show the potentiating effect of glycine on responses to NMDA and surmised that extracellular glycine levels were sufficient to activate fully this modulatory site. It now appears, however, that in some situations this is not the case (see Thompson 1989 for a review) and hence there is the possibility that NMDA receptor function can be enhanced by increased tissue levels of glycine, as originally suggested by Johnson and Ascher (1987). Indeed, the hyper-irritability and seizures in a 4-month-old patient with non-ketotic hyperglycinaemia with cerebrospinal fluid (CSF) levels of 240 μM glycine showed considerable improvement on treatment with ketamine (Ohya *et al.* 1991).

It has, however, been easier to demonstrate the importance of this site by the use of selective glycine antagonists. In fact probing of this site led to the elucidation of the modes of action of HA-966 (Fletcher and Lodge 1988) and kynurenate (Kessler *et al.* 1989), two early NMDA antagonists with previously unclear modes of action. The antagonism by HA-966 and kynurenate, but not that of competitive NMDA antagonists or of PCP-like compounds, could be readily reversed by exogenous glycine or D-serine. The action of the pyrrolidone HA-966 now appears to be that of a weak partial agonist (Foster and Kemp 1989), a feature of some other cyclic analogues of glycine (Fletcher *et al.* 1990, Thompson 1990). Quinoxalinediones, better known as competitive antagonists at non-MNDA receptors, also proved to have actions at the glycine modulatory site (Birch *et al.* 1988) which explained their poor selectivity between responses induced by non-NMDA and NMDA agonists (Honore *et al.* 1988).

These three structures — pyrrolidones, quinoxalinediones and kynurenates — have been used as the base for synthesis of putative glycine antagonists. The (+)-isomer of HA-966 is the more active but the addition of a methyl group at the 4-position ((+)-*cis*-4-methyl-HA-966 or L-687414) produces a glycine antagonist ten-fold more potent than (+)-HA-966 but still with some weak agonist activity (Kemp *et al.* 1991). This compound, although intrinsically less potent than some of the kynurenates mentioned below, is able to cross the blood–brain barrier; for example, it is anticonvulsant in a variety of seizure models at doses of 5-25 mg/kg (Saywell *et al.* 1991).

The potential for development based on the quinoxaline moiety has recently been demonstrated (see chapter 11); 5,7-dinitro-quinoxalinedione (MNQX) is 30-fold more potent than the 6,7-analogue (DNQX) at the glycine site and five-fold less potent at AMPA/kainate receptors (Sheardown *et al.* 1989), whereas the 6-nitro-7-sulphamoylbenzo(*f*)-analogue (NBQX) is four-fold more potent at the AMPA receptor and considerably less potent at the glycine site (Sheardown *et al.* 1990).

Various halide substitutions, mostly chloride or fluoride, in the 5–8-positions of kynurenic acid also provide NMDA antagonist activity, largely by displacing endogenous glycine from its modulatory site (Kleckner and Dingledine 1989). Unlike the parent compound, which is both weak and non-selective for NMDA responses, the first such analogue described, 7-chlorokynurenate, was a potent, selective and full NMDA antagonist, acting competitively at the glycine site (Kemp *et al.* 1988). This is the most potent of the single chloride substitutions tested but the 6,7- and 5,7-dichlorokynurenate and 6,7-dichloro-3-hydroxyquinoxaline-2-carboxylate (see Fig. 2) are similarly potent (McNamara *et al.* 1990). More importantly, however, of these only 5,7-dichlorokynurenate exhibits real selectivity for the glycine site; it is 500-fold more potent against responses of NMDA than those of kainate, whereas the equivalent figure for 7-chlorokynurenate is 40 (McNamara *et al.* 1990). Fluoride and bromide substitutions have thus far been less potent and results for iodide substitutions are awaited.

Other glycine antagonists include 1-amino-cyclobutane- and -cyclopentane-carboxylates and indole-2-carboxylate, although these are considerably weaker than the chlorokynurenates and MNQX (see Lodge and Johnson 1990, McNamara *et al.* 1990). The two-point binding site and associated pockets for glycine agonists and antagonists have recently been modelled (Fletcher *et al.* 1990).

Several of the above glycine antagonists have been shown to have anticonvulsant and neuroprotective properties and hence share the therapeutic potential of other types of NMDA antagonists. They may, however, have advantages over competitive NMDA antagonists in not being displaced by high extracellular concentrations of glutamate and over PCP-like compounds by inducing less marked behavioural effects; competitive NMDA antagonists have a profile of behavioural effects different from PCP-like compounds (Willetts *et al.* 1990) and (+)-HA-966 reduces some of stimulant effects of PCP (Bristow *et al.* 1991). Partial agonists may also have their use here; complete abolition of NMDA function may not be desirable even in such an acute and life-threatening situation as cerebral ischaemia. Certainly some of these compounds should be tested in hyperglycinaemic patients.

If, as seems possible, glycine levels are not fully saturating (Thomson 1989, 1990), glycine site agonists may have benefits in promoting the more physiological aspects of NMDA receptor-mediated events. Indeed, D-cycloserine, a weak glycine site agonist, and milacemide, a precursor of glycine, have been promoted as possible enhancers of memory (Handelmann *et al.* 1988) and, if established as such, could find many suitable patients!

POLYAMINE SITE

Polyamines, such as spermine and spermidine (see Fig. 3), enhance the binding of PCP-like compounds to the NMDA receptor–channel complex, suggesting positive allosteric modulation of this complex at a site distinct from that of NMDA and glycine (Ransom and Stec 1988, Williams *et al.* 1989, Reynolds 1990, Sacaan and Johnson 1990). That this is receptor-mediated rather than a non-specific membrane effect is suggested by two facts. First, there is a clear SAR; being effective at submicromolar concentrations, *N,N′*-bis(3-aminopropyl)-1,3-propanediamine is one of the most potent agonists yet described (see Lodge and Johnson 1990). Secondly, several antagonists, e.g. putrescine, have been described which displace the spermine dose–response curves in a parallel fashion. Of these, arcaine, with a K_i of about 10 μM, is the most potent. It also appears that activation of this site is necessary for normal NMDA receptor function; polyamine antagonists block the NMDA receptor-stimulated binding of PCP-like compounds (Reynolds 1990, Sacaan and Johnson 1990), although an inverse agonist role cannot be excluded (Williams *et al.* 1990).

Certain arthropod toxins (see chapter 13), e.g. argiotoxin and philanthotoxin (see Fig. 3), have long polyamine tails and hence it is not unreasonable to propose that this polyamine site is responsible for some of their NMDA blocking properties (Jackson

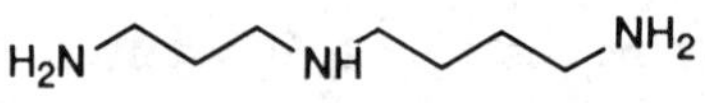

spermidine

N,N'-bis(3-aminopropyl)-1,3-propanediamine

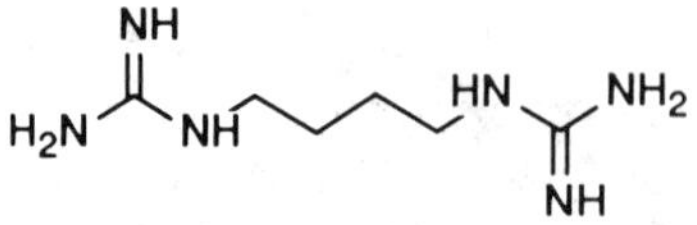

arcaine

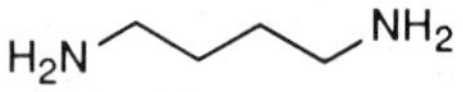

putrescine

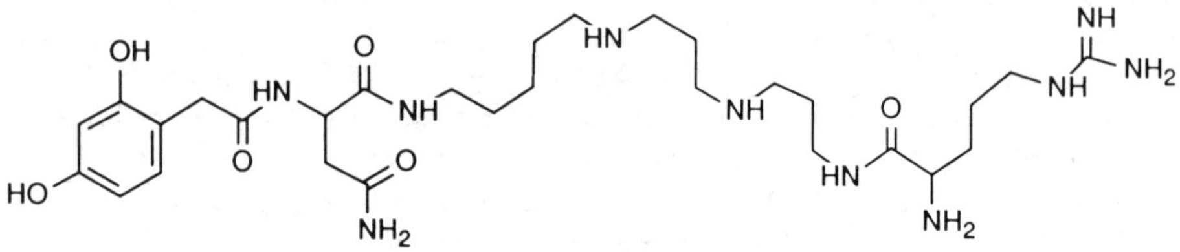

argiotoxin 636

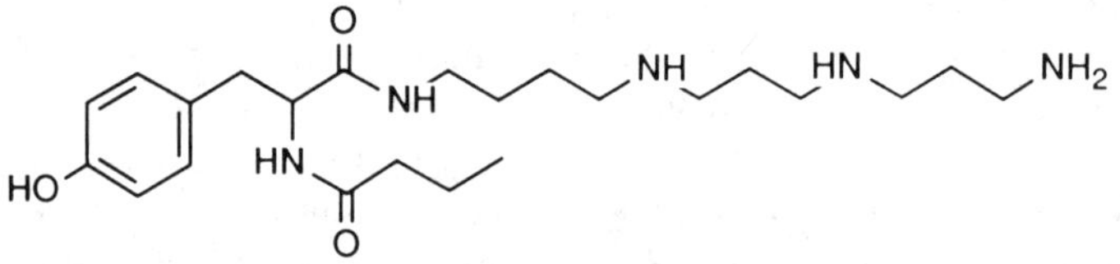

philanthotoxin 433

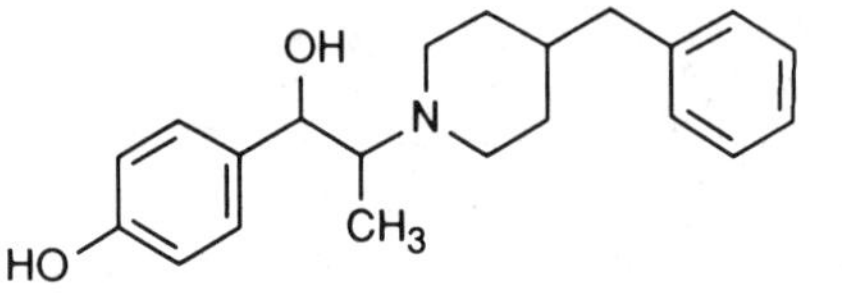

ifenprodil

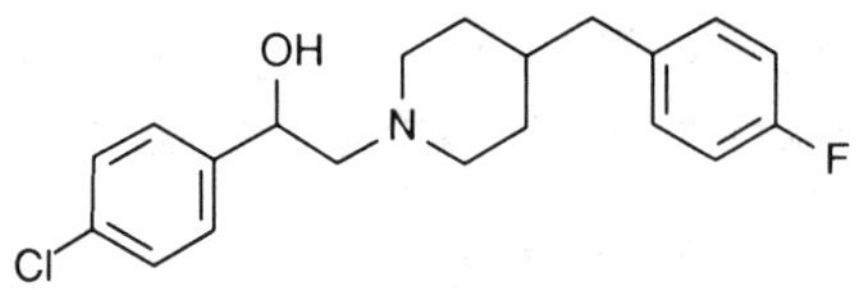

SL 82.0715

Fig. 3 — Putative modulators of NMDA via the polyamine site. The top two rows are agonists and antagonists, respectively. Argiotoxin and philanthotoxin have NMDA antagonist properties more akin to channel blockers. Ifenprodil and SL 82.0715 competitively antagonize actions of spermidine.

and Usherwood 1988, Priestley *et al.* 1989, Anis *et al.* 1990), although it should be mentioned that *in vivo* philanthotoxin reduces AMPA/kainate responses to a greater extent than those of NMDA (Jones *et al.* 1990). With respect to this hypothesis, however, spermidine did not alter the potency of argiotoxin$_{636}$ in MK-801 binding studies but the toxin did reduce the effects of magnesium on MK-801 binding (Reynolds 1991). Previously, it had been shown that argiotoxin$_{636}$ binding was inhibited by magnesium (Mena *et al.* 1989), that like magnesium, argiotoxin$_{636}$

increased glycine binding (Gullack *et al.* 1988) and that polyamines share a site of action with magnesium as inhibitors of PCP site binding (Sacaan and Johnson 1991, Reynolds 1990). Interestingly, the 17 amino acid peptide toxin conantokin-G, a novel non-competitive NMDA antagonist without action on the PCP site, also enhances glycine binding (Mena *et al.* 1990). Whatever the site of action of this peptide toxin, a complex interaction between magnesium, polyamines and arthropod toxins is emerging, which perhaps is not surprising since they all carry considerable positive charges.

The physiological significance of the polyamine site is not yet known. Very few physiological experiments have been attempted; the location, intracellular or extracellular, of the site is not established and the endogenous levels of the various polyamine agonists and antagonists and their interactions are still the subject of debate. Thus, NMDA receptor stimulation activates ornithine decarboxylase, the rate-limiting enzyme for polyamine synthesis, resulting in the possibility of a positive (Siddiqui *et al.* 1988) or negative feedback depending on the resultant polyamine mix; some polyamines antagonize the neurotoxic actions of glutamate (Gilad and Gilad 1988).

Ifenprodil and the related SL 82.0715 (see Fig. 3) may produce their neuroprotective actions via the polyamine site (Carter *et al.* 1989). Both drugs produce a shift to the right of dose–response curves for spermidine enhancement of PCP site binding, although in the case of ifenprodil it is not a parallel shift. Indeed, Reynolds and Miller (1989) argue that the kinetics of ifenprodil's action on MK-801 binding is different from that of polyamine antagonists. Nevertheless, this compound has novel non-competitive NMDA antagonist properties, is neuroprotective but does not act at the NMDA, glycine or PCP sites (Carter *et al.* 1988).

OTHER MODULATORY SITES

Ethanol has recently received considerable attention as a modulator of NMDA receptor function (Gonzales 1990). Although earlier studies suggested a non-specific antagonism of EAA responses (Teichberg *et al.* 1984), recently Lovinger *et al.* (1989) showed that 5–50 mM concentrations of ethanol were selective for responses of hippocampal neurons to NMDA. Whether these effects of ethanol are directed at the NMDA macromolecular complex or the lipid–protein interface is not yet known. These observations have largely been substantiated in several other preparations (see Gonzales 1990), although in many cases doses above 50 mM are required to produce robust effects. In contrast to these *in vitro* studies, the injection of up to 4 g/kg of ethanol i.v. did not change systematically responses of spinal neurons to NMDA or other EAAs (Jones and Lodge, unpublished observations). A possible explanation for this discrepancy is related to the ability of micromolar concentrations of glycine to reverse the effects of ethanol and *vice versa* (see Gonzales 1990). Such levels of glycine are likely to be present *in vivo*, although other glycine antagonists do block NMDA responses in this preparation (Fletcher *et al.* 1990). These apparent contrary observations require more study, although it is interesting to note that prenatal exposure to ethanol results in reduced NMDA responses in the pups before and after weaning (Morrisett *et al.* 1989, Noble and Ritchie 1989). This seems opposite to the normal antagonist-induced up-regulation but is presumably due to interference with neuronal development.

A novel synthetic cannabinoid, HU-211 (see Fig. 4), acts as a non-competitive NMDA antagonist in TCP binding studies at a site distinct from others described above (Feigenbaum *et al.* 1989) but is free of psychotropic cannabinoid effects (Mechoulam *et al.* 1988). Such a drug may have useful therapeutic advantages.

The acid–base status of the brain is an important determinant of NMDA receptor function (Aram and Lodge 1987, Grantyn and Lux 1988, Church and McLennan 1989, Vyklicky *et al.* 1990b, Traynelis and Cull-Candy 1990). Hydrogen ions reduce and bicarbonate ions facilitate NMDA-induced currents. Oxidation and reduction of disulphide bridges also leads respectively to marked enhancement or inhibition of NMDA receptor-mediated responses (see Aizenman *et al.* 1990 for references). ATP prevents run-down of NMDA receptor currents in whole-cell patch-clamp experiments, indicating an intracellular site for phosphorylation of the NMDA receptor–channel complex (Mody *et al.* 1988). Although these sites may become the targets for drug development in the future, it is interesting to note that they are all likely to limit NMDA receptor function in periods of hypoxaemia or ischaemia when the pH and ATP levels fall and when free radicals increase during reperfusion.

The plethora of modulatory sites on extra-, inter- and intra-membranal domains of the NMDA receptor complex will continue to be a challenge to scientists for a considerable time. In view of the involvement of NMDA receptors in normal development and physiology, and their role in neurology and neuropathology, the benefits will undoubtedly be worth the effort.

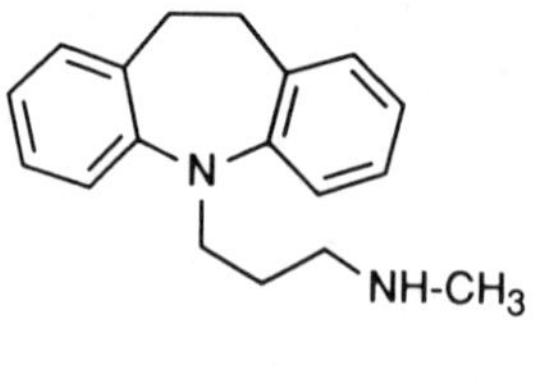

N NH-CH_3

desipramine

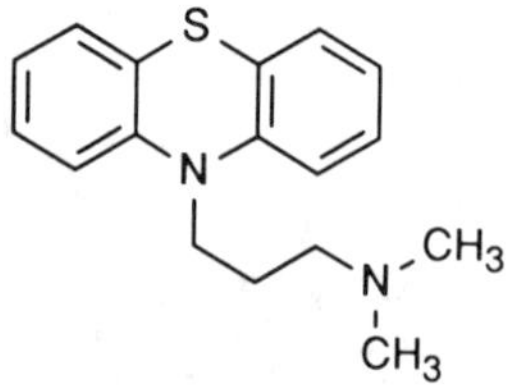

S N N CH_3 CH_3

promazine

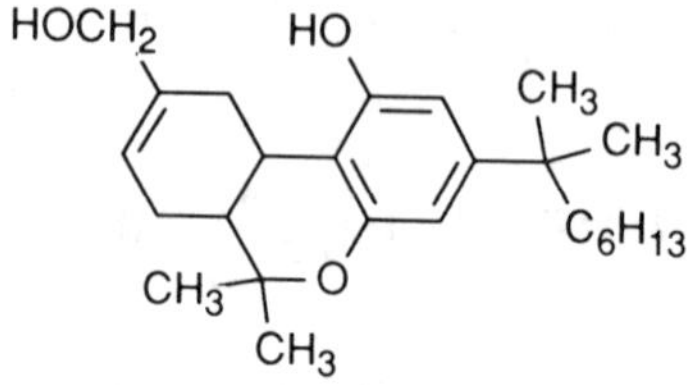

$HOCH_2$ HO CH_3 CH_3 C_6H_{13} CH_3 O CH_3

HU-211

CH_3CH_2OH

ethanol

Fig. 4 — Other putative NMDA antagonists with unknown or novel sites of action.

REFERENCES

Aizenman, E., Hartnett, K. A. and Reynolds, I. J. (1990) *Neuron* **5** 841–846.

Anis, N. A., Sherby, S., Goodnow, R., Jr., Niwa, M., Konno, K., Kallimopoulos, T., Bukownik, R., Nakanishi, K., Usherwood, P., Eldefrawi, A. and Eldefrawi, M. (1990) *J. Pharmacol. Exp. Ther.* **254** 764–773.

Aram, J. A. and Lodge, D. (1987) *Neurosci. Lett.* **83** 345–350.

Ault, B., Evans, R. H., Francis, A. A., Oakes, D. J. and Watkins, J. C. (1979) *Br. J. Pharmacol.* **67** 591–603.

Ault, B., Evans, R. H., Francis, A. A., Oakes, D. J. and Watkins, J. C. (1980) *J. Physiol.* **307** 413–428.

Birch, P. J., Grossman, C. J. and Hayes, A. G. (1988) *Eur. J. Pharmacol.* **156** 177–180.

Bristow, D. R., Bowery, N. G. and Woodruff, G. N. (1986) *Eur. J. Pharmacol.* **126** 303–307.

Bristow, L. J., Baucutt, L. J., Hutson, P. H., Thorn, L. and Tricklebank, M. D. (1991) *Br. J. Pharmacol. Proc.* (in press).

Carter, C., Benavides, J., Legendre, P., Vincent, J. D., Noel, F., Thuret, F., Lloyd, K. G., Arbilla, S., Zivkovic, B., Mackenzie, E. T., Scatton, B. and Langer, S. Z. (1988) *J. Pharmacol. Exp. Ther.* **247** 1222–1232.

Carter, C., Rivy, J.-P. and Scatton, B. (1989)*Eur. J. Pharmacol.* **164** 611–612.

Church, J. and McLennan, H. (1989) *J. Physiol.* **415** 85–108.

Church, J., Millar, J. D., Jones, M. G. and Lodge, D. (1991) *Eur. J. Pharmacol.* **192** 337–342.

Collingridge, G. L. and Singer, W. (1990) *Trends Pharmacol. Sci.* **11** 290–296.

Dingledine, R., McBain, C. J. and McNamara, J. O. (1990) *Trends Pharmacol. Sci.* **11** 334–338.

Evans, R. H., Francis, A. A. and Watkins, J. C. (1977) *Experientia* **33** 489–491.

Evans, R. H., Francis, A. A. and Watkins, J. C. (1978) *Brain Res.* **148** 536–542.

Fagg, G. E. (1987) *Neurosci. Lett.* **76** 221–227.

Feigenbaum, J. F., Bergmann, F., Richmond, S. A., Mechoulam, R., Nadler, V., Keogg, Y. and Sokolovsky, M. (1989) *Proc. Nat. Acad. Sci. USA* **86** 9584–9587.

Fletcher, E. J. and Lodge, D. (1988) *Eur. J. Pharmacol.* **15** 161–162.

Fletcher, E. J., Beart, P. M. and Lodge, D. (1990) In: Ottersen, O. P. and Storm-Mathisen, J. (eds) *Glycine Neurotransmission.* Wiley, Chichester, pp. 193–218.

Foster, A. C. and Wong, E. H. F. (1987) *Br. J. Pharmacol.* **91** 403–409.

Foster, A. C. and Kemp, J. A. (1989) *J. Neurosci.* **9** 2191–2196.

Gilad, G. M. and Gilad, V. H. (1988) *Soc. Neurosci. Abstr.* **14** 519.

Gonzales, R. A. (1990) *Trends Pharmacol. Sci.* **11** 137–139.

Grantyn, R. and Lux, H. D. (1988) *Neurosci. Lett.* **89** 198–203.

Gullack, M. F., Pagnozzi, M. J., Richter, K. E. and Mena, E. E. (1988) *Soc. Neurosci. Abstr.* **14** 168.

Handelmann, G. E., Mueller, L. L. and Cordi, A. A. (1988) *Soc. Neurosci. Abstr.* **14** 249.

Headley, P. M. and Grillner, S. (1990) *Trends Pharmacol. Sci.* **11** 205–211.

Honore, T., Davies, S. N., Drejer, J., Fletcher, E. J., Jacobsen, P., Lodge, D. and Nielsen, F. E. (1988) *Science* **241** 701–703.

Jackson, H. and Usherwood, P. N. R. (1988) *Trends Neurosci.* **11** 279–283.

Johnson, J. W. and Ascher, P. (1987) *Nature* **325** 529–531.

Jones, M. G., Yuliang, L., Lodge, D. and Millar, J. A. (1989) *Br. J. Pharmacol.* **97** 376P.

Jones, M. G., Anis, N. A. and Lodge, D. (1990) *Br. J. Pharmacol.* **101**, 968–970.

Kemp, J. A., Foster, A. C. and Wong, E. H. F. (1987) *Trends Neurosci.* **10** 294–298.

Kemp, J. A., Foster, A. C., Leeson, P. D., Priestley, T., Tridgett, R. and Iversen, L. L. (1988) *Proc. Nat. Acad. Sci. USA* **85** 6547–6550.

Kemp, J. A., Priestley, T., Marshall, G. R., Leeson, P. D. and Williams, B. J. (1991) *Br. J. Pharmacol. Proc.* (in press).

Kessler, M., Terrammani, T., Lynch, G. and Baudry, M. (1989) *J. Neurochem.* **52** 1319–1328.

Kleckner, N. W. and Dingledine, R. (1988) *Science* **241** 835–837.

Kleckner, N. W. and Dingledine, R. (1989) *Molec. Pharmacol.* **36** 430—436.

Kloog, Y., Haring, R. and Sokolovsky, M. (1988) *Biochemistry* **27** 843–848.

Lam, K. H. K., Palmer, A. J., Millar, J. D. and Lodge, D. (1991) *Br. J. Pharmacol. Proc.* (in press).

Legendre, P. and Westbrook, G. L. (1990) *J. Physiol.* **429** 429–449.

Lodge, D. and Anis, N. A. (1982) *Eur. J. Pharmacol.* **77** 203–204.

Lodge, D. and Johnson, K. M. (1990) *Trends Pharmacol. Sci.* **11** 81–86.

Lodge, D. and Johnston, G. A. R. (1985) *Neurosci. Lett.* **56** 373–375.

Lodge, D., Aram, J. A. Church, J., Davies, S. N., Martin, D., Millar, J. and Zeman, S. (1988) In: Lodge, D. (ed.) *Excitatory Amino Acids in Health and Disease*. Wiley, Chichester, pp. 237–259.

Loo, P., Braunwalder, A., Lehmann, J. and Williams, M. (1986) *Eur. J. Pharmacol.* **123** 467–468.

Lovinger, D. M., White, G. and Weight, F. F. (1989) *Science* **243** 1721–1723.

MacDonald, J. F. and Nowak, L. M. (1990) *Trends Pharmacol. Sci.* **11** 167–172.

Maragos, W. F., Penney, J. B. and Young, A. B. (1988) *J. Neurosci.* **8** 493–501.

Martin, D. and Lodge, D. (1989) *Pharmacol. Biochem. Behav.* **31** 279–286.

Mayer, M. L. and Vyklicky, L., Jr (1989) *J. Physiol.* **415** 351–365.

Mayer, M. L., Westbrook, G. L. and Guthrie, P. B. (1984) *Nature* **309** 261–263.

Mayer, M. L. and Vyklicky, L., Jr and Westbrook, G. L. (1989) *J. Physiol.* **415** 329–350.

McNamara, D., Smith, E. C. R., Calligaro, D. O., O'Malley, P. J., McQuaid, L. A. and Dingledine, R. (1990) *Neurosci. Lett.* **120** 17–20.

Mechoulam, R., Feigenbaum, J. J., Lander, N., Segal, M., Jarbe, T. U. S., Hiltunen, A. J. and Consroe, P. (1988) *Experientia* **44** 762–764.

Meldrum, B. and Garthwaite, J. (1990) *Trends Pharmacol. Sci.* **11** 379–387.

Mena, E. E., Gullack, M. F., Pagnozzi, M. J., Phillips, D. and Saccomano, N. (1989) *Soc. Neurosci. Abstr.* **15** 1168.

Mena, E. E., Gullack, M. F., Pagnozzi, M. J., Richter, K. E., Rivier, J., Cruz, L. J. and Olivers, B. M. (1990) *Neurosci. Lett.* **118** 241–244.

Mody, I., Salter, M. W. and MacDonald, J. F. (1988) *Neurosci. Lett.* **93** 73–78.

Morrisett, R. A., Martin, D., Wilson, W. A., Savage, D. D. and Swartzwelder, H. S. (1989) *Alcohol* **6** 415–420.

Nadler, V., Kloog, Y. and Sokolovsky, M. (1990) *Eur. J. Pharmacol.* **188** 97–104.

Noble, E. P. and Ritchie, T. (1989) *Life Sci.* **45** 803–810.

Nowak, L., Bregestovski, P., Ascher, P., Herbert, A. and Prochiantz, A. (1984) *Nature* **307** 462–465.

Ohya, Y., Ochi, N., Mizutani, N. and Watanabe, K. (1991) In: Kameyama, T., Nabeshima, T. and Domino, E. F. (eds) NPP Books, Ann Arbor, MI (in press).

Olney, J. W., Labruyere, J. and Price, M. T. (1989) *Science* **244** 1360–1362.

Priestley, T., Woodruff, G. N. and Kemp. J. A. (1989) *Br. J. Pharmacol.* **97** 1315–1323.

Ransom, R. W. and Stec, N. L. (1988) *J. Neurochem.* **51** 830–836.

Reynolds, I. J. (1990) *Eur. J. Pharmacol.* **177** 215–216.
Reynolds, I. J. (1991) *Br. J. Pharmacol.* (in press).
Reynolds, I. J. and Miller, R. J. (1988) *Br. J. Pharmacol.* **95** 95–102.
Reynolds, I. J. and Miller, R. J. (1989) *Molec. Pharmacol.* **36** 758–765.
Reynolds, I. J., Murphy, S. N. and Miller, R. J. (1987) *Proc. Nat. Acad. Sci. USA* **84** 7744–7748.
Sacaan, A. I. and Johnson, K. M. (1990) *Molec. Pharmacol.* **37** 572–577.
Sacaan, A. I. and Johnson, K. M. (1991) *Neurosci. Lett.* **121** 219–222.
Sather, W., Johnson, J. W., Henderson, G. and Ascher, P. (1990) *Neuron* **4** 725–731.
Saywell, K., Singh, L., Oles, R. J., Vass, C., Leeson, P. D., Williams, B. J. and Tricklebank, M. D. (1991) *Br. J. Pharmacol. Proc.* (in press).
Sernagor, E., Kuhn, D., Vyklicky, L., Jr and Mayer, M. L. (1989) *Neuron* **2** 1221–1227.
Shannon, H. E. (1983) *J. Pharmacol. Exp. Ther.* **255** 144–152.
Sheardown, M. J., Drejer, J., Jensen, L. H., Stidsen, C. E. and Honore, T. (1989) *Eur. J. Pharmacol.* **174** 197–204.
Sheardown, M. J., Nielsen, E. O., Hansen, A. J., Jacobsen, P. and Honore, T. (1990) *Science* **247** 571–574.
Siddiqui, F., Iqbal, Z. and Koenig, H. (1988) *Soc. Neurosci. Abstr.* **14** 1048.
Snell, L. D. and Johnson, K. M. (1985) *J. Pharmacol. Exp. Ther.* **235** 50–57.
Snell, L. D., Morter, R. S. and Johnson, K. M. (1987) *Neurosci. Lett.* **83** 313–317.
Teichberg, V. I., Tal, N., Goldberg, O. and Luini, A. (1984) *Brain Res.* **291** 285–292.
Thomson, A. M. (1989) *Trends Neurosci.* **12** 349–353.
Thomson, A. M. (1990) *Prog. Neurobiol.* **35** 3–74.
Traynelis, S. F. and Cull-Candy, S. G. (1990) *Nature* **345** 347–350.
Vyklicky, L., Jr, Beneviste, M. and Mayer, M. L. (1990a) *J. Physiol.* **428** 313–331.
Vyklicky, L., Jr, Vlachova, V. and Krusek, J. (1990b) *J. Physiol.* **430** 497–517.
Willetts, J., Balster, R. L. and Leander, J. D. (1990) *Trends Pharmacol. Sci.* **11** 423–428.
Williams, K., Romano, C. and Molinoff, P. B. (1989) *Molec. Pharmacol.* **36** 575–581.
Williams, K., Dawson, V. L., Romano, C., Dichter, M. A. and Molinoff, P. B. (1990) *Neuron* **5** 199–208.
Wong, E. H. F., Kemp, J. A., Priestley, T., Knight, A. R., Woodruff, G. N. and Iversen, L. L. (1986) *Proc. Nat. Acad. Sci. USA* **83** 7104–7108.
Xie, X. and Smart, T. G. (1991) *Nature* **349** 521–524.
Zukin, S. R. and Zukin, R. S. (1979) *Proc. Nat. Acad. Sci. USA* **76** 5372–5376.

10

AMPA receptor agonists: structural, conformational and stereochemical aspects

Jan J. Hansen, Flemming S. Jørgensen, Trine M. Lund, Birgitte Nielsen, Annemarie Reinhardt, Irene Breum, Lotte Brehm and **Povl Krogsgaard-Larsen**
PharmaBiotec Research Centre, Department of Organic Chemistry, The Royal Danish School of Pharmacy, DK-2100 Copenhagen, Denmark

ABSTRACT

The non-selective excitatory amino acid receptor agonist ibotenic acid has been converted into a number of selective (*RS*)-2-amino-3-(3-hydroxy-5-methylisoxazol-4-yl)propanoic acid (AMPA) receptor agonists by systematic variations of structural, conformational and stereochemical features. In general, these isoxazole amino acids show a good correlation between excitatory potency and affinity towards binding sites for AMPA, supporting the view that [^{3}H]AMPA binding represents the recognition sites of physiological AMPA receptors. Thus, like several other heterocyclic groups, the 3-hydroxyisoxazole moiety of compounds derived from ibotenic acid can act as a bioisostere of the γ-carboxyl group of (*S*)-glutamic acid (Glu), the presumed endogeneous ligand at AMPA receptors. Although Glu is a very flexible compound, its structural element can be incorporated into highly rigid structures without loss of activity, and the resulting compounds, 5-HPCA and 7-HPCA, are, like AMPA, highly selective agonists at AMPA receptors. Enzymatic and chromatographic resolution techniques have furnished the enantiomers of several isoxazole amino acids acting as agonists or antagonists at AMPA receptors. *In vivo* and *in vitro* studies of these enantiomers demonstrate a pronounced stereoselectivity of AMPA receptors, which are mainly activated by agonists with the L-configuration, and a eudismic ratio of approximately 4000 is observed for AMPA. It must be emphasized, however, that a precise determination of receptor stereoselectivity necessitates an accurate determination of very small amounts of enantiomeric impurities.

INTRODUCTION

In the decade since its synthesis (Hansen and Krogsgaard-Larsen 1980), (*RS*)-2-amino-3-(3-hydroxy-5-methylisoxazol-4-yl)propanoic acid (AMPA; Fig. 1) has be-

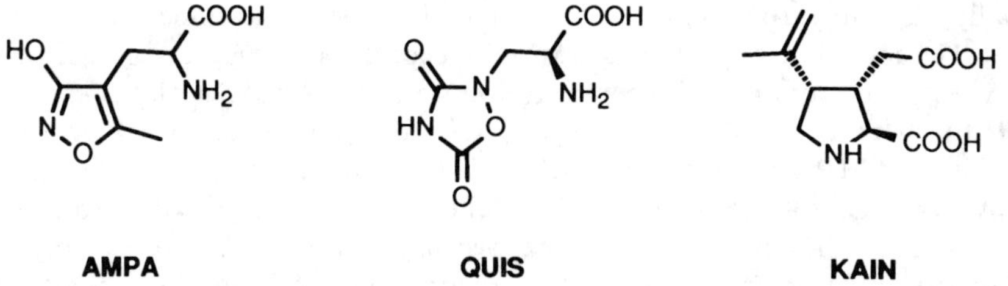

Fig. 1 — Structures of AMPA, QUIS and KAIN. Like most other structures in the following figures, these compounds are, for simplicity, depicted in their uncharged form although they are fully charged at physiological pH.

come the preferred ligand for characterizing the particular subtype of excitatory amino acid (EAA) receptors which now is termed the AMPA receptor but earlier was named after its prototypical ligand (*S*)-quisqualic acid (QUIS) as the QUIS receptor and, transitionally, as the QUIS/AMPA receptor. This recent change in nomenclature (Watson and Abbott 1989) reflects the high selectivity of AMPA and the predominant use of [^{3}H]AMPA as radioligand for this receptor subtype in contrast to the multiple effects of QUIS (see below) within the central nervous system (CNS). The identification of different subtypes of physiological EAA receptors is mainly based on electrophysiological and biochemical studies using selective agonists and antagonists, and the AMPA receptor, as defined in chapter 2, shows a distinct pharmacology by such *in vivo* and *in vitro* studies on intact animals, tissue preparations and cell cultures. Especially, binding studies on CNS tissues and synaptic membrane preparations using radiolabelled ligands have been useful for elucidating the anatomical distribution within the CNS and quantitating ligand affinities and other binding characteristics of AMPA receptors.

The AMPA receptor, like the receptor named after *N*-methyl-(*R*)-aspartic acid (NMDA), appears to constitute physiological synaptic receptors (Davies and Watkins 1983, Fagg *et al.* 1986) for an endogenous excitatory neurotransmitter, for which (*S*)-glutamic acid (Glu) is the most likely candidate (Fonnum 1984, Robinson and Coyle 1987). In autoradiographic binding studies the AMPA receptor and the NMDA receptor show very similar anatomical distributions in the brain (Monaghan and Cotman 1986, Young and Fagg 1990). Both subtypes appear to be postsynaptically located (Greenamyre and Young 1989) and may to a large extent be co-localized on individual synapses (Bekkers and Stevens 1989). AMPA receptors seem to mediate the normal fast synaptic activity in the CNS (Mayer and Westbrook 1987) and may act as a 'tripping wire' for activation of NMDA receptors, which then in turn participate in the synaptic plasticity (Collingridge and Singer 1990) necessary for memory and learning processes (Izquierdo 1991, Morris *et al.* 1986).

The presence of potent and selective antagonists at NMDA receptors (see chapter 8) allows a clear identification of this subclass of EAA receptors. AMPA receptors and (2*S*,3*S*,4*S*)-kainic acid (KAIN; Fig. 1) receptors, on the other hand, are difficult to distinguish due to a lack of antagonists which selectively discriminate between them, and these two subtypes of EAA receptors are often collectively referred to as non-NMDA receptors. The recently developed quinoxaline antagonists 6-cyano-7-nitroquinoxaline-2,3-dione (CNQX), 6,7-dinitroquinoxaline-2,3-dione (DNQX) and, especially, 2,3-dihydroxy-6-nitro-7-sulphamoylbenzo(*f*)quinoxaline (NBQX) (see chapter 11) act selectively at non-NMDA receptors, but they do not distinguish AMPA receptors from KAIN receptors (Honoré *et al.* 1988, Sheardown *et al.* 1990). GDEE, the diethyl ester of Glu, has been used for this purpose as a weak but selective antagonist at AMPA receptors with little effect on excitations induced by KAIN or NMDA (Krogsgaard-Larsen *et al.* 1980, McLennan and Lodge 1979). However, the usefulness of GDEE has been questioned, because it is susceptible to hydrolysis and its selectivity is not observed in all systems (Turner and Meldrum 1991, Watkins and Evans 1981).

Other main criteria for the differentiation of AMPA receptors and KAIN receptors include the presence of a relatively pure population of KAIN receptors at C-fibre afferents in the spinal cord of immature rats (Agrawal and Evans 1986, Evans *et al.* 1987) and binding studies, where the binding sites for AMPA and KAIN show different ligand affinity profiles (Foster and Fagg 1984, Nielsen *et al.* 1986) and, by autoradiography, different regional distributions in the CNS (Greenamyre *et al.* 1985, Young and Fagg 1990).

The binding sites for [^{3}H]AMPA are generally believed to represent the recognition site of physiological AMPA receptors (Hansen and Krogsgard-Larsen 1990, Monaghan *et al.* 1989, Young and Fagg 1990, Watkins *et al.* 1990). [^{3}H]AMPA binding (Honoré *et al.* 1982) is markedly increased by the addition of SCN^- and other chaotropic ions (Honoré and Nielsen 1985), and [^{3}H]AMPA binding studies are therefore usually performed in the presence of 100 mM thiocyanate (Honoré and Drejer 1988, Murphy *et al.* 1987, Olsen *et al.* 1987). Both high- and low-affinity binding components can be observed and have been suggested (see chapter 11) to represent two interconvertible conformations of the AMPA receptor which exist, possibly in association with a modulatory protein (Honoré and Nielsen 1985), in a conformational equilibrium that is affected by thiocyanate ions (Honoré and Drejer 1988).

The physiological role of [^{3}H]KAIN binding, on the other hand, and indeed that of KAIN receptors, is presently not clear (see chapter 2), possibly due to lack of selectivity of KAIN for a single subtype of EAA receptors. The pharmacological profile of the traditional KAIN receptor can be studied in spinal C-fibre afferents (see chapter 12), where the excitatory potency for a number of compounds correlates reasonably well with their neurotoxicity (Coyle 1983) and ability to displace [^{3}H]KAIN from rat brain membranes (Agrawal and Evans 1986). However, KAIN also appears to have presynaptic actions (Ferkany and Coyle 1983, Virgili *et al.* 1986) and, in addition, shows a moderate affinity towards AMPA receptors (Honoré *et al.* 1982). In fact, accumulating data from recent electrophysiological and biochemical work on mammalian neurons suggest that the excitatory effects of KAIN may, at least partially, be mediated via its activation of AMPA receptors (Honoré 1989, Honoré *et al.* 1989, Kiskin *et al.* 1986, Pin *et al.* 1989, Sheardown *et al.* 1990). This suggestion is also supported by studies of EAA receptors expressed in *Xenopus* oocytes after injection with rat brain mRNA where responses to KAIN and AMPA appear to be mediated with different efficacies by the same non-NMDA receptor, AMPA possibly acting as a partial agonist (Rassendren *et al.* 1989, Verdoorn and Dingledine 1988; see, however, Sugiyama *et al.* 1989).

A functional EAA-gated ion channel has recently been cloned (see chapter 3) and expressed in *Xenopus* oocytes (Hollmann *et al.* 1989) and human kidney cells (Keinänen *et al.* 1990). Additional cDNA clones, each in two alternative mRNA splicing forms (Sommer *et al.* 1990), were subsequently isolated from a rat brain cDNA library (Keinänen *et al.* 1990, Boulter *et al.* 1990). The protein expressed from these clones show about 70% sequence homology and are believed to represent subunits of physiological EAA receptors. However, their size (862–889 amino acid residues) and lack of homology with other ionotropic receptors indicate a distant or non-existent evolutionary relationship with the $GABA_A$ receptor and the nicotinic acetylcholine receptor (Barnard and Henley 1990). Very recently also a human cDNA has been isolated, encoding a protein with 97% sequence homology with one of the rat EAA receptor subunits (Puckett *et al.* 1991).

While the pharmacological profile and mRNA distribution in rat brain for these cloned EAA receptor subunits are characteristic of AMPA receptors (Keinänen *et al.* 1990), their sensitivity to both AMPA and KAIN support the notion of a unitary non-NMDA receptor. The functional expression of both homo-oligomeric and hetero-oligomeric subunit compositions of the cloned receptors is similar to but distinct from the ionotropic EAA receptors expressed in *Xenopus* oocytes after injection of rat brain mRNA, possibly reflecting different subunit compositions. In analogy with the present situation for $GABA_A$ receptors (Lüddens and Wisden 1991), a variety of subunits for EAA receptors can now be expected to be identified, probably with different ligand affinity profiles. However, the number of their potential combinations into receptors will probably vastly exceed that which can be distinguished by current pharmacological techniques.

Glu IS A FLEXIBLE CHIRAL MOLECULE

As already stated, Glu appears to be the endogenous ligand for AMPA receptors. However, Glu shows little selectivity for EAA receptor subtypes and participates in a number of metabolic processes in addition to its role as a precursor for the inhibitory transmitter 4-aminobutanoic acid (GABA). Glu contains four carbon–carbon bonds with more or less free rotation and would, therefore, be expected to adopt a number of low-energy conformations. Hence, a possible explanation for the lack of selectivity of Glu at EAA receptors could be that different conformations of Glu activate different receptor subtypes. If so, conformational restriction of the structural element of Glu by incorporation into rigid or semi-rigid structures might lead to compounds which selectively activate only one subtype of EAA receptors.

In order to study the conformational flexibility of Glu we have performed a detailed conformational analysis of the fully charged molecule by use of the MM2(85) molecular mechanics program (Burkert and Allinger 1982). In this analysis the two torsional angles labelled τ_1 and τ_2 in Fig. 2 were varied in small steps and the Glu molecule was geometry optimized by MM2(85) for all possible combinations of the two torsional angles. This so-called double-drive around the two C_{sp^3}–C_{sp^3} bonds allowed us to construct a potential energy surface, describing the steric energy of Glu as a function of the torsional angles τ_1 and τ_2. This potential energy surface, shown in Fig. 2, indicates that Glu is able to adopt at least six different conformations which differ in energy by less than 2 kcal/mol, confirming from a theoretical point of view the expected high molecular flexibility of Glu.

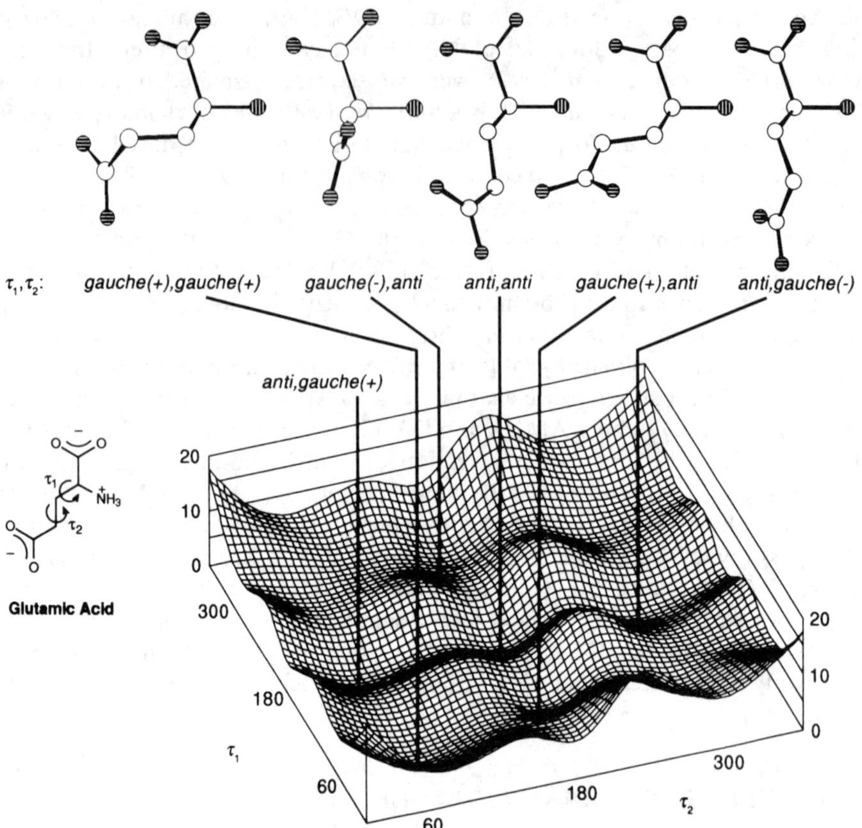

Fig. 2 — Potential energy surface calculated by MM2(85) for a two-bond drive of Glu with different solid state conformations of Glu derivatives from the Cambridge Structural Database.

This conclusion is supported by experimental data obtained from the Cambridge Structural Database (CSD). The CSD presently contains approximately 90 000 X-ray structures of organic molecules, and the presence of a number of differently substituted derivatives of a compound may provide information about its conformational mobility (Allen *et al.* 1983). A search of the CSD for compounds incorporating the structural element of Glu revealed the presence of several Glu derivatives in the database (F. S. Jørgensen, unpublished). After exclusion of Glu-containing peptides and coordination compounds involving transition metal complexes, the remaining Glu derivatives could be classified into five groups, each corresponding to a different conformation of the Glu structure. A representative of each of the five conformational groups is shown in Fig. 2, from which it appears that the five experimentally observed conformations of the Glu structure all correspond to low-energy conformations on the calculated potential energy surface. The depicted conformations of Glu show considerable variation, and the torsional angles τ_1 and τ_2 (see Fig. 2), as well as the distances between the charged groups, are listed in Table 1. The close agreement between the experimentally observed and theoretically determined low-energy confor-

Table 1 — Torsional angles and distances between the charged groups for the Glu conformations shown in Fig. 2.

Compound	CSD ref. code	τ_1	τ_2	d_1	d_2
(*S*)-Glu	LGLUAC03	59°	68°	3.8 Å	4.5 Å
(*S*)-Glu	LGLUAC11	– 171°	– 73°	4.7 Å	3.8 Å
(*S*)-Glu, HCl	LGLUTA	171°	– 173°	4.9 Å	4.5 Å
Ca^{2+}-(*S*)-Glu, $3H_2O$	CAGLCL10	67°	170°	4.6 Å	5.0 Å
(*RS*)-Glu, H_2O	CADVUY	– 55°	– 165°	4.5 Å	4.3 Å

τ_1 = C1–C2–C3–C4, τ_2 = C2–C3–C4–C5, d_1 = the distance between C1 and C5, d_2 = the distance between C5 and N. The distance between C1 and N is 2.4 Å for all five compounds.

mations of Glu (Fig. 2) supports the applicability of molecular mechanics calculations and documents the conformational flexibility of Glu from both an experimental and a theoretical point of view (see also Bhat and Vijayan 1977).

In addition to this conformational flexibility, also the chirality of Glu may well be essential for its complex synaptic activities. Therefore, a rational approach for studying the molecular mechanisms of Glu at different synaptic sites will involve design, synthesis and structure–activity studies of analogues with a systematic variation of structural, conformational and stereochemical features.

STRUCTURAL AND CONFORMATIONAL ASPECTS

The naturally occurring 3-hydroxyisoxazole amino acid ibotenic acid (IBO; Fig. 3), isolated from the fly agaric, *Amanita muscaria*, (Eugster 1969), has been used as a lead structure for the design of potent and selective agonists and antagonists at specific subtypes of synaptic receptors for Glu (Krogsgaard-Larsen *et al.* 1980, Watkins *et al.* 1990; see chapter 2). IBO is an analogue of Glu which is partially restricted in conformational flexibility, and the planar 3-hydroxyisoxazole moiety acts as a bioisostere of the γ-carboxyl group of Glu. Like Glu, IBO is fully charged at physiological pH and the 3-hydroxyisoxazole moiety shows the same type of delocalization of the negative charge as a carboxylate group.

A semi-empirical CNDO/2 molecular orbital study indicated that the conformation depicted in Fig. 4 as IBO-II is the preferred conformation of IBO in the

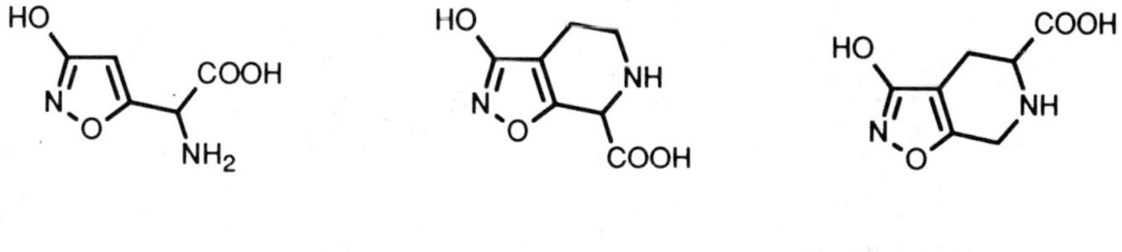

IBO **7-HPCA** **5-HPCA**

Fig. 3 — Structures of IBO, 7-HPCA and 5-HPCA.

'conservative state' (Borthwick and Steward 1976). However, allowance for an aqueous medium predicts a change into the conformation shown (Fig. 4) as IBO-I (Borthwick and Steward 1976). A more recent study (Brehm *et al.* 1990) using the semi-empirical AM1 method (Dewar and Stewart 1986) investigated the variation of potential energy as a function of rotation around the C_{sp^3}–C_{sp^2} bond connecting the amino acid moiety with the 3-hydroxyisoxazole group in the IBO zwitterion (Fig. 4). Rotation around this bond is the cause of the major part of the conformational

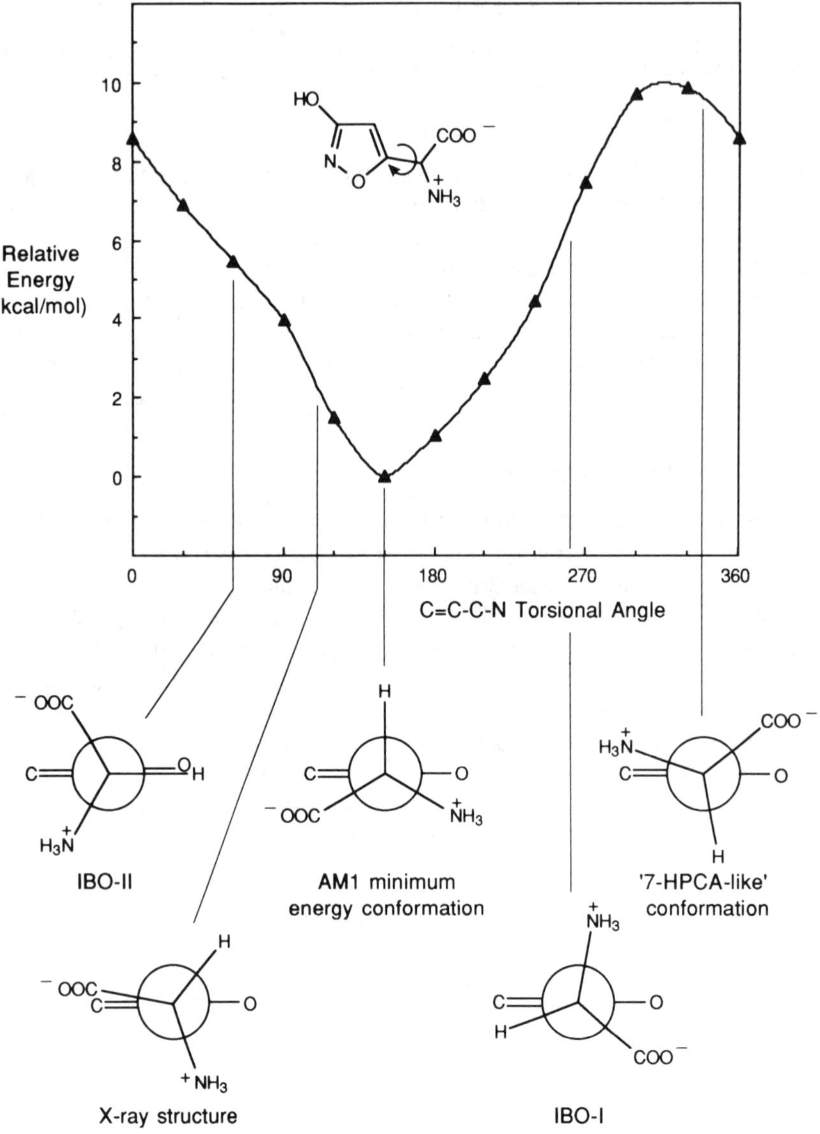

Fig. 4 — Potential energy profile calculated by AM1 for rotation around the C5–αC bond of IBO with Newman projections of different conformations mentioned in the text.

flexibility of IBO, and the global minimum energy conformation of IBO was found to be a conformation with the carboxylate and the ammonium groups placed on the same side of the plane of the isoxazole ring and the α-hydrogen atom placed orthogonal to the plane (Brehm *et al.* 1990). As shown in Fig. 4, this conformation, with the ammonium group close to the oxygen atom of the isoxazole ring, is quite similar to the conformation of IBO found in the solid state by X-ray crystallography (Brehm *et al.* 1988).

The strong excitatory effects of IBO (Curtis *et al.* 1979, Johnston *et al.* 1968) are mainly caused by its activation of NMDA receptors (McLennan and Lodge 1979, Krogsgaard-Larsen *et al.* 1980), but IBO also interacts potently with metabotropic EAA receptors (Qp receptors) (Schoepp *et al.* 1990; see chapter 15) and weakly with AMPA receptors in addition to being a moderately potent inhibitor of [^{3}H]KAIN binding (Krogsgaard-Larsen *et al.* 1980). However, by molecular engineering this non-selective agonist has been converted into a number of isoxazole amino acids with potent and selective effects at NMDA receptors (see chapter 6) and non-NMDA receptors (Hansen and Krogsgaard-Larsen 1990).

To further restrict its conformational flexibility the amino acid side-chain of IBO was incorporated in a six-membered ring to give the bicyclic compound (*RS*)-3-hydroxy-4,5,6,7-tetrahydroisoxazolo[5,4-*c*]pyridine-7-carboxylic acid (7-HPCA; Fig. 3), which is an analogue of Glu with considerable conformational restrictions. Surprisingly, 7-HPCA turned out to be a potent and selective agonist at AMPA receptors (Table 2) with no effect at NMDA receptors *in vivo* and no affinity towards [^{3}H]KAIN binding sites (Krogsgaard-Larsen *et al.* 1984). In the crystalline state the 7-HPCA zwitterion adopts a conformation (Fig. 5) with the carboxylate group in a pseudo-equatorial position (Krogsgaard-Larsen *et al.* 1985). A similar conformation, termed '7-HPCA-like' in Fig. 4, is accessible to IBO, and it seems likely that 7-HPCA, possibly with a pseudo-axial carboxylate group, reflects the conformation of IBO when acting at AMPA receptors. However, the energy requirement for the conformational change of IBO from the global minimum energy conformation to the '7-HPCA-like' conformation is approximately 9 kcal/mol (Brehm *et al.* 1990), which is close to the energy barrier for rotation around the C5–αC bond indicated in Fig. 4. Thus, the presence in IBO of only a small population of '7-HPCA-like' conformers may account for the low activity of IBO at AMPA receptors. Similarly, the inability of 7-HPCA to assume conformations recognized by the NMDA receptor or, alternatively, its lack of conformational flexibility may prevent 7-HPCA from activating NMDA receptors.

Like 7-HPCA, the isomeric compound (*RS*)-3-hydroxy-4,5,6,7-tetrahydroisoxazolo[5,4-*c*]pyridine-5-carboxylic acid (5-HPCA; Fig. 3) is a bicyclic compound with strong restrictions on its conformational flexibility. 5-HPCA is equipotent with 7-HPCA as an excitant and shows a similar high affinity and selectivity for AMPA receptors with no effect on NMDA receptors and no inhibition of [^{3}H]KAIN binding (Krogsgaard-Larsen *et al.* 1985). The preferred conformation of 5-HPCA, as determined by ^{1}H-NMR spectroscopy in aqueous solution (Nielsen *et al.* 1986), contains the six-membered ring in a half-chair conformation with the carboxylate group in an equatorial position, and the resulting structure is very similar to that found for 7-HPCA in the solid state (Fig. 5). Both compounds adopt an almost planar conformation, and their quite rigid structures indicate that molecular flexibility is not a prerequisite for activation of AMPA receptors. The solvent-accessible surfaces of the

Table 2 — Electrophysiological and binding studies on isoxazole amino acids

Compound	Relative excitatory potency[a]		$[^3H]$AMPA binding IC_{50} (μM)	$[^3H]$KAIN binding IC_{50} (μM)	$[^3H]$Glu binding[b] IC_{50} (μM)
AMPA	+ + + + + +	(3.5)	0.04	> 100	⩾ 100
5-HPCA	+ + + + +		1.9	> 100	> 100
7-HPCA	+ + + + +		0.60	> 100	> 100
ABPA	+ + + + +		0.03	> 100	> 100
ATPA	+ + + +		6.2	> 100	> 100
APPA[c]		(340)	35	> 100	> 100
p-Cl-APPA[d]			58	> 100	> 100
4-AHCP[e]	+ + + + + + +	(10)	0.74	> 100	> 100
HIBO	+(+)		3.7	> 100	0.16
Br-HIBO	+ + + + +		0.60	> 100	0.20
Me-HIBO	+ + +(+)	(20)	0.61	> 100	0.18
HE-HIBO[f]		(90)	2.0	> 100	0.14
Bu-HIBO[f]		(40)	2.0	> 100	0.18
Oc-HIBO[f]			> 100	> 100	0.02
6-HPCA	0		> 100	> 100	> 100
N-Me-AMPA[g]		(2000)	50	> 100	> 100
O-Me-AMPA	0		> 100		
IBO	+ + + +		44	12	0.19
Glu	+ +(+)		0.50	0.27	0.17
QUIS	+ + + + + + +		0.02	0.29	0.21
KAIN	+ + + + + + +(+)		6.0	0.02	> 100
NMDA	+ + + + + +		> 100	> 100	> 100

All compounds, except for the last four, were tested as racemates, and the data are essentially as listed in Hansen *et al.* (1989a) and Nielsen *et al.* (1986).
[a]By iontophoresis on cat spinal neurons or (in parentheses) EC_{50} values (μM) in the rat cortical wedge preparation.
[b]In the presence of 2.5 mM $CaCl_2$.
[c]Christensen *et al.* (1989).
[d]A. Reinhardt *et al.* (unpublished).
[e]Lund *et al.* (1991).
[f]Christensen *et al.* (submitted).
[g]Madsen *et al.* (1990a).

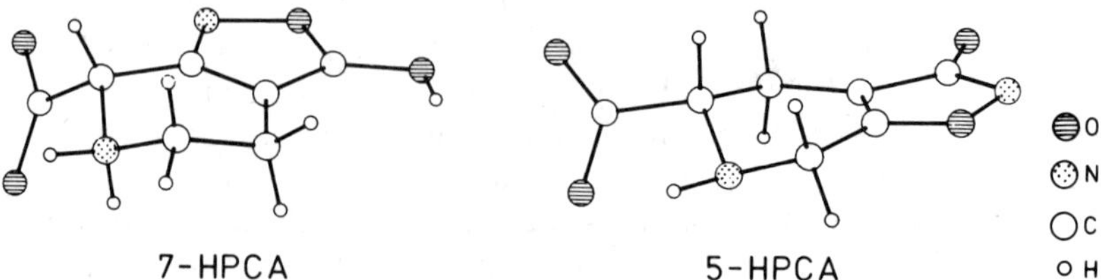

Fig. 5 — Conformations of 7-HPCA zwitterion in the crystalline state, as determined by X-ray crystallography, and 5-HPCA, as determined by ^{1}H-NMR in D_2O at pH 10.

fully ionized molecules of 5-HPCA and 7-HPCA are superimposed in Fig. 6, showing that the two molecules have very similar shapes (Brehm *et al.* 1990). In addition to these steric characteristics, 5-HPCA and 7-HPCA also display nearly identical charge distributions due to delocalization of the negative charge on the 3-hydroxyisoxazole moieties.

Opening of the piperidine ring in the structure of 5-HPCA regains some of the conformational flexibility of Glu, and the resulting compound, AMPA, shows an increased affinity for AMPA receptors as compared with 5-HPCA. In contrast to IBO and QUIS, AMPA does not activate metabotropic EAA receptors (Recasens *et al.* 1988, Schoepp *et al.* 1990), and AMPA is very potent and selective as an agonist at AMPA receptors, with no effect on NMDA receptors and no affinity towards [^{3}H]KAIN binding sites.

In order to study the structural requirements for activation of AMPA receptors several structural analogues of AMPA have been prepared (Fig. 7). (*RS*)-2-Amino-3-(3-hydroxy-5-*tert*-butylisoxazol-4-yl)propanoic acid (ATPA) contains the bulky *tert*-butyl group, which causes a decrease in excitatory potency and affinity, but not in selectivity, for AMPA receptors (Lauridsen *et al.* 1985). Another analogue, (*RS*)-2-amino-3-(3-hydroxy-5-bromomethylisoxazol-4-yl)propanoic acid (ABPA), contains the chemically reactive bromomethyl group and was designed as an analogue capable of irreversibly alkylating the receptor (Krogsgaard-Larsen *et al.* 1985). No such effect was observed, however, and ABPA is equipotent with AMPA as a selective agonist at AMPA receptors. The results obtained with ATPA and ABPA indicate that the substituent in the 5-position of the isoxazole ring does not come in close contact with the receptor, suggesting that the AMPA receptor may contain a 'cavity' where this substituent is located (see Fig. 8 and chapter 2) or, alternatively, that the substituent is directed away from the recognition site of the receptor during binding.

The solid state conformation of AMPA, as determined by X-ray crystallography (Honoré and Lauridsen 1980), is quite different from the conformation of the same structural element in 5-HPCA, as found by ^{1}H-NMR in aqueous solution (Fig. 5).

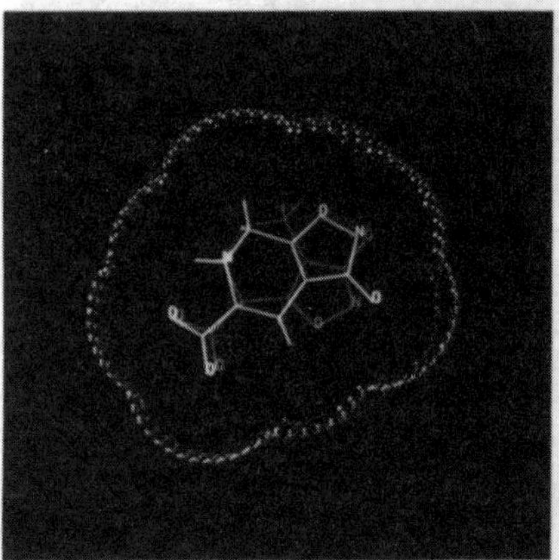

Fig. 6 — (**See colour section.**) Structural similarities of 7-HPCA (red) and 5-HPCA (green) as shown by superposition of their solvent-accessible surfaces.

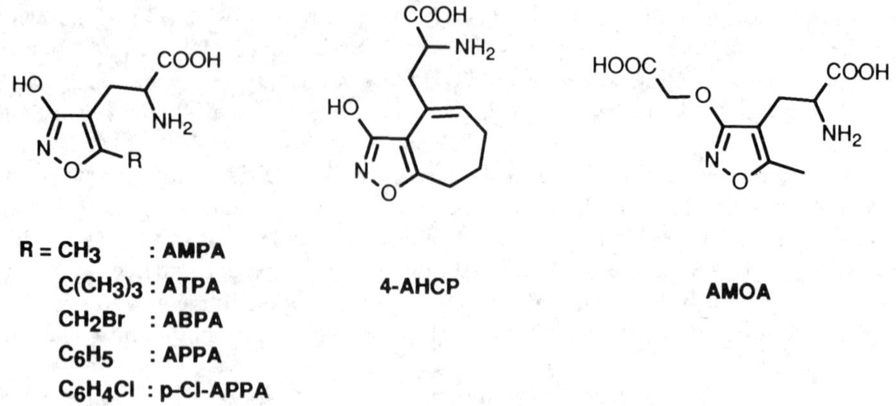

COOH
HO
NH_2
N O R

$R = CH_3$: **AMPA**
$C(CH_3)_3$: **ATPA**
CH_2Br : **ABPA**
C_6H_5 : **APPA**
C_6H_4Cl : **p-Cl-APPA**

COOH
NH_2
HO
N O

4-AHCP

HOOC
O
COOH
NH_2
N O

AMOA

Fig. 7 — Structures of AMPA analogues, 4-AHCP and the non-NMDA receptor antagonist AMOA.

However, AMPA is a relatively flexible molecule and other conformations than the one found in crystals may be accessible to AMPA.

The conformational flexibility of AMPA, ABPA and ATPA has been studied (Lauridsen *et al.* 1985) by MM2 calculations (Burkert and Allinger 1982), which indicate that these isoxazole amino acids all have similar minimum energy conformations. Rotation around the two carbon–carbon bonds closest to the isoxazole ring in the amino acid side-chain is associated with energy barriers of about 5 kcal/mol and 7 kcal/mol, respectively, for both AMPA and ABPA, whereas ATPA shows a considerably higher rotational barrier (Lauridsen *et al.* 1985). A more recent conformational analysis (Lund *et al.* 1991), using a combination of molecular mechanics (Tripos force field) and quantum mechanics (AM1) calculations, has shown that AMPA may adopt three low-energy conformations, none of which corresponds to the solid state conformation of AMPA found by X-ray crystallography (Honoré and Lauridsen 1980).

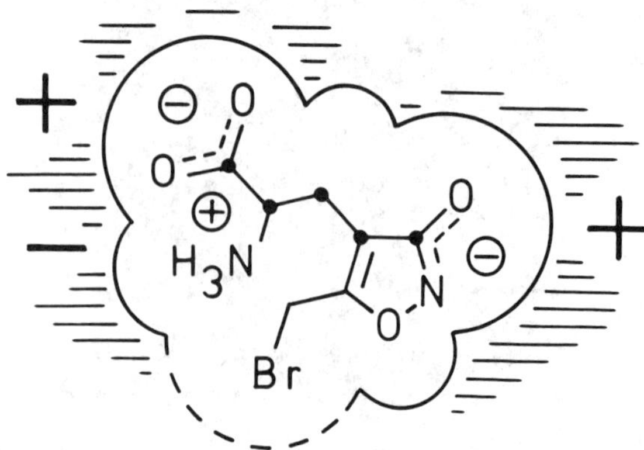

Fig. 8 — Hypothetical model of the AMPA receptor with ABPA occupying the binding site.

These conformational analyses indicate that AMPA, ABPA and, especially, ATPA do not easily assume the same conformation as that found in the conformationally restricted analogue 5-HPCA because of steric repulsion between the ammonium group and the 5-alkyl substituent of the isoxazole ring. However, a molecular modelling analysis has shown that the charged groups of the more flexible compounds AMPA and ATPA could be superimposed on those of the relatively rigid analogues 5-HPCA and 7-HPCA, as shown in Fig. 9 for AMPA, ATPA and 5-HPCA (Brehm *et al.* 1988). The shape of the compounds is illustrated by their solvent-accessible surfaces, and although AMPA and ATPA occupy more space than 5-HPCA, this preliminary pharmacophore model suggests that AMPA and ATPA, by adopting a folded conformation, can obtain an arrangement of the charged groups similar to that of the conformationally restricted compounds 5-HPCA and 7-HPCA.

A quantitative model for agonist activity at AMPA receptors has been derived from these superpositions by placing opposite charges, reflecting hypothetical receptor functionalities, around the bicyclic analogues 5-HPCA and 7-HPCA (Jørgensen *et al.* 1988).

Substitution of the methyl group of AMPA with a phenyl group gives (*RS*)-2-amino-3-(3-hydroxy-5-phenylisoxazol-4-yl)propanoic acid (APPA; Fig. 7), which displays a markedly lower affinity and efficacy at AMPA receptors and appears to be the first unambiguous partial agonist at these receptors (Christensen *et al.* 1989). The corresponding *para*-chlorophenyl derivative (*p*-Cl-APPA) shows approximately the same weak affinity as APPA towards [^{3}H]AMPA binding sites, but *p*-Cl-APPA was inactive both as an agonist and as an antagonist in the rat cortical wedge preparation (A. Reinhardt *et al.* unpublished).

N-Methylation of the amino acid group of AMPA causes a considerable decrease in excitatory potency but does not affect the selectivity for AMPA receptors (Krogsgaard-Larsen *et al.* 1984, Madsen *et al.* 1990a). *O*-Methylation of the 3-hydroxy group on the isoxazole ring, on the other hand, completely eliminates all neuronal effects and the resulting compound is inactive both as an excitant and an inhibitor of [^{3}H]AMPA binding (Hansen *et al.* 1983). Similarly, esterification of the

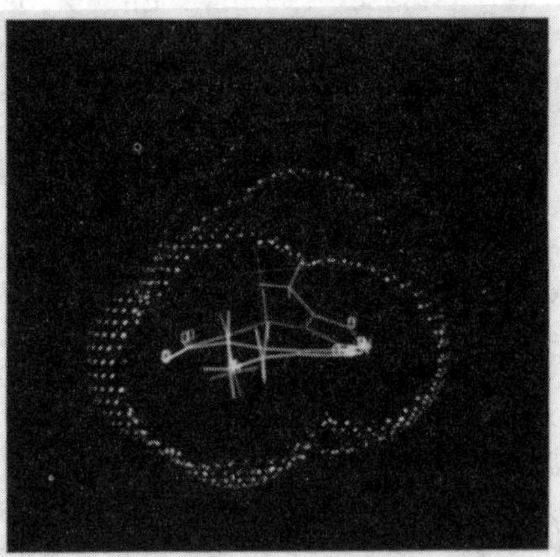

Fig. 9 — (**See colour section.**) The solvent-accessible surfaces of ATPA (red), AMPA (blue) and 5-HPCA (green).

carboxyl group of AMPA abolishes the neuronal activity (Madsen *et al.* 1990b). It is, in fact, generally observed for these 3-hydroxyisoxazole amino acids that protection of either or both acidic functions results in weak or inactive compounds with no significant agonist or antagonist properties. However, substitution of the 3-hydroxy group of specific AMPA receptor agonists like AMPA by the acidic carboxymethoxy group may cause a conversion into useful non-NMDA receptor antagonists (Krogsgaard-Larsen *et al.* 1991) as evidenced by the AMPA derivative (*RS*)-2-amino-3-(3-(carboxymethoxy)-5-methylisoxazol-4-yl)propanoic acid (AMOA; Fig. 7).

The bicyclic compound (*RS*)-2-amino-3-(3-hydroxy-7,8-dihydro-6*H*-cyclohepta[1,2-*d*]isoxazol-4-yl)propanoic acid (4-AHCP; Fig. 7) can be regarded as a homologue of AMPA and Glu, and 4-AHCP is at least as potent as AMPA as a selective GDEE- and NBQX-sensitive excitant on cat spinal neurons *in vivo*, although somewhat less potent in the rat cortical wedge preparation and as an inhibitor of [^{3}H]AMPA binding (Krogsgaard-Larsen *et al.* 1984, Lund *et al.* 1991). Further homologization of AMPA, however, causes a significant decrease in both excitatory potency and affinity for AMPA receptors (Christensen *et al.* 1989).

The remarkable potency of 4-AHCP as an excitant at AMPA receptors *in vivo* prompted a study of the conformational flexibility of this compound and a comparison of its low-energy conformations with those of AMPA and ATPA (Lund *et al.* 1991). The conformational analysis, using molecular mechanics (Tripos force field) and subsequent quantum mechanics (AM1) calculations, was performed using a build-up procedure by first considering the conformational flexibility of the parent bicyclic ring system and subsequently adding the amino acid side-chain. For each ring conformation were generated the possible low-energy conformations of the side-chain, and finally a complete geometry optimization of each of the resulting 4-AHCP conformations was effected. This conformational analysis revealed that 4-AHCP is a quite flexible compound with four low-energy conformations, none of which resemble the solid state conformation of AMPA. However, one of the low-energy conformations of 4-AHCP shows a striking similarity both with one of the three low-energy conformations of AMPA and with one of the three low-energy conformations found for ATPA, allowing the charged groups of each of these three compounds to occupy the same positions in space. In this superposition (Fig. 10), the methylene bridge of 4-AHCP occupies the same position as the 5-alkyl substituents of AMPA and ATPA which at the receptor site may be situated in a 'cavity' as depicted for ABPA in Fig. 8.

Like AMPA, IBO is an analogue of Glu, and homologization of IBO gives homoibotenic acid ((*RS*)-2-amino-3-(3-hydroxyisoxazol-5-yl)propanoic acid; HIBO), which interacts weakly, but selectively, with AMPA receptors. Introduction of substituents in the 4-position of the isoxazole ring causes a considerable increase in excitatory potency, and a series of 4-substituted homoibotenic acids (Fig. 11) have been prepared. Like 4-AHCP, these homologues of IBO show strong excitatory properties at AMPA receptors despite being analogues of the NMDA receptor antagonists 2-aminoadipic acid (α-AA) and 2-amino-5-phosphonopentanoic acid (AP5). 4-Bromohomoibotenic acid (Br-HIBO) and the corresponding, pharmacologically very similar, 4-methyl derivative (Me-HIBO), which is a structural isomer of AMPA, are the most potent members of this series (Krogsgaard-Larsen *et al.* 1980) and their excitatory effects can be antagonized by GDEE and CNQX but not by α-AA or AP5. Interestingly, the polar 2-hydroxyethyl group and the hydrophobic *n*-butyl

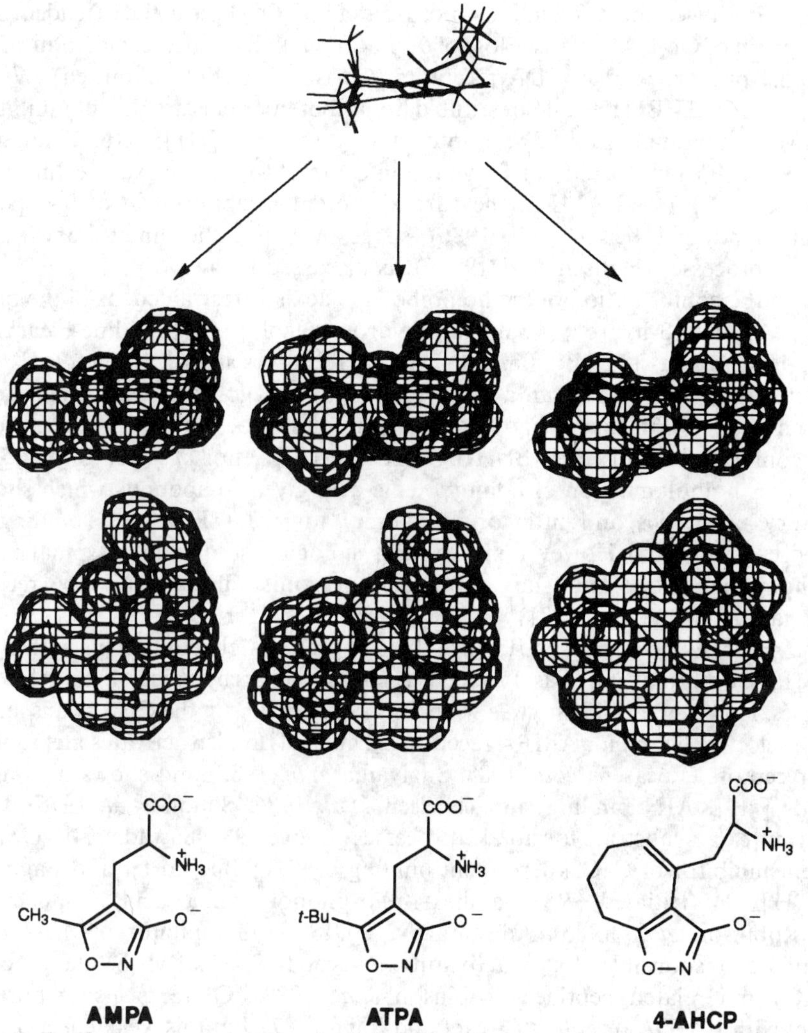

AMPA **ATPA** **4-AHCP**

Fig. 10 — Superposition of selected low-energy conformations of AMPA, ATPA and 4-AHCP with van der Waals volumes of the compounds from two different views, of which the upper row corresponds to that of the superposition.

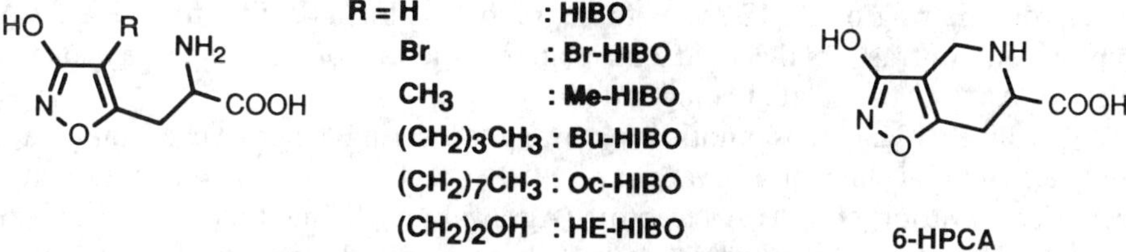

Fig. 11 — Structures of IBO homologues.

group are both well tolerated in the 4-position of HIBO whereas the extended, apolar *n*-octyl chain of Oc-HIBO causes loss of excitatory activity and, instead, induces weak antagonist properties at NMDA receptors (Christensen *et al.* submitted). With the exception of Oc-HIBO these 4-substituted homoibotenic acids are potent inhibitors of [^{3}H]AMPA binding (Table 2) without affinity towards [^{3}H]KAIN binding sites (Krogsgaard-Larsen *et al.* 1980). However, in contrast to other isoxazole amino acids derived from IBO and AMPA, they are all potent inhibitors of $CaCl_2$-dependent [^{3}H]Glu binding (Nielsen *et al.* 1986), suggesting that they interfere with EAA transport processes (Hansen *et al.* 1989b), as discussed below.

When the conformation of the homoibotenic acids is restrained, as in the bicyclic analogue (*RS*)-3-hydroxy-4,5,6,7-tetrahydroisoxazolo[4,5-*c*]pyridine-6-carboxylic acid (6-HPCA; Fig. 11), all affinity to AMPA receptors is lost. In fact, 6-HPCA is completely inactive both as an agonist and as an antagonist at cat spinal neurons (Madsen *et al.* 1986), suggesting that this compound does not reflect the receptor-active conformations of either Br-HIBO and Me-HIBO nor of α-AA and AP5.

Further homologization of homoibotenic acid gives compounds which show no excitatory properties and little or no affinity toward [^{3}H]AMPA binding sites (Lauridsen *et al.* 1985). However, in electrophysiological studies on cat spinal neurons these homologues of HIBO show moderate antagonist effects at NMDA receptors (Krogsgaard-Larsen *et al.* 1981).

The naturally occurring QUIS (Fig. 1) is approximately equipotent with AMPA at AMPA receptors and was formerly used as the archetypical ligand for these receptors. However, QUIS shows a wide range of activities in the CNS and is far from being a selective ligand for AMPA receptors. Thus, QUIS also activates metabotropic EAA receptors (Recasens *et al.* 1988, Sugiyama *et al.* 1987) and shows high affinity towards [^{3}H]KAIN binding sites (Nielsen *et al.* 1986, Simon *et al.* 1976). QUIS, furthermore, is a substrate for uptake processes (Lodge 1981, Sawada *et al.* 1985) and a potent inhibitor of $CaCl_2$-dependent binding of [^{3}H]Glu (Foster and Fagg 1984) and [^{3}H]AP4 (tritiated (*RS*)-2-amino-4-phosphonobutanoic acid) (Butcher *et al.* 1987, Robinson *et al.* 1985). Additionally, QUIS is an inhibitor of the NAALA dipeptidase responsible for inactivating *N*-acetyl-(*S*)-aspartyl-(*S*)-glutamic acid (NAAG) and related peptides (Robinson *et al.* 1987). QUIS is also reported to interact with NMDA receptors (Foster and Fagg 1987), but this weak effect of QUIS may be due to the presence of small amounts of Glu- and aspartic acid (Asp)-like impurities in most QUIS preparations (Cha *et al.* 1989).

X-ray crystallography and semi-empirical calculations of QUIS and some isosteric derivatives have indicated that the heterocyclic ring nitrogen atom, where the amino acid side-chain is attached, is pyramidal and rapidly interconverting (Boden *et al.* 1986, Jackson *et al.* 1988). This conformational flexibility and lack of coplanarity of the heterocyclic ring of QUIS seems to be important for its ability to activate EAA receptors, in contrast to the rigid and planar structure of the 3-hydroxyisoxazole moiety of AMPA and related compounds.

Willardiine (Fig. 12) is another naturally occurring heterocyclic amino acid, mainly tested in racemic form (Evans *et al.* 1980), which appears to exert its excitatory effects via activation of AMPA receptors (Agrawal and Evans 1986, Sugiyama *et al.* 1989). However, the substituted derivative 5-bromowillardiine seems mainly to act at KAIN receptors based on its potent excitatory effect at spinal afferent C-fibres

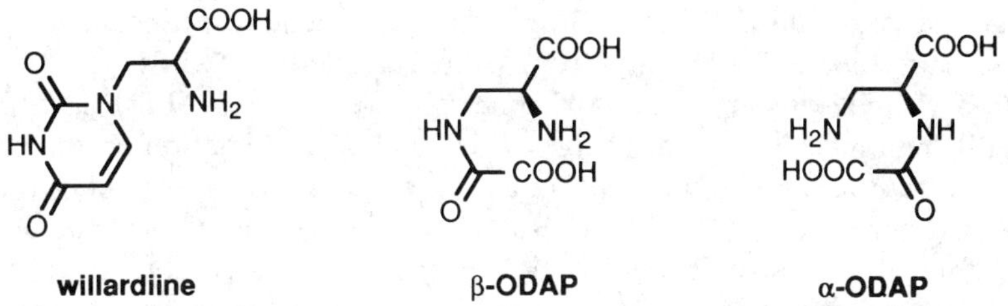

willardiine **β-ODAP** **α-ODAP**

Fig. 12 — Structures of willardiine, β-ODAP and α-ODAP.

(Agrawal and Evans 1986, Davies *et al.* 1982; see chapter 12).

The lathyrus constituent (*S*)-2-amino-3-(*N*-oxalylamino)propanoic acid (β-ODAP, often also termed BOAA; Fig. 12) may be the causative agent in human neurolathyrism, a relatively rare neurological disorder observed after excessive ingestion of seeds of the chickling pea, *Lathyrus sativus*, and characterized by irreversible spastic paralysis of the legs (Ludolph *et al.* 1987). β-ODAP is a potent and stereoselective excitant, preferentially acting at AMPA receptors but also showing some affinity to [^{3}H]KAIN binding sites and weak interactions with NMDA receptors (MacDonald and Morris 1984, Bridges *et al.* 1989).

While β-ODAP, like 4-AHCP and HIBO, can be considered a homologue of Glu, the isomeric 3-amino-(*S*)-2-(*N*-oxalylamino)propanoic acid (α-ODAP; Fig. 12) has the same chain length as Glu between the acidic groups. However, α-ODAP no longer contains a free α-amino group and is considerably weaker than β-ODAP both as an excitant (MacDonald and Morris 1984) and as an inhibitor of [^{3}H]AMPA binding (Bridges *et al.* 1989). The homologues of β-ODAP, 2-amino-4-(*N*-oxalylamino)butanoic acid (γ-ODAB), δ-*N*-oxalylornithine and ε-*N*-oxalyllysine (Fig. 13), exhibit little affinity for [^{3}H]AMPA and [^{3}H]KAIN binding sites but show weak antagonist properties at NMDA receptors (Bridges *et al.* 1989, Evans *et al.* 1982), in analogy with the results obtained by homologization of HIBO (Krogsgaard-Larsen *et al.* 1981, Lauridsen *et al.* 1985).

A related amino acid, (*S*)-2-amino-3-methylaminopropanoic acid (β-methylamino-L-alanine; BMAA; Fig. 13), is a constituent of the seeds of the false sago palm, *Cycas circinalis*, which were formerly used for food and medicinal purposes on the

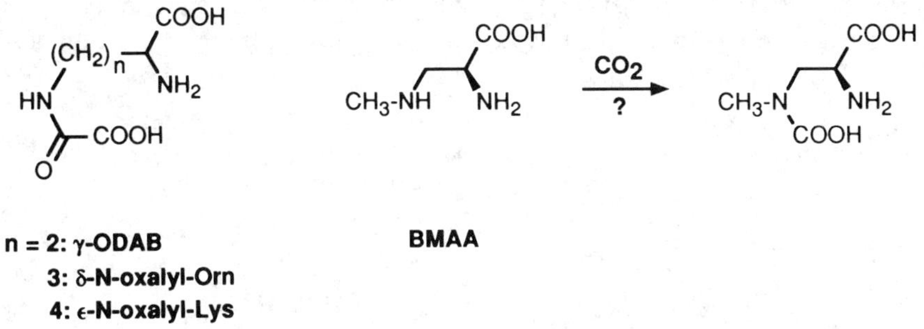

n = 2: γ-ODAB
3: δ-N-oxalyl-Orn
4: ε-N-oxalyl-Lys

BMAA

Fig. 13 — Structures of γ-ODAB, δ-*N*-oxalyl-Orn, ε-*N*-oxalyl-Lys and BMAA with a hypothetical carbamate which may be formed in bicarbonate-containing buffers.

Marianas Islands, including Guam. BMAA has been suggested (Spencer *et al.* 1987) to be the causative agent in amyotrophic lateral sclerosis (ALS) which, often in association with Parkinsonism and Alzheimer-type dementia (GALS-PD), showed an abnormally high incidence in the indigenous Chamarro population on these islands. Ingestion of BMAA causes neurodegenerative changes in primates similar to those of GALS-PD (Spencer *et al.* 1987), but BMAA, curiously, lacks the ω-acidic group normally found in excitotoxic amino acids. However, in mouse cortical cell cultures the neurodegenerative and excitatory properties of BMAA could only be observed in bicarbonate-containing buffers (Weiss and Choi 1988), suggesting that these effects were caused by BMAA in combination with HCO_3^- via an ionic or covalent bond. The effects of BMAA are partially prevented by NMDA receptor antagonists (Spencer *et al.* 1987, Weiss and Choi 1988), suggesting an interaction with NMDA receptors, but BMAA also appears to interact with non-NMDA receptors (Smith and Meldrum 1990, Weiss *et al.* 1989) and metabotropic EAA receptors (Copani *et al.* 1990).

A number of excitatory sulphur-containing amino acids (Fig. 14) may have a physiological function as regulators of the levels of excitability at excitatory and inhibitory pathways (Dunlop and Griffiths 1990) and some of them, particularly (*S*)-homocysteic acid (L-HCA), have been proposed as endogenous transmitter candidates (Cuenod *et al.* 1990, Knöpfel *et al.* 1987, Pullan *et al.* 1987; see chapter 7). Like Glu, these acidic amino acids are mixed agonists and interact in varying degrees with AMPA receptors in addition to their predominant interaction with NMDA receptors and generally weak inhibition of [^{3}H]KAIN binding (Murphy and Williams 1987, Pullan *et al.* 1987).

Among the sulphur-containing amino acids, (*R*)-cysteine-*S*-sulphonic acid (L-CSS), being equi-active with Glu, is the most potent inhibitor of [^{3}H]AMPA binding (Murphy and Williams 1987, Pullan *et al.* 1987). L-CSS is a neurotoxic metabolite of cysteine, formed in abnormally high amounts in a metabolic disorder termed sulphite oxidase deficiency, and may be responsible for the brain damage observed in this rare disease (Olney *et al.* 1975).

In summary, by structural and conformational modifications of selective AMPA receptor agonists a reasonably good correlation between affinity towards [^{3}H]AMPA binding sites and excitatory potency at AMPA receptors has been obtained (see Table 2), consistent with the notion that [^{3}H]AMPA binding sites represent the recognition site of AMPA receptors. High affinity and potent excitatory effects at AMPA receptors necessitate the presence of an α-amino acid structure containing a positive and two negative charges. Substituents on the positive ammonium group are not

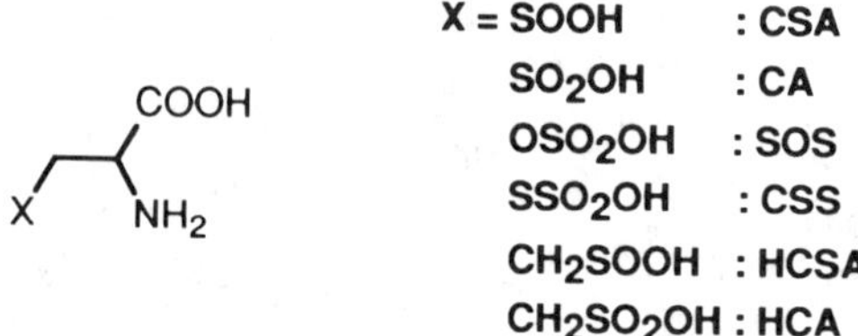

Fig. 14 — Structures of excitatory sulphur-containing amino acids.

always tolerated, and the chain of atoms, not necessarily carbon atoms, separating the two negative charges may be longer than that required for potent agonist activity at NMDA receptors (see chapter 6). Maximal agonist activity at AMPA receptors is observed with three or four atoms separating the two acidic groups, while further elongation of the chain confers antagonist properties or loss of activity. As with NMDA receptor agonists, the α-carboxyl group seems to be mandatory, whereas more diversity is allowed for the ω-acidic group and several acidic heterocycles, sometimes containing bulky substituents, can act as bioisosteres of the γ-carboxyl group of Glu. A high conformational flexibility of agonists is not necessary for high affinity and excitatory potency at AMPA receptors, and compounds with quite rigid structures may effectively activate these receptors.

STEREOCHEMICAL ASPECTS

Despite the curious steric requirements of the NMDA receptors (Hansen and Krogsgaard-Larsen 1990, Watkins and Olverman 1988) and the warning served by α-AA where the antagonist effect of the D-enantiomer (McLennan and Hall 1978) was masked by the agonist properties of the L-form (Curtis and Watkins 1960), delaying for several years the discovery of a useful antagonist, stereochemical aspects of structure–activity studies at EAA receptors are often neglected because of the difficulties in obtaining the enantiomeric components of synthetic racemates. Thus, with the exception of the mixed-action sulphur-containing amino acids (Mewett *et al.* 1983, Murphy and Williams 1987, Pullan *et al.* 1987), relatively few of the AMPA receptor agonists have been investigated with respect to the pharmacological properties of individual enantiomers.

The L-form of Glu is only slightly more potent than its D-enantiomer as an excitant at cat spinal neurons (Curtis and Watkins 1960), despite having two orders of magnitude higher affinity towards the binding sites for AMPA (Table 3) and an equally high stereoselectivity at NMDA receptors (Watkins and Olverman 1988) and KAIN binding sites (Simon *et al.* 1976). Similarly, the D-forms of HCA and, especially, homocysteine sulphinic acid (HCSA; Fig. 14) are at least as potent neuronal excitants as their corresponding L-enantiomers, probably due to a lack of efficient uptake mechanisms for these D-amino acids (Mewett *et al.* 1983). However, with the exception of HCA, where the enantiomers are approximately equipotent as inhibitors of [^{3}H]AMPA binding, the L-forms of the sulphur-containing amino acids, including (*S*)-HCSA and (*R*)-CSS, display somewhat higher affinity than their corresponding D-enantiomers for [^{3}H]AMPA binding sites (Murphy and Williams 1987, Pullan *et al.* 1987).

QUIS is an L-amino acid but also the D-enantiomer of QUIS shows significant excitatory effects, although weaker than the L-form, in the locust neuromuscular junction (Boden *et al.* 1986). Molecular modelling studies indicate that the excitatory activity of the D-enantiomer may be explained by an inversion of the pyramidal ring nitrogen atom where the amino acid side-chain is attached, allowing the D-form to display considerable similarity with the more potent L-enantiomer (Bycroft and Jackson 1988).

The D-form of β-ODAP does not exhibit the excitatory and neurotoxic properties of the L-enantiomer, and only the L-enantiomer shows affinity towards [^{3}H]AMPA and [^{3}H]KAIN binding sites although both enantiomers display weak affinity towards NMDA receptors (Bridges *et al.* 1989).

From these studies it appears, with the exception of HCA, that AMPA receptors are preferentially activated by the L-enantiomers of agonists. In order to test the extent and generality of this stereoselectivity we have engaged in a systematic effort to investigate the enantiomers of a number of isoxazole amino acids acting as agonists or antagonists at AMPA receptors. Several of these have now been prepared by kinetic resolution using enzymatic techniques or by direct chromatographic resolution by high-performance liquid chromatography (HPLC) on chiral stationary phases. Compared to asymmetric synthesis these resolution procedures have the advantage of providing, in a single set of operations, both enantiomers for biological testing.

The resolution of 4-bromohomoibotenic acid (Br-HIBO), using the enzyme amino acylase in covalently immobilized form, is shown in Fig. 15 (Hansen *et al.* 1989b). The *N*-acetylated *O*-methyl derivative of racemic Br-HIBO was used as a substrate in the enzymatic step, and the resolution was monitored by quantitating the hydrolysis of the acetylamino group via reaction, in aliquots, of the liberated amino group with ninhydrin. The reaction progressed only until 50% conversion of the racemic substrate (Fig. 16), suggesting a highly stereoselective hydrolysis of the substrate. After separation of the free *O*-methyl amino acid from unreacted *N*-acetyl-*O*-methyl amino

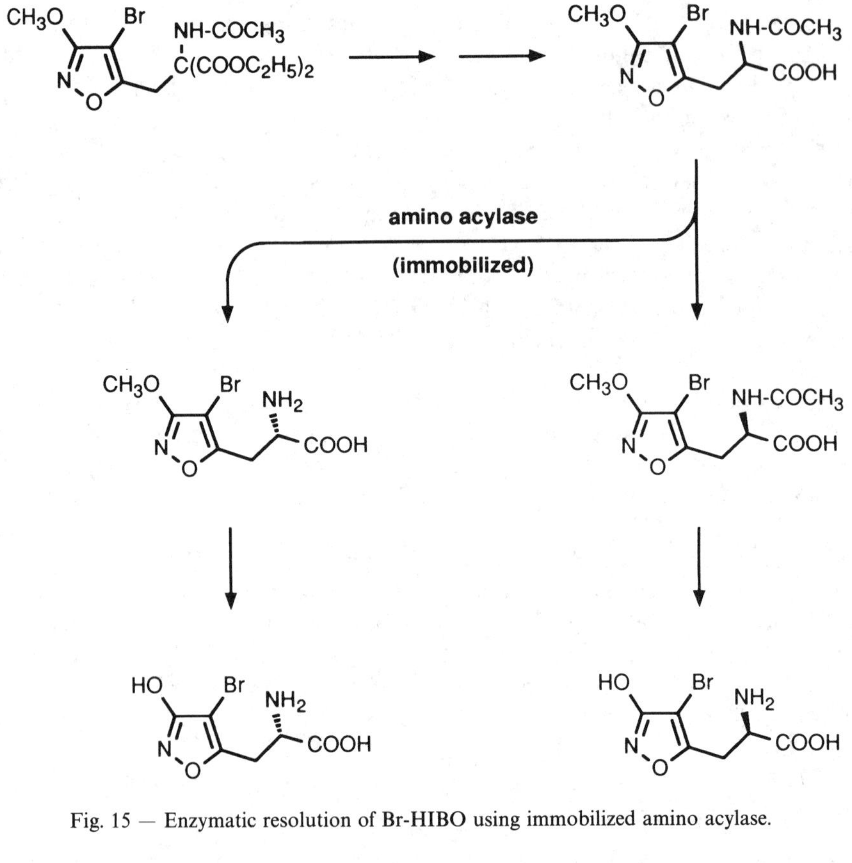

CH_3O Br NH-$COCH_3$ C($COOC_2H_5$)$_2$ → → CH_3O Br NH-$COCH_3$ COOH

amino acylase

(immobilized)

CH_3O Br NH_2 COOH | CH_3O Br NH-$COCH_3$ COOH

HO Br NH_2 COOH | HO Br NH_2 COOH

Fig. 15 — Enzymatic resolution of Br-HIBO using immobilized amino acylase.

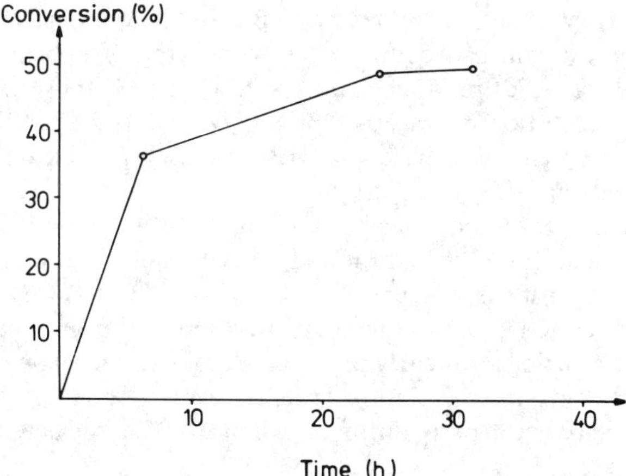

Fig. 16 — Progression of the amino acylase-catalysed hydrolysis of the *N*-acetyl-*O*-methyl derivative of racemic Br-HIBO.

acid by ion-exchange chromatography, the enantiomers of Br-HIBO were obtained after deprotection. A similar resolution (B. Nielsen *et al.* unpublished), where the amino acylase was immobilized by physical entrapment in hollow fibres, was used to prepare the enantiomers of Me-HIBO (Fig. 11).

In addition to means for monitoring the progression of the resolution, a resolution procedure, ideally, should also provide a determination of the absolute configuration and enantiomeric purity of the products. The absolute configuration of the enantiomers of Br-HIBO was unequivocally established by X-ray crystallography (Fig. 17), confirming the expected preference of amino acylase for hydrolysis of the L-form of the substrate (Hansen *et al.* 1989b).

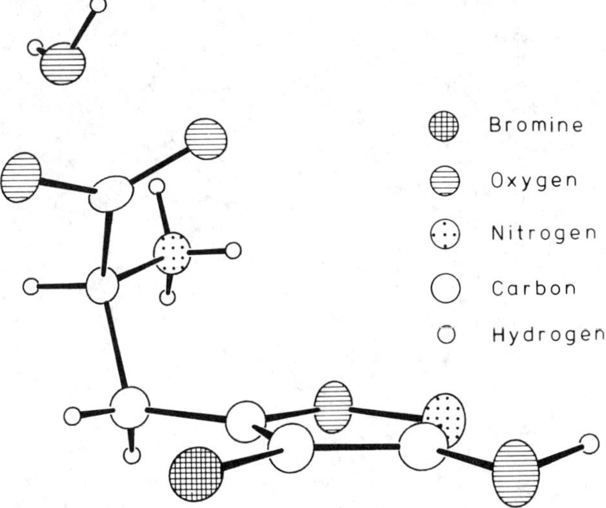

Fig. 17 — X-ray structure of L-Br-HIBO monohydrate.

The optical purity of the enantiomers of Br-HIBO was determined by ligand-exchange HPLC on a chiral stationary phase consisting of optically active proline covalently bound to silica and chelated with Cu^{2+} (Gübitz *et al.* 1981). The chromatographic resolution of racemic Br-HIBO and the analytical separations obtained with the two enantiomers of Br-HIBO by use of a chiral stationary phase containing D-proline are shown in Fig. 18. The L-form of Br-HIBO elutes before the D-enantiomer on this stationary phase, allowing quantitation of the 0.6% enantiomeric impurity present in this preparation of D-Br-HIBO (Fig. 18B). In L-Br-HIBO the enantiomeric impurity is obscured by the tail of the dominant peak, and the chromatogram (Fig. 18C) does not allow quantitation of the amount of D-Br-HIBO present. However, by using a similarly prepared chiral stationary phase containing L-proline instead of D-proline the elution order is reversed, and the presence of 0.1% enantiomeric impurity in the preparation of L-Br-HIBO can be determined (chromatogram not shown).

AMPA was previously resolved using immobilized amino acylase by a procedure (Hansen *et al.* 1983) analogous to the resolution shown in Fig. 15 for Br-HIBO. Recently, however, an improved resolution, using a solution of the cheaper and more readily available enzyme α-chymotrypsin, has been performed (B. Nielsen *et al.* unpublished). This resolution (Fig. 19), which is faster than the original procedure, utilizes the esterase activity of the enzyme with the ethyl ester of racemic *N*-acetyl-*O*-ethyl-AMPA as substrate. The enzymatic reaction was performed in a pH-stat maintained at pH 7.8 by continuous addition of base to neutralize the carboxylic acid released during hydrolysis. This procedure allows a direct monitoring of the progression of the reaction by following the consumption of base. As in the amino acylase-catalysed reactions, the enzymatic hydrolysis stopped after 50% conversion of the racemic substrate, and after separation of the free carboxylic acid from the unreacted ester the two enantiomers of AMPA were obtained after deprotection. The enantiomeric purities of the products were higher than 99.6% enantiomeric excess (e.e.) as

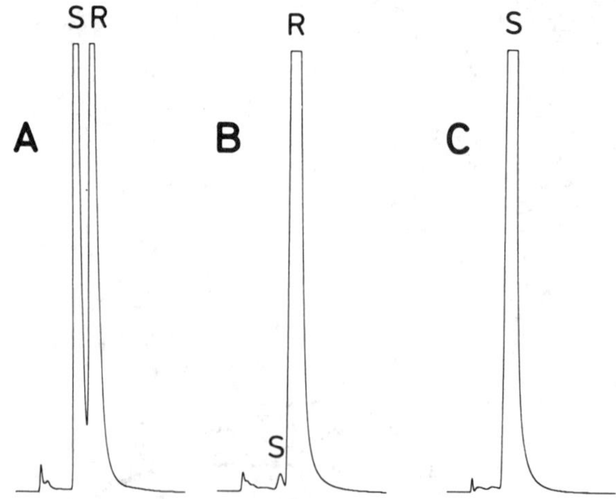

Fig. 18 — Chiral ligand-exchange HPLC of (A) the racemate, (B) the D- and (C) the L-enantiomer of Br-HIBO on a stationary phase containing D-proline.

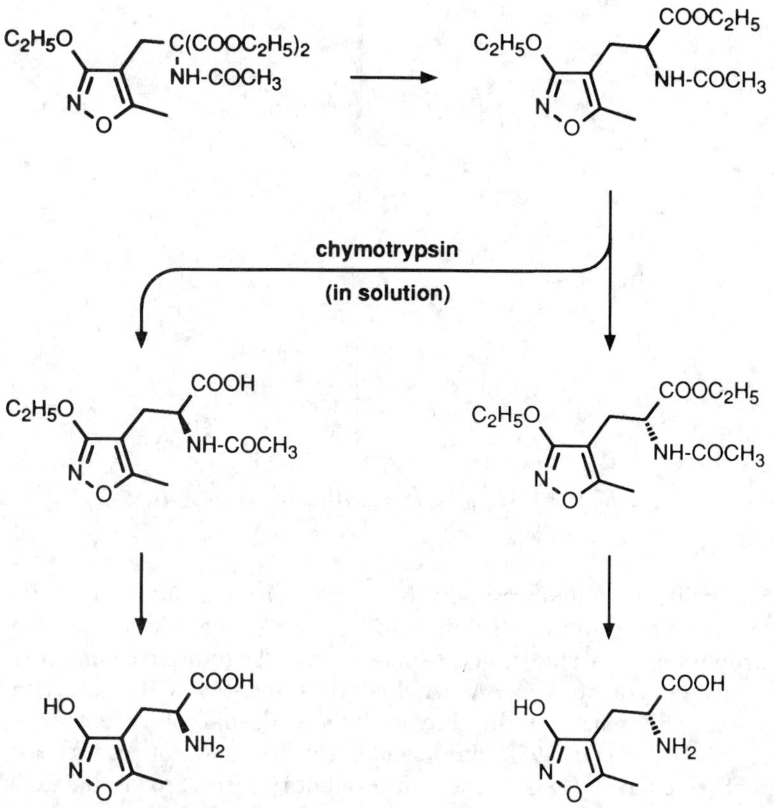

Fig. 19 — Enzymatic resolution of AMPA using α-chymotrypsin in solution.

determined by chiral ligand-exchange HPLC, analogous to the separations shown in Fig. 18 for the enantiomers of Br-HIBO. The AMPA derivative hydrolysed by α-chymotrypsin (Fig. 19) had the same absolute configuration as that of the AMPA derivative (*N*-acetyl-*O*-methyl-AMPA) hydrolysed by amino acylase (Hansen *et al.* 1983). From the well-established stereoselectivty of these two enzymes (Barrett 1985), especially that of amino acylase (Chenault *et al.* 1989), the hydrolysed derivatives are assumed to be of the L-configuration.

The same general procedure as described in Fig. 19 for AMPA was applied to the resolution of 5-HPCA using α-chymotrypsin in immobilized form (A. Reinhardt *et al.* unpublished). This procedure gave the enantiomers of 5-HPCA with enantiomeric purities higher than 98.8% e.e. as determined by chiral ligand-exchange HPLC (Fig. 20). The absolute configurations of the enantiomers are presently unknown, however, because α-chymotrypsin cannot be trusted to hydrolyse preferentially the L-enantiomer of a conformationally restrained substrate (I. Breum *et al.* unpublished, Hein and Niemann 1961). Hence, a determination of the absolute configuration of the enantiomers of 5-HPCA by X-ray crystallographic techniques is in progress (L. Brehm *et al.* unpublished).

No significant hydrolysis of the 3-alkoxyisoxazole moiety was observed in these

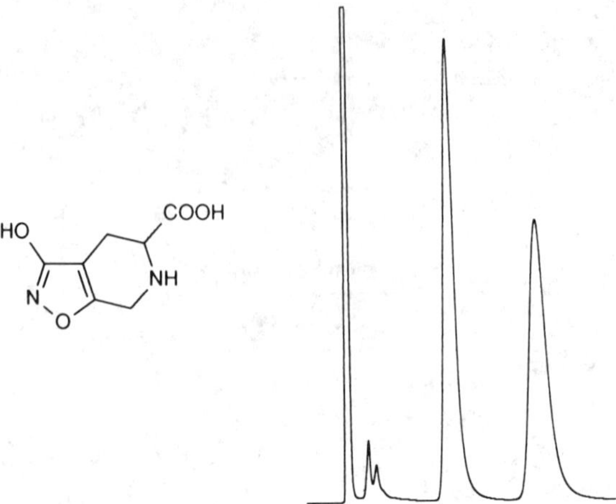

Fig. 20 — Chiral ligand-exchange HPLC of racemic 5-HPCA.

and other similar α-chymotrypsin-catalysed reactions, indicating that this potential bioisostere of a carboxyl ester is not a substrate for the enzyme.

Electrophysiological studies in cat spinal cord of the resolved enantiomers showed that the excitatory actions of AMPA, Br-HIBO and Me-HIBO reside in their L-enantiomers, while the D-forms are essentially devoid of excitatory properties (Krogsgaard-Larsen *et al.* 1982, Hansen *et al.* 1989b). A pronounced stereoselectivity was also observed for 5-HPCA, where the excitatory effects are located exclusively in the (+)-enantiomer (A. Reinhardt *et al.* unpublished). Surprisingly, the excitatory potency of the racemates of Br-HIBO and Me-HIBO was significantly stronger than that of their L-enantiomers. Correspondingly, the excitatory potency of L-Br-HIBO was increased by co-administration of the excitatorily inactive D-enantiomer (Hansen *et al.* 1989b), indicating some kind of cooperative effect of the enantiomers. No such effect was observed for AMPA or 5-HPCA.

A pronounced stereoselectivity was also observed in [^{3}H]AMPA binding studies on rat brain synaptic membranes, where the L-form of AMPA displays 3800 times higher affinity than its D-enantiomer (Table 3). Of the compounds listed in Table 3 AMPA shows the highest affinity for AMPA receptors. Its high eudismic ratio is, therefore, consistent with Pfeiffer's rule (Pfeiffer 1956) which, however, has recently been questioned (Barlow 1990). A limited stereoselectivity might, perhaps, have been expected for 5-HPCA owing to its almost planar structure (see Fig. 5). However, as for Br-HIBO, Me-HIBO and Glu, the [^{3}H]AMPA binding sites also show considerable stereoselectivity towards 5-HPCA, and the observed eudismic ratio is $\geqslant 100$ for all four compounds. This stereoselectivity, observed for isoxazole amino acids at [^{3}H]AMPA binding sites, corresponds well with the stereoselective excitatory effects of these compounds in electrophysiological studies.

With the exception of Glu, none of the compounds listed in Table 3 show any affinity towards NMDA receptors or [^{3}H]KAIN binding sites, nor do they inhibit D-[^{3}H]Asp binding which conventionally is taken to represent the sodium-dependent high-affinity uptake system assumed to be responsible for inactivation of EAAs (Cross

Table 3 — Stereoselectivity of EAA binding sites

	$[^3H]$AMPA binding IC_{50} (μM)	$[^3H]$Glu binding[a] IC_{50} (μM)	D-$[^3H]$Asp binding IC_{50} (μM)
DL-AMPA	0.04	$\geqslant 100$	> 100
L-AMPA	0.02	$\geqslant 100$	
D-AMPA	76	> 100	
DL-Br-HIBO	0.60	0.20	> 100
L-Br-HIBO	0.34	0.22	
D-Br-HIBO	32	0.15	
DL-Me-HIBO	0.61	0.18	> 100
L-Me-HIBO	0.32	0.17	
D-Me-HIBO	> 100	0.19	
DL-5-HPCA	1.9	> 100	> 100
(+)-5-HPCA	0.88		
(−)-5-HPCA	> 100		
L-Glu	0.50	0.17	7.5
D-Glu	47	2.5	> 100

[a]In the presence of 2.5 mM $CaCl_2$.

et al. 1986). However, the enantiomers of Br-HIBO and Me-HIBO, in contrast to those of AMPA and 5-HPCA, displace $[^3H]$Glu binding in the presence of chloride and calcium ions, and the D-enantiomers of these two compounds are at least as potent as their corresponding L-forms as inhibitors of this $CaCl_2$-dependent $[^3H]$Glu binding.

The physiological relevance of the $CaCl_2$-dependent binding of $[^3H]$Glu to synaptic membranes has been the subject of considerable discussion (Fagg and Lanthorn 1985). However, these binding sites now seem to be related to EAA transport processes (Hansen *et al.* 1989b) and probably represent a chloride-linked reuptake of Glu into resealed membrane vesicles (Kessler *et al.* 1987a, Pin *et al.* 1984, Zaczek *et al.* 1987). Synaptic $CaCl_2$-dependent $[^3H]$Glu binding is inhibited by AP4, QUIS and several sulphur-containing amino acids, including cystine and cysteic acid (CA; Fig. 14) (Kessler *et al.* 1987b, Koyama *et al.* 1989). A related chloride-dependent transport system, albeit insensitive to AP4, appears to be present on glial membranes (Bridges *et al.* 1987).

Although the functional significance of these chloride-dependent transport systems is unknown, they may explain the paradoxical excitatory effects observed with the racemate and the enantiomers of Br-HIBO and Me-HIBO. If the enantiomers bind to a transport site, they may be substrates for the transport process. The presence of the excitatorily inactive D-enantiomers could then inhibit the removal of the excitatorily active L-enantiomers, thereby prolonging their presence near the synapse and thus increasing their excitatory potency (Hansen *et al.* 1989b).

The observed IC_{50} values (Table 3) of 0.34 μM and 32 μM, respectively, for the more potent enantiomer (eutomer) and the less potent enantiomer (distomer) of Br-HIBO as inhibitors of [^{3}H]AMPA binding give an observed eudismic ratio of 94 for Br-HIBO at AMPA receptors. However, a considerable proportion of the observed affinity of the distomer is caused by the eutomer present as an enantiomeric impurity. If the amounts of enantiomeric impurity present in the distomer and eutomer samples are known, the observed affinities (proportional to the inverse IC_{50} values) may be corrected for the contribution of the enantiomeric impurities by assuming that the observed affinity of the distomer sample (Affinity_d) is the sum of the corrected affinity of the distomer (affinity_d^c) times its molar proportion in the distomer sample (y_d) and the corrected affinity of the eutomer (Affinity_e^c) times its molar proportion in the distomer sample ($1 - y_d$):

$$\text{Affinity}_d = y_d \times \text{Affinity}_d^c + (1 - y_d) \times \text{Affinity}_e^c$$

where $y_d = \frac{1}{2}(1 + \text{enantiomeric purity of distomer sample})$. A similar equation can be set up for the observed affinity of the eutomer sample, giving two equations with two unknowns (Affinity_d^c and Affinity_e^c) from which these two corrected affinities can be calculated.

Thus, when the 0.1% and 0.6% enantiomeric impurities present in the eutomer and distomer samples, respectively, of Br-HIBO are taken into account, the corrected IC_{50} values are calculated to be 0.34 μM and 73 μM for the L- and D-form, respectively,

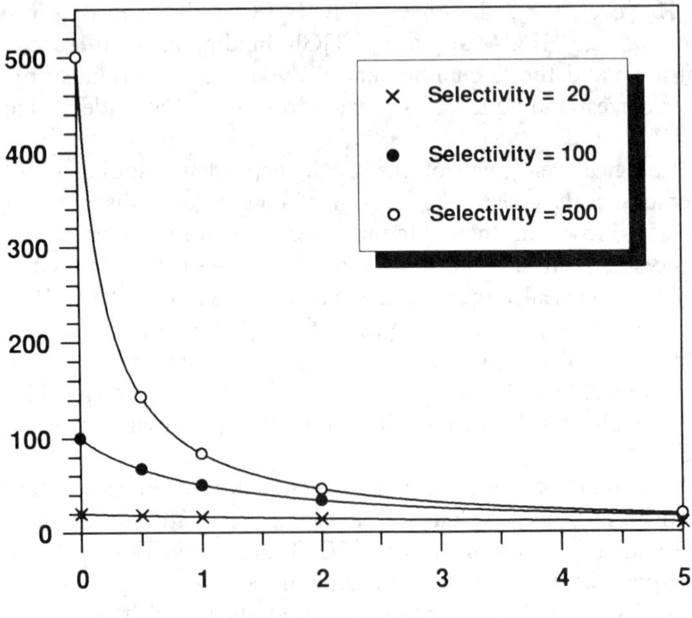

Fig. 21 — Observed eudismic ratio as a function of the proportion of enantiomeric impurity present in the distomer for three different 'true' receptor stereoselectivities.

resulting in a corrected eudismic ratio of 220. This value is more than twice as high as the observed eudismic ratio of 94, demonstrating, as earlier noted by Barlow *et al.* (1972), that observed eudismic ratios will cause a general underestimation of receptor stereoselectivity if not corrected for enantiomeric impurities.

The affinity of the eutomer is little affected by the presence of a small amount of enantiomeric impurity, and the observed affinity for a eutomer sample is normally a good approximation of the corrected value, as also found above for L-Br-HIBO. For a distomer sample, however, the affinity can be greatly influenced by the presence of an enantiomeric impurity, resulting in a large effect on the eudismic ratio. Assuming that the observed affinity of the eutomer is equal to its corrected affinity Fig. 21 shows the observed eudismic ratio as a function of the amount of enantiomeric impurity present in the distomer sample for three different 'true' receptor stereoselectivities. It is apparent from Fig. 21 that the presence of even small amounts of an enantiomeric impurity in the distomer has a dramatic effect on the observed eudismic ratio at highly stereoselective receptors. Consequently, a lack of information about the enantiomeric purity, as is usually the case in the pharmacological literature, may lead to a serious underestimation of the stereoselectivity of receptors.

ACKNOWLEDGEMENTS

We thank the Danish Biotechnology Research Initiative, the Danish Medical and Technical Research Councils and the Lundbeck Foundation for financial support.

REFERENCES

Agrawal, S. G. and Evans, R. H. (1986) *Br. J. Pharmacol.* **87** 345–355.

Allen, F. H., Kennard, O. and Taylor, R. (1983) *Acc. Chem. Res.* **16** 146–153.

Barlow, R. (1990) *Trends Pharmacol. Sci.* **11** 148–150.

Barlow, R. B., Franks, F. M. and Pearson, J. D. M. (1972) *J. Pharm. Pharmacol.* **24** 753–761.

Barnard, E. A. and Henley, J. M. (1990) *Trends Pharmacol. Sci.* **11** 500–507.

Barrett, G. C. (1985) In: Barrett, G. C. (ed.) *Chemistry and Biochemistry of the Amino Acids.* Chapman and Hall, London, pp. 338–353.

Bekkers, J. M. and Stevens, C. F. (1989) *Nature* **341** 230–233.

Bhat, T. N. and Vijayan, M. (1977) *Acta Cryst.* **B33** 1754–1759.

Boden, P., Bycroft, B. W., Chhabra, S. R., Chiplin, J., Crowley, P. J., Grout, R. J., King, T. J., McDonald, E., Rafferty, P. and Usherwood, P. N. R. (1986) *Brain Res.* **385** 205–211.

Borthwick, P. W. and Steward, E. G. (1976) *J. Mol. Struct.* **33** 141–144.

Boulter, J., Hollmann, M., O'Shea-Greenfield, A., Hartley, M., Deneris, E., Maron, C. and Heinemann, S. (1990) *Science* **249** 1033–1037.

Brehm, L., Jørgensen, F. S., Hansen, J. J. and Krogsgaard-Larsen, P. (1988) *Drug News Perspect.* **1** 138–144.

Brehm, L., Christensen, I. T., Nielsen, B., Reinhardt, A., Ebert, B., Madsen, U., Jørgensen, F. S., Hansen, J. J. and Krogsgaard-Larsen, P. (1990) In: Jensen, B., Jørgensen, F. S. and Kofod, H. (eds) *Frontiers in Drug Research: Crystallographic and Computational Methods.* Munksgaard, Copenhagen, pp. 92–105.

Bridges, R. J., Nieto-Sampedro, M. and Cotman, C. W. (1987) *J. Neurochem.* **48** 1709–1715.

Bridges, R. J., Stevens, D. R., Kahle, J. S., Nunn, P. B., Kadri, M. and Cotman, C. W. (1989) *J. Neurosci.* **9** 2073–2079.

Burkert, U. and Allinger, N. L. (1982) *Molecular Mechanics.* American Chemical Society, Washington DC.

Butcher, S. P., Roberts, P. J. and Collins, J. F. (1987) *Brain Res.* **419** 294–302.

Bycroft, B. W. and Jackson, D. E. (1988) In: Lunt, C. C. (ed.) *Neurotox'88: Molecular Basis of Drug and Pesticide Action.* Elsevier, Amsterdam, pp. 461–467.

Cha, J.-H. J., Hollingsworth, Z. R., Greenamyre, J. T. and Young, A. B. (1989) *J. Neurosci. Meth.* **27** 143–148.

Chenault, H. K., Dahmer, J. and Whitesides, G. M. (1989) *J. Am. Chem. Soc.* **111** 6354–6364.

Christensen, I. T., Reinhardt, A., Nielsen, B., Ebert, B., Madsen, U., Nielsen, E. Ø., Brehm, L. and Krogsgaard-Larsen, P. (1989) *Drug Des. Delivery* **5** 57–71.

Christensen, I. T., Ebert, B., Madsen, U., Nielsen, B., Brehm, L. and Krogsgaard-Larsen, P., submitted to *J. Med. Chem.*

Collingridge, G. L. and Singer, W. (1990) *Trends Pharmacol. Sci.* **11** 290–296.

Copani, A., Canonico, P. L. and Nicoletti, F. (1990) *Eur. J. Pharmacol.* **181** 327–328.

Coyle, J. T. (1983) *J. Neurochem.* **41** 1–11.

Cross, A. J., Skan, W. J. and Slater, P. (1986) *Neurosci. Lett.* **63** 121–124.

Cuenod, M., Do, K. Q. and Streit, P. (1990) *Trends Pharmacol. Sci.* **11** 477–478.

Curtis, D. R. and Watkins, J. C. (1960) *J. Neurochem.* **6** 117–141.

Curtis, D. R., Lodge, D. and McLennan, H. (1979) *J. Physiol.* **291** 19–28.

Davies, J. and Watkins, J. C. (1983) *Exp. Brain Res.* **49** 280–290.

Davies, J., Evans, R. H., Jones, A. W., Smith, D. A. S. and Watkins, J. C. (1982) *Comp. Biochem. Physiol.* **72C** 211–224.

Dewar, M. S. J. and Stewart, J. J. P. (1986) *QCPE Bull.* **6** 24 (QCPE Program 506).

Dunlop, J. and Griffiths, R. (1990) *Biochem. Soc. Trans.* **18** 378–381.

Eugster, C. H. (1969) *Fortschr. Chem. Org. Naturst.* **27** 261–321.

Evans, R. H., Jones, A. W. and Watkins, J. C. (1980) *J. Physiol.* **308** 71P–72P.

Evans, R. H., Nunn, P. B. and Pearson, S. (1982) *J. Physiol.* **327** 81P.

Evans, R. H., Evans, S. J., Pook, P. C. and Sunter, D. C. (1987) *Br. J. Pharmacol.* **91** 531–537.

Fagg, G. E. and Lanthorn, T. H. (1985) *Br. J. Pharmacol.* **86** 743–751.

Fagg, G. E., Foster, A. C. and Ganong, A. H. (1986) *Trends Pharmacol. Sci.* **7** 357–363.

Ferkany, J. W. and Coyle, J. T. (1983) *J. Pharmacol. Exp. Ther.* **225** 399–406.

Fonnum, F. (1984) *J. Neurochem.* **42** 1–11.

Foster, A. C. and Fagg, G. E. (1984) *Brain Res. Rev.* **7** 103–164.

Foster, A. C. and Fagg, G. E. (1987) *Eur. J. Pharmacol.* **133** 291–330.

Greenamyre, J. T. and Young, A. B. (1989) *Neurosci. Lett.* **101** 133–137.

Greenamyre, J. T., Olson, J. M. M., Penney, J. B. and Young, A. B. (1985) *J. Pharmacol. Exp. Ther.* **233** 254–263.

Gübitz, G., Jellenz, W. and Santi, W. (1981) *J. Liq. Chromatogr.* **4** 701–712.

Hansen, J. J. and Krogsgaard-Larsen, P. (1980) *J. Chem. Soc., Perkin Trans. I* 1826–1833.

Hansen, J. J. and Krogsgaard-Larsen, P. (1990) *Med. Res. Rev.* **10** 55–94.

Hansen, J. J., Lauridsen, J., Nielsen, E. and Krogsgaard-Larsen, P. (1983) *J. Med. Chem.* **26** 901–903.

Hansen, J. J., Madsen, U., Nielsen, E. Ø., Nielsen, B., Brehm, L. and Krogsgaard-Larsen, P. (1989a) In: Makriyannis, A. (ed.) *New Methods in Drug Research*, Vol. 3. Prous Science Publishers, Barcelona, pp. 73–101.

Hansen, J. J., Nielsen, B., Krogsgaard-Larsen, P., Brehm, L., Nielsen, E. Ø. and Curtis, D. R. (1989b) *J. Med. Chem.* **32** 2254–2260.

Hein, G. E. and Niemann, C. (1961) *Proc Nat. Acad. Sci. USA* **47** 1341–1355.

Hollmann, M., O'Shea-Greenfield, A., Rogers, S. W. and Heinemann, S. (1989) *Nature* **342** 643–648.

Honoré, T. (1989) *Med. Res. Rev.* **9** 1–23.

Honoré, T. and Drejer, J. (1988) *J. Neurochem.* **51** 457–461.

Honoré, T. and Lauridsen, J. (1980) *Acta Chem. Scand.* **B34** 235–240.

Honoré, T. and Nielsen, M. (1985) *Neurosci. Lett.* **54** 27–32.

Honoré, T., Lauridsen, J. and Krogsgaard-Larsen, P. (1982) *J. Neurochem.* **38** 173–178.

Honoré, T., Davies, S. N., Drejer, J., Fletcher, E. J., Jacobsen, P., Lodge, D. and Nielsen, F. E. (1988) *Science* **241** 701–703.

Honoré, T., Drejer, J., Nielsen, E. Ø. and Nielsen, M. (1989) *Biochem. Pharmacol.* **38** 3207–3212.

Izquierdo, I. (1991) *Trends Pharmacol. Sci.* **12** 128–129.

Jackson, D. E., Bycroft, B. W. and King, T. J. (1988) *J. Comp.-Aided Mol. Des.* **2** 321–328.

Johnston, G. A. R., Curtis, D. R., de Groat, W. C. and Duggan, A. W. (1968) *Biochem. Pharmacol.* **17** 2488–2489.

Jørgensen, F. S., Byberg, J. R., Krogsgaard-Larsen, P. and Snyder, J. P. (1988) In: Lunt, C. C. (ed.) *Neurotox'88: Molecular Basis of Drug and Pesticide Action.* Elsevier, Amsterdam, pp. 497–506.

Keinänen, K., Wisden, W., Sommer, B., Werner, P., Herb, A., Verdoorn, T. A., Sakman, B. and Seeburg, P. H. (1990) *Science* **249** 556–560.

Kessler, M., Petersen, G., Vu, H. M., Baudry, M. and Lynch, G. (1987a) *J. Neurochem.* **48** 1191–1200.

Kessler, M., Baudry, M. and Lynch, G. (1987b) *Neurosci. Lett.* **81** 221–226.

Kiskin, N. I., Krishtal, O. A. and Tsyndrenko, A. Y. (1986) *Neurosci. Lett.* **63** 225–230.

Knöpfel, T., Zeise, M. L., Cuénod, M. and Zieglgänsberger, W. (1987) *Neurosci. Lett.* **81** 188–192.

Koyama, Y., Baba, A. and Iwata, H. (1989) *Brain Res.* **487** 113–119.

Krogsgaard-Larsen, P., Honoré, T., Hansen, J. J., Curtis, D. R. and Lodge, D. (1980) *Nature* **284** 64–66.

Krogsgaard-Larsen, P., Honoré, T., Hansen, J. J., Curtis, D. R. and Lodge, D. (1981) In: DiChiara, G. and Gessa, G. L. (eds) *Glutamate as a Neurotransmitter.* Raven Press, New York, pp. 285–294.

Krogsgaard-Larsen, P., Hansen, J. J., Lauridsen, J., Peet, M. J., Leah, J. D. and Curtis, D. R. (1982) *Neurosci. Lett.* **31** 313–317.

Krogsgaard-Larsen, P., Nielsen, E. Ø. and Curtis, D. R. (1984) *J. Med. Chem.* **27** 585–591.

Krogsgaard-Larsen, P., Brehm, L., Johansen, J. S., Vinzents, P., Lauridsen, J. and Curtis, D. R. (1985) *J. Med. Chem.* **28** 673–679.

Krogsgaard-Larsen, P., Ferkany, J. W., Nielsen, E. Ø., Madsen, U., Ebert, B., Johansen, J. S., Diemer, N. H., Bruhn, T., Beattie, D. T. and Curtis, D. R. (1991) *J. Med. Chem.* **34** 123–130.

Lauridsen, J., Honoré, T. and Krogsgaard-Larsen, P. (1985) *J. Med. Chem.* **28** 668–672.

Lodge, D. (1981) In: DeFeudis, F. V. and Mandel, P. (eds) *Amino Acid Neurotransmitters.* Raven Press, New York, pp. 327–332.

Lüddens, H. and Wisden, W. (1991) *Trends Pharmacol. Sci.* **12** 49–51.

Ludolph, A. C., Hugon, J., Dwivedi, M. P., Schaumburg, H. H. and Spencer, P. S. (1987) *Brain* **110** 149–165.

Lund, T. M., Madsen, U., Ebert, B., Jørgensen, F. S. and Krogsgaard-Larsen, P. (1991) *Med. Chem. Res.* **1** 136–141.

MacDonald, J. F. and Morris, M. E. (1984) *Exp. Brain Res.* **57** 158–166.

Madsen, U., Schaumburg, K., Brehm, L., Curtis, D. R. and Krogsgaard-Larsen, P. (1986) *Acta Chem. Scand.* **B40** 92–97.

Madsen, U., Ferkany, J. W., Jones, B. E., Ebert, B., Johansen, T. N., Holm, T. and Krogsgaard-Larsen, P. (1990a) *Eur. J. Pharmacol. Mol. Pharmacol. Sect.* **189** 381–391.

Madsen, U., Nielsen, E. Ø., Curtis, D. R., Beattie, D. T. and Krogsgaard-Larsen, P. (1990b) *Acta Chem. Scand.* **44** 96–102.

Mayer, M. L. and Westbrook, G. L. (1987) *Prog. Neurobiol.* **28** 197–276.

McLennan, H. and Hall, J. G. (1978) *Brain Res.* **149** 541–545.

McLennan, H. and Lodge, D. (1979) *Brain Res.* **169** 83–90.

Mewett, K. N., Oakes, D. J., Olverman, H. J., Smith, D. A. S. and Watkins, J. C. (1983) In: Mandel, P. and DeFeudis, F. V. (eds) *CNS Receptors: From Molecular Pharmacology to Behaviour.* Raven Press, New York, pp. 163–174.

Monaghan, D. T. and Cotman, C. W. (1986) In: Roberts, P. J., Storm-Mathisen, J. and Bradford, H. F. (eds) *Excitatory Amino Acids.* Macmillan, London, pp. 279–299.

Monaghan, D. T., Bridges, R. J. and Cotman, C. W. (1989) *Annu. Rev. Pharmacol. Toxicol.* **29** 365–402.

Morris, R. G. M., Anderson, E., Lynch, G. S. and Baudry, M. (1986) *Nature* **319** 774–776.

Murphy, D. E. and Williams, M. (1987) In: Hicks, T. P., Lodge, D. and McLennan, H. (eds) *Excitatory Amino Acid Transmission.* Liss, New York, pp. 63–66.

Murphy, D. E., Snowhill, E. W. and Williams, M. (1987) *Neurochem. Res.* **12** 775–782.

Nielsen, E. Ø., Madsen, U., Schaumburg, K., Brehm, L. and Krogsgaard-Larsen, P. (1986) *Eur. J. Med. Chem.* **21** 433–437.

Olney, J. W., Misra, C. H. and de Gubareff, T. (1975) *J. Neuropathol. Exp. Neurol.* **34** 167–176.

Olsen, R. W., Szamraj, O. and Houser, C. R. (1987) *Brain Res.* **402** 243–254.

Pfeiffer, C. C. (1956) *Science* **124** 29–30.

Pin, J.-P., Bockaert, J. and Recasens, M. (1984) *FEBS Lett.* **175** 31–36.

Pin, J.-P., van Vliet, B. J. and Bockaert, J. (1989) *Eur. J. Pharmacol. Mol. Pharmacol. Sect.* **172** 81–91.

Puckett, C., Gomez, C. M., Korenberg, J. R., Tung, H., Meier, T. J., Chen, X. N. and Hood, L. (1991), *Proc. Nat. Acad. Sci. USA* **88** 7557–7561.

Pullan, L. M., Olney, J. W., Price, M. T., Compton, R. P., Hood, W. F., Michel, J. and Monahan, J. B. (1987) *J. Neurochem.* **49** 1301–1307.

Rassendren, F.-A., Lorry, P., Pin, J.-P., Bockaert, J. and Nargeot, J. (1989) *Neurosci. Lett.* **99** 333–339.

Recasens, M., Guiramand, J., Nourigat, A., Sassetti, I. and Devilliers, G. (1988) *Neurochem. Int.* **13** 463–467.

Robinson, M. B. and Coyle, J. T. (1987) *FASEB J.* **1** 446–455.

Robinson, M. B., Crooks, S. L., Johnson, R. L. and Koerner, J. F. (1985) *Biochemistry* **24** 2401–2405.

Robinson, M. B., Blakely, R. D., Couto, R. and Coyle, J. T. (1987) *J. Biol. Chem.* **262** 14498–14506.

Sawada, S., Higashima, M. and Yamamoto, C. (1985) *Exp. Brain Res.* **60** 323–329.

Schoepp, D., Bockaert, J. and Sladeczek, F. (1990) *Trends Pharmacol. Sci.* **11** 508–515.

Sheardown, M. J., Nielsen, E. Ø., Hansen, A. J., Jacobsen, P. and Honoré, T. (1990) *Science* **247** 571–574.

Simon, J. R., Contrera, J. F. and Kuhar, M. J. (1976) *J. Neurochem.* **26** 141–147.

Smith, S. E. and Meldrum, B. S. (1990) *Neurochem. Int.* **16** (Suppl. 1) 69.

Sommer, B., Keinänen, K., Verdoorn, T. A., Wisden, W., Burnashev, N., Herb, A., Köhler, M., Takagi, T., Sakmann, B. and Seeburg, P. H. (1990) *Science* **249** 1580–1585.

Spencer, P. S., Nunn, P. B., Hugon, J., Ludolph, A. C., Ross, S. M., Roy, D. N. and Robertson, R. C. (1987) *Science* **237** 517–522.

Sugiyama, H., Ito, I. and Hirono, C. (1987) *Nature* **325** 531–533.

Sugiyama, H., Watanabe, M., Taji, H., Yamamoto, Y. and Ito, I. (1989) *Neurosci. Res.* **7** 164–167.

Turner, J. P. and Meldrum, B. S. (1991) *Br. J. Pharmacol.* **104** 445–451.

Verdoorn, T. A. and Dingledine, R. (1988) *Mol. Pharmacol.* **34** 298–307.

Virgili, M., Poli, A., Contestabile, A., Migani, P. and Barnabei, O. (1986) *Neurochem. Int.* **9** 29–33.

Watkins, J. C. and Evans, R. H. (1981) *Annu. Rev. Pharmacol. Toxicol.* **21** 165–204.

Watkins, J. C. and Olverman, H. J. (1988) In: Lodge, D. (ed.) *Excitatory Amino Acids in Health and Disease.* Wiley, New York, pp. 13–45.

Watkins, J. C., Krogsgaard-Larsen, P. and Honoré, T. (1990) *Trends Pharmacol Sci.* **11** 25–33.

Watson, S. and Abbott, A. (1989) *Trends Pharmacol. Sci.* **10** (Receptor Nomenclature Suppl.) 1–30.

Weiss, J. H. and Choi, D. W. (1988) *Science* **241** 973–975.

Weiss, J. H., Koh, J.-y. and Choi, D. W. (1989) *Brain Res.* **497** 64–71.

Young, A. B., and Fagg, G. E. (1990) *Trends Pharmacol. Sci.* **11** 126–133.

Zaczek, R., Balm, M., Arlis, S., Drucker, H. and Coyle, J. T. (1987) *J. Neurosci. Res.* **18** 425–431.

11

Quinoxalinediones as AMPA receptor antagonists: mechanism and structure–activity studies

Poul Jacobsen, Lars Naerum, Flemming E. Nielsen and **Tage Honoré**
Novo Nordisk A/S, CNS Division, Novo Nordisk Park, DK-2760, Maaloev, Denmark

ABSTRACT

The discovery of the quinoxalinediones, a series of potent and selective antagonists at non-*N*-methyl-D-aspartate (NMDA) excitatory amino acid receptors, has greatly facilitated the study of the structure and function of the 2-amino-3-(3-hydroxy-5-methylisoxazol-4-yl)propanoate (AMPA) and kainate receptor subtypes. The quinoxalinediones CNQX, DNQX and NBQX bind selectively to non-NMDA sites as defined by inhibition of [^{3}H]AMPA and [^{3}H]kainate binding to rat cortical membranes. Furthermore, binding studies using [^{3}H]CNQX as radioligand show that CNQX binds with high affinity to both high- and low-affinity [^{3}H]AMPA binding sites in rat brain. Classical structure–activity studies on substituted quinoxaline-2,3-diones and computer-based quantitative structure–activity relationship analyses on 1-hydroxyquinoxaline-2,3-diones show that the aromatic substitution pattern determines both the potency and the selectivity of the compounds.

INTRODUCTION

At least four different subtypes of excitatory amino acid (EAA) receptors are known. Three of these are coupled to ionophores: *N*-methyl-D-aspartate (NMDA), 2-amino-3-(3-hydroxy-5-methylisoxazol-4-yl)propanoate (AMPA) and kainate receptors, whereas the fourth, metabotropic quisqualate receptors, is coupled via G-protein to inositol phosphate turnover.

The NMDA receptor complex consists of at least three entities: the recognition site, which binds glutamate and NMDA as well as the competitive antagonists 2-amino-5-phosphonopentanoate (AP5) and 3-(2-carboxypiperazin-4-yl)propyl-1-phosphonoate (CPP); the modulatory glycine site, which is insensitive to strychnine, but binds glycine and D-serine as well as the antagonists HA-966 and 7-chlorokynurenic acid; and the ionophore site, which binds phencyclidine (PCP) and MK-801 (see chapter 9).

The AMPA- and kainate-induced responses have many similarities, including the sensitivities to antagonists (Blake *et al.* 1988, Davies *et al.* 1982, Fletcher *et al.* 1988), and in some cases kainate effects may be mediated by AMPA receptor activation (see Chapter 2). However, some effects of kainate appear independent of AMPA receptors. Of these, depolarization of mammalian C-fibres is the most important (Evans *et al.* 1987).

It has recently been described that quinoxalinediones such as 6,7-dinitroquinoxaline-2,3-dione (DNQX), 6-cyano-7-nitroquinoxaline-2,3-dione (CNQX), and 2,3-dihydroxy-6-nitro-7-sulphamoylbenzo(*f*)quinoxaline (NBQX) (Fig. 1) are potent non-NMDA antagonists (Honoré *et al.* 1988b, Sheardown *et al.* 1990). Shortly after the appearance of DNQX and CNQX it was determined that these compounds, in addition to their potent non-NMDA antagonism, had moderate to weak antagonistic effects on the recently described strychnine-insensitive glycine sites coupled to NMDA receptors (Birch *et al.* 1988, Drejer *et al.* 1989, Sheardown *et al.* 1989). The two effects are, however, not coupled to the same structural elements (Table 1), as it is possible to optimize either the glycine antagonism (5,7-dinitroquinoxaline-2,3-dione (MNQX); Fig. 1) or the non-NMDA antagonism (NBQX; Fig. 1).

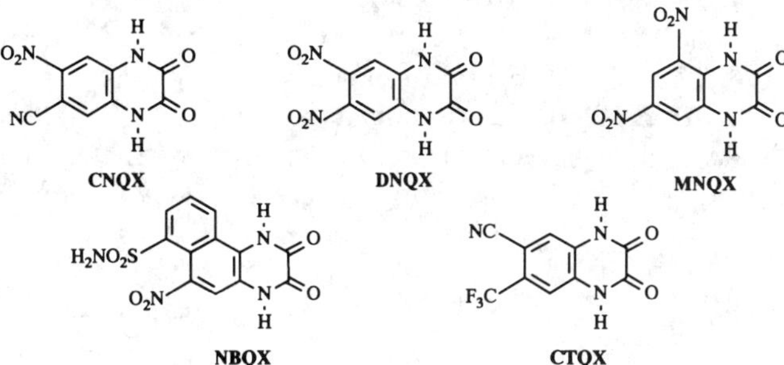

Fig. 1 — Structures of antagonistic quinoxalinediones.

Table 1 — Inhibition of binding to excitatory amino acid receptors on rat cortical membranes

Compound	[^{3}H]AMPA	IC$_{50}$ (μM) [^{3}H]Kainate	[^{3}H]CPP	[^{3}H]Glycine
MNQX	2.4	11	44	1.0
CTQX	1.5	17.6	23.9	10.0
DNQX	0.5	2.3	40	9.5
CNQX	0.3	1.7	28	14
NBQX	0.2	4.8	> 100	> 300

Investigations of a series of analogues of DNQX and CNQX showed that their ability to inhibit [^{3}H]AMPA binding to rat cortical membranes correlates very well with their ability to antagonize quisqualate-induced spreading depression in chicken retina (Fig. 2, $p < 0.001$ Student's t-test) (unpublished; Drejer *et al.* 1989). From these results, it is suggested that quisqualate effects are mediated via the AMPA receptors, and that the antagonism in chicken retina is due to blockade of AMPA receptors.

Here we describe the molecular pharmacology of AMPA receptors, as well as structure–activity studies of a series of quinoxalinediones.

AMPA RECEPTORS

The best radioligands for examining AMPA receptors are [^{3}H]AMPA and [^{3}H]CNQX (Honoré *et al.* 1982, Honoré *et al.* 1989, Honoré and Nielsen 1985, Murphy *et al.* 1987, Olsen *et al.* 1987). Autoradiographic studies have shown that [^{3}H]AMPA binding sites have the highest densities in the dentate gyrus and stratum radiatum of CA1 in the hippocampus (Monaghan *et al.* 1984, Nielsen *et al.* 1988, Rainbow *et al.* 1984). [^{3}H]AMPA binds with high affinity (125 nM) to binding sites in rat cortical membranes (Honoré and Drejer 1988). The molecular weight of the [^{3}H]AMPA binding site, as determined by high-energy irradiation techniques, is 51.6 kDa (Honoré and Nielsen 1985). Furthermore, radiation inactivation of [^{3}H]AMPA binding indicates the involvement of a 128-kDa modulatory protein.

[^{3}H]AMPA binding to rat cortical membranes is dramatically increased by the presence of chaotropic ions. The most efficient ion, SCN^-, increases the specific binding of [^{3}H]AMPA at least fourfold (Honoré and Drejer 1988). Furthermore, in the presence of SCN^- ions, the saturation binding study gave curvilinear Scatchard plots (Honoré and Drejer 1988). If it is assumed that curvilinearity reflects multiplicity of binding sites, the curves can be resolved into two binding sites with $K_{D1} = 14$ nM and $K_{D2} = 235$ nM; SCN^- ions apparently increased the density of binding sites.

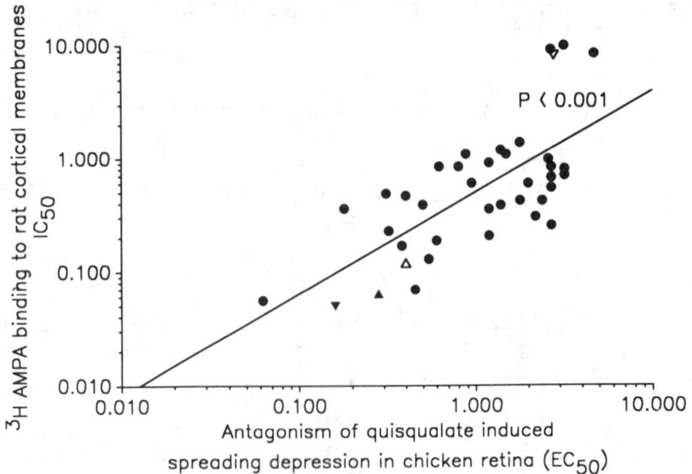

Fig. 2 — Correlation of binding to AMPA receptors and antagonism of quisqualate-induced spreading depression. ▼, NBQX; ▲, CNQX; △, DNQX; ▽, 6-Cl-QX.

However, kinetic, studies of [^{3}H]AMPA binding were not in agreement with the saturation study and multiplicity of binding sites (Honoré and Drejer 1988). Furthermore, dissociation of [^{3}H]AMPA initiated by unlabelled ligand was similar to dissociation after dilution, indicating no cooperative interactions (Honoré and Drejer 1988). A better model to explain the obtained results is a two-state model (Fig. 3) (Honoré and Drejer 1988). In this model, one conformation (R_1) has high affinity for [^{3}H]AMPA whereas a second conformation (R_2) has low affinity for [^{3}H]AMPA.

In contrast to [^{3}H]AMPA, [^{3}H]CNQX apparently binds to a homogeneous population of receptors as judged from the Scatchard analysis (Fig. 4) (Honoré *et al.* 1989). However, when AMPA is used as inhibitor of [^{3}H]CNQX binding, the inhibition curve is biphasic (Fig. 4) (Honoré *et al.* 1989). These results indicate that [^{3}H]CNQX binds with the same affinity to the two conformations of AMPA receptors. The highest density of [^{3}H]CNQX binding is in the stratum radiatum of CA1 in the hippocampus, similar to [^{3}H]AMPA binding, but a substantial density of [^{3}H]CNQX binding sites is also found in the CA3 region of hippocampus (Nielsen *et al.* 1990).

The hypothesis that the two populations of AMPA binding sites are two conformations of the same site is further substantiated by molecular target site

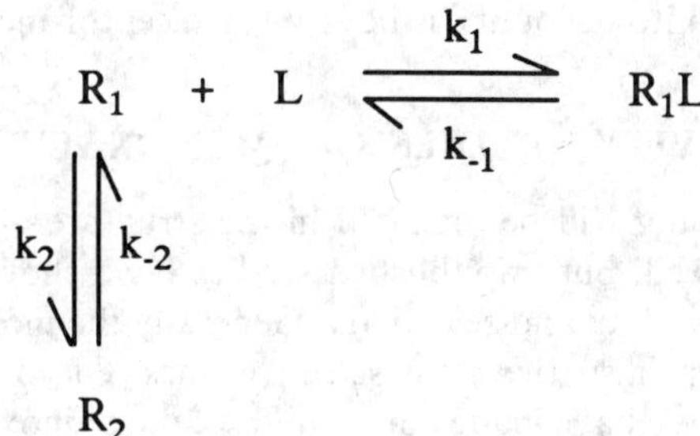

Fig. 3 — Two-state model of [^{3}H]CNQX binding to rat cortical membranes.

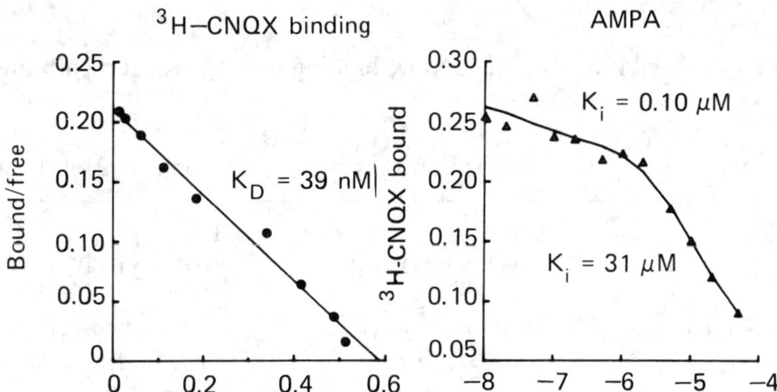

Fig. 4 — Left: Scatchard plot of [^{3}H]CNQX binding to rat cortical membranes. Right: Inhibition of [^{3}H]CNQX binding by AMPA.

analysis of [^{3}H]CNQX binding, which shows the same molecular size (51.8 kDa) for the two sites (Honoré *et al.* 1989). Furthermore, the monoexponential decay curve obtained in the radiation inactivation of [^{3}H]CNQX binding suggests that the previously mentioned modulatory site does not influence the binding of antagonists.

When a series of standard non-NMDA agonists (see chapters 10 and 12) were tested as inhibitors of [^{3}H]CNQX binding to rat cerebral cortex, the affinity for the two conformations of AMPA receptors could be estimated (Table 2) (Honoré *et al.* 1989 and unpublished). AMPA and quisqualate bind with highest affinity to the conformation named R_1 (K_i = 0.3 and 0.21 μM, respectively). Domoate and kainate, on the other hand, bind with the highest affinity to the conformation named R_2 (K_i = 0.32 and 12 μM, respectively). Thus, the rank order of inhibitor potency of the standard excitatory amino acids is different for the two conformations; conformation R_1: quisqualate > AMPA > domoate > kainate; conformation R_2: domoate > quisqualate > kainate > AMPA. Evaluation of rank orders obtained with these agonists in functional assays may give important information concerning which AMPA receptor population is involved in mediating the responses.

Some evidence from patch-clamp studies suggest that the channels associated with kainate and AMPA receptors have different properties (Ascher and Nowak 1986, Cull-Candy and Usowicz 1987, Jahr and Stevens 1987, Mayer 1987). However, the possibility that the difference is due to two conformations of the same channel exists. This property is taken into account in the AMPA receptor model shown in Fig. 5.

STRUCTURE—ACTIVITY STUDIES OF QUINOXALINEDIONES

Only [^{3}H]AMPA binding will be included in the structure–activity considerations. Selectivity is not involved, but unpublished studies have shown a good correlation between [^{3}H]AMPA binding and selectivity. Generally the most potent inhibitors of [^{3}H]AMPA binding are also the most selective ones. Of course, there are several exceptions to this rule. We have made compounds of the quinoxalinedione type which at the same time are potent AMPA antagonists and potent glycine antagonists.

The unsubstituted quinoxaline-2,3-dione (Fig. 6) has quite weak activity as an inhibitor of [^{3}H]AMP binding and totally lacks selectivity. Monosubstitution in the

Table 2 — Inhibition of [^{3}H]CNQX binding to rat cortical membranes

	K_i (μM)[a]	
	Conformation A	Conformation B
NBQX	0.02 (2)	0.02 (2)
CNQX	0.039 ± 0.002 (3)	0.039 ± 0.002 (3)
AMPA	0.3 ± 0.2 (4)	22 ± 9 (4)
Quisqualate	0.21 ± 0.02 (3)	5.9 ± 0.2 (3)
Domoate	9.6 ± 2.3 (3)	0.32 ± 0.19 (3)
Kainate	240 ± 70 (3)	12 ± 4 (3)

[a] Values are mean ± SEM. Number of experiments is given in parentheses.

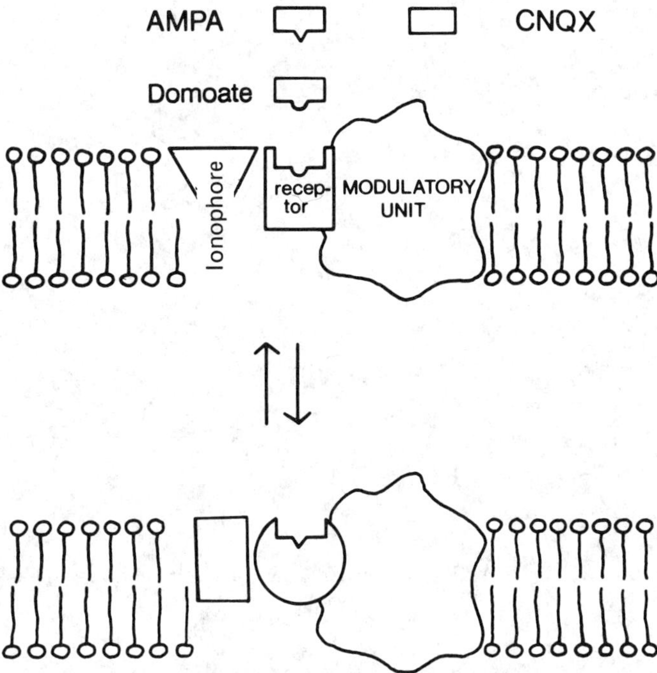

Fig. 5 — Hypothetical AMPA receptor model.

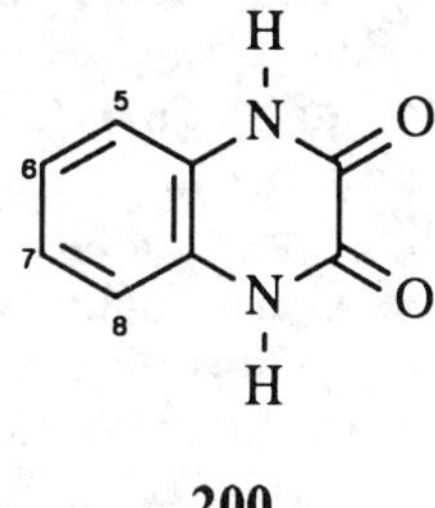

200

Fig. 6 — Structure and [^{3}H]AMPA binding (IC_{50}, μM) of the unsubstituted quinoxaline-2,3-dione.

6-position gives, as a general rule, an increase in activity whereas substitution in the 5-position either has no effect or gives a decrease in activity (Fig. 7) Other examples of substituents are trifluoromethyl, chloro and bromo, which in the 5-position are without effect but which in the 6-position give an increase in activity. A large series of different 6-substituted quinoxaline-2,3-diones has been prepared, and a selection of the most interesting ones is shown in Table 3. As can be seen, compounds with nitro, cyano, or trifluoromethyl have the highest [^{3}H]AMPA binding activity. It is, however, difficult to see what determines the difference in activity of the many substituents. Electronic properties are obviously important but are surely not the whole truth. Steric properties, at least, are also of some importance.

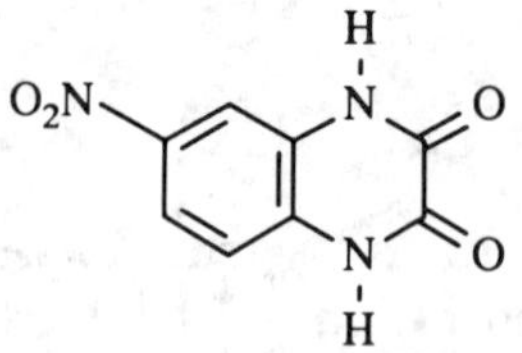

5.8 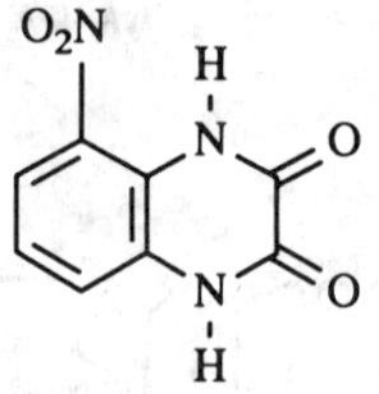 **340**

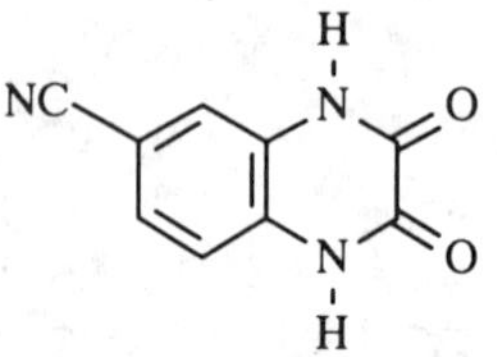

8.0 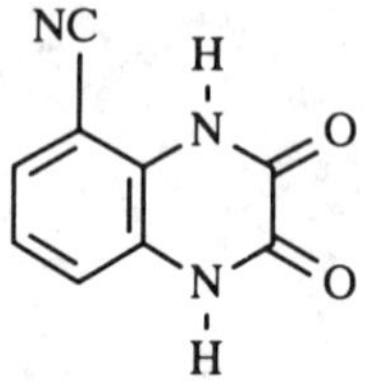 **> 1000**

Fig. 7 — Structures and [^{3}H]AMPA binding (IC_{50}, μM) of 5- or 6-substituted quinoxaline-2,3-diones.

Disubstitution of the quinoxaline-2,3-dione follows the same rules as were established for the monosubstituted analogues, namely that 6,7-disubstitution gives rise to a much larger increase in activity than 5,6- or 5,7-substitution (Fig. 8). A large series of 6,7-disubstituted quinoxaline-2,3-diones has been prepared (Honoré *et al.* 1988a), and analysis of the binding data showed that the increases in activity for each of the two substituents are cumulative. To get the potency of a disubstituted compound, one can simply multiply the increases in activity coming from each substituent (Fig. 9). By doing this for a series of 6,7-disubstituted analogues, it can be seen that the calculated IC_{50} values fit reasonably well with the observed potencies (Table 4). From the data on 6-monosubstituted compounds it can be predicted that CNQX and DNQX are the most potent compounds obtainable in this series of substituted quinoxaline-2,3-diones.

QUANTITATIVE STRUCTURE—ACTIVITY RELATIONSHIP (QSAR) OF 1-HYDROXYQUINOXALINEDIONES

In order to reduce symmetry problems in the analysis, 1-hydroxyquinoxaline-2,3-diones (Honoré *et al.* 1990) were used as model compounds (Fig. 10). It has previously been shown (unpublished) that *N*-hydroxylation does not influence the affinity for [^{3}H]AMPA binding sites significantly; 6-cyano-7-trifluoromethylquinoxaline-2,3-dione (CTQX; Fig. 1) and its 1-hydroxy analogue, for example, have IC_{50} = 1.5 and 0.6 μM, respectively.

Sixteen 1-hydroxyquinoxaline-2,3-diones were selected from a group of 48 compounds to define the model (Fig. 10). The selection and the QSAR models were performed using the SIMCA program (Wold and Sjöström 1977), which uses principal component (PC) and partial least-squares (PLS) statistics. The analyses

Table 3 — [^{3}H]AMPA binding of 6-substituted quinoxaline-2,3-diones. The increase in potency as compared to the unsubstituted compound is given in parenthesis

R_1	IC_{50} (μM)	
NO_2	5.8	(× 35)
CN	8.0	(× 25)
CF_3	13.0	(× 15)
$COCH_3$	28.9	(× 7)
Br	29.8	(× 7)
SO_2NH_2	32.8	(× 6)
Cl	41.7	(× 5)
$COCF_3$	46.5	(× 4)
CH_3	48.3	(× 4)
OCH_3	54.5	(× 4)
CH_2CN	79.6	(× 3)
SO_2CF_3	88.4	(× 2)
SO_2NHCH_3	91.4	(× 2)
SO_2CH_3	154	(× 1.5)
OH	208	(× 1)
H	204	
NH_2	> 300	
COOH	> 300	
$NHCOCH_3$	> 300	

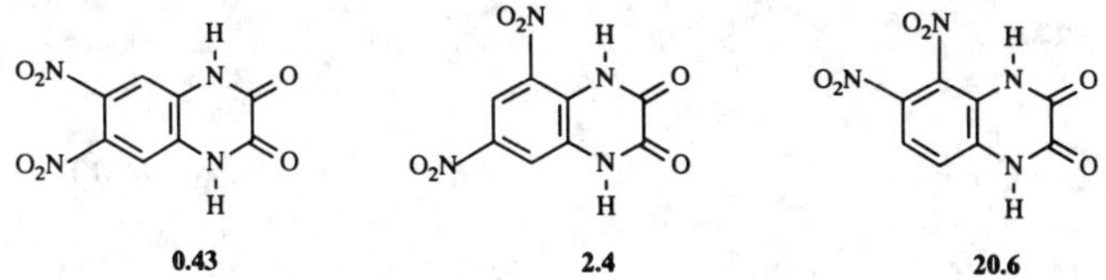

Fig. 8 — Structures of [^{3}H]AMPA binding (IC_{50}, μM) of disubstituted quinoxaline-2,3-diones.

were based on ten classical substituent parameters: PI, lipophilic aromatic substituent constant; ES, Taft steric parameters; VW, van der Waals volume; MR, molar refractivity; L, second-generation STERIMOL length parameter; B1, second-generation STERIMOL minimum width parameter; B5, second-generation STERIMOL maximum width parameter; F, field/inductive parameter; R, resonance parameter; MU, group dipole moment for aromatic substituent (Van de Waterbeemd and Testa 1987).

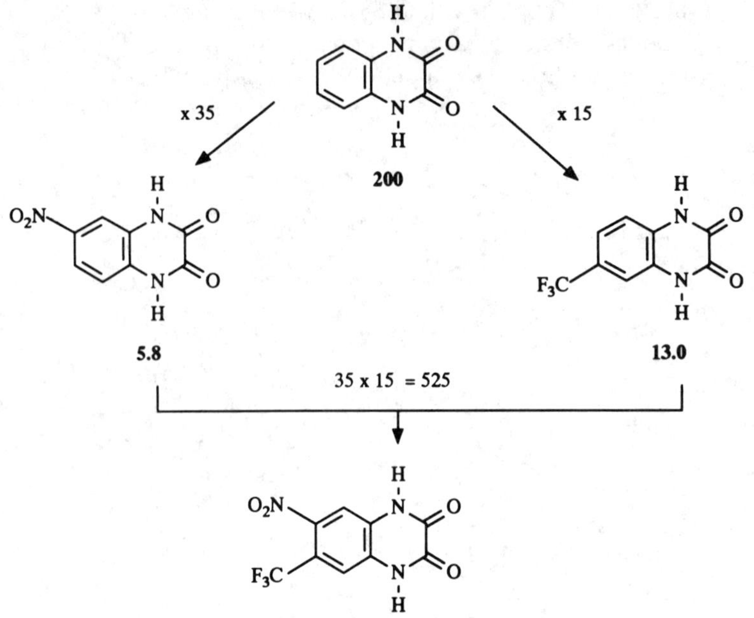

Theor. IC_{50} = 0.38 μM

(Determined IC_{50} = 0.51 μM)

Fig. 9 — Calculation of theoretical IC_{50} values of disubstituted quinoxaline-2,3-diones.

Table 4 — [^{3}H]AMPA binding (calculated IC_{50} values in parentheses) of 6,7-disubstituted quinoxaline-2,3-diones.

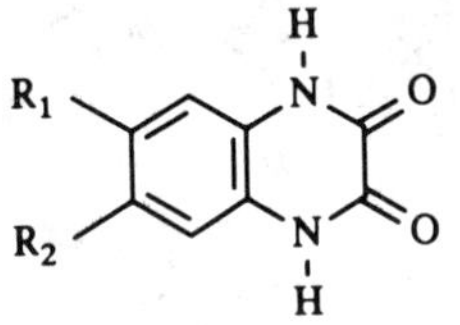

	R_1	R_2	IC_{50} (μM)
CNQX	CN	NO_2	0.26 (0.23)
DNQX	NO_2	NO_2	0.43 (0.17)
	CF_3	NO_2	0.51 (0.38)
	CN	CN	0.89 (0.32)
	Br	NO_2	1.50 (0.82)
	CF_3	CN	1.52 (0.53)
	Cl	NO_2	2.01 (1.2)
	Cl	CN	3.69 (1.6)
	Cl	Cl	8.19 (8.0)
	SO_2CH_3	Cl	21.4 (26.7)

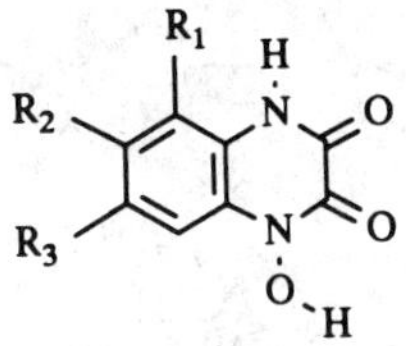

R_1	R_2	R_3
H	H	CN
H	OCH_3	OCH_3
CF_3	H	Cl
H	H	SO_2CF_3
H	CH_3	H
H	CN	H
H	H	SO_2CH_3
H	CF_3	SO_2NH_2
H	CF_3	CN
Cl	H	H
CH_3	H	Br
CH_3	CH_3	H
H	H	C_6H_5
H	H	NH_2
F	H	H
H	H	C_4H_9

Fig. 10 — Selected model compounds.

PC analysis reduced the description of the compounds to two dimensions as depicted in Fig. 11. From this plane 16 compounds representing the plane were selected.

Fig. 12 shows the correlation between calculated affinity versus observed affinity for [^{3}H]AMPA binding sites ($R^2 = 0.95$) (filled circles). The open circles correspond to the 32 compounds not used in the model. The correlation between calculated and measured affinity for [^{3}H]AMPA binding sites indicates that the model is valid.

Antagonism of NMDA-induced [^{3}H]GABA release in cultured cortical neurons from mice is a model for both NMDA and glycine antagonism of quinoxalinediones. In this case, a correlation ($R^2 = 0.67$) between calculated and measured activity was found. The slightly lower correlation may be due to the biological model representing two different receptors. However, as judged from inhibition of the binding of the

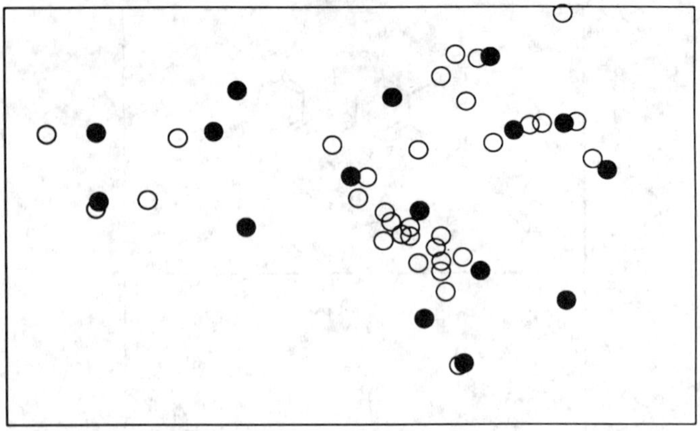

Fig. 11 — PC analysis in two dimensions of the selected compounds. The filled circles show the selected compounds.

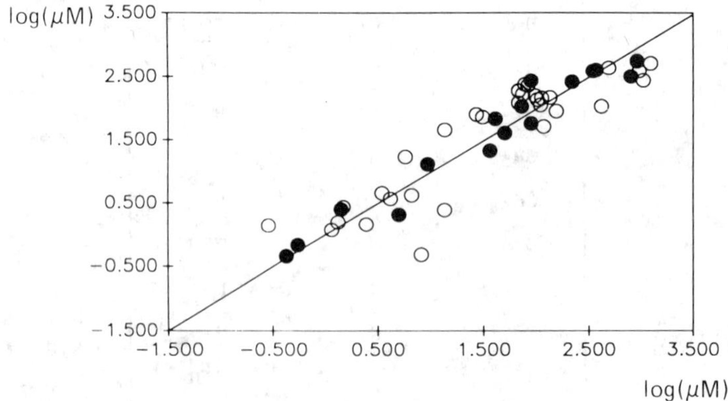

Fig. 12 — Correlation of [^{3}H]AMPA binding, calculated versus observed affinity (IC_{50}) of model compounds (filled circles) and 32 test compounds (open circles).

NMDA antagonist [^{3}H]CPP, the model compounds are mainly glycine antagonists (unpublished).

These results suggest that the model based on the 16 selected compounds can be used for further analysis of the influence of substitution pattern on potency and selectivity of the compounds for both AMPA and glycine receptors.

With respect to AMPA binding, substituents in position 5 and 6 gave deviation from zero in eight out of ten coefficients (from the calculated QSAR equation), indicating that these positions are important for affinity to AMPA receptors. A closer look at the coefficients showed that the best substituents in position 5 are very small ones (like hydrogen) whereas the best substituents in position 6 are electronegative ones (like cyano).

On the other hand, when the correlation was made to glycine antagonism, the substituents in positions 6 and 7 gave deviation from zero in the analysis, indicating the importance of these positions in the interaction with glycine receptors.

Based on the abovementioned results, it is suggested that substituents in position 6 determine the affinities of quinoxalinediones for both AMPA and glycine receptors. Positions 5 and 7 in the 1-hydroxyquinoxalinediones determine the selectivity of the compounds for either AMPA or glycine receptors. Glycine selectivity of a given compound is obtained when the 7-substituent interacts positively with the glycine receptor and the 5-substituent interacts negatively with the AMPA receptor.

THREE-DIMENSIONAL QSAR (3D-QSAR)

The structure–activity relationship of the 1-hydroxyquinoxalinediones can also be analysed by using 3D-QSAR based on the relatively new comparative molecular field analysis (CoMFA) approach (Cramer *et al.* 1988). CoMFA is based on the assumption that differences in non-covalent fields surrounding the molecules can be related to the differences in activity.

The compounds studied are non-flexible, which is an advantage in placing each compound in its proposed active conformation. The computer-drawn molecules are placed in a suitable manner against each other and the fields are calculated as interaction energies between the molecules and some isolated carbon atoms with a charge of + 1. These probe atoms are placed around the molecules in the empty space areas. Around all the model compounds two fields are now generated: one which is describing the steric potential (Lennard-Jones) and one which is describing the electrostatic potential (Coulomb) of the molecules.

The calculated fields and the activity of the model compounds are analysed by using PLS statistics. The result can be shown by contouring the model coefficients in 3D. However, it is clear that only space areas which have been structurally varied in the model compound will be able to model differences in activity of the model compounds. One therefore should bear in mind that the interpretation of the results is not equivalent to real receptor maps as the analysis is only based on rather few, relatively similar compounds which cannot explain all possible compound–receptor interactions. Another problem is that the CoMFA used does not include entropically related interactions such as hydrophobicity.

The QSAR module in the SYBYL program (Tripos Associate Inc., USA, ver. 5.3) was used for the CoMFA analyses. Geometries and charges were calculated with the semi-empirical MNDO method (program AMPAC, QCPE No. 506).

A CoMFA model based on AMPA binding gave approximately the same good correlation between observed and calculated activity as the earlier-mentioned substituent paramcter based model. Fig. 13 shows the model coefficients (stdev*coefficients) regarding the steric fields. In this case the steric field is the most important (80%) as compared to the electrostatic (20%) fields. The results show that the substituents should be placed close to the blue/green areas and away from the magenta/yellow areas in order to obtain an increase in activity. This result is basically identical to the result generated from the substituent parameter-based analyses of the AMPA binding data.

A CoMFA analysis of the NMDA activity gave a poor correlation as compared to the substituent parameter-based method.

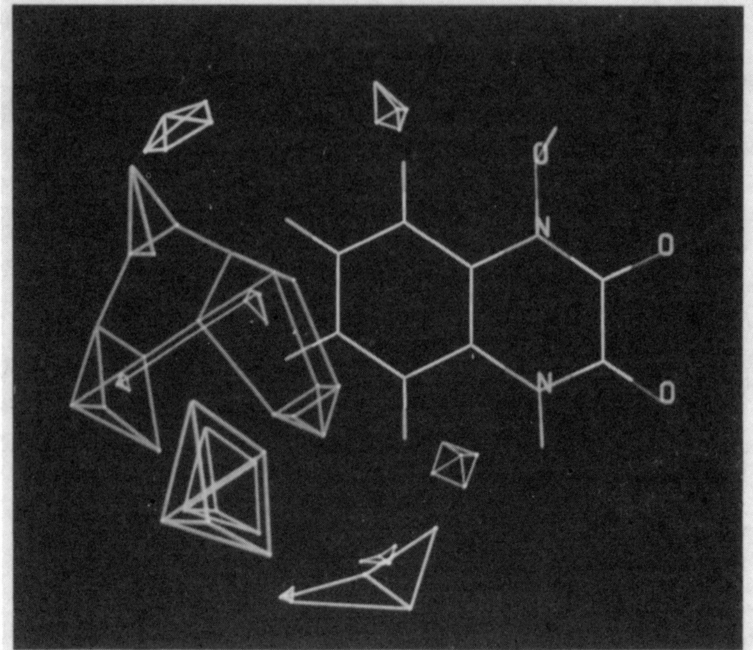

Fig. 13 — (**See colour section.**) Steric 3D coefficient map.

The 3D-QSAR-based method has some advantages over the substituent parameter-based methods. The results can be analysed in a suitable graphical manner, and because the method is independent of substituent parameters, all compounds, including compounds with different ring systems, can be studied.

CONCLUSION

During the last decade the knowledge of the molecular pharmacology of interaction with non-NMDA receptors, in particular the AMPA receptor subtype, has improved considerably.

Binding studies using [^{3}H]AMPA and the antagonist [^{3}H]CNQX have revealed information about the receptors, which are shown to exist in two states with different selectivities for standard compounds. The syntheses of a growing number of quinoxalinediones have made QSAR possible on these structures, and the first studies have given information about the effect of the substitution pattern on potency and selectivity of AMPA/glycine antagonism.

Today it is not known if the agonists, such as AMPA and quisqualate, and the antagonists, such as CNQX, DNQX and NBQX, interact with exactly the same amino acid sequence in the receptor, but it is expected from biological studies that the compounds interact with the same protein. Medicinal chemistry studies in combination with the increasing amount of information from molecular neurobiology studies (Barnard *et al.* 1990, Gregor *et al.* 1989, Hollmann *et al.* 1989, Keinänen *et al.* 1990, Sommer *et al.* 1990, Wada *et al.* 1989) of non-NMDA receptors give promises for the future.

REFERENCES

Ascher, P. and Nowak, L. (1986) *Adv. Exp. Med. Biol.* **203** 507–511.

Barnard, E. A., Ambrosini, A., Henley, J. M. and Usherwood, P. N. R. (1990) *Excitatory Amino Acids* (in press).

Birch, P. J., Grossman, C. J. and Hayes, A. G. (1988) *Eur. J. Pharmacol.* **156** 177–180.

Blake, J. F. Brown, M. W. and Collingridge, G. L. (1988) *Neurosci. Lett.* **89**, 182–186.

Cramer, R. D., Patterson, D. E. and Bunce, J. D. (1988) *J. Am. Chem. Soc.* **110** 5961–5967.

Cull-Candy, S. G. and Usowicz, M. M. (1987) *Nature* **325** 525–528.

Davies, J., Evans, R. H., Jones, A. W., Smith, D. A. S. and Watkins, J. C. (1982) *Comp. Biochem. Pharmacol.* **72** 211–224.

Drejer, J., Sheardown, M., Nielsen, E. O. and Honoré, T. (1989) *Neurosci. Lett.* **98** 333–338.

Dunn, W. J. (1977) *Eur. J. Chem.* **12** 109–112.

Evans, R. H., Evans, S. J., Pook, P. C. K. and Sunter, D. C. (1987) *Br. J. Pharmacol.* **91** 531–537.

Fletcher, E. J., Martin, D., Aram, J. A., Lodge, D. and Honoré, T. (1988) *Br. J. Pharmacol.* **95** 585–597.

Gregor, P., Mano, I., Maoz, I., McKeown, M. and Teichberg, V. I. (1989) *Nature* **342** 689–692.

Hollmann, M., O'Shea-Greenfield, A., Rogers, S. W. and Heinemann, S. (1989) *Nature* **342** 643–648.

Honoré, T. and Drejer, J. (1988) *J. Neurochem.* **51** 457–461.

Honoré, T. and Nielsen, M. (1985) *Neurosci. Lett.* **54** 27–32.

Honoré, T., Lauridsen, J. and Krogsgaard-Larsen, P. (1982) *J. Neurochem.* **38** 173–178.

Honoré, T., Drejer, J., Jacobsen, P. and Nielsen, F. E. (1988a) *EP Appl.* publication no. 260467.

Honoré, T., Davies, S. N., Drejer, J., Fletcher, E. J., Jacobsen, P., Lodge, D. and Nielsen, F. E. (1988b) *Science* **241** 701–703.

Honoré, T., Drejer, J., Nielsen, E. O. and Nielsen, M. (1989) *Biochem. Pharmacol.* **38** 3207–3212.

Honoré, T., Jacobsen, P., Nielsen, F. E. and Naerum, L. (1990) *EP Appl.* publication no. 377112.

Jahr, C. E. and Stevens, C. F. (1987) *Nature* **325** 522–525.

Keinänen, K., Wisden, W., Sommer, B., Werner, P., Herb, A., Verdoorn, T. A., Sakman, B. and Seeburg, P. H. (1990) *Science* **249** 556–560.

Mayer, M. (1987) *Nature* **325** 480–481.

Monaghan, D. T., Yao, D. and Cotman, C. W. (1984) *Brain Res.* **324** 160–164.

Murphy, D. E., Snowhill, E. W. and Williams, M. (1987) *Neurochem. Res.* **12**, 775–782.

Nielsen, E. O., Cha, J.-H. J., Honoré, T., Penney, J. B. and Young, A. B. (1988) *Eur. J. Pharmacol.* **157** 197–203.

Nielsen, E. O., Drejer, J., Cha, J.-H. J., Young, A. B. and Honoré, T. (1990) *J. Neurochem.* **54** 686–695.

Olsen, R. W., Szamraj, O. and Houser, C. R., (1989) *Brain Res.* **402** 243–254.

Rainbow, T. C., Wieczorek, C. M. and Halpain, S. (1984) *Brain Res.* **309** 173–177.

Sheardown, M. J., Drejer, J., Jensen, L. H., Stidsen, C. E. and Honoré, T. (1989) *Eur. J. Pharmacol.* **174** 197–204.

Sheardown, M. J., Nielsen, E. O., Hansen, A. J., Jacobsen, P. and Honoré, T. (1990) *Science* **247** 571–574.

Sommer, B., Keinänen, K., Verdoorn, T. A., Wisden, W., Burnashev, U., Herb, A., Köhler, A., Tagaki, T., Sakman, B. and Seeburg, P. H. (1990) *Science* **249** 1580–1585.

Van de Waterbeemd, H. and Testa, B. (1987). In: Testa, B. (ed.) *Advances in Drug Research*, Vol. 16, Academic Press, London, pp. 87–228.

Wada, K., Dechesne, C. J., Shimasaki, S., King, R. G., Kusano, K., Buonanno, A., Hampson, D. R., Banner, C., Wenthold, R. J. and Nakatani, Y. (1989) *Nature* **342** 684–689.

Wold, S. and Sjöström, M. (1977) *ACS Symposium Series* **52** 243.

12

Kainic acid receptor agonists

Haruhiko Shinozaki
The Tokyo Metropolitan Institute of Medical Science, 3-18-22 Honkomagome, Bunkyo-ku, Tokyo 113, Japan

ABSTRACT

Powerful excitatory actions of kainic acid (KAIN) on mammalian central neurons gave rise to the excitotoxic concept that glutamate (Glu) destroys neurons by excessive activation of excitatory receptors. In this chapter neuropharmacological actions of several KAIN receptor agonists, including newly synthesized kainoids, are described. Acromelic acid A (ACRO A), a naturally occurring kainoid, demonstrates the highest depolarizing activity among known excitatory amino acids (EAAs) in the newborn rat spinal motor neuron. Systemic administration of ACRO A to the rat exhibits characteristic behavioural signs quite distinct from those of KAIN or domoic acid despite the fact that they contain the common chemical moiety of KAIN, and there is regional difference in the distribution of neuron damage between ACRO A and KAIN. The most pronounced behavioural change seen after systemic administration of ACRO A is the persistent spastic paraplegia which appeared on the day following injection, with selective neuron damage confined to the lower spinal cord. One of the newly synthesized kainoids showed interesting KAIN-like properties with a higher depolarizing activity than ACRO A. The difference in the rank order of depolarizing activity of these kainoids between the motor neuron and the C-fibre is discussed.

INTRODUCTION

A seaweed, *Digenea simplex*, one of the red algae, *Rhodomelacea*, has been used as an effective drug for killing intestinal worms in Japan. In 1953, Takemoto and his colleagues succeeded in isolating L-α-kainic acid (KAIN) and L-α-allo-KAIN (possessing a reversed orientation of the isopropenyl side-chain) as the effective ascaricidal components of *Digenea simplex*, and identified their structures by chemical synthesis. The author first demonstrated the marked excitatory actions of KAIN on mammalian central neurons sensitive to L-glutamate (Glu) (Shinozaki and Konishi 1970). At that time, discovery of potent Glu agonists and antagonists had been expected with great interest for pharmacological identification of Glu as an excitatory neurotransmitter in the mammalian central nervous system (CNS). It was reasonable to suppose that KAIN caused marked muscle contraction in ascarides through activation of their presumed glutamatergic neuromuscular junction (NMJ), because KAIN contains the chemical moiety of Glu (Fig. 1). The molecules involved in chemical transmission and the mechanism of their action on membranes are generally similar in vertebrates and invertebrates. Therefore, potent excitatory compounds in invertebrates attracted notice in view of their possible actions on the mammalian CNS. Successively, the author first discovered the powerful excitatory action of L-quisqualic acid (QUIS) at the crayfish NMJ, based on the structural similarity of QUIS to Glu (Shinozaki and Shibuya 1974a). The electron density on the nitrogen atom in the structure —CO—NH—CO— of the dioxo-oxadiazolidine moiety of QUIS is reduced by the two adjacent carbonyl groups. As a result, the hydrogen atom in this moiety functions as an acidic hydrogen. This acidic hydrogen is separated from the α-amino group by a distance similar to that separating the γ-carboxyl group from the α-amino group in Glu. QUIS can therefore be regarded as a Glu analogue. Both KAIN and QUIS are conformationally restricted Glu analogues. The conformation of chemical substances is one of the most interesting factors affecting their pharmacological activity. The introduction of these potent naturally occurring excitatory amino acids (EAAs) to neuroscience led neurobiologists to new opportunities for elucidation of Glu excitatory transmitter function (Shinozaki 1978, McGeer *et al.* 1978). Classification of the Glu receptor subtypes in the mammalian CNS (Davies and Watkins 1979, Watkins 1981) was one of these opportunities, and electrophysiological and biochemical analysis of responses of QUIS or ibotenic acid (IBO) led to the discovery of a novel Glu receptor subtype, the metabotropic EAA receptor. The powerful excitatory actions of KAIN on mammalian central neurons gave rise to the excitotoxic concept that Glu destroys neurons by excessive activation of excitatory receptors on the dendrosomal surface of neurons (Olney *et al.* 1974, Olney 1978). Thus, KAIN and QUIS, in addition to *N*-methyl-D-aspartic acid (NMDA), have had a marked influence upon the development of EAA research in neuroscience. KAIN and other excitotoxic amino acids have already been seized on by many laboratories as lesioning tools which may offer selectivities completely different in type but similar in usefulness (McGeer *et al.* 1978). In recent years, considerable interest has focused on the neurotoxic properties of EAAs, in particular KAIN and its analogues (the kainoids), and their possible relevance for the study of human neurodegenerative disorders. The recent development of selective agonists and antagonists of EAA receptors has also

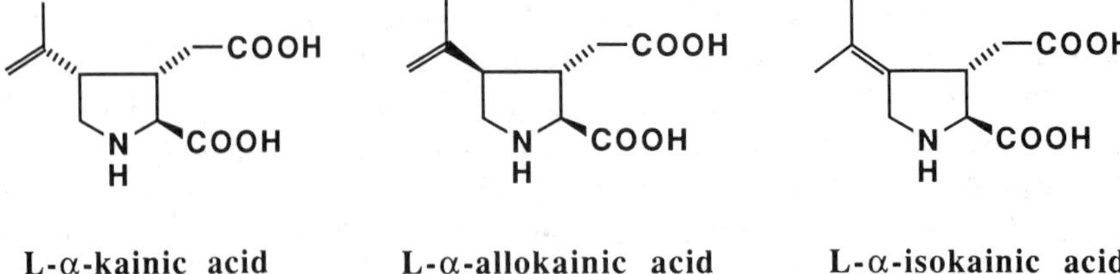

Fig. 1 — Chemical structures of KAIN and its isomers isolated from the seaweed *Digenea simplex*. They contain the common chemical moiety of Glu.

aided in the elucidation of excitotoxic mechanisms. Pharmacological blockade of excitotoxicity may constitute a novel therapeutic strategy for the treatment of these disease states. In this chapter neuropharmacological and neurological aspects of our recent work on the kainoids are reviewed with the aim of finding pharmacological tools.

EXCITOTOXIC ACTIONS OF KAINOIDS ON MAMMALIAN CENTRAL NEURONS

Kainic acid

KAIN, like QUIS, is one of the most potent excitants of mammalian central neurons. Iontophoretic administration of KAIN to rat cortical neurons induces marked spike discharges which are effectively blocked by GABA (Shinozaki and Konishi 1970). Also L-α-allo-KAIN causes spike discharges, albeit with a weaker activity than KAIN. In terms of the frequency of spike discharges produced by iontophoretic currents of the same strength, KAIN seems to be more potent than Glu. Large amounts of KAIN cause excessive depolarization of rat cortical neurons, reducing the amplitude of spike discharges in spite of the continued presence of KAIN. Spike discharges induced by KAIN persist for a much longer time after cessation of the iontophoretic application than those evoked by Glu. Such prolonged generation of spike discharges is one of the characteristics of KAIN-induced responses, probably reflecting slow dissociation of KAIN from receptors.

The iontophoretic technique, which has been most widely employed in mammalian CNS experiments (Stone 1985), has the major disadvantage that concentrations of drugs at the receptors are unknown, limiting its use in quantitative studies of drug–receptor interactions. Thus, the spinal cord of newborn rats has been used for *in vitro* electrophysiological experiments in the mammalian CNS because of the ease of changing drug concentration in the bathing solution. In this preparation, KAIN and other EAAs induce depolarization of spinal motor neurons in a concentration-dependent manner. The depolarizing activity of KAIN is much higher than that of Glu on a molar basis, but slightly lower than that of QUIS. Among the factors which affect depolarizing activities are concentrations of agonists, their conformational change and their uptake from the vicinity of receptors by the surrounding neurons and glia. When potential changes in the spinal motor neuron of the newborn rat were recorded extracellularly from ventral roots, KAIN showed a relatively slower depolarizing response (prolonged time-course of action) than did Glu, QUIS or NMDA. The half-decay time of the depolarizing response to KAIN seemed to be almost constant when its concentration was changed, while peak amplitudes of the depolarizing responses were increased in a dose-dependent manner. Selective NMDA blockers, such as 3-((±)-2-carboxypiperazin-4-yl)propyl-1-phosphonic acid (CPP), D-(−)-2-amino-5-phosphonopentanoic acid (D-AP5), (+)-5-methyl-10,11-dihydro-5*H*-dibenzo[*a,d*]cyclohepten-5,10-imine maleate (MK-801) and Mg^{2+} ions, did not affect the KAIN-induced depolarization. 6-Cyano-7-nitroquinoxaline-2,3-dione (CNQX) or 6,7-dinitroquinoxaline-2,3-dione (DNQX) blocked effectively the KAIN-induced depolarization, so that at high concentrations the depolarization was completely abolished, suggesting that the inhibition is in a competitive manner (Fletcher *et al.* 1988). However, these quinoxalinediones are not always selective for responses to KAIN because they depress responses to 2-amino-3-(3-hydroxy-5-methylisoxazol-4-yl)propanoic acid (AMPA), QUIS and NMDA as well as those to KAIN (Laster *et al.* 1989). The NMDA-blocking action of these quinoxalinediones is considered to be due to their antagonism of the co-agonist activity of glycine at the NMDA receptor-channel complexes. The *Xenopus* oocyte provides a system for expression of exogenous mRNA that permits detailed study of structure and function of receptors, and some agents are reported to be potent non-competitive antagonists of human brain KAIN receptors expressed in *Xenopus* oocytes (Umbach and Gundersen 1989, Randle 1990). However, these agents are not always effective at KAIN-induced depolarization in the newborn rat spinal cord preparation.

KAIN is a powerful neurotoxin that produces selective neuronal lesions in the mammalian CNS. Excitotoxic actions of KAIN have been well documented, and KAIN has been a useful tool for the study of the mechanism of neuron damage. A particularly significant feature of lesions made by EAAs is their sparing of fibres that pass to or through the zone of neuronal degeneration. The local administration of KAIN in the rat causes damage of almost all neurons in the injected region, but axons of passage or of termination are spared. Accordingly, excitotoxic amino acids, especially KAIN, are being employed as tools to destroy postsynaptic elements in the brain while preserving presynaptic structures. KAIN can be used as a tool in neurobiology to produce an animal model of Huntington's chorea (Coyle *et al.* 1978), and the unilateral microinjection of KAIN into the rat pre-pyriform cortex produces generalized motor seizures in a dose-dependent manner. Intracerebral or intraventricular injection of KAIN produces neuronal loss selectively in the CA3/CA4 cell fields of the hippocampus (Nadler *et al.* 1978) which correlate well with the high density of KAIN receptors (Foster *et al.* 1981, Monaghan and Cotman 1982). It seems likely that intrahippocampal KAIN-induced destruction of CA3/CA4 pyramidal cells is mediated by non-NMDA receptors and that KAIN-induced loss of these cells is associated with the neurobehavioural effect of intrahippocampal administration of KAIN, because KAIN-induced CA3/CA4 damage is not affected by MK-801 pretreatment, although high concentrations of MK-801 reduce the amount of KAIN-induced granule cell loss and, to some extent, CA1 pyramidal cell damage in the hippocampus (Rogers and Tilson 1990). Systemic administration of KAIN to the rat causes characteristic behaviour and selective neuron damage. The most pronounced behavioural changes in the rat seen after systemic injection of KAIN are immobility, increased incidence of 'wet-dog-shakes' (WDS), severe 'rearing, praying, salivating' limbic motor seizures, severe catalepsy, and tonic–clonic convulsions (Collins *et al.* 1980, Schwob *et al.* 1980, Lothman and Collins 1981, Sloviter and Damiano 1981, Worm *et al.* 1981, Lassman *et al.* 1984, Ben-Ari 1985), inducing selective neuron damage in the hippocampus, pyriform cortex, amygdala, several thalamic nuclei and lateral septum (Ben-Ari 1985). It seems likely that the number of WDS is an index of the progression of limbic seizures toward generalization (Rondouin *et al.* 1987), and WDS is eliminated by bilateral destruction of ventral dentate granule cells but unaffected by that of dorsal dentate granule cells (Grimes *et al.* 1990). In the hippocampus, neuronal cell loss produced by systemic KAIN is most extensively demonstrated in the CA1 region (Schwob *et al.* 1980, Lothman and Collins 1981, Lassman *et al.* 1984) where there is a low density of KAIN receptors (Foster *et al.*

1981, Ben-Ari 1985). Although the mechanism of KAIN and Glu neurotoxicity has not yet been fully elucidated, the CA1 damage caused by systemic KAIN closely resembles that produced by cerebral ischaemia (Brown and Brierly 1972). Glu neurotoxicity has several specific characteristics consistent with an important role in the pathogenesis of ischaemic brain damage (Choi and Rothman 1990). It is of great interest that the distribution of the lesions induced by KAIN does not always correlate with the regions of high density of KAIN receptors. It is reported that the extent of neuron damage resulting from systemic administration of KAIN is dependent on the extent of seizure activity (O'Shaughnessy and Herber 1986) and that the toxicity of KAIN is uniquely dependent on neuronally released Glu (McGeer and Zhu 1990). The extent of seizure activity seems to be affected by other transmitter systems as well. For example, partial depletion of noradrenaline and dopamine in the rat brain by treatment with the tyrosine hydroxylase inhibitor α-methyl-*p*-tyrosine markedly potentiated KAIN-induced epileptic symptoms (Baran *et al.* 1989).

Glu neurotoxicity is mediated through Glu excitatory receptors and entails sustained depolarization of postsynaptic dendrosomal membranes, increased membrane permeability and impaired ion homeostasis, probably by a toxic influx of extracellular Ca^{2+} (Choi 1987, 1988b). Only NMDA receptors open a membrane channel that is highly permeable to Ca^{2+} (MacDermott *et al.* 1986). The NMDA receptor-activated channel is the major route by which Glu induces a toxic Ca^{2+} influx, although additional Ca^{2+} entry probably occurs through voltage-activated Ca^{2+} channels, the Na^{+}–Ca^{2+} exchanger, and non-specific membrane leakage (Choi 1988a). Additionally, Ca^{2+} is released from intracellular stores by the action of Glu on metabotropic EAA receptors (Sladeczek *et al.* 1985, Nicoletti *et al.* 1986). Sheardown *et al.* (1990) have hypothesized that Glu released during ischaemia may trigger a post-ischaemic release of Glu or another endogenous substance, which acts at non-NMDA receptors, and they have recently reported that 2,3-dihydroxy-6-nitro-7-sulphamoyl-benzo(*f*)quinoxaline (NBQX), which preferentially blocks responses to QUIS and AMPA, is an effective neuroprotectant for cerebral ischaemia (Sheardown *et al.* 1990).

Domoic acid

Domoic acid (DOMO), like KAIN, is one of the potent neurotoxic kainoids of natural origin (Fig. 2). DOMO was first isolated from a seaweed, *Chondria armata*, the same family of algae as *Digenea simplex*, and recently some DOMO derivatives such as domoilactones and isodomoic acids were also isolated from the seaweed. The common name of *Chondria armata* in Japanese is *Domoi* or *Hanayanagi*, and the seaweed has excellent ascaricidal properties. The effective ascaricidal component, therefore, was named domoic acid, which also has been noted to have fly-killing properties. Ohfune and Tomita (1982) have succeeded in the chemical synthesis of DOMO. DOMO demonstrates a potent depolarizing action in a manner similar to KAIN, being more active than KAIN (Patneau and Mayer 1990), and DOMO is one of the potentially valuable EAAs (Biscoe *et al.* 1975, 1976, Shinozaki and Shibuya 1976, Debonnel *et al.* 1989a, 1989b, Huettner 1990). DOMO was much more potent in the CA3 than in the CA1 regions of the rat hippocampal slice, whereas no such regional difference could be detected with QUIS and NMDA (Debonnel *et al.* 1989a). The differential regional depolarizing response of hippocampal CA1 and CA3 pyramidal neurons to KAIN is attributable to the extremely high density of KAIN

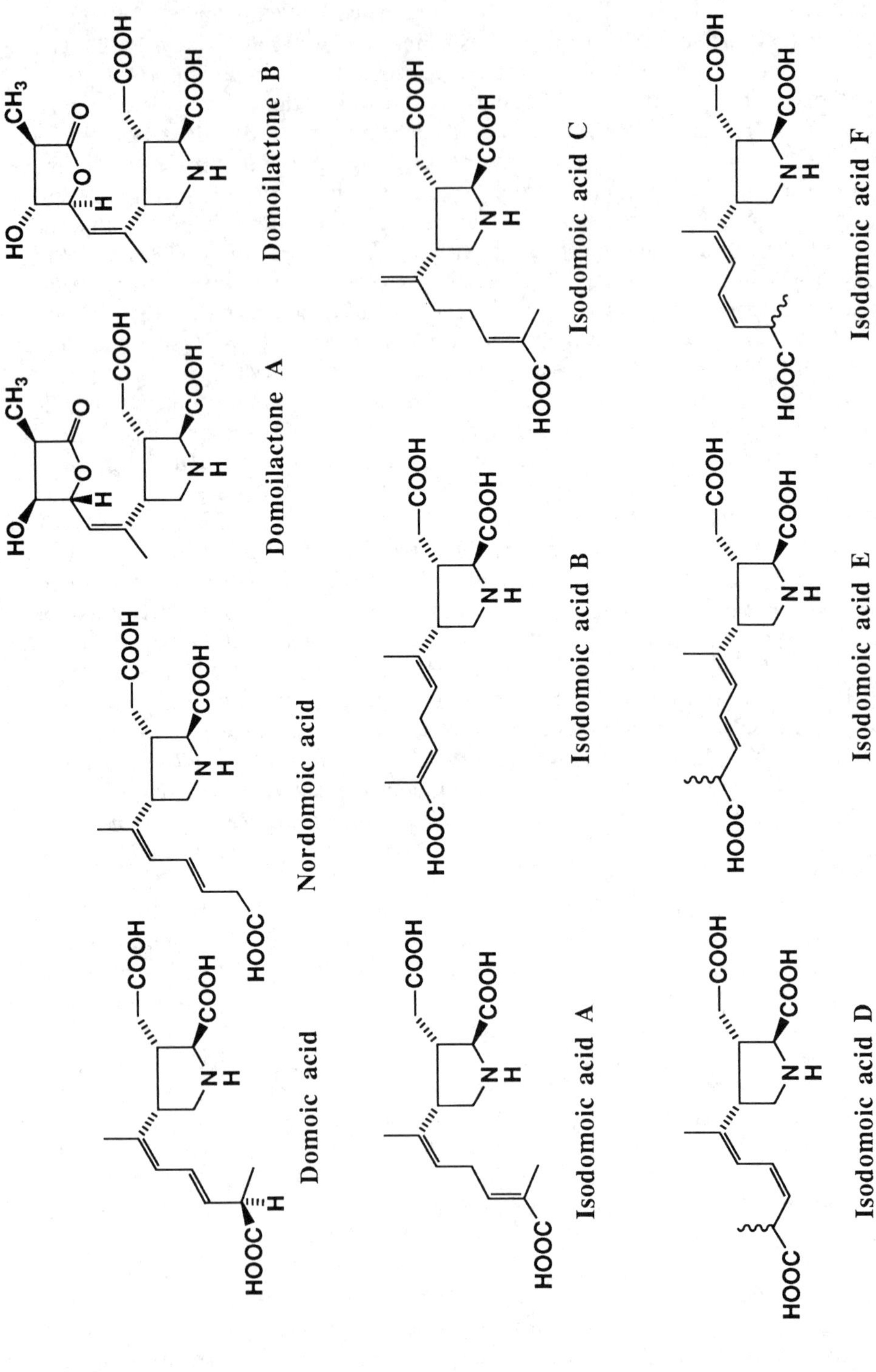

Domoic acid

Isodomoic acid A

Isodomoic acid D

Nordomoic acid

Isodomoic acid B

Isodomoic acid E

Domoilactone A

Isodomoic acid C

Isodomoic acid F

Domoilactone B

Fig. 2 — Chemical structures of DOMO and its isomers isolated from the seaweed *Chondria armata.*

receptors in the CA3 region (Foster *et al.* 1981, Monaghan and Cotman 1982). Electrophysiological experiments failed to reveal the difference in qualitative properties between DOMO and KAIN (Debonnel *et al.* 1989a). In the newborn rat spinal motor neuron, there were no apparent differences in electrophysiological and pharmacological features between KAIN and DOMO. Reponses to KAIN and DOMO did not show appreciable desensitization in cultured embryonic hippocampal neurons, whereas those to QUIS, AMPA and NMDA showed rapid desensitization (Mayer and Vyklicky 1989). A plant lectin, concanavalin A (Con A), reduced desensitization at AMPA receptors, with no effect on responses to KAIN or NMDA. In passing, reduction of desensitization of Glu receptors by Con A has been well documented at the invertebrate NMJ (Mathers and Usherwood 1976, Mathers 1981, Shinozaki and Ishida 1979).

From its structural similarity to KAIN, it has been predicted that DOMO may produce its neurotoxic effects through activation of KAIN receptors (Tryphonas and Iverson 1990). Systemic administration of DOMO to the mouse induces a characteristic syndrome, essentially similar to that of KAIN, including sluggishness, scratching stereotypy, convulsions and death (Glavin *et al.* 1990); however, particularly prominent is involuntary scratching of the shoulders with the hind legs. There was no apparent regional difference in neuron damage induced by DOMO and KAIN. Mussel (*Mytilus edulis*) intoxication in Prince Edward Island in 1987 has been proven to be due to DOMO. Isodomoic acids have also been detected in the mussels (Wright *et al.* 1990). The patients had a rapid onset of confusion, disorientation, and memory loss within 24 hours of eating mussels from the island. Four post-mortem anatomopathological examinations of patients intoxicated with the mussels revealed extensive damage of the hippocampus, as well as spotty damage of thalamic and forebrain regions (Perl *et al.* 1990, Teitelbaum *et al.* 1990). Some of the survivors apparently sustained permanent brain damage as they have continued to show profound memory impairment. A disproportionately high percentage of severely affected individuals were elderly. As Olney (1990) described, the apparently heightened vulnerability of elderly individuals to the neuropsychological consequences of DOMO poisoning is of particular interest, as this suggests that the KAIN receptor is a potentially sensitive mediator of excitotoxic neuropathology in old age.

Acromelic acid

Acromelic acid (ACRO) has been isolated from a poisonous Japanese mushroom, *Clitocybe acromelalga*, which is one of the most poisonous ones found in Japan (Konno *et al.* 1983, 1986, 1988). ACRO has two structural isomers, ACRO A and ACRO B (Fig. 3). Recently, another kainoid, ACRO C, was isolated from the mushroom (Fushiya *et al.* 1990). ACRO C is not an isomer of ACRO A or B but has the structure of decarboxylated ACRO B with weaker toxicity than ACRO A and B. On account of the limited availability of ACRO B and C, ACRO A has been used exclusively, and is here referred to simply as ACRO, unless otherwise noted. Recently a method for chemical synthesis of ACRO has been designed (Hashimoto *et al.* 1986, Takano *et al.* 1987, 1989, Konno *et al.* 1983, 1986, 1988, Shirahama *et al.* 1988, 1990, Baldwin and Li 1988, Hashimoto and Shirahama 1990), and small samples of ACRO and some KAIN derivatives have become available for neurobiological experiments (Shinozaki 1988, Shinozaki and Ishida 1988a, 1988b, Shinozaki *et al.* 1989a, 1990,

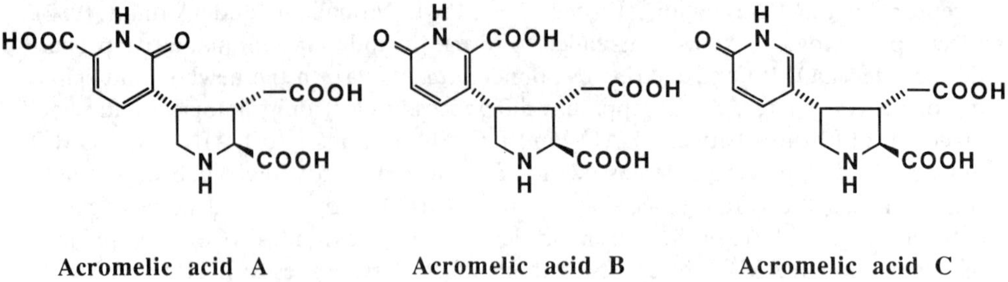

Fig. 3 — Chemical structures of ACRO A, B and C isolated from the mushroom *Clitocybe acromelalga*.

Kwak *et al.* 1990). ACRO shows a yellow coloration in ninhydrin tests and acts as a stronger acid than KAIN in ion-exchange chromatography and paper electrophoresis (Konno *et al.* 1988). When ACRO is dissolved in distilled water, it shows a strong blue fluorescence (excitation 320 nm, emission 390 nm).

ACRO demonstrates the highest deplarizing activity of the three kainoids (KAIN, DOMO and ACRO) in the newborn rat spinal cord (Ishida and Shinozaki 1988, Shinozaki *et al.* 1991) and the frog spinal cord (Maruyama and Takeda 1989), followed by DOMO and KAIN in that order. Their threshold concentrations to cause depolarization are almost identical in the newborn rat spinal cord. There are some differences in the electrophysiological features of depolarizing responses among these three kainoids, including the slope of the dose–response curves and the time-course of depolarizing responses (Fig. 4). ACRO induces depolarizing responses with the shortest time-course, and the tail of depolarizing respones to ACRO decreases more quickly in amplitude than that of responses to KAIN or DOMO. This short time-course is one of the characteristics of ACRO-induced depolarization in the isolated newborn rat spinal cord, in striking contrast to that of KAIN. Several factors, such as desensitization of receptors, equilibrium constants of agonist-receptor complexes, and differences in rates of uptake of agonists, would affect the time-course and the peak amplitude of depolarizing responses but have not yet been sufficiently analysed. In order to avoid an underestimate of the more slowly developing responses relative to the fast responses, responses were compared after prolonged application of EAAs, but ACRO was the most potent, regardless of the duration of EAA application. The dose–response curves demonstrate that ACRO is the most potent of the known EAAs, including NMDA, QUIS and KAIN, in the newborn rat spinal cord (Fig. 5). The excitatory actions of ACRO are expected to be similar in other regions of the mammalian CNS as well, because ACRO exhibits significant excitatory actions on rat cortical neurons when administrered iontophoretically.

CPP, D-AP5, MK-801 and Mg^{2+} ions, which are known to be selective NMDA blockers, do not affect ACRO-induced depolarizations in the newborn rat spinal cord, but the quinoxalinediones, such as CNQX and DNQX, effectively depress depolarizing responses to ACRO in a concentration-dependent manner, suggesting that ACRO is a potent non-NMDA-type agonist. At present, a highly selective antagonist for KAIN receptors is not available; therefore, it is difficult to know pharmacologically whether ACRO acts selectively on KAIN receptors. In our recent studies, CNQX

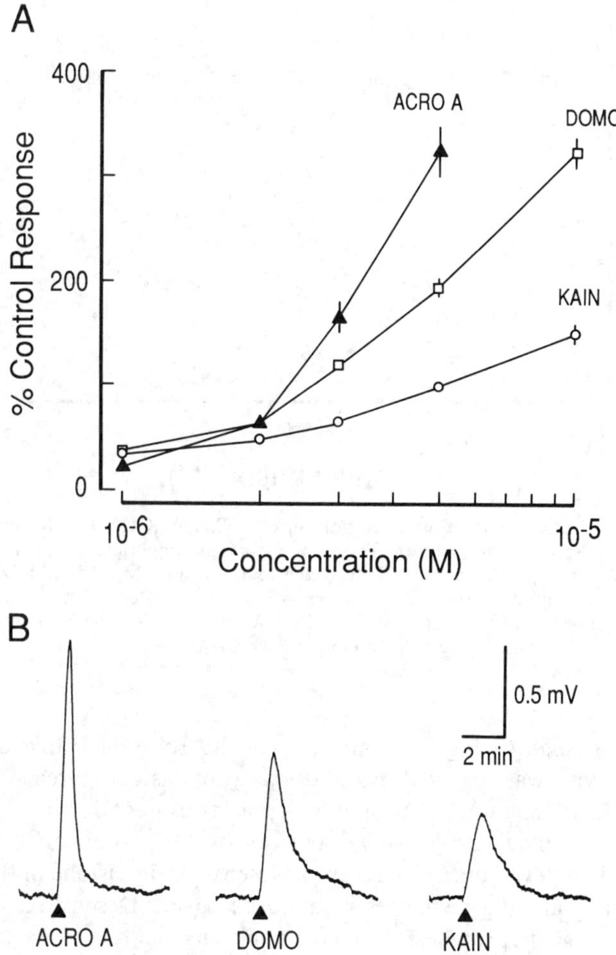

Fig. 4 — A: Depolarization-concentration relationships for ACRO A, DOMO and KAIN. The responses (peak amplitude) to each kainoid were recorded extracellularly from the newborn rat ventral root in Mg^{2+}-free, tetrodotoxin(TTX)-containing solution. Responses (10-s application) from individual preparations have been normalized, so that results are expressed as a percentage of the control depolarization to KAIN (5 μM). Vertical bars represent the standard errors of means (SEM) (*n* at least 4). B: Sample recordings of depolarizing responses to KAIN, DOMO and ACRO A at a concentration of 3 μM.

depressed depolarizing responses to ACRO as well as AMPA more effectively than those to KAIN in the newborn rat spinal motor neuron (Ishida and Shinozaki 1991). It is known that C-fibres in the dorsal root isolated from immature rats are depolarized selectively by KAIN, whereas QUIS and AMPA are much less active, and NMDA does not cause depolarization of the dorsal root even at high concentrations (Agrawal and Evan 1986, Evans *et al.* 1987). Therefore, the dorsal root C-fibre may be used for differentiating KAIN-type agonists from other non-NMDA-type ones. ACRO causes depolarization of the newborn rat dorsal root in a dose-dependent manner. The depolarizing activity of ACRO is lower than that of DOMO in the

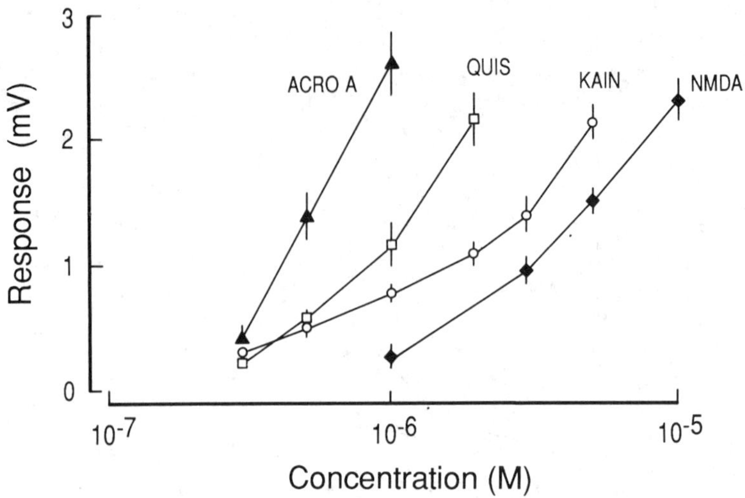

Fig. 5 — Depolarization–concentration relationships for NMDA, KAIN, QUIS and ACRO A in the newborn rat spinal motor neuron. The responses (peak amplitude) to each EAA (2-min application) were recorded extracellularly from the ventral root in Mg^{2+}-free, TTX-containing solution. Peak amplitudes of depolarizing responses were plotted against concentrations. Vertical bars represent SEM (*n* at least 5). ACRO A demonstrated the highest depolarizing activity among known EAAs.

dorsal root, but slightly higher than that of KAIN. The rank order of their depolarizing activity was reversed in the dorsal roots as compared with that of the motor neuron. As in the motor neuron, the depolarization of dorsal roots induced by ACRO is effectively blocked by CNQX, but not by Mg^{2+} ions or selective NMDA blockers. A marked development of receptor desensitization is one of the characteristic features in depolarizing responses in the dorsal root. Desensitization to various pharmacological agents provides means for analysing the role of particular types of receptors in drug effects. In particular, crossed desensitization between two excitatory compounds has been an effective way to show their identity or to examine their physiological and pharmacological properties. When KAIN was applied to the dorsal root for a prolonged time, initial depolarization was not maintained but decreased to a sustained plateau level, demonstrating the development of desensitization of KAIN receptors. Under these conditions, ACRO and DOMO did not cause any further depolarization of the dorsal root. From these results, ACRO seems to be a KAIN-type agonist (see below).

It seems reasonable to predict that ACRO would demonstrate excitotoxic actions quite similar to KAIN or DOMO, because all of them contain the common chemical moiety of KAIN. The local administration of ACRO to the rat induces neuron damage, and the minimal lethal dose of subcutaneous ACRO was less than half that of KAIN. Systemic administration of ACRO to the rat causes abnormal behaviour quite distinct from that of KAIN or DOMO, and there is a striking difference in the regional distribution of neuron damage (Shinozaki *et al.* 1989a, 1990, Kwak *et al.* 1990, 1992a). The most pronounced behavioural changes in the rat seen after a single systemic injection of ACRO are the tonic extension of the hindlimbs, their transient flaccid paralysis, and the persistent spastic paraplegia, which appear in consecutive order. These symptoms are quite distinct from those induced by systemic administration of KAIN or DOMO, although most rats receiving effective doses of ACRO died from severe tonic–clonic convulsions in a manner similar to that observed with systemic KAIN or DOMO. The transient flaccid paralysis of the hindlimbs seemed to be a premonitory symptom towards the development of persistent spastic paraplegia, because the surviving rats that had experienced heavy transient flaccid paralysis of the hindlimbs developed a severe spastic paraplegia on the day following a single injection of ACRO. The spastic paraplegia, which was significantly reinforced by sensory stimuli such as touching and poking, persisted until the rat was sacrificed for histological analysis (at least three months after the single application of ACRO). Systemic KAIN caused neither spastic paraplegia nor tonic extension of the hindlimb at any doses; instead, it caused severe limbic motor seizures (Collins *et al.* 1980, Schwob *et al.* 1980, Lothman and Collins 1981, Lassman *et al.* 1984, Ben-Ari 1985). ACRO never caused limbic seizures in rats, and there were only a few incidences of WDS even at lethal doses. These characteristic symptoms induced by ACRO suggested that the spinal cord was exquisitely sensitive to ACRO in the first stage after the systemic administration, but the subsequent tonic–clonic convulsions probably reflected actions at the brain rather than the lower spinal cord levels. The distribution of neuron damage induced by systemic administration of ACRO is confined to the lower spinal cord in the rat and, in particular, to the lumbar and sacral segments, as predicted from behavioural signs (Fig. 6). No apparent pathological changes were found in hippocampal neurons, in striking contrast to the vulnerability of hippocampal pyramidal cells to KAIN (McGeer *et al.* 1978, Nadler *et al.* 1978, Schwob *et al.* 1980, Nadler 1981, Lothman and Collins 1981, Lassman *et al.* 1984, Ben-Ari 1985). However, a mild glial reaction has been detected in the CA4 area of the hippocampus by immunohistochemical staining for glial fibrillary acidic protein (GFAP) (Kwak *et al.* 1990, 1992a). Severe rigidospastic paraparesis with extensive damage of spinal interneurons has been known to be induced after spinal ischaemia by ligation of the thoracic aorta in the dog, cat, rabbit and rat (Gelfan and Tarlov, 1959, 1963, Hadzovic 1970). The distribution of neuron damage induced by ACRO closely resembles that of spinal neuron damage by ligation of the thoracic aorta. The similarity of neuron damage induced by brain and spinal ischaemia to those induced by KAIN and ACRO, respectively, may suggest a possibility for a common mechanism underlying neuron damage by ischaemia and kainoid neurotoxicity.

Since *c-fos*, an immediate early gene, is a transcriptional factor gene which is induced by various stimulations, including EAAs, CNS stimulants, dibutyryl cyclic AMP, Ca^{2+} influx, and phorbol ester, it is thought to be one of the useful markers of neuronal activity in the CNS (Nakajima *et al* 1989). It is known that KAIN produces a seizure syndrome associated with a transient rise in the *c-fos* protein within the hippocampus in the rat (Popovici *et al.* 1988, Le Gal La Salle 1988). Before the generalized tonic–clonic convulsions appeared, ACRO significantly induced *c-fos* mRNA only in granule cells of the dentate gyrus, lateral septum and hypothalamic nucleus, in striking constrast to KAIN, which induced *c-fos* mRNA in a variety of brain areas, including the CA1, CA3 and CA4 pyramidal cells of hippocampus, lateral septum, hypothalamic nucleus, amygdala complex, entorhinal cortex, and pyriform cortex (Shinozaki *et al.* 1991). The expression of *c-fos* mRNA was not observed when

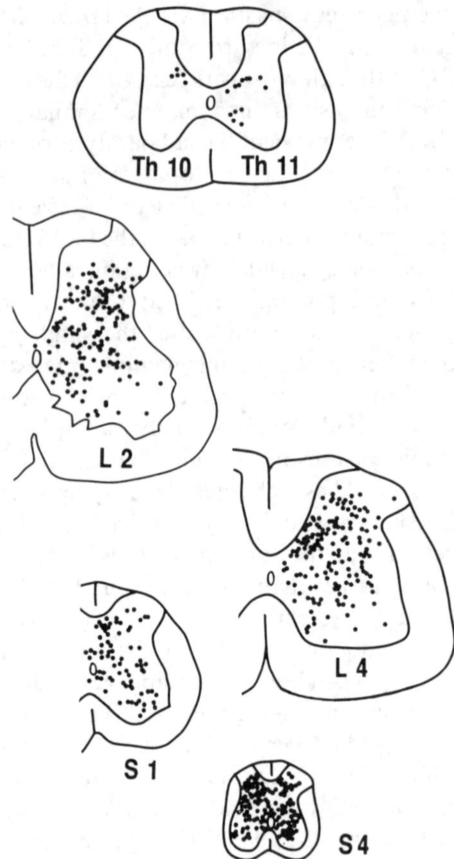

Fig. 6 — The distribution of degenerated neurons in various segments of the spinal cord of the spastic paraplegic rat seven days after a single injection of ACRO A (5 mg/kg, s.c.). The distribution of degenerated neurons closely resembled that induced by spinal ischaemia. Th, thoracic; L, lumbar; S, sacral.

the mouse was under the initial sedative stage after subcutaneous injection of ACRO. These data indicate that the neurons activated by ACRO may be different from those activated by KAIN. ACRO is superior to KAIN in causing depolarization of newborn rat spinal motor neurons (Ishida and Shinozaki 1988) and has been considered to activate KAIN receptors in the frog spinal cord (Maruyama and Takeda 1989). However, the abnormal behavioural symptoms and pathology revealed by systemic administration of ACRO are quite distinct from those induced by KAIN. In addition, ACRO is a weak displacer of both [^{3}H]KAIN and [^{3}H]AMPA binding to adult rat spinal cord synaptic membranes (Table 1) (Kwak *et al.* 1992b). At present there is no clear-cut explanation for the above discrepancy, but the idea that there are two or more subtypes of KAIN receptors is very attractive. Smith and Mallhinney (1992) reported that in the absence of chaotropic ions, ACRO is the most potent displacer of AMPA binding yet described, and they suggested that ACRO distinguishes two

Table 1 — The binding ability of KAIN derivatives to KAIN and AMPA receptors

Compound	$[^3H]KAIN^a$ K_i (μM)	$[^3H]KAIN^b$ IC_{50} (μM)	$[^3H]KAIN^c$ IC_{50} (μM)	$[^3H]AMPA^c$ IC_{50} (μM)
DOMO	0.013	0.65	0.0035	0.47
HFPA			0.0058	0.13
KAIN	0.021	0.35	0.012	1.9
MFPA			0.014	0.052
MKPA	0.43			
QUIS	0.61	35	0.097	0.0019
ACRO B			0.12	
Glu	1.52	> 1000	0.26	0.12
ACRO A		0.003	0.33	0.06
CNQX			2.3	
DihydroKAIN	7.4			
Allo-KAIN	> 10			
AMPA			99	0.0086
NMDA			> 100	> 10

[a]Rat cerebellum neurons (Slevin *et al.* 1983).
[b]Frog spinal cord (Maruyama and Takeda 1989).
[c]Rat spinal cord (Aizawa and Kwak 1991).

KAIN binding sites in rat brain synaptic plasma membranes. Indeed, whole-cell voltage-clamp recording data from cultured rat hippocampal neurons reveal the presence of two KAIN responses which differ in their rectification properties and their permeability to Ca^{2+} (Iino *et al.* 1990). Recent data on cloned Glu receptor molecules suggest a multiplicity of KAIN receptors in the rat brain (see chapter 3). KAIN or AMPA receptors are composed of heteromeric subunits, and genes encoding each subunit express their transcripts differently in brain areas (Boulter *et al.* 1990, Keinanen *et al.* 1990, Sommer *et al.* 1990).

Erroneous ingestion of the mushroom *Clitocybe acromelalga*, from which ACRO was isolated, causes a sharp pain, and a marked reddish oedema (erythromelalga) appears in the hand and foot about a week later and lasts for about a month. The striking features of intoxication by this poisonous mushroom are the lack of fever, gastrointestinal and renal disturbances and abnormal signs of the CNS, which are quite different from those of DOMO intoxication. Other compounds, such as clitidine, clithioneine and 4-aminoquinolinic acid, have so far been isolated from the mushroom (Konno *et al.* 1982, 1984, Hirayama *et al.* 1989), but it is not yet known whether these compounds, including ACRO, cause the symptoms of poisoning. According to a Japanese domestic medical journal, axonal damage has been suggested in the peripheral nerves of a patient poisoned by the mushroom.

The selective loss of interneurons in the spinal cord is one of the characteristic pathological features of human 'stiffman syndrome', a progressive neurological disease characterized by slowly ascending rigidospasticity of the axial and limb musculature with painful muscle spasms (Kasperek and Zebrowski 1971, Martinelli *et al.* 1978, Howell *et al.* 1979). Although there is no direct evidence that ACRO *per se* induces the stiffman syndrome, the close similarity of clinical and neuropathological features of this disorder with those of the rat with spastic paraplegia would deserve further investigation for the mechanism of neurodegenerative disorders. In fact, an aetiological role of EAAs has been claimed in several diseases affecting predominantly the spinal cord, including amyotrophic lateral sclerosis (Spencer *et al.* 1987, Young 1990) and neurolathyrism (Rao *et al.* 1964, Rao 1978, Spencer *et al.* 1984, Bridges *et al.* 1989) (see chapter 10). Elucidation of a mechanism of excitotoxic neuron damage may provide a valuable clue to the selective neuronal death in neurodegenerative disorders. ACRO will probably prove a useful addition to known EAAs as a tool for neuroscience research.

PHARMACOLOGICAL ACTIONS OF KAINOIDS ON THE INVERTEBRATE NMJ

The invertebrate neuromuscular system presents some advantages, especially in the use of certain microtechniques on individual cells that are difficult to perform in vertebrates. The invertebrate NMJ has some properties in common with mammalian central synapses, and in addition to providing us with an insight into the chemical transmission process at many synapses, it can be used as a model for studying the mechanism of drug action on synaptic transmission in the mammalian CNS (Katz 1966, Shinozaki 1980, 1988). Potentiation of Glu responses is one of the characteristic actions of the kainoids at the crayfish NMJ (Shinozaki and Shibuya 1974b, Takeuchi and Onodera 1975, Onodera and Takeuchi 1980, King and Wheal 1984). KAIN has a powerful excitatory action on mammalian central neurons, but its depolarizing action on crayfish or locust muscle fibres is not particularly potent. However, the ability of the kainoids to potentiate the Glu response in the crayfish opener muscle is almost proportional to their depolarizing activity in the mammalian central neuron.

Both ACRO A and B caused a marked depolarization of the crayfish opener muscle fibre at very low concentrations. ACRO A and B are of almost equal potency, but other kainoids are not particularly active at the invertebrate glutamatergic NMJ (Shinozaki *et al.* 1986). ACRO A and B, unlike Glu or QUIS, did not cause receptor desensitization of the crayfish NMJ. The depolarizing potency of bath-applied ACRO, in terms of its threshold concentration, was approximately 100 times higher than that of KAIN and about ten times higher than that of DOMO or methylketone KAIN (Fig. 7) (Shinozaki *et al.* 1986). The threshold concentration of DOMO and methylketone KAIN required for causing depolarization was five to ten times lower than that of Glu. Brief iontophoretic application of ACRO produced a significant depolarization of the crayfish muscle fibre, in striking contrast to that of KAIN or DOMO (Shinozaki 1988). Iontophoretically applied DOMO produced a slow and small depolarization, in contrast to the faster and larger depolarization produced by Glu or QUIS (Shinozaki and Shibuya 1976). The ACRO- and DOMO-sensitive areas overlapped Glu-sensitive ones, but no depolarization was observed when KAIN was iontophoretically applied to any area, including a Glu-sensitive spot. There was no apparent pharmacological difference among ACRO, DOMO and KAIN, except for the higher depolarizing potency of ACRO at the crayfish NMJ.

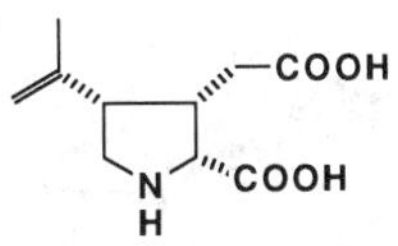

L-β-kainic acid 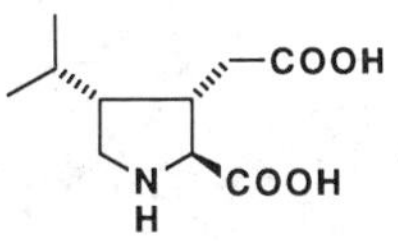 L-α-dihydro kainic acid 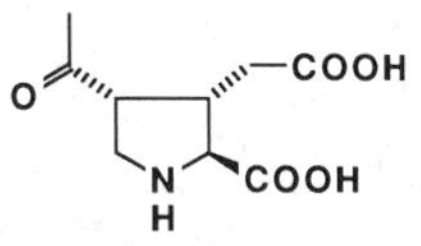 L-α-methylketone kainic acid

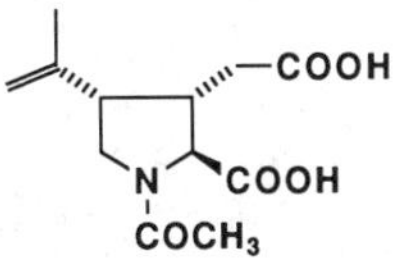

L-α-N-acetyl kainic acid 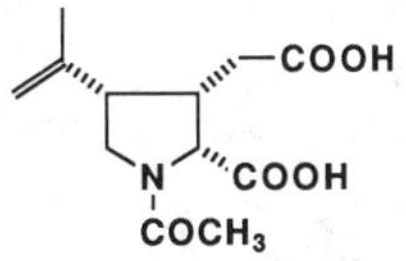 L-β-N-acetyl kainic acid 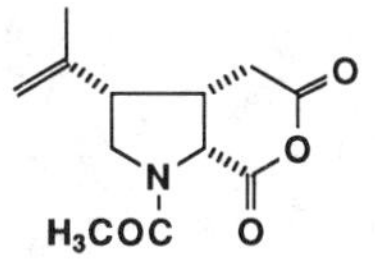 L-β-N-acetyl kainic anhydride

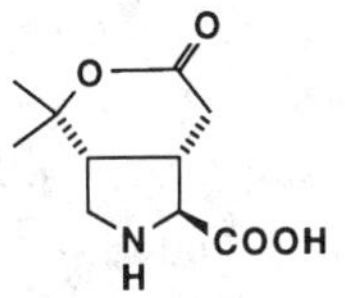

L-α-kainic acid lactone 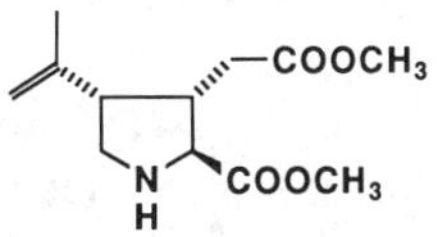 L-α-kainic acid dimethylester 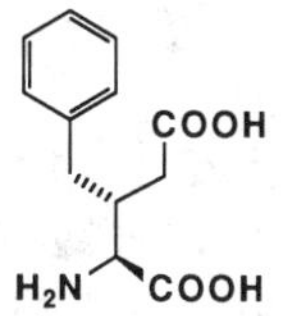 (2S,3R)-3-benzyl-glutamic acid

Fig. 7 — KAIN derivatives (I).

On the other hand, esterification or *N*-acetylation of KAIN caused a complete loss of both its depolarizing and potentiating activities at the crayfish NMJ, and neither allo- or dihydro-KAIN nor the *β*-isomer of KAIN had any effect on the crayfish NMJ (Fig. 7). Recent data show that conformationally restricted spirocyclo analogues give a loss of affinity for the receptor recognition site (Kozikowski and Fauq 1990). These results suggest that the isopropenyl side-chain (or π-electrons) plays a key role for depolarizing activity. Removal of the 5-carbon of KAIN leads to 3-alkyl-Glu derivatives (Yanagida *et al.* 1989), for example (2*S*,3*R*)-3-benzyl-Glu where the stereochemistry is quite similar to that of KAIN. This compound (Fig. 7) did not cause depolarization, but potentiated the Glu response in low concentrations at the crayfish NMJ, being about ten times less potent than KAIN.

Some naturally occurring kainoids and their derivatives (Fig. 8) have been examined predominantly in invertebrates, but none of them are particularly more active than DOMO or KAIN, except for nordomoic acid (nor-DOMO) (Maeda *et al.* 1987). According to Maeda *et al.* (1987), low concentrations of isodomoic acids and nor-DOMO (see Fig. 2) enhanced the Glu sensitivity of the insect NMJ, and DOMO affected Ca^{2+} transport and stimulated a Ca^{2+}-dependent process that regulated the release of Glu from presynaptic nerve endings. In addition, DOMO activated a Ca^{2+}-dependent proteinase, leading to generation of Glu receptors from the latent state on the neuromuscular postsynaptic membrane.

Extremely low concentrations of the kainoids, such as ACRO, DOMO, methylketone KAIN and KAIN, well below those causing depolarization of the muscle fibre, caused marked potentiation of responses to Glu. When potentials evoked by long

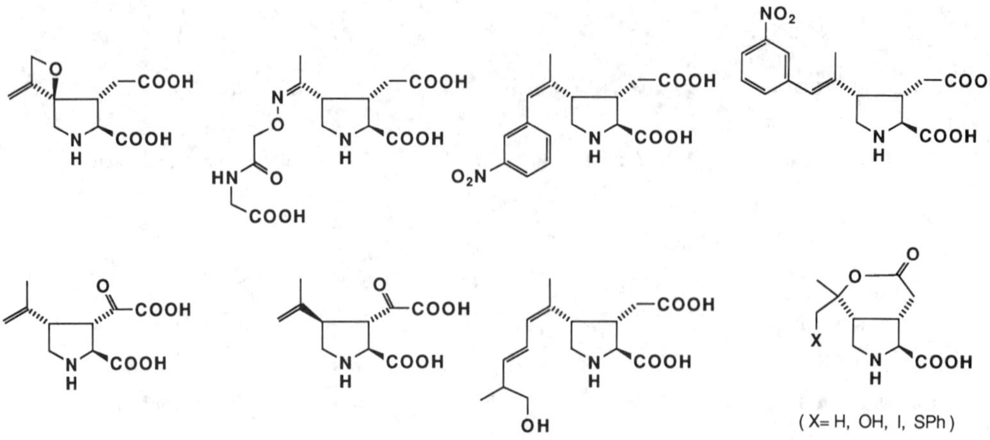

Fig. 8 — KAIN derivatives (II).

iontophoretic pulses of Glu slowly subsided to a sustained low plateau level due to receptor desensitization, despite the continued presence of Glu additional iontophoretic application of the kainoids began to depolarize the muscle fibre as if they inhibited the development of desensitization. On the other hand, these kainoids depressed depolarizing responses to QUIS at the crayfish NMJ in a dose-dependent manner (Shinozaki and Ishida 1976; Shinozaki 1988) in spite of the fact that QUIS and Glu act on common receptors at this junction (Shinozaki and Shibuya 1974a, Shinozaki and Ishida 1981) and although these kainoids *per se* are excitants. The threshold concentration of ACRO to depress depolarizing responses to QUIS was far lower than that of KAIN or DOMO. On the other hand, the response to low concentrations of QUIS was slightly augmented by extremely low concentrations of ACRO. Thus facilitation and depression of the test Glu response (Takeuchi and Takeuchi 1964, Shinozaki and Shibuya 1974a) are observed following a conditioning application of ACRO in a similar way to KAIN and DOMO. In passing, CNQX and DNQX did not affect responses to Glu and QUIS at the crayfish NMJ; furthermore, AMPA did not cause depolarization of the muscle fibre at all. Such pharmacological differences between vertebrates and invertebrates lead to the question of whether separate receptors and channels exist for QUIS and KAIN in the mammalian CNS as well, and, if so, whether binding to one of these receptors would modify the response elicited through the other. According to Perouansky and Grantyn (1989), QUIS acts as a low-affinity competitive antagonist at the KAIN site and as a high-affinity agonist at its own receptor in cultured neurons from the rat superior colliculus. A relatively higher affinity of KAIN towards low-affinity AMPA binding sites has also been suggested (Watkins *et al.* 1990) (see chapters 2 and 11). These findings may explain the discrepancy in neuropharmacological properties between invertebrates and vertebrates.

KAINATE DERIVATES

The three representative kainoids, KAIN, DOMO and ACRO, thus revealed significant difference in their depolarizing activities and regional neuron damage in the mammalian CNS. Such differences may indicate that more active and/or pharmacologically useful excitants are present among the unknown kainoids. Here, the depolarizing actions of several KAIN derivatives were examined in the isolated newborn rat spinal cord with the purpose of finding new active compounds. Almost all KAIN derivates shown in Fig. 9 caused depolarization of the rat spinal motor neuron, albeit with a large variation in their activities (Ishida and Shinozaki 1991). The methoxyphenyl derivative (MFPA) showed the highest activity, but several derivatives caused depolarization much more effectively than KAIN. Since ACRO so far has been the most potent of the known EAAs, it should be noted that MFPA was more potent than ACRO (Fig. 10). The depolarizing responses to the compounds shown in Fig. 9 could be roughly divided into two groups showing fast and slow responses. KAIN produced a relatively prolonged time-course of action (slow responses) in rat spinal motor neurons, but KAIN derivatives did not always show slow responses, and ACRO A and B, QUIS and Glu showed faster responses than KAIN. The time-course of depolarizing responses to MFPA closely resembled that of KAIN or DOMO. Although the slope of dose–response curves differed, mean equimolar potency ratios for the depolarization of newborn rat spinal motor neurons were tentatively obtained by plotting peak amplitudes of responses against the concentrations of agonists. By this procedure, the rank order of depolarizing activities of KAIN derivatives in the newborn rat spinal motor neuron was: MFPA > ACRO A > ACRO B ≏ DOMO > HFPA > HMPPA ≏ CPPA > KAIN > MPPA ≏ FPA > CNOPA > MKPA > Glu > dihydro KAIN (see Fig. 10). This order did not change when the duration of drug application was prolonged, although the value of mean equimolar potency sometimes varied slightly. However, it seemed to be technically difficult to determine exactly the relative potency of excitation, induced by

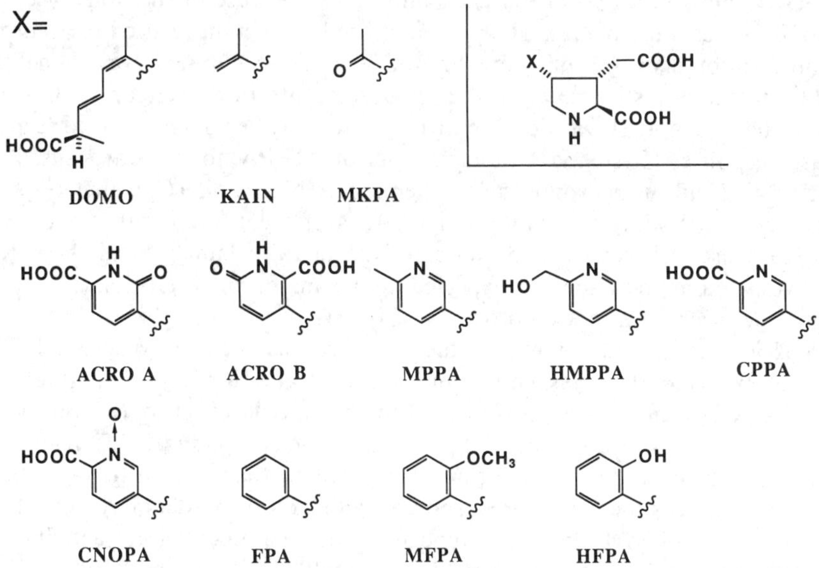

Fig. 9 — KAIN derivatives (III).

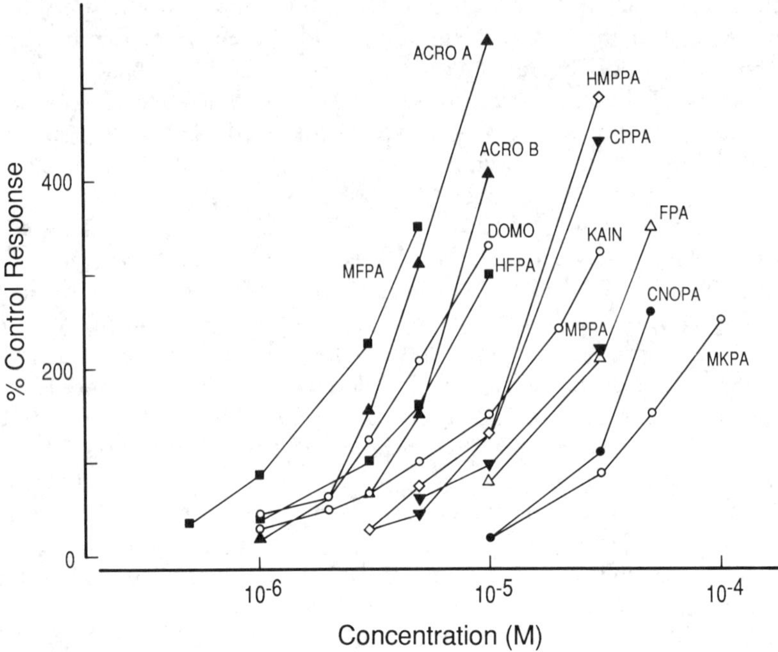

Fig. 10 — Comparison of depolarizing responses to various KAIN derivatives. The responses (peak amplitude) to each KAIN derivative were recorded extracellularly from the newborn rat ventral root in Mg^{2+}-free, TTX-containing solution. Responses from individual preparations have been normalized, so that results are expressed as a percentage of the control depolarization to KAIN (5 μM) (*n* at least 4). Abbreviations are the same as those in Fig. 9.

these KAIN derivatives, from the peak amplitude, because of the difference in the slope of dose–response curves. The peak amplitude sometimes varied in accordance with the duration of drug application (Shinozaki *et al.* 1989c), and it was often difficult to obtain maximal responses even when high concentrations were applied. Another reason is that the more slowly developing responses may be underestimated relative to the fast responses. Systemic administration of MFPA to the rat caused both characteristic syndromes produced by systemic KAIN and ACRO, which, respectively, caused quite distinct behavioural signs. It is of great interest that each KAIN derivative causes a selective and characteristic neuron damage, and these KAIN derivatives would, therefore, be expected to be major pharmacological tools for studying the mechanism of excitotoxicity of EAAs.

(2*S*,3*R*)-3-Benzyl-Glu potentiated the Glu response at the crayfish NMJ. Therefore, it is expected that this compound demonstrates KAIN-like actions in the mammalian CNS, and it is of great interest to examine the effects of the four possible stereoisomers of this 3-substituted Glu. Among these compounds, (2*S*,3*R*)-3-benzyl-Glu was shown to be about three times more potent than Glu in the newborn rat spinal cord, whereas the other three isomers had slight or no depolarizing activity.

The β-isomer of KAIN has been reported to have anticonvulsant activity with a profile suggesting EAA antagonism (Collins *et al.* 1984, Turkski *et al.* 1985, 1987). β-KAIN preferentially antagonizes myoclonic seizures induced by NMDA and quinolinate (QUIN) (Turski *et al.* 1985), but β-KAIN does not antagonize any of the major agonists at EAA receptors (Stone and Collins 1986). However, β-KAIN depressed the K^+-evoked release of endogenous Glu at relatively high concentrations, and this effect may underlie the anticonvulsant properties of β-KAIN (Connick and Stone 1986).

Decarboxylation of IBO converts its pharmacological properties as an excitatory agonist into those of an inhibitory GABA agonist (muscimol). However, unlike IBO, 2-decarboxylated KAIN was only a weak inhibitor of Na^+-independent binding of GABA to rat brain membranes, suggesting that decarboxylated Glu agonists do not always interact with GABA receptors (Allan 1978). The bicyclic [2*S*-(2α,3β,4β)]-2-carboxy-4-(1-hydroxy-1-methylethyl)-3-pyrrolidineacetic acid δ-lactone and its 4-[1-hydroxy-1-(iodomethyl)ethyl], 4-[1-hydroxy-1-(hydroxymethyl)ethyl] and 4-[1-hydroxy-1-[(phenylthio)methylethyl] analogues (Fig. 8), which were chemically derived from KAIN, inhibited the stimulation of Na^+ fluxes induced by KAIN and NMDA in rat brain slices and were effective in protecting mice from picrotoxin-induced convulsions when administered intracerebroventricularly (Goldberg and Teichberg 1985). These bicyclic δ-lactones of KAIN, however, only inhibited the stimulation of Na^+ fluxes induced by KAIN and NMDA in rat brain slices at considerably high concentrations (Goldberg *et al.* 1983). Bovine serum albumin conjugated with kainylaminooxyacetylglycine has been reported as a novel ligand of KAIN receptors in the chick, goldfish, and rat brain (Eshhar *et al.* 1989). In addition to these experimental results, various actions of KAIN and its derivatives, such as their receptor binding affinity, their ability to elevate cyclic GMP concentration, and their inhibitory actions on uptake of EAAs, have been well documented (Watkins and Evans 1981, Slevin *et al.* 1982, 1983, Conway *et al.* 1984, Takeuchi *et al.* 1984, Anand *et al.* 1986a, 1986b).

KAINATE-LIKE ACTIONS OF NON-KAINOIDS

It is reasonable to assume that an acyclic transmitter molecule has different conformations and is capable of fitting to different types of receptors. The Glu molecule is relatively flexible (see Chapter 10), but the interaction of Glu with its receptors is not well understood with respect to their preference for particular conformers of the Glu molecule. 2-(Carboxycyclopropyl)glycine (CCG) is a conformationally restricted Glu analogue with eight stereoisomers in which the cyclopropyl group fixes the Glu chain in an extended or folded form (Fowden *et al.* 1969, Kurokawa and Ohfune 1985, Yamanoi *et al.* 1988). The carbon chain of Glu is able to conform almost completely with that of CCG. CCG isomers have provided useful information about the relationship between the conformation of the Glu molecule and activation of Glu receptor subtypes (Fig. 11). In the newborn rat spinal motor neurons, the eight CCG stereoisomers demonstrated a large variety of depolarizing activities (Shinozaki *et al.* 1989b, 1989c). The (2*S*,3*S*,4*S*) and (2*S*,3*R*,4*R*) isomers of CCG (L-CCG-I and L-CCG-II, respectively) activated metabotropic EAA receptors, although L-CCG-II was much less potent than L-CCG-I. The depolarization induced by L-CCG-I was not affected by selective NMDA receptor antagonists or CNQX, suggesting that L-CCG-I is neither a NMDA- nor non-NMDA-type agonist (Shinozaki *et al.* 1989c). L-CCG-I induced oscillatory chloride responses in *Xenopus* oocytes injected with mRNA obtained from rat brain (Ishida *et al.* 1990) and increased phosphoinositide

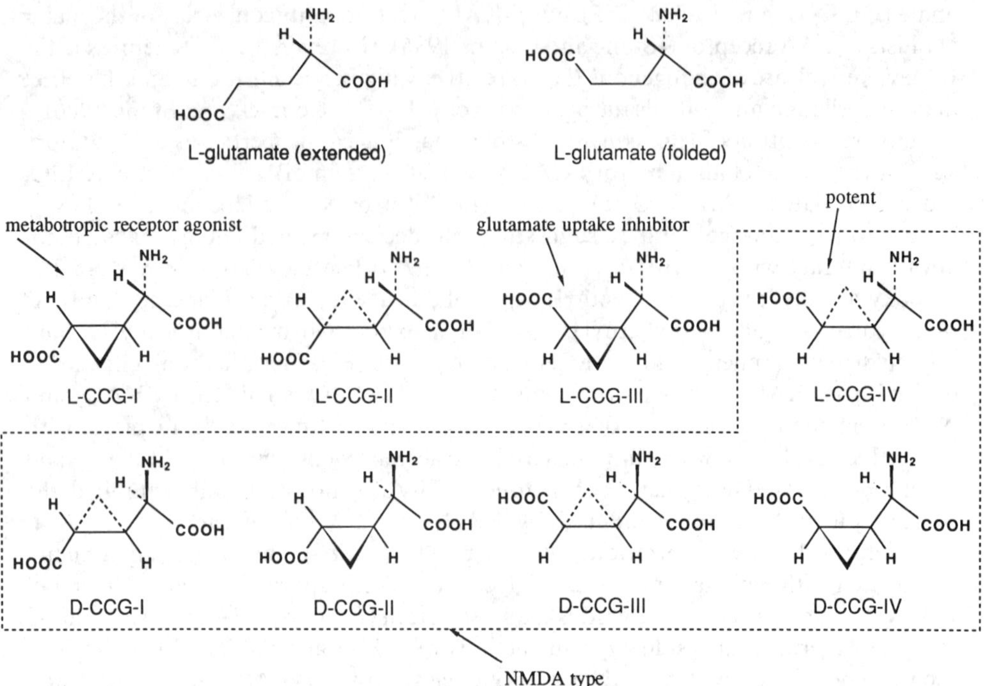

Fig. 11 — Chemical structures of 2-(carboxycyclopropyl)glycine (CCG) and its structural isomers and their pharmacological properties in the newborn rat spinal cord preparation. With respect to the detailed pharmacological actions and depolarizing activity, see Shinozaki *et al.* (1989c).

hydrolysis in rat hippocampal cells (Nakagawa *et al.* 1990), showing it to be a potent metabotropic EAA receptor agonist. L-CCG-I is almost equal in potency to (1*S*,3*R*)-1-amino-cyclopentyl-1,3-dicarboxylate (1*S*,3*R*-ACPD) in causing depolarization of newborn rat spinal motor neurons and increasing phosphoinositide hydrolysis. Also the (2*S*,3*R*,4*S*) isomer of CCG (L-CCG-IV) exhibited significant depolarizing potency, and it was superior to NMDA (more than five times as active) in the newborn rat spinal cord preparation (Shinozaki *et al.* 1988b, 1989b, 1989c, 1990, Ishida and Shinozaki 1990) and the depolarization induced by L-CCG-IV was effectively depressed by selective NMDA antagonists. It has been known that Ca^{2+} influx is increased through activation of NMDA receptors (MacDermott *et al.* 1986). Like NMDA, L-CCG-IV significantly activated influx of extracellular Ca^{2+} in the mammalian central neuron and its potency to cause an increase in intracellular Ca^{2+} concentration, $[Ca^{2+}]_i$, was about 100 times higher than that of NMDA and about ten times higher than that of Glu in cultured rat hippocampal neurons (Shinozaki 1990). KAIN and QUIS also produced large increases in $[Ca^{2+}]_i$ that were abolished by removal of external Ca^{2+} (Murphy and Miller 1989), but the increase in $[Ca^{2+}]_i$ produced by L-CCG-IV was effectively blocked by selective NMDA antagonists in a concentration-dependent manner, suggesting that the increase in $[Ca^{2+}]_i$ was through activation of NMDA receptors. These experimental results have been supported by receptor binding assay (Table 2) (Kawai *et al.* 1992). It should be noted

Table 2 — The binding ability of CCG isomers to various EAA receptors in rat cortical neurons

Compound	IC$_{50}$ (μM)			
	[^{3}H]CPP	[^{3}H]AMPA	[^{3}H]KAIN	[^{3}H]Glycine
Glu	0.327	0.426	0.362	> 1000
NMDA	15.0			
QUIS	21.7	0.020	0.131	
KAIN	327	8.89	0.0096	
L-CCG-1	53.3	> 100	> 100	> 100
L-CCG-II	> 100	> 100	> 100	> 100
L-CCG-III	> 100	> 100	26.1	> 100
L-CCG-IV	0.019	2.37	1.67	> 100
D-CCG-I	14.2	48.1	36.1	> 100
D-CCG-II	0.412	> 100	> 100	> 100
D-CCG-III	76.9	> 100	> 100	> 100
D-CCG-IV	0.215	29.4	42.1	> 100

Data from Kawai *et al.* (1992).

that L-CCG-IV demonstrated a powerful binding to the NMDA receptor. The structure of L-CCG-IV closely resembles the folded conformation of Glu, and the carbon chain of Glu is able to conform almost completely with that of L-CCG-IV. From these observations, the active conformation of Glu at the NMDA-type receptor is strongly suggested to be a folded form. Despite the fact that the (2*S*,3*S*,4*R*) isomer (L-CCG-III) of CCG has a folded structure similar to L-CCG-IV and its configurations at C3 and C4 are the only differences between them, the actions of L-CCG-III are in striking contrast to those of L-CCG-IV (Shinozaki *et al.* 1989c). L-CCG-III markedly enhanced the responses to some EAAs, in particular Glu and Asp, for a period of several hours, with much lower depolarizing activity (NMDA-type) than L-CCG-IV in the newborn rat spinal motor neuron. Therefore, it seems likely that the cyclopropyl group of CCG isomers exerts significant influence upon their neuropharmacological properties.

The C6-substituted L-CCG-IV derivatives shown in Fig. 12 were much more potent than Glu but less potent than L-CCG-IV or NMDA in causing depolarization of the newborn rat spinal motor neuron (Ishida *et al.* 1991). (6*R*)-MOM-CCG was about twice as potent as (6*S*)-MOM-CCG while (6*R*)-BOM-CCG was about ten times less potent than (6*S*)-MOM-CCG and about 20 times less potent than L-CCG-IV. The rank order of depolarizing activity in the newborn rat motor neuron was: QUIS > AMPA > KAIN = L-CCG-IV > NMDA > (6*R*)-MOM-CCG > (6*S*)-MOM-CCG > (6*R*)-BOM-CCG > Glu. Responses to L-CCG-IV and its (6*S*)-MOM derivative were effectively blocked by CPP in a concentration-dependent manner, being completely depressed by high concentrations of CPP. On the other hand, responses to (6*R*)-substituted CCG derivatives were almost insensitive to

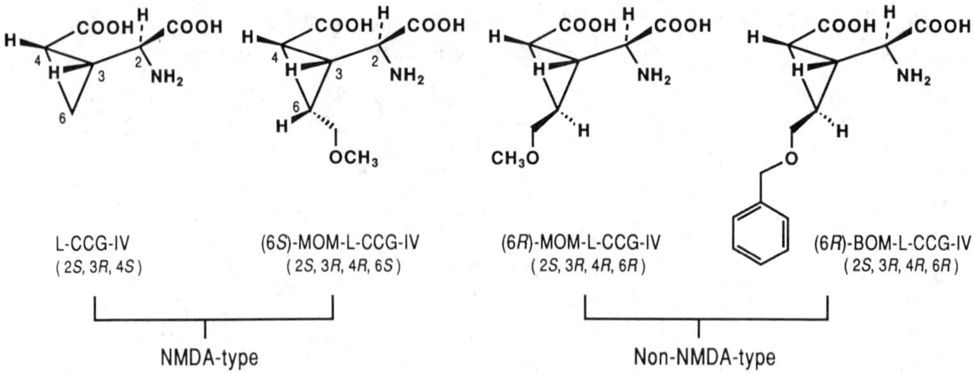

L-CCG-IV (2*S*, 3*R*, 4*S*)

(6*S*)-MOM-L-CCG-IV (2*S*, 3*R*, 4*R*, 6*S*)

(6*R*)-MOM-L-CCG-IV (2*S*, 3*R*, 4*R*, 6*R*)

(6*R*)-BOM-L-CCG-IV (2*S*, 3*R*, 4*R*, 6*R*)

NMDA-type

Non-NMDA-type

Fig. 12 — Structures of C6-substituted L-CCG-IV derivatives and their receptor preference for Glu receptor subtypes.

relatively high concentrations of CPP, but quite sensitive to CNQX. Responses to (6*R*)-BOM-CCG were slightly decreased by a high concentration of CPP, but their residual amplitudes in the presence of high concentrations of CPP were completely depressed by the addition of low concentrations of CNQX. These results suggest that (6*R*)-substituted CCG derivatives prefer to bind to non-NMDA-type receptors in the rat spinal motor neuron, unlike the parent compound L-CCG-IV and its (6*S*) derivatives. Although (6*R*)-substitution of L-CCG-IV led to a considerable loss of depolarizing potency, the receptor preference was converted from the NMDA-type to the non-NMDA-type. According to experiments in the dorsal root C-fibre, which is depolarized selectively by KAIN derivatives, the (6*R*)-substituted CCG derivatives seem to be KAIN-type agonists (see below) (Ishida *et al.* 1991). This change of receptor preference is of great interest from the view-point of design of new agonists and antagonists.

KAIN RECEPTORS ON THE C-FIBRE

The dorsal root isolated from immature rats is useful for differentiating KAIN agonists from other non-NMDA agonists. Glu and almost all KAIN derivatives shown in Fig. 9 cause depolarization of the dorsal root in a concentration-dependent manner (Fig. 13), and there was no significant difference in the slope of dose–response curves, except for the case of Glu responses, probably because of a marked development of receptor desensitization. DOMO demonstrated a particularly high depolarizing activity, while AMPA was much less active than KAIN, and NMDA did not cause depolarization of the dorsal root even at high concentrations. The kainoids tested here had a very short time-course of depolarizing responses, and it was quite easy to determine their relative potency because of the similar slopes of dose–response curves. The rank order of depolarizing activities in the dorsal root fibres was considerably different from that in spinal motor neurons, being as follows: DOMO (34) > ACRO B (13) > 5-Br-willardiine (5.9) > MFPA (5.0) > ACRO A (1.7) > HFPA (1.3) > KAIN (1.0) > CPPA (0.8) > HMPPA (0.5) > CNOPA (0.3) > FPA (0.2) > MKPA (0.1) ≏ Glu (0.1) > QUIS (0.03) > 3-Benzyl-Glu (0.02) > AMPA

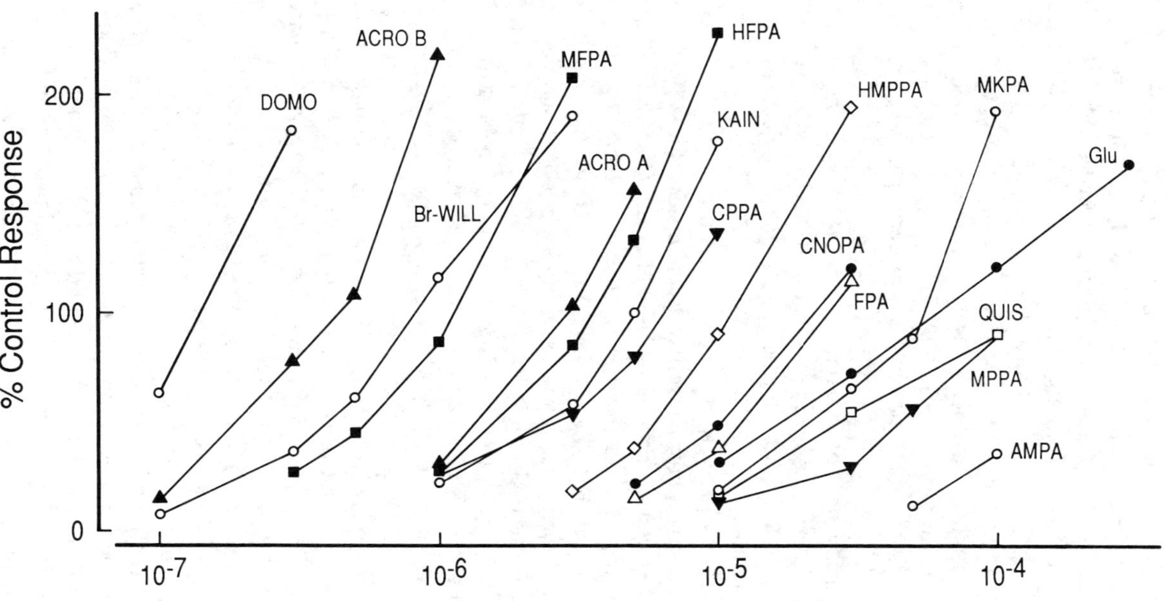

Fig. 13 — Depolarization–concentration relationships for various KAIN derivatives in the newborn rat spinal dorsal root. The central end of dorsal roots was placed inside the suction electrode, which was positioned 2–3 mm from the peripheral end, which was bathed with Ringer and drug solutions. Responses from individual preparations have been normalized, so that results are expressed as a percentage of the control depolarization to KAIN (5 μM) (n at least 5). Abbreviations are the same as those in Fig. 9.

(0.01) > dihydro KAIN > L-α-*N*-acetyl KAIN. The figures in parentheses represent the potency relative to KAIN (5 μM). Both DOMO and MFPA were quite sensitive to CNQX in the rat dorsal root and their pharmacological properties seemed to be almost similar except for their relative potency of depolarizing activity.

It seems likely that there are some differences in neuropharmacological properties of KAIN receptors between motor neurons and C-fibres (Evans *et al.* 1987). Although these differences has not been examined satisfactorily, analysis of them could provide useful information for elucidating the pharmacological properties of KAIN receptors in the mammalian CNS (Davies *et al.* 1979). One of these differences is the extent of the development of receptor desensitization observed at the C-fibre, in striking constrast to the rat spinal motor neuron. After development of receptor desensitization induced by prolonged application of KAIN to the dorsal roots, GABA still causes a marked depolarization, but additional KAIN, AMPA, and QUIS do not cause any further depolarization. Crossed desensitization was observed between KAIN, Glu, ACRO A, DOMO, and MFPA (Fig. 14). On the other hand, in the rat spinal motor neuron desensitization hardly developed even after prolonged application of high concentrations of KAIN. After prolonged application of KAIN to the spinal motor neurons, amplitudes of the ACRO-induced depolarization did not decrease but rather increased in an additive manner.

(6*R*)-Substituted L-CCG-IV derivatives, which caused non-NMDA-type depolarization in the newborn rat spinal motor neuron, caused considerable depolarization of dorsal roots in relatively low concentrations. (6*R*)-MOM- and (6*R*)-BOM-CCG were of almost equal potency in the dorsal root. On the other hand, the (6*S*)-MOM derivative was much less potent than AMPA. L-CCG-IV also was a weak excitant in the dorsal root and its depolarizing activity was slightly lower than that of Glu despite the fact that L-CCG-IV was one of the most potent NMDA-type excitants in newborn rat spinal motor neurons. L-CCG-III also caused a slight depolarization at high concentrations, but other CCGs did not cause any appreciable depolarization even at high concentrations. The depolarization of the dorsal root caused by (6*R*)-substituted CCG derivatives and L-CCG-IV was effectively reduced by CNQX, in a manner similar to that of QUIS, AMPA, DOMO and Glu, but was not affected by CPP. GABA is known to cause significant picrotoxin-sensitive depolarization of the dorsal roots (Agrawal and Evans 1986), but picrotoxin did not affect the depolarization induced by (6*R*)-substituted CCG derivatives. The rank order of the depolarizing activity of CCG-related compounds in the rat dorsal root was as follows: DOMO (34) > KAIN (1.0) > (6*R*)-BOM-CCG (0.39) ≏ (6*R*)-MOM-CCG (0.26) > Glu (0.11) > L-CCG-IV (0.04) ≏ QUIS (0.03) > AMPA (0.01) ⩾ (6*S*)-MOM-CCG > NMDA. The figures in parentheses represent the potency relative to KAIN (5 μM).

(6*R*)-Substituted CCG derivatives do not contain the chemical moiety of KAIN but, nevertheless, activate KAIN receptors in the rat dorsal root. In no other case do non-kainoids cause appreciable depolarization of the dorsal roots, except for Glu and 5-Br-willardiine (Agrawal and Evans 1986). Willardiine has a structure quite similar to QUIS and prefers to activate AMPA receptors (see Chapter 10). 5-Br-willardiine is about fives times more potent than KAIN in causing depolarization of the dorsal root (see Fig. 13) where willardiine does not cause depolarization. With regard to receptor desensitization, there is an apparent difference between KAIN and 5-Br-willardiine. Responses to 5-Br-willardiine and Glu showed similar rapid desensitization to AMPA

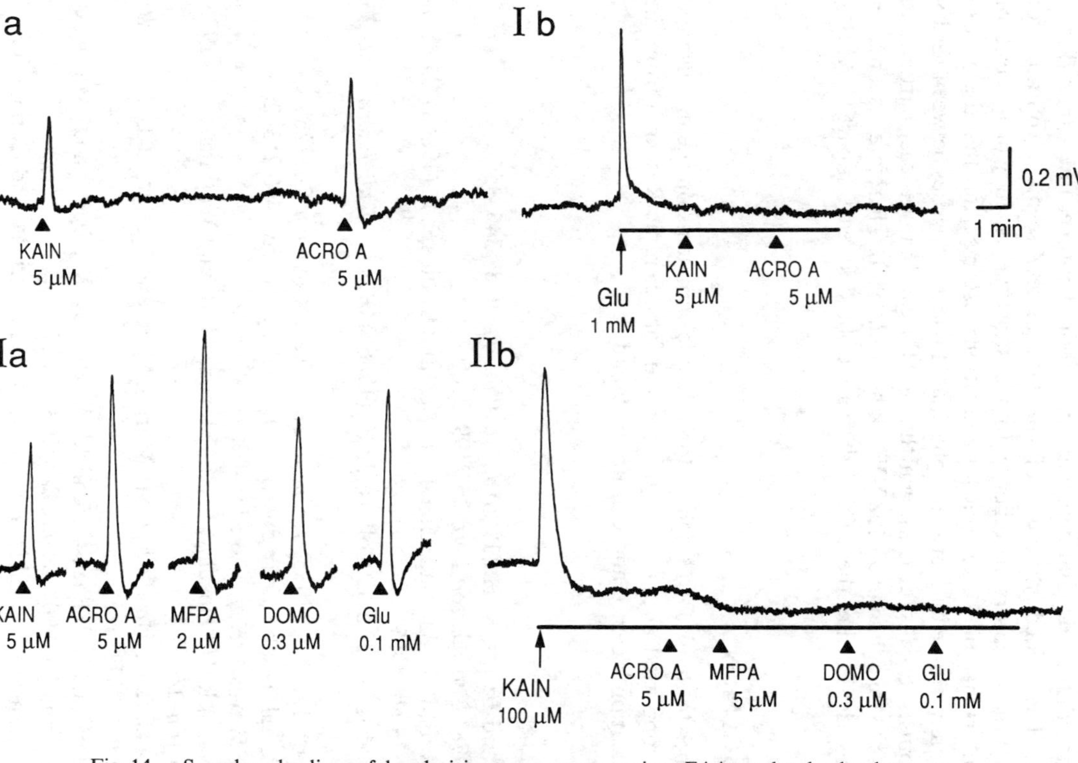

Fig. 14 — Sample redordings of depolarizing responses to various EAAs under the development of receptor desensitization in the newborn rat dorsal root. Numerals under each trace represent the concentration. Ia: control responses to KAIN and ACRO A (10-s application). Ib: Effects of prolonged application of Glu on depolarizing responses to KAIN and ACRO A (10-s application). Glu was applied for a period indicated by the horizontal bar. IIa: Control responses (10-s application). IIb: ACRO A, MFPA, DOMO and Glu (10-s application) did not cause any depolarization of the dorsal root during prolonged application of KAIN.

and QUIS in cultured embryonic hippocampal neurons, but KAIN and DOMO did not develop appreciable desensitization (Mayer and Vyklicky 1989). In addition, neither 5-Br-willardiine nor (6*R*)-CCG derivatives were potent displacers of high-affinity [^{3}H]KAIN binding to rat spinal cord synaptic membranes (Kwak *et al.* unpublished). It is of great interest that (6*R*)-substituted CCG derivatives stimulate KAIN receptors more markedly than does the parent compound L-CCG-IV in the dorsal root, although their depolarizing activity is about ten times lower than that of 5-Br-willardiine. The folded structure of the Glu moiety seems favourable for activation of KAIN receptors together with C6 substituents, in which steric or stereoelectric factors may play a key role for activation. It seems likely that these CCG derivatives are intermediate in their neuropharmacological properties between KAIN and AMPA agonists, because it is known that KAIN can bind at least partially to AMPA receptors (Monaghan *et al.* 1989, Watkins *et al.* 1990). Therefore, as a matter of course, it is not denied that these CCG derivatives bind at least in part to AMPA receptors.

ACKNOWLEDGEMENT

The author wishes to thank Dr M. Ishida for valuable comments on the manuscript. This work was supported in part by a grant-in-aid for Scientific Research from the Ministry of Education, Science and Culture of Japan.

REFERENCES

Agrawal, S. G. and Evans, R. H. (1986) *Br. J. Pharmacol.* **87** 345–355.

Allan, R. D. (1978) *Tetrahedron Lett.* **25** 2199–2200.

Anand, H., Roberts, P. J., Badman, G., Dixon, A. J. and Collins, J. F. (1986a) *Biochem. Pharmacol.* **35** 409–415.

Anand, H., Roberts, P. J. and Collins, J. F. (1986b) *Biochemical. Pharmacol.* **35**, 1055–1057.

Baldwin, J. E. and Li, C.-S. (1988) *J. Chem. Soc. Chem. Commun.* 261–263.

Baran, H., Hortnagl, H. and Hornykiewicz, O. (1989) *Brain res.* **495** 253–260.

Ben-Ari, Y. (1985) *Neuroscience* **14** 375–403.

Biscoe, T. J., Evans, R. H., Headley, P. M., Martin, M. R. and Watkins, J. C. (1975) *Nature* **255** 166–167.

Biscoe, T. J., Evans, R. H., Headley, P. M., Martin, M. R. and Watkins, J. C. (1976) *Br. J. Pharmacol.* **58** 373–382.

Boulter, J., Hollmann, M., O'Shea-Greenfield, A., Hartley, M., Deneris, D., Maron, C. and Heinemann, S. (1990) *Science* **249** 1033–1037.

Bridges, R. J., Stevens, D. R., Kahle, J. S., Nunn, P. B., Kadri, M. and Cotman, C. W. (1989) *J. Neurosci.* **9** 2073–2079.

Brown, A. W. and Brierly, I. B. (1972) *J. Neurol. Sci.* **16** 59–54.

Choi, D. W. (1987) *J. Neurosci.* **7** 369–379.

Choi, D. W. (1988a) *Trends Neurosci.* **11** 465–469.

Choi, D. W. (1988b) *Neuron* **1** 623–634.

Choi, D. W. and Rothman, S. M. (1990) *Annu. Rev. Neurosci.* **13** 171–182.

Collins, R. C., McLean, M. and Olney, J. W. (1980) *Life Sci.* **27** 855–862.

Collins, J. F., Dixon, A. J., Badman, G., de Sarro, G., Chapman, A. G., Hart, G. P. and Meldrum, B. S. (1984) *Neurosci. Lett.* **51**, 371–376.

Connick, J. H. and Stone, T. W. (1986) *Biochem. Pharmacol.* **35** 3631–3635.

Connway, G. A., Park, J. S., Maggiora, L., Mertes, M. P., Galton, N. and Michaelis, E. K. (1984) *J. Med. Chem.* **27** 52–56.

Coyle, J. T., McGeer, E. G., McGeer, P. L. and Schwarcz, R. (1978) In: McGeer, E. G., Olney, J. W. and McGeer, P. L. (eds) *Kainic Acid as a Tool in Neurobiology.* Raven Press, New York, pp. 139–159.

Davies, J. and Watkins, J. C. (1979) *J. Physiol.* **297**, 621–636.

Davies, J., Evans, R. H., Francis, A. A. and Watkins, J. C. (1979) *J. Physiol.* (Paris) **75**, 641–654.

Debonnel, G., Beauchesne, L. and de Montigny, C. (1989a) *Can. J. Physiol. Pharmacol.* **67** 29–33.

Debonnel, G., Weiss, M. and de Montigny, C. (1989b) *Can. J. Physiol. Pharmacol.* **67** 904–908.

Eshhar, N., Lederkremer, G., Beaujean, M., Goldberg, O., Gregor, P., Ortega, A., Triller, A. and Teichberg, V. I. (1989) *Brain Res.* **476** 57–70.

Evans, R. H., Evans, S. J. Pook, P. C. and Sunter, D. C. (1987) *Br. J. Pharmacol.* **91**, 531–537.

Fletcher, E. J., Martin, D., Aram, J. A., Lodge, D. and Honoré, T. (1988) *Br. J. Pharmacol.* **95** 585–597.

Foster, A. C., Mena, E. E., Monaghan, D. T. and Cotman, C. W. (1981) *Nature* **289** 73–75.

Fowden, L., Smith, A., Millington, D. S. and Sheppard, R. C. (1969) *Phytochemistry* **8**, 437–443.

Fushiya, S., Sato, S., Kanazawa, T., Kusano, G. and Nozoe, S. (1990) *Tetrahedron Lett.* **31** 3901–3904.

Gelfan, S. and Tarlov, I. M. (1959) *J. Physiol.* **146**, 594–617.

Gelfan, S. and Tarlov, I. M. (1963) *Am. J. Physiol.* **205** 606–616.

Glavin, G. B., Pinsky, C. and Bose, R. (1990) *Brain Res. Bull.* **24** 701–703.

Goldberg, O. and Teichberg, V. I. (1985) *J. Med. Chem.* **28** 1947–1958.

Goldberg, O., Luini, A. and Teichberg, V. I. (1983) *J. Med. Chem.* **26** 39–42.

Grimes, L. M., Earnhardt, T. S., Mitchell, C. L., Tilson, H. A. and Hong, J. S. (1990) *Brain Res.* **514** 167–170.

Hadzovic, S. (1970) *Jugoslav. Physiol. Pharmacol. Acta* **6** 97–100.

Hashimoto, K. and Shirahama, H. (1990) In: Lubec, G. and Rosenthal, G. A. (eds) *Amino Acids: Chemistry, Biology and Medicine.* ESCOM, Leiden, pp. 566–572.

Hashimoto, K., Konno, K., Shirahama, H. and Matsumoto, T. (1986) *Chem. Lett.* 1399–1400.

Hirayama, F., Konno, K., Shirahama, H. and Matsumoto, T. (1989) *Phytochemistry* **28** 1133–1135.

Hollmann, M., O'Shea-Greenfield, A., Rogers, S. W. and Heinemann, S. (1989) *Nature* **342** 643–648.

Howell, D. A., Lees, A. J. and Toghill, P. J. (1979) *J. Neurol. Neurosurg. Psychiatr.* **42** 733–785.

Huettner, J. E. (1990) *Neuron* **5** 255–266.

Iino, M., Ozawa, S. and Tsuzuki, K. (1990) *J. Physiol.* **424** 151–165.

Ishida, M. and Shinozaki, H. (1988) *Brain Res.* **474** 386–389.

Ishida, M. and Shinozaki, H. (1990) In: Lubec, G. and Rosenthal, G. A. (eds) *Amino Acids: Chemistry, Biology and Medicine.* ESCOM, Leiden, pp. 398–409.

Ishida, M. and Shinozaki, H. (1991) *Br. J. Pharmacol.* **104** 873–878.

Ishida, M., Akagi, H., Shimamoto, K., Ohfune, Y. and Shinozaki, H. (1990) *Brain Res.* **537** 311–314.

Ishida, M., Ohfune, Y., Shimada, Y., Shimamoto, K. and Shinozaki, H. (1991) *Brain Res.* **550** 152–156.

Kasperek, S. and Zebrowski, S. (1971) *Arch. Neurol.* **24** 22–30.

Katz, B. (1966) *Nerve, Muscle, and Synapse.* McGraw-Hill, New York.

Kawai, M., Horikawa, Y., Ishihara, T., Shimamoto, K. and Ohfune, Y. (1992) *Eur. J. Pharmacol.* (in press).

King, A. E. and Wheal, H. V. (1984) *Eur. J. Pharmacol.* **102** 129–134.

Konno, K., Hayano, K., Shirahama, H., Saito, H. and Matsumoto, T. (1982) *Tetrahedron* **38** 3281–3284.

Konno, K., Shirahama, H. and Matsumoto, T. (1983) *Tetrahedron Lett.* **21** 939–942.

Konno, K., Shirahama, H. and Matsumoto, T. (1984) *Phytochemistry* **23**, 1003–1006.

Konno, K., Hashimoto, K., Ohfune, Y., Shirahama, H. and Matsumoto, T. (1986) *Tetrahedron Lett.* **27** 607–610.

Konno, K., Hashimoto, K., Ohfune, Y., Shirahama, H. and Matsumoto, T. (1988) *J. Am. Chem. Soc.* **110** 4807–4815

Kozikowski, A. P. and Fauq, A. H. (1990) *Tetrahedron Lett.* **31** 2967–2970.

Kwak, S., Aizawa, H., Ishida, M., Gotoh, Y. and Shinozaki, H. (1990) In: Lubec G. and Rosenthal, G. A. (eds) *Amino Acids, Chemistry: Biology and Medicine.* ESCOM, Leiden, pp. 318–326.

Kwak, S., Aizawa, H., Ishida, M. and Shinozaki, H. (1992a) *Exp. Neurol.* (in press).

Kwak, S., Aizawa, H., Ishida, M. and Shinozaki, H. (1992b) *Neurosci. Lett.* (in press).

Kurokawa, N. and Ohfune, Y. (1985) *Tetrahedron Lett.* **26** 83–84.

Lassman, H., Petsche, V., Kitz, K., Baran, G. and Sperk, G. (1984) *Neuroscience* **13** 691–704.

Laster, R. A., Quarum, M. L., Parker, J. D., Weber, E. and Jahr, C. E. (1989) *Mol. Pharmacol.* **35** 565–570.

Le Gal La Salle, G. (1988) *Neurosci. Lett.* **88** 127–130.

Lothman, E. W. and Collins, R. C. (1981) *Brain Res.* **218** 299–318.

MacDermott, A., Mayer, M. L., Westbrook, G. L., Smith, S. J. and Barker, J. C. (1986) *Nature* **321** 519–522.

Maeda, M., Kodama, T., Saito, M., Tanaka, T., Yoshizumi, H., Nomoto, K. and Fujita, T. (1987) *Pestic. Biochem. Physiol.* **28** 85–92.

Martinelli, P., Pazzaglia, P., Montagna, P., Coccagna, G., Rizzuto, N., Simonati, S. and Lugaresi, E. (1978) *J. Neurol. Neurosurg.* **41** 458–462.

Maruyama, M. and Takeda, K. (1989) *Brain Res.* **504** 328–331.

Mathers, D. A. (1981) *J. Physiol.* **312** 1–8.

Mathers, D. A. and Usherwood, P. N. R. (1976) *Nature* **259** 409–411.

Mayer, M. L. and Vyklicky, L. Jr, (1989) *Proc. Nat. Acad. Sci. USA* **86** 1411–1415.

McGeer, E. G. and Zhu, S. G. (1990) *Neurosci. Lett.* **112**, 348–351.

McGeer, E. G., Olney, J. W. and McGeer, P. L. (1978) *Kainic Acid as a Tool in Neurobiology.* Raven Press, New York.

Monaghan, D. T. and Cotman, C. W. (1982) *Brain Res.* **252**, 91–100.

Monaghan, D. T., Bridges, R. J. and Cotman, C. W. (1989) *Annu. Rev. Pharmacol. Toxicol.* **29** 365–402.
Murphy, S. N. and Miller, R. J. (1989) *J. Pharmacol. Exp. Ther.* **249** 184–193.
Nadler, J. V. (1981) *Life Sci.* **29** 2031–2042.
Nadler, J. V., Perry, B. W. and Cotman, C. W. (1978) *Nature* **271** 676–677.
Nakagawa, Y., Saitoh, K., Ishihara, T., Ishida, M. and Shinozaki, H. (1990) *Eur. J. Pharmacol.* **184** 205–206.
Nakajima, T., Daval, J., Gleiter, C. H., Deckert, J., Post, R. M. and Marangos, P. (1989) *Epilepsy Res* **4** 156–159.
Nicoletti, F., Wroblewski, J. T., Novelli, A., Alho, H., Guidotti, A. Costa, E. (1986) *J. Nurosci.* **6** 1905–1911.
Ohfune, Y. and Tomita, M. (1982) *J. Am. Chem. Soc.* **104** 3511–3518.
Olney, J. W. (1978) In: McGeer, E. G., Olney, J. W. and McGeer, P. L. (eds) *Kainic Acid as a Tool in Neurobiology*. Raven Press, New York, pp. 95–121.
Olney, J. W. (1990) *Annu. Rev. Pharmacol. Toxicol.* **30** 47–71.
Olney, J. W., Rhee, V. and Ho, O. L. (1974) *Brain Res.* **77** 507–512.
Onodera, K. and Takeuchi, A. (1980) *J. Physiol.* **306** 233–250.
O'Shaughnessy, D. and Herber, G. J. (1986) *Neurotoxicology* **7** 187–202.
Patneau, D. K. and Mayer, M. L. (1990) *J. Neurosci.* **10** 2385–2399.
Perouansky, M. and Brantyn, R. (1989) *J. Neurosci.* **9** 70–80.
Perl, T. M., Bedard, L., Kosatssky, T., Hockin, J. C., Todd, E. C. D. and Remis, R. S. (1990) *N. Engl. J. Med.* **322** 1775–1780.
Popovici, T., Barbin, G. and Ben-Ari, Y. (1988) *Eur. J. Pharmacol.* **150** 405–406.
Randle, J. C. (1990) *Can. J. Physiol. Pharmacol.* **68** 1069–1078.
Rao, S. L. N. (1978) *J. Neurochem.* **30** 1467–1470.
Rao, S. L. N., Adiga, P. R. and Sarma, P. S. (1964) *Biochemistry* **3**, 432–436.
Rogers, B. C. and Tilson, H. A. (1990) *Neurosci. Lett.* **109** 335–340.
Rondouin, G., Lerner-Natori, M. and Hashizume, A. (1987) *Exp. Neurol.* **95** 500–505.
Schwob, J. E., Fuller, T., Price, J. L. and Olney, J. W. (1980) *Neuroscience* **5** 991–1014.
Sheardown, M. J., Nielsen, E. O. Hansen, A. J., Jacobsen, P. and Honoré (1990) *Science* **247** 571–574.
Shinozaki, H. (1978) In: McGeer, E. G., Olney, J. W. and McGeer, P. L. (eds) *Kainic Acid as a Tool in Neurobiology*. Raven Press, New York, pp. 17–35.
Shinozaki, H. (1980) *Prog. Neurobiol.* **14** 121–155.
Shinozaki, H. (1988) *Prog. Neurobiol.* **30** 399–435.
Shinozaki, H. (1990) *Jap. J. Pharmacol.* **52** 39P.
Shinozaki, H. and Ishida, M. (1976) *Brain Res.* **109** 435–439.
Shinozaki, H. and Ishida, M. (1979) *Brain Res.* **161** 493–501.
Shinozaki, H. and Ishida, M. (1981) *J. Pharm. Dyn.* **4** 42–48.
Shinozaki, H. and Ishida, M. (1988a) In: Lunt, G. G. (eds) *Neurotox '88: Molecular Basis of Drugs and Pesticide Action*. Elsevier, Amsterdam, pp. 91–104.
Shinozaki, H. and Ishida, M. (1988b) In: Kanazawa, I. (ed) *Neurotransmitters: Focus on Excitatory Amino Acids*. Excerpta Medica, Tokyo, pp. 17–30.
Shinozaki, H. and Konishi, S. (1970) *Brain Res.* **24** 368–371.
Shinozaki, H. and Shibuya, I. (1974a) *Neuropharmacology* **13** 665–672.
Shinozaki, H. and Shibuya, I. (1974b) *Neuropharmacology* **13** 1057–1065.
Shinozaki, H. and Shibuya, I. (1976) *Neuropharmacology* **15** 145–147.

Shinozaki, H., Ishida, M. and Okamoto, T. (1986) *Brain Res.* **399** 395–398.

Shinozaki, H., Ishida, M., Gotoh, Y. and Kwak, S. (1989a) *Brain Res.* **503** 330–333.

Shinozaki, H., Ishida, M., Shimamoto, K. and Ohfune, Y. (1989b) *Brain Res.* **480** 355–359.

Shinozaki, H., Ishida, M., Shimamoto, K. and Ohfune, Y. (1989c) *Br. J. Pharmacol.* **98** 1213–1224.

Shinozaki, H., Ishida, M., Gotoh, Y. and Kwak, S. (1990) In: Lubec, G. and Rosenthal, G. A. (eds) *Amino Acids: Chemistry, Biology and Medicine.* ESCOM, Leiden, pp. 281–293.

Shinozaki, H., Ishida, M., Kwak, S. and Nakajima, T. (1991) In: Conn, M. (ed) *Methods in Neuroscience.* Academic Press, New York, pp. 38–57.

Shirahama, H., Konno, K., Hashimoto, K. and Matsumoto, T. (1988) In: Lunt, G. G. (ed) *Neurotox '88: Molecular Basis of Drugs and Pesticide Action.* Elsevier, Amsterdam, pp. 105–122.

Shirahama, H., Hashimoto, K., Yanagida, M. and Hirokawa, M. (1990) In: Lubec, G. and Rosenthal, G. A. (eds) *Amino Acids: Chemistry, Biology and Medicine.* ESCOM, Leiden, pp. 39–46.

Sladeczek, F., Pin, J. P., Recasens, M., Bockaert, J. and Weiss, S. (1985) *Nature* **317** 717–719.

Slevin, J. T., Collins, J. F., Lindsley, K. and Coyle, J. T. (1982) *Brain Res.* **249** 353–360.

Slevin, J. T., Collins, J. F. and Coyle, J. T. (1983) *Brain Res.* **265** 169–172.

Sloviter, R. S. and Damiano, B. P. (1981) *Neuropharmacology* **20** 1003–1011.

Smith, A. L. and Mallhinney, R. A. J. (1992) *Br. J. Pharmacol.* **105** 83–86.

Sommer, B., Keinanen, K., Verdoorn, T. A., Wisden, W., Bunashev, N., Herb, A., Kohler, M., Takagi, T., Sakmann, B. and Seeberg, P. H. (1990) *Science* **249** 1580–1585.

Spencer, P. S., Schaumburg, H. H., Cohn, D. F. and Seth, P. K. (1984) In: Rose, F. C. (ed) *Research Progress in Motor Neuron Disease.* Pitman, London, pp. 312–327.

Spender, P. S., Nunn, P. B., Hugon, J., Ludolph, A. C., Ross, S. M., Roy, D. N. and Robertson, R. C. (1987) *Science* **237** 517–522.

Stone, T. W. (1985) *Microiontophoresis and Pressure Ejection.* IBRO Handbook Series, Vol. 8. Wiley, Chichester.

Stone, T. W. and Collins, J. F. (1986) *Neuroscience* **17** 629–633.

Takano, S., Iwaguchi, Y. and Ogasawara, K. (1987) *J. Am. Chem. Soc.* **109** 5523–5524.

Takano, S., Tomita, S., Iwaguchi, Y. and Ogasawara, K. (1989) *Heterocycles* **29** 1473–1476.

Takeuchi, A. and Onodera, K. (1975) *Neuropharmacology* **14** 619–625.

Takeuchi, A. and Takeuchi, N. (1964) *J. Physiol.* **170** 269–317.

Takeuchi, H., Watanabe, K., Nomoto, K., Ohfune, Y. and Takemoto, T. (1984) *Eur. J. Pharmacol.* **102** 325–332.

Teitelbaum, J. S., Zatotte, R. J., Carpenter, S., Gendron, D., Evans, A. C., Gjedde, A. and Cash, N. R. (1990) *N. Engl. J. Med.* **322** 1781–1787.

Tryphonas, L. and Iverson, R. (1990) *Toxicol. Pathol.* **18** 165–169.

Turski, L., Meldrum, B. S. and Collins, J. F. (1985) *Brain Res.* **336** 162–166.

Turski, L., Klockgether, T., Schwarz, M., Sontag, K. H. and Meldrum, B. S. (1987) *Neuroscience* **20** 285–292.

Umbach, J. A. and Gundersen, C. B. (1989) *Mol. Pharmacol.* **36** 582–588.

Watkins, J. C. (1981) In: Roberts, P. J., Storm-Mathisen, J. and Johnstone, G. A. R. (eds) *Glutamate: Transmitter in the Central Nervous System.* Wiley, New York, pp. 1–24.

Watkins, J. C. and Evans, R. H. (1981) *Annu. Rev. Pharmacol. Toxicol.* **21** 165–204.

Watkins, J. C., Krogsgaard-Larsen, P. and Honoré, T. (1990) *Trends Pharmacol. Sci.* **11** 25–33.

Worm, P., Willigen, M.-T. and Lloyd, K. G. (1981) *Life Sci.* **29** 2215–2225.

Wright, J. L. L., Falk, M., McInnes, A. and Walter, J. A. (1990) *Can. J. Chem.* **68** 22–25.

Yamanoi, K., Ohfune, Y., Watanabe, K., Li, P. N. and Takeuchi, H. (1988) *Tetrahedron Lett.* **29** 1181–1184.

Yanagida, M., Hashimoto, K., Ishida, M., Shinozaki, H. and Shirahama, H. (1989) *Tetrahedron Lett.* **30** 3799–3802.

Young, A. B. (1990) *Ann. Neurol.* **28** 9–11.

13

Polyamine toxins as excitatory amino acid receptor ligands

Nobufumi Kawai
Department of Physiology, Jichi Medical School, Minamikawachi-machi Kawachi-gun, Tochigi-ken, 329-04 Japan

ABSTRACT

A variety of toxins which affect the excitatory amino acid receptors have been found in the venoms of spiders and wasps. In the venoms of Araneid spiders, Joro spider toxin (JSTX) from *Nephila clavata*, NSTX from *Nephila maculata*, and argiopin (argiotoxin) from *Argiope lobata* act postsynaptically on quisqualate-type receptors. These toxins share a common structure of a phenolic moiety connected to a polyamine. Purified natural toxins and chemically synthesized spider toxins have been used in the characterization of the excitatory amino acid receptors. Labelling of synthesized JSTX-3 has enabled morphological and biochemical studies of the glutamate receptors in crustacean muscle and mammalian brain. In the wasp venom, another antagonist for the glutamate receptor has been found. One component, δ-philanthotoxin, contains, similarly to the spider toxins, a polyamine chain in the structure. The structure–activity relationships of the synthesized analogues of these spider and wasp toxins have served in the elucidation of the function of the ionotropic excitatory amino acid receptors.

INTRODUCTION

Toxins from natural sources have been valuable tools for characterizing receptors and ion channel proteins. The venoms of arthropods such as wasps and spiders have attracted interest in the field of neurobiology because of their potent biological actions (Tu 1984, Geren 1986, Kawai and Nakajima 1990). Since wasps and spiders catch and paralyse insects as their prey, they are expected to possess neuroactive substances in their venoms. In spiders, until the 1980s, the venoms of only a few species such as black widow spiders and Sydney funnel web spiders were known to possess neuroactive toxins. Extensive studies were carried out on black widow spider (*Latrodectus mactans*) venom, revealing a unique presynaptic effect (Clark *et al.* 1972). The main component of the venom, α-latrotoxin, is a high-molecular-weight (ca. 130 000 Da) protein which acts on the nerve terminal, causing massive release of transmitter by opening cation channels in the presynaptic membrane (Hurlbut and Ceccarelli 1979).

In the early 1980s, a novel type of neurotoxin from 'Joro' spider, *Nephila clavata*, was found (Kawai *et al.* 1982). The toxin acts postsynaptically on glutamate receptors, blocking synaptic transmission in invertebrates (Abe *et al.* 1983, Kawai *et al.* 1983b, Saito *et al.* 1985a, Miwa *et al.* 1987) and mammalian brain (Akaike *et al.* 1987, Saito *et al.* 1985b, 1989). Chemical characterization of the *Nephila* toxins (JSTXs, NSTXs) was carried out by Nakajima and colleagues (Aramaki *et al.* 1986), and established a variety of related compounds with a unique structure containing a phenolic group connected to a polyamine. Toxins which block glutamate receptors have subsequently been found in other Araneid spiders, such as *Argiope lobata*, *Argiope trifasciata* and *Araneus gemma* ((Michaelis *et al.* 1984, Usherwood *et al.* 1984, Grishin *et al.* 1986, Adams *et al.* 1987). So far, toxins from more than 20 species of orb-web spiders have been isolated (Jackson and Usherwood 1988, Jackson and Parks 1989).

As another source for polyamine toxins, some solitary wasps of Hymenoptera contain a number of components which have specific actions on the glutamate receptors (Piek and Spanjer 1986). The active component of *Philanthus trangulum* venom was isolated and named philanthotoxin 433 (Eldefrawi *et al.* 1988).

This chapter describes recent studies of the polyamine toxins from natural sources and structure–activity relationships of synthesized toxins with the aim of understanding the function of the excitatory amino acid receptors.

TOXINS OF *NEPHILA CLAVATA* — JSTXs

In the venom of the spider *Nephila clavata* which inhabits Japan and East Asia, a group of closely homologous toxic substances was found (Kawai *et al.* 1982). The toxins, collectively named Joro spider toxin (JSTX), turned out to be of low molecular weight and to act postsynaptically to block the glutamate receptors (Abe *et al.* 1983, Kawai *et al.* 1983b, Aramaki *et al.* 1986). The blocking mode of JSTX was studied using the neuromuscular synapses of spiny lobster (*Palinurus japonicus*), which are convenient for studying glutamatergic transmission, since excitatory postsynaptic potentials (EPSPs) can be evoked stably for hours by stimulation of a single axon (Kawai and Niwa 1982) and are thus well suited for testing the toxin activity. Purified JSTX irreversibly blocks EPSPs without affecting inhibitory postsynaptic potentials (IPSPs). No recovery from the action of JSTX was found over more than 10 hours, indicating that the toxin has a practically irreversible action. The EC_{50} of JSTX for suppression of the EPSP has been estimated to be as low as 10 nM (Abe *et al.* 1983, Kawai *et al.* 1984). JSTX has no effect on the presynaptic action potential or on the resting conductance of the postsynaptic membrane. Pharmacological studies of the effect of JSTX were done using iontophoretic application of a variety of agonists which selectively activate subtypes of glutamate receptors (Watkins and Evans 1981, Shinozaki 1989, Collingridge and Lester 1990). JSTX was found preferentially to block quisqualate-type receptors, while aspartate-induced responses were insensitive to the toxin. Using macro patch-clamp recording from the synaptic sites, excitatory postsynaptic currents (EPSCs) were recorded to further analyse the action of JSTX (Miwa *et al.* 1987). The peak amplitude of EPSCs increased linearly with hyperpolarization in the presence as well as in the absence of JSTX, suggesting that the toxin has a voltage-independent action. Furthermore, in the presence of JSTX the decay phase of EPSCs could be fitted with a single exponential function, with no biphasic components observed over a wide range of membrane potentials. The time constants for the decay phases of miniature EPSCs were slightly reduced in the presence of JSTX but the rate of reduction by hyperpolarization was virtually the same with or without toxin. These results suggest that the active site of the toxin is remote from the ionic channels subject to membrane potential changes.

The effect of JSTX was also examined on the squid giant synapse, another invertebrate glutamatergic synapse (Miledi 1967). In this preparation, simultaneous intracellular recordings were easily feasible from pre- and postsynaptic axons. JSTX blocked the postsynaptic potentials while the presynaptic and antidromic action potentials remained unaltered, showing that voltage-dependent Na^+ channels are left intact (Kawai *et al.* 1983b). Spontaneous miniature EPSPs in the postsynaptic membrane were blocked by JSTX, a further indication that the action of JSTX is confined to the postsynaptic membrane (Saito *et al.* 1985a). A pharmacological study of the effects of JSTX on agonist-induced responses in the squid giant synapse gave similar results to those obtained at the lobster neuromuscular synapse.

The effect of JSTX on mammalian brain has been studied in the guinea-pig hippocampal slice (Saito *et al.* 1985b, 1989). Pressure-applied JSTX blocked EPSPs in CA1 pyramidal neurons in response to stimulation of Schaffer collateral/commissural fibres, but had no effect on the antidromic action potentials evoked by alveus stimulation. JSTX suppressed both EPSPs and iontophoretically induced glutamate potentials in a CA1 pyramidal neuron in a similar manner. Agonist-induced potentials showed that JSTX blocked glutamate preferentially to aspartate, and quisqualate preferentially to *N*-methyl-D-aspartate (NMDA). Using the single electrode voltage-clamp method, Sahara *et al.* (1988, 1991) showed that EPSCs in some pyramidal neurons are composed of NMDA- as well as non-NMDA-type currents. By analysis of current–voltage relationships, it was shown that JSTX blocked a component of the EPSC which increased linearly with hyperpolarization but failed to block a non-linear-type current (NMDA-type). The effect of JSTX on the behaviour of mice was studied by direct injection of JSTX-3 into the ventricles (Himi *et al.* 1990a, 1990b). They reported that JSTX-3 at a dose of 4.7 nmol per brain, which did not produce any abnormality *per se*, prevents convulsions induced by quisqualate. Injection of JSTX-2 (22.2 pmol per brain) into the lateral ventricles inhibited memory retrieval in a step-through test, but had no effect on the acquisition or consolidation of memory.

STRUCTURE–ACTIVITY RELATIONSHIP OF JSTXs

The chemical structures of the JSTXs were determined by Nakajima and colleagues (Aramaki *et al.* 1986, 1987) (Fig. 1). The JSTXs proved to consist of several closely similar compounds. Most notably, a 2,4-dihydroxyphenylacetyl asparaginyl cadaverine moiety connected to a polyamine is common to all these toxins. Following the characterization of the structures, a major component of Joro spider toxin, JSTX-3, was synthesized (Hashimoto *et al.* 1987). The blocking activity of the synthetic spider toxin on glutamate receptors was confirmed at the lobster neuromuscular synapse

JSTX-1

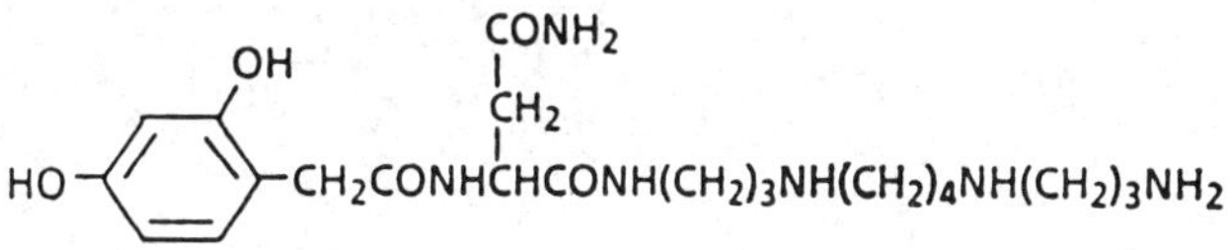

$CONH_2$ | CH_2 | (on the CH)

HO–(ring, OH)–$CH_2CONHCHCONH(CH_2)_3NH(CH_2)_4NH(CH_2)_3NH_2$

JSTX-2

NH_2 | $(CH_2)_3$ | ; $CONH_2$ | CH_2 |

HO–(ring, OH)– $CH_2CONHCHCONHCHCONH(CH_2)_5NHCO(CH_2)_2NH(CH_2)_4NHCO(CH_2)_2NH(CH_2)_4NH_2$

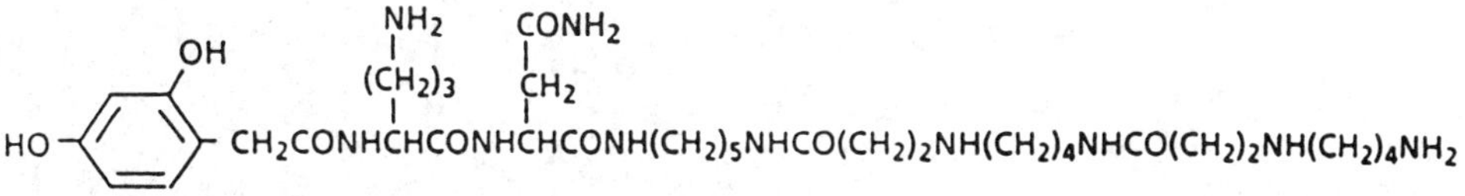

JSTX-3

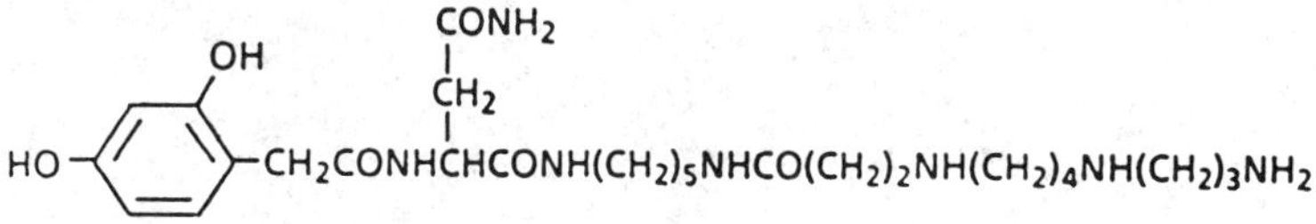

$CONH_2$ | CH_2 |

HO–(ring, OH)–$CH_2CONHCHCONH(CH_2)_5NHCO(CH_2)_2NH(CH_2)_4NH(CH_2)_3NH_2$

JSTX-4

$CONH_2$ | CH_2 | ; NH_2 | $(CH_2)_3$ | ; NH_2 | ; NH ||

HO–(ring, OH)–$CH_2CONHCHCONH(CH_2)_5NHCOCHNHCOCH(CH_2)_3NHCNH_2$

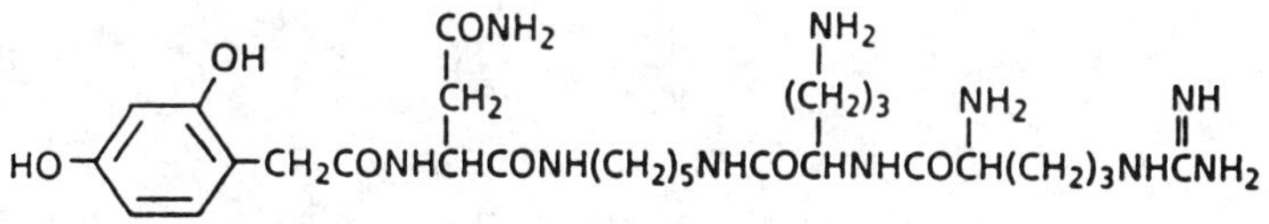

Fig. 1 — Structure of JSTXs.

(Shudo *et al.* 1987). The structure–activity relationship was studied using a range of other synthesized analogues, and Fig. 2 compares the suppressive activities of synthesized analogues of JSTX-3. 2,4-Dihydroxyphenylacetic acid (DHP) (1) and *N*-(2,4-dihydroxyphenylacetyl)asparagine (DHP-Asn) (2) have no suppressive action. Connection of a polyamine chain to DHP-Asn produces the suppressive activity. *N*-(2,4-Dihydroxyphenylacetyl)asparaginyl cadaverine (3) and *N*-(2,4-dihydroxyphenyl-acetyl)asparaginyl spermine (4) caused suppression with potencies of approximately 0.01 and 0.1 times that of JSTX-3 (5).

Fig. 3 shows the effect of synthetic analogues compared to that of JSTX-3 on the EPSPs of the lobster neuromuscular synapse. Following application of C-1 (No. 3 in Fig. 2), ESPSs were gradually suppressed, while the resting potential and the conductance of the postsynaptic membrane remained unaffected. The amplitude of EPSPs was reduced to approximately 30% of the control 5 min after the application. When the preparation was washed, EPSPs recovered slowly and complete recovery was obtained after washing for 20 minutes. C-2 (No. 4 in Fig. 2) gave rise to a more potent suppression of EPSPs than C-1. After C-2 was applied, EPSPs were rapidly suppressed, then completely blocked within a few minutes. The recovery of EPSPs from C-2 effects was very slow. At 30 minutes after C-2 application, the amplitude of EPSPs was only 5% of the control and at 70 min, 17% of the control. Complete

(1) HO–C_6H_3(OH)–CH_2COOH

(2) HO–C_6H_3(OH)–$CH_2CONHCH(CH_2CONH_2)COOH$

(3) HO–C_6H_3(OH)–$CH_2CONHCH(CH_2CONH_2)CONH(CH_2)_5NH_2$

(4) HO–C_6H_3(OH)–$CH_2CONHCH(CH_2CONH_2)CONH(CH_2)_3NH(CH_2)_4NH(CH_2)_3NH_2$

(5) HO–C_6H_3(OH)–$CH_2CONHCH(CH_2CONH_2)CONH(CH_2)_5NHCO(CH_2)_2NH(CH_2)_4NH(CH_2)_3NH_2$

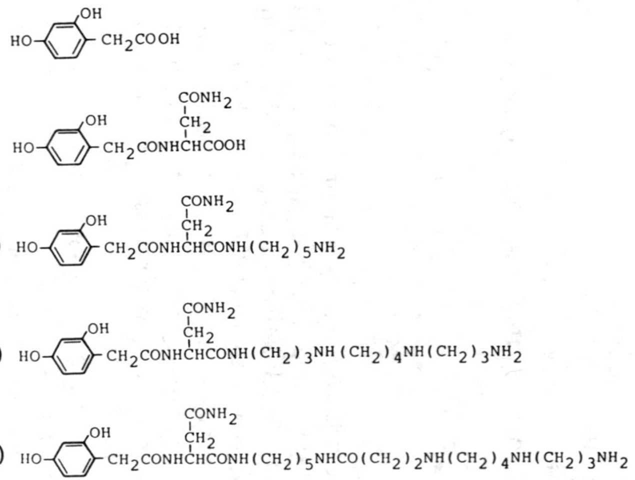

No.	Activity	Effect
(1)	—	—
(2)	—	—
(3) = C 1	0.01	R
(4) = C 2	0.1	R
(5) = JSTX	1	IR

Fig. 2 — Structure–activity relationship of JSTX and its analogues. Activity indicates the relative potency of the compounds on suppression of EPSPs in the lobster neuromuscular synapse. R, reversible effects; IR, irreversible effect.

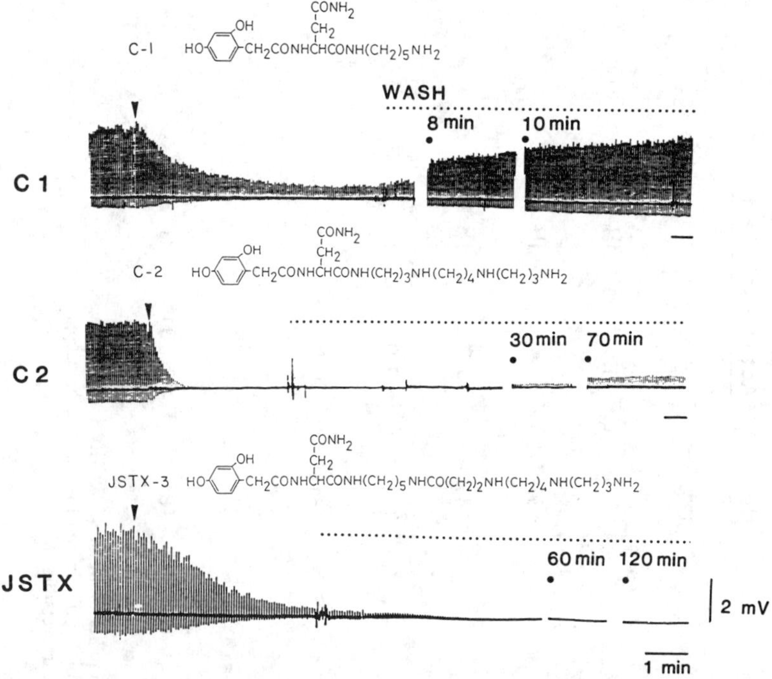

Fig. 3 — Comparison of the effects of synthesized analogues (C1 and C2) and natural spider toxin (JSTX) on EPSPs in the lobster neuromuscular synapse. Concentrations were 8.3 μM (C1), 22 μM (C2) and 0.02 μM (JSTX). Numbers in minutes are the time after applying the toxins as indicated by arrows.

recovery could not be obtained even after washing the preparation for 2 hours. From these results, it can be deduced that the polyamine part acts to enhance the toxic activity. Analogues in which aromatic or aliphatic compounds replaced the (2,4-dihydroxyphenylacetyl)asparaginyl part in JSTX-3 still showed some suppressive action similar to JSTX but with less potency (Asami *et al.* 1989).

STRUCTURE–ACTIVITY RELATIONSHIP OF NSTXs

A number of toxins which are structurally similar to JSTX were isolated from the venom of a Papua New Guinean spider (*Nephila maculata*) and named NSTXs (Aramaki *et al.* 1986). A major component of the venom, NSTX-3, was synthesized (Teshima *et al.* 1987), and the structure–activity relationship of NSTX-3 and synthesized analogues were studied (Teshima *et al.* 1990). The blocking activity on EPSPs of the lobster neuromuscular synapse was compared for seven analogues of NSTX-3 (Fig. 4). Asparaginyl-cadaverinyl-putranine (Asp-Cad-Pua) (1) was entirely inactive. A des-arginine derivative of NSTX-3, i.e. (2,4-dihydroxyphenylacetyl)Asn-Cad-Pua (2), showed an irreversible block of EPSPs with a relative potency one-tenth of NSTX-3. However, an acetyl derivative (7) of des-Arg-NSTX-3 was extremely weak and reversible in its action. A comparison was made of analogues in which the

	Structure	No.	Activity	Effect
(1)	$NH_2CH(CH_2CONH_2)CONH(CH_2)_5NHCO(CH_2)_2NH(CH_2)_4NHCOCH(NH_2)(CH_2)_3NHC(NH_2)=NH$	(1)	—	—
(2)	$HO-C_6H_3(OH)-CH_2CONHCH(CH_2CONH_2)CONH(CH_2)_5NHCO(CH_2)_2NH(CH_2)_4NH_2$	(2)	0.1	IR
(3)	$HO-C_6H_3(OH)-CH_2CONHCH(CH_2CONH_2)CONH(CH_2)_5NHCO(CH_2)_2NH(CH_2)_4NHCOCH(NH_2)(CH_2)_3NHC(NH_2)=NH$	(3)=NSTX	1	IR
(4)	$HO-C_6H_3(OH)-CH_2CONHCH(CH_2CONH_2)CONH(CH_2)_5NHCO(CH_2)_2NH(CH_2)_4NHCOCH(NH_2)(CH_2)_4NH_2$	(4)	0.2	IR
(5)	$HO-C_6H_3(OH)-CH_2CONHCH(CH_2CONH_2)CONH(CH_2)_5NHCO(CH_2)_2NH(CH_2)_4NHCOCH(NH_2)CH_3$	(5)	0.05	IR
(6)	$HO-C_6H_3(OH)-CH_2CONHCH(CH_2CONH_2)CONH(CH_2)_5NHCO(CH_2)_2NH(CH_2)_4NHCOCH(NH_2)CH_2COOH$	(6)	>0.001	R
(7)	$HO-C_6H_3(OH)-CH_2CONHCH(CH_2CONH_2)CONH(CH_2)_5NHCO(CH_2)_2NH(CH_2)_4NHCOCH_3$	(7)	>0.001	R
(8)	$HO-C_6H_3(OH)-CH_2CONHCH(CH_2CONH_2)CONH(CH_2)_5NHCO(CH_2)_2NH(CH_2)_4NHCOCH(NH_2)(CH_2)_3NHC(NH_2)-NHCOCH_3$	(8)	>0.001	R

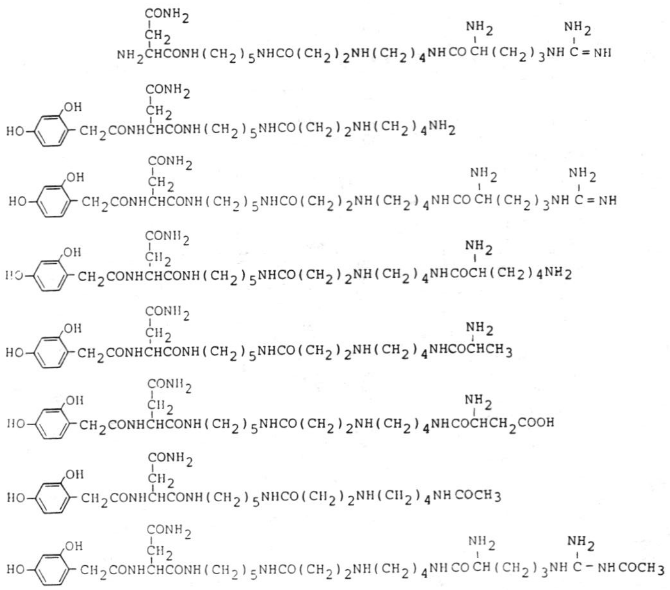

Fig. 4 — Structure–activity relationship of NSTX and its analogues. Activity indicates the relative potency of the compounds on suppression of EPSPs in the lobster neuromuscular synapse. R, reversible effect; IR, irreversible effect.

arginine of NSTX-3 was replaced with acidic (aspartic acid), neutral (alanine) and basic (lysine) amino acids. The alanine (5) and lysine (4) derivative showed irreversible suppression of EPSPs with relative potencies of 0.05 and 0.2 of NSTX-3, respectively, while the aspartic acid derivative (6) showed much weaker activity. These results suggest that a positive charge at the position of Arg in the structure of NSTX-3 plays a key role in the suppressive activity. In addition, it appears that the α-amino group in arginine is also a requisite for blocking activity, since the acetyl-arginine derivative (8) which possesses only one positive charge in its guanidino group failed to show suppressive activity.

LABELLED JSTX AND BINDING TO GLUTAMATE RECEPTORS

Labelling of synthesized JSTX-3 has been performed for histological study of glutamate receptors. ^{125}I-labelled JSTX-3 was found to produce an irreversible block on EPSPs in the lobster neuromuscular synapse similarly to the unlabelled toxin (Hagiwara *et al.* 1988). After treatment with [^{125}I]JSTX-3, the distribution of radioactivity on the lobster muscle fibre was observed by autoradiography. Under light microscopic observation, highly concentrated spots were found sparsely distributed on the surface of muscle fibres. Electron microscopy of adjoining ultrathin sections revealed that the spots corresponded to structures characteristic of the synaptic profile, with spherical synaptic vesicles in the nerve terminal. These findings indicate that [^{125}I]JSTX-3 had accumulated exclusively at synaptic areas with negligible radioactivity in non-synaptic regions of muscle fibres. Reconstruction of the radioactive spots on the muscle surface revealed that their distribution pattern was similar to that of the nerve terminals as stained by methylene blue (Shimazaki *et al.* 1988).

Another labelled version of the toxin, biotinylated JSTX-3, was synthesized for staining glutamate receptors in mammalian brain. Using the avidin–biotin complex method, the binding of biotinylated JSTX-3 to rat cerebellum and hippocampus was studied histochemically (Shimazaki *et al.* 1990). In the hippocampus, biotinylated JSTX-3 bound mainly to the pyramidal cells, the stratum oriens and stratum lucidum of the CA3 region. The pyramidal cell layer was more densely stained in the CA3 region than in other regions of Ammon's horn. In the cerebellum, a high concentration of biotinylated JSTX-3 binding was seen in the Purkinje cell and molecular layer, with a much lower concentration in the granular cell layer. In the molecular layers, a high concentration of binding was observed on the dendrites of Purkinje cells (Fig. 5). Thus the distribution of biotinyl-JSTX-3 binding corresponds to the distribution of the quisqualate-preferring subtype of glutamate receptors as assessed electrophysiologically (Ito *et al.* 1982) and biochemically (Foster and Fagg 1984).

Biotin-labelled JSTX-3 was also used for a biochemical study directed towards the purfication of the glutamate receptor (Shimazaki *et al.* 1989a, 1989b). A crude synaptic membrane fraction from rat hippocampus was solubilized by Triton X-100 and incubated with [^{3}H]biotinyl-JSTX-3. Affinity purification of the JSTX binding protein was performed using iminobiotinyl-JSTX-3. At pH 8.5, JSTX-3-bound protein was trapped on an avidin–agarose column by the molecular interaction between iminobiotin and avidin. The trapped proteins eluted with iminobiotinyl-JSTX-3 at pH 4.0 since iminobiotin interacts poorly with avidin at acidic pH. The presence of several

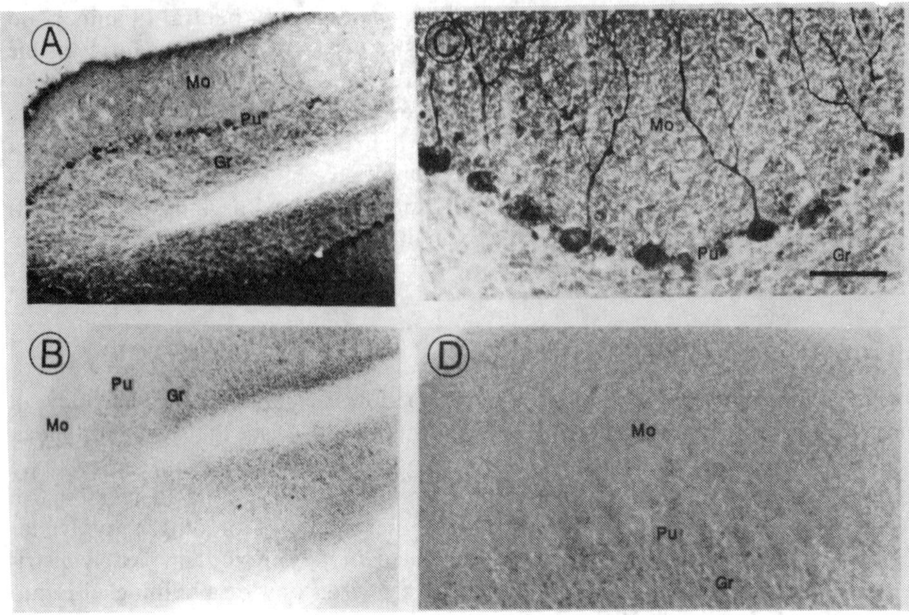

Fig. 5 — Distribution of JSTX binding sites in the rat cerebellum. A: Positive structures after treatment with biotinyl-JSTX-3. B: Enlarged photograph of Purkinje cell layer. C and D: control preparations of A and B, treated with an excess of JSTX-3. Mo, molecular layer; Pu, Purkinje cell layer; Gr, granular cell layer. Bar = 250 μm (A and C) and 50 μm (B and D).

polypeptides which bound to JSTX-3 was demonstrated by SDS–polyacrylamide gel electrophoresis. At least four discrete protein bands visualized by the silver stain method appeared in the lane containing the sample treated with iminobiotinyl JSTX-3.

^{3}H-labelled NSTX was used for characterization of the binding sites for the toxin in rat brain (Ino *et al.* 1990). In the cerebellum, high-affinity ($K_d = 7.75$ nM, $B_{max} = 0.37$ pmol/mg) and low-affinity ($K_d = 202$ nM, $B_{max} = 5.54$ pmol/mg) binding sites were found. The binding was not inhibited by glutamate agonists, suggesting that the binding site is different from the agonist recognition site on the receptor molecule.

TOXINS OF *ARGIOPE*

Antagonists of glutamate receptors similar to the JSTXs have been found in the venom of other spiders of the Araneidae family. Kawai *et al.* (1983a) reported block of lobster neuromuscular synapse by crude venom of *Araneus ventricosus* and *Neoscona nautica.* Tashmukhamedov and colleagues found that a venom-gland extract of *Argiope lobata* blocked glutamate responses in the locust muscle, and at much higher doses inhibited acetylcholine potentials in frog muscle (Tashmukhamedov *et al.* 1985).

The effective components in the venom of *Argiope lobata* was isolated by Grishin *et al.* (1986) and named argiopin (Fig. 6). Usherwood and colleagues reported non-competitive antagonists of the glutamate receptor at the locust neuromuscular

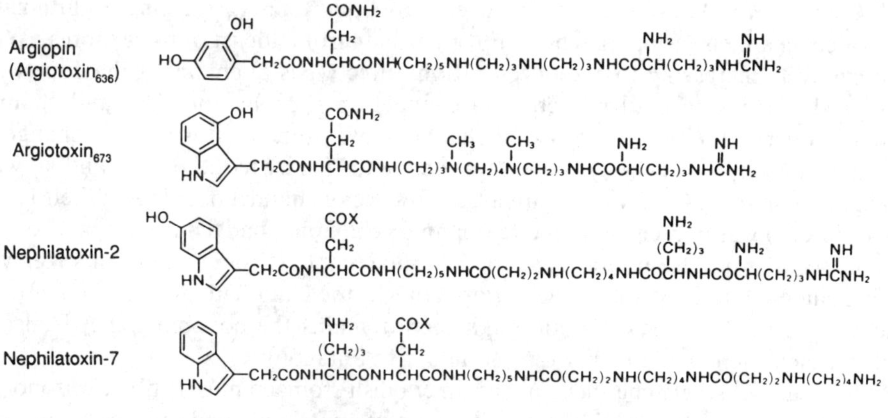

Argiopin (Argiotoxin$_{636}$)

HO–(OH-phenyl)–$CH_2CONHCH(CH_2CONH_2)CONH(CH_2)_5NH(CH_2)_3NH(CH_2)_3NHCOCH(NH_2)(CH_2)_3NHC(=NH)NH_2$

Argiotoxin$_{673}$

(4-OH-indolyl)–$CH_2CONHCH(CH_2CONH_2)CONH(CH_2)_3N(CH_3)(CH_2)_4N(CH_3)(CH_2)_3NHCOCH(NH_2)(CH_2)_3NHC(=NH)NH_2$

Nephilatoxin-2

(HO-indolyl)–$CH_2CONHCH(CH_2COX)CONH(CH_2)_5NHCO(CH_2)_2NH(CH_2)_4NHCOCH((CH_2)_3NH_2)NHCOCH(NH_2)(CH_2)_3NHC(=NH)NH_2$

Nephilatoxin-7

(indolyl)–$CH_2CONHCH((CH_2)_3NH_2)CONHCH(CH_2COX)CONH(CH_2)_5NHCO(CH_2)_2NH(CH_2)_4NHCO(CH_2)_2NH(CH_2)_4NH_2$

Fig. 6 — Structure of argiopin, argiotoxins and nephilatoxins.

synapse in the venoms of *Argiope trifasciata*, *Argiope florida* and *Araneus gemma* (Usherwood *et al.* 1984, Bateman *et al.* 1985, Budd *et al.* 1988). A 636-Da toxin isolated from the venom of a group of *Argiope* spiders, argiotoxin$_{636}$, was found to be identical to argiopin. Argiopin (argiotoxin$_{636}$) shows a large degree of homology with NSTXs and JSTXs in having a chromophore group coupled to asparagine and aliphatic polyamines (Grishin *et al.* 1989). Adams *et al.* (1987) isolated three types of argiotoxins from the venom of *Argiope aurantia.* These types have common features: a hydrophilic, basic domain consisting of arginine, a polyamine and an asparagine connected to an aromatic moiety, either 4-hydroxyindole-3-acetic acid (argiotoxin$_{659}$ and argiotoxin$_{673}$) or 2,4-dihydroxyphenylacetic acid (argiotoxin$_{636}$) (Fig. 6). Some of these toxins were synthesized and confirmed to have a paralytic effect on insects.

Nakajima and colleagues isolated several compounds different from the JSTXs from the venom of *Nephila clavata*, and named them nephilatoxins (Toki *et al.* 1988a, 1988b) (Fig. 6). Nephilatoxins are structurally similar to each other and have a 6-hydroxyindole-3-acetyl or indole-3-acetyl moiety linked to the N-terminus of a polycationic part composed of basic amino acids and polyamines. These toxins induce histamine release from rat peritoneal mast cells and also have an insecticidal action against wriggler larvae of mosquitoes. Nephilatoxins possessing a 6-hydroxyindole moiety had a less potent blocking action on crustacean glutamate receptors than those possessing an indole moiety, though both types of toxins caused an irreversible block.

EFFECT OF ARGIOPIN (ARGIOTOXIN$_{636}$) ON GLUTAMATE-ACTIVATED CURRENTS

A considerable number of electrophysiological studies on the block of glutamate receptors by toxins in *Argiope* (argiopin or argiotoxin$_{636}$) have been reported in locust muscle fibres (Usherwood *et al.* 1984, Bateman *et al.* 1985, Budd *et al.* 1988), blowfly muscle (Grishin *et al.* 1986, Magazanik *et al.* 1987), frog spinal cord (Antov *et al.* 1987), and hippocampal CA1 pyramidal neurons of rat (Ashe *et al.* 1989).

Kerry *et al.* (1988) studied the effects of argiotoxin$_{636}$ on single glutamate-activated channels using patch-clamp recording from locust muscle fibres. With progressively increasing toxin concentration, three types of channel behaviour were observed. Type 1 entailed reduction in channel open probability (P_o) and channel event frequency (f) by an increase in channel closed time (m_c) with little change in mean channel open time (m_o). Type 2 was characterized by reduced P_o, f, and m_o with an increase in m_c. Type 3 was an apparent absence of channel openings. When 10^{-12} M argiotoxin was in the patch pipette, channel behaviour changed from type 1 to type 2 in 60 seconds and from type 2 to type 3 in 120 seconds. From these results Kerry *et al.* concluded that in insect muscle argiopin blocks the glutamate-activated channel at the level of the open channel, although there remains the possibility that it also is either a closed channel blocker or a competitive antagonist.

In a study of single-channel currents in crayfish stomach muscle fibre, Antonov *et al.* (1989) reported that argiopin blocks the glutamate-activated current by at least two mechanisms. One is a potential-dependent block caused by 10^{-7} M or less of argiopin. The channels can be recovered from the block by extreme hyperpolarizations to 190 mV below the resting potential. The potential-dependent block of argiopins does not affect the opening and closing kinetics, but the number of channels available for opening is reduced in a concentration-dependent manner. The second type of block is a flickering block which takes place at an argiopin concentration of 10^{-6} M without any voltage dependence. This block reduces the open time per burst of the channel to about half of the control, with the blocking and unblocking rates in the range of 10^3 per second. The observed potential dependence appeared only with large hyperpolarizations. Since the spider toxin is a polar molecule with the negative group believed to participate in binding, strong hyperpolarization would result in destabilization of the negative binding site.

In cultured rat cortical neurons, argiotoxin$_{636}$ effectively reduced NMDA-induced whole cell currents at hyperpolarized membrane potentials in an agonist- and voltage-dependent manner (Priestley *et al.* 1989) (see chapter 9). The potency of argiotoxin$_{636}$ in blocking NMDA-activated channels was reported to be more than 30 times higher than for quisqualate- and kainate-activated ion channels. In this study, however, contrary to the results in the crayfish muscle described above, membrane hyperpolarization promoted the block of inward currents.

INTERACTION OF SPIDER TOXIN WITH METAL IONS

Some polyamine spider toxins show an unusual interaction with metal ions. Yoshioka *et al.* (1988) reported the development of colour by purified JSTX with Fe^{3+}, Hg^{2+} or Ce^{3+}, possibly as a result of chelating reactions. Another interesting feature is that zinc ions enhance the potency of the blocking action of synthetic analogues of JSTX (Kawai *et al.* 1989). Fig. 7 shows the relative potency of synthesized analogues which contain aza-crowns substituting for the polyamine part. N-2 has no suppressive action on EPSPs even at concentrations as high as 1 mM. N-4, which possesses an N-4 aza-crown type polyamine, produced a moderate suppressive action, which was, however, greatly potentiated by addition of Zn^{2+} (Fig. 8). N-6 aza-crown analogue had a significant suppressive action but showed no further potentiation upon addition of

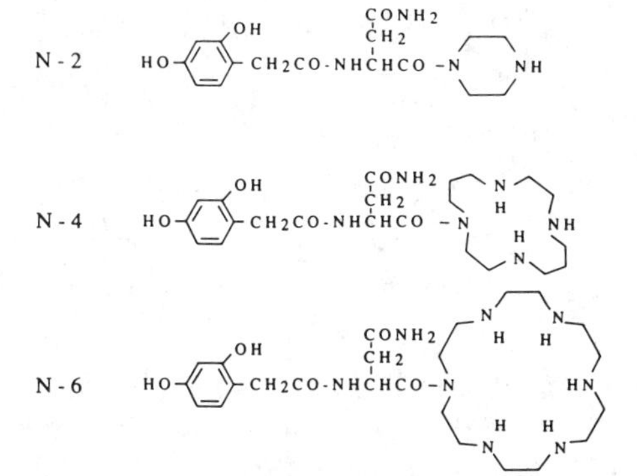

N-2 HO-C6H3(OH)-CH2CO-NHCH(CH2CONH2)CO-N(piperazine)NH

N-4 HO-C6H3(OH)-CH2CO-NHCH(CH2CONH2)CO-N(aza-crown, NH)

N-6 HO-C6H3(OH)-CH2CO-NHCH(CH2CONH2)CO-N(aza-crown, HN)

JSTX-3 HO-C6H3(OH)-CH2CO-NHCH(CH2CONH2)CO-NH(CH2)5NHCO(CH2)2NH(CH2)4NH(CH2)3NH2

	Inhibition (%)	
	alone	$+Zn^{2+}$(100 μM)
N-2 (100 μM)	0	0
N-4 (100 μM)	12	52
N-6 (100 μM)	47	43
JSTX (1 μM)	100	100

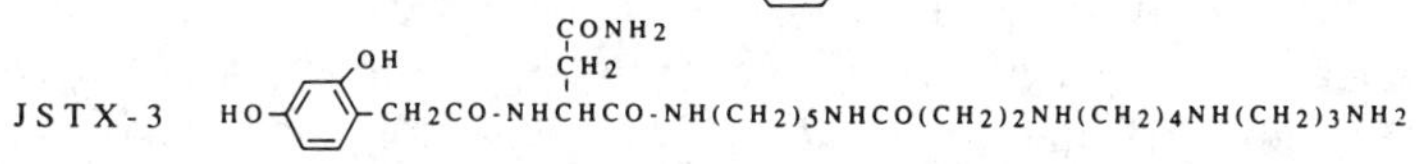

Fig. 7 — Structure of JSTX-3 and its aza-crown analogues. The table shows percentage inhibition of EPSPs in lobster neuromuscular synapse after a single dose of compound (alone) and with 100 μM Zn^{2+}.

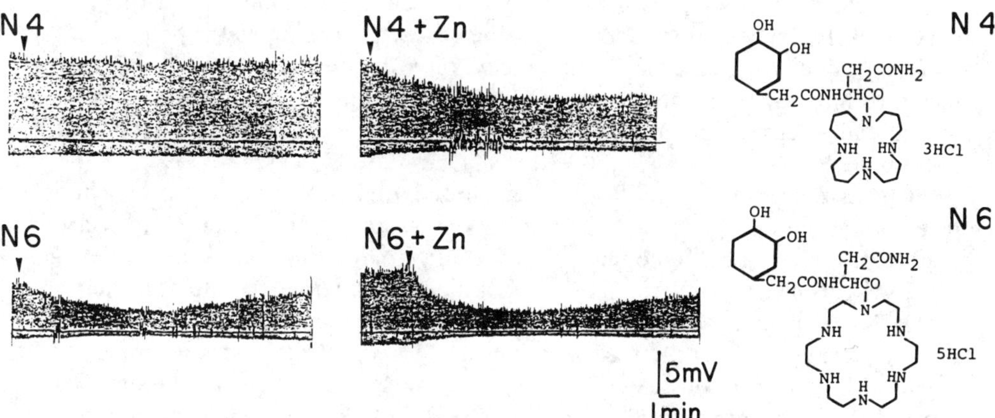

Fig. 8 — Effect of Zn^{2+} in combination with aza-crown analogues of spider toxins on lobster neuromuscular synapse. All compounds were applied at a concentration of 100 μM and applied at the arrow.

Zn^{2+}. Since metal ions regulate the conformation of aza-crown compounds, Zn^{2+} may stabilize a restricted conformation of the N-4 type of spider toxin analogues.

A WASP TOXIN — PHILANTHOTOXIN

The venom of parasitic wasps contains a number of toxins which paralyse insects (Piek *et al.* 1986, 1989). One group of these toxins, collectively named δ-philanthotoxin (δ-PhTX) (Fig. 9), was isolated from the venom of *Philanthus triangulum* (Spanjer *et al.* 1982) and the main component, philanthotoxin-433 (PhTX-433), was

δ-Philanthotoxin

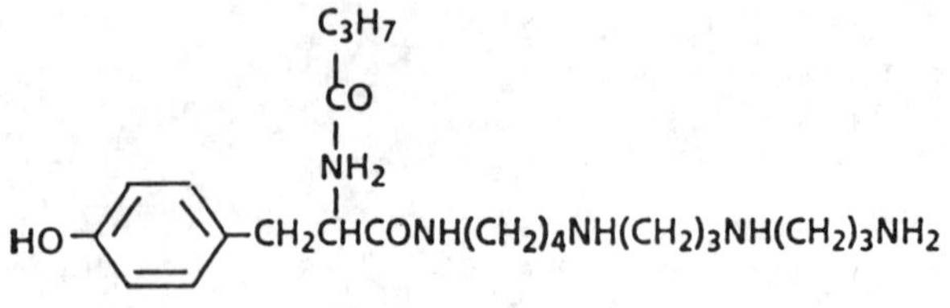

$$\mathrm{HO{-}C_6H_4{-}CH_2CH(NH{-}CO{-}C_3H_7)CONH(CH_2)_4NH(CH_2)_3NH(CH_2)_3NH_2}$$

Fig. 9 — Structure of δ-philanthotoxin (PhTX-433).

structurally characterized (Eldefrawi *et al.* 1988). PhTX-433 (Fig. 9) and two structural analogues (PhTX-334 and PhTX-343) non-competitively antagonize quisqualate receptors in locust muscle (Clark *et al.* 1982, Eldefrawi *et al.* 1988). In central neurons, PhTX suppressed firings induced by quisqualate and kainate, with little effect on NMDA-induced firings (Jones *et al.* 1989), suggesting that this toxin is a selective non-NMDA antagonist.

STRUCTURE-ACTIVITY RELATIONSHIP OF PHILANTHOTOXINS

Using a series of analogues of PhTX, Bruce *et al.* (1990) reported the structure–activity relationships of wasp toxins on a nerve–muscle preparation of locust leg. They concluded that one or more of the following characteristics raises the potency of the blocking action: increased hydrophobicity of aromatic and tyrosyl regions; an increased number of protonated groups in the polyamine region; a guanidine instead of a spermine terminal amino moiety; and incorporation of a butyl side-group in the polyamine.

Synthesized analogues of PhTX were found to inhibit nicotinic acetylcholine receptors as well as glutamate receptors (Anis *et al.* 1990). The structure–activity relationships of 36 analogues were studied by the binding activity of [^{3}H]perhydrohistrionicotoxin to the nicotinic acetylcholine receptor and also by [^{3}H]MK-801 binding to the NMDA receptor. The potencies of the PhTX analogues on both receptors generally rose with increasing lipophilicity and polyamine chain length. Modification of the amine terminal of PhTX affects inhibition of the acetylcholine receptor more than that of the NMDA receptor, whereas the tyrosine moiety was of greater importance for inhibition of the NMDA receptor than of the acetylcholine receptor. The complex nature of PhTX interactions with the two receptors suggests that PhTX may bind to two sites, one being an external polyamine binding site and the other a channel binding site.

REFERENCES

Abe, T., Kawai, N. and Miwa, A. (1983) *J. Physiol.* **339** 243–252.

Adams, M. E., Carney, R. L., Enderlin, F. E., Fu, E. T., Jarema, M. A., Li, J. P., Miller, C. A., Schooley, D. A., Shapiro, M. J. and Venema, V. J. (1987) *Biochem. Biophys. Res. Commun.* **148** 678–683.

Akaike, N., Kawai, N., Kishin, N. I., Kljuchko, E. M., Krishtal, O. A. and Tsyndrenko, A. Y. (1987) *Neurosci. Lett.* **79** 326–330.

Anis, N., Sherby, S., Goodnow, R., Niwa, M., Konno, K., Kallimopoulos, T., Bukownik, R., Nakanishi, K., Usherwood, P. N. R., Eldefrawi, A. and Eldefgrawi, M. (1990) *J. Pharmacol. Exp. Ther.* **254** 764–773.

Antonov, S. M., Grishin, E. V., Magazanik, L. G., Shupliakov, O. V., Vesslkin, N. P. and Volkova, T. M. (1987) *Neurosci. Lett.* **83** 179–184.

Antonov, S. M., Dudel, J., Franke, C. and Hatt, H. (1989) *J. Physiol.* **419** 569–587.

Aramaki, Y., Yasuhara, T., Higashijima, T., Yoshioka, M., Miwa, A., Kawai, N. and Nakajima, T. (1986) *Proc. Japan Acad.* **62 Ser B** 359–362.

Aramaki, Y., Yasuhara, T., Shimazaki, K., Kawai, N. and Nakajima, T. (1987) *Biomed. Res.* **8** 241–245.

Asami, T., Kagechika, H., Hashimoto, Y., Shudo, K., Miwa, A., Kawai, N. and Nakajima, T. (1989). *Biomed. Res.* **10** 185–189.

Ashe, J. H., Cox, C. L. and Adams, M. E. (1989) *Brain Res.* **480** 234–241.

Bateman, A., Boden, P., Dell, A., Duce, I. R., Quick, D. L. J. and Usherwood, P. N. R. (1985) *Brain Res.* **339** 237–244.

Bruce, M., Bukownik, R., Eldefrawi, A. T., Eldefrawi, M. E., Goodnow, R., Jr, Kallimopoulos, T., Konno, K., Nakanishi, K., Niwa, M. and Usherwood, P. N. R. (1990) *Toxicon* **28** 1333–1346.

Budd, T., Clinton, P., Dell, A., Duce, I. R., Johnson, S. J., Quicke, D. L. J., Taylor, G. W., Usherwood, P. N. R. and Usoh, G. (1988) *Brain Res.* **448** 30–39.

Clark, A. W., Hurlbut, W. P. and Mauro, A. (1972) *J. Cell Biol.* **52** 1–14.

Clark, R. B., Donaldson, P. L., Gration, K. A. F., Lambert, J. J., Piek, T., Ramsey, R. L., Spanjer, W. and Usherwood, P. N. R. (1982) *Brain Res.* **21** 105–114.

Collingridge, G. L., and Lester, R. A. J. (1990) *Physiol. Rev.* **40** 145–210.

Eldefrawi, A. T., Eldefrawi, M. E., Konno, K., Mansour, N. A., Makanishi, K., Oltz, E. and Usherwood, P. N. R. (1988) *Proc. Nat. Acad. Sci. USA.* **85** 4910–4913.

Foster, A. C. and Fagg, G. (1984) *Brain Res. Rev.* **7** 103–164.

Geren, C. R. (1986) *J. Toxicol.-Toxin Rev.* **5** 161.

Grishin, E. V., Volkova, T. M., Arseniev, A. S., Reshetova, O. S., Onoprienko, V. V., Magazanik, L. G., Antonov, S. M. and Fedrov, I. M. (1986) *Bioorg. Khim.* **12** 1121–1124.

Grishin, E. V., Volkova, T. M. and Arseniev, A. S. (1989) *Toxicon* **27** 541–549.

Hagiwara, K., Aramaki, Y., Shimazaki, K., Kawai, N. and Nakajima, T. (1988) *Chem. Pharm. Bull.* **36** 1233–1236.

Hashimoto, Y., Yasuhara, T., Endo, Y., Shudo, K., Aramaki, Y., Kawai, N. and Nakajima, T. (1987) *Tetrahedron Lett.* **28** 3511–3514.

Himi, T., Saito, H., Kawai, N. and Nakajima, T. (1990a) *J. Neural Transm.* **80** 95–104.

Himi, T., Saito, H. and Nakajima, T. (1990b) *J. Neural Transm.* **80** 79–89.

Hurlbut, W. P. and Ceccarelli, B. (1979) In: Ceccarelli, B. and Clementi, F. (eds) *Neurotoxins: Tools in Neurobiology*. Raven Press, New York, pp. 87–115.

Ino, H., Nakade, S., Niinobe, M., Ikenaka, K., Teshima, T., Wakamiya, T., Matsumoto, T., Shiba, T., Kawai, N. and Mikoshiba, K. (1990) *Neurosci. Res.* **8** 29–39.

Ito, M. Sakurai, M. and Tongroach, P. (1982) *J. Physiol.* **324** 113–134.

Jackson, H. and Parks, T. N. (1989) *Annu. Rev. Neurosci.* **12** 405–414.

Jackson, H. and Usherwood, P. N. R. (1988) *Trends Neurosci.* **11** 278–283.

Jones, M. G., Anis, N. A. and Lodge, D. (1989) *Neurosci. Lett.* (Suppl.) **36** S.

Kawai, N. and Nakajima, T. (1990) *J. Toxicol.-Toxin Rev.* **5** 203–223.

Kawai, N. and Niwa, A. (1982) *J. Physiol.* **305** 73–85.
Kawai, N., Miwa, A. and Abe, T. (1982) *Brain Res.* **247** 169–171.
Kawai, N., Miwa, A. and Abe, T. (1983a) *Toxicon* **21** 438–440.
Kawai, N., Miwa, A. and Abe, T. (1983b) In: Mandel, P. and DeFeudis, E. V. (eds) *CNS Receptors: From Molecular Pharmacology to Behaviour*. Raven Press, New York, pp. 221–227.
Kawai, N., Yamagishi, S., Saito, M. and Furuya, K. (1983b) *Brain Res.* **278** 346–349.
Kawai, N., Miwa, A., Saito, M., Pan-Hou, H. S. and Yoshioka, M. (1984) *J. Physiol.* (Paris) **79** 228–231.
Kawai, N., Miwa, A., Hashimoto, Y., Shudo, K., Asami, T. and Nakajima, T. (1989) *Neurosci. Res.* **6** 358–362.
Kerry, C. J., Ramsey, R. L., Sansom, M. S. P. and Usherwood, P. N. R. (1988) *Brain Res.* **459** 312–327.
Magazanik, L. G., Antonov, S. M., Federova, I. M., Volkova, T. M. and Grishin, E. V. (1987) In: Ovchinikov and Hucho, F. (eds) *Receptors and Ion Channels.* De Gruyter, New York, pp. 305–312.
Michaelis, E. K., Galton, N. and Earley, S. L. (1984) *Proc. Nat. Acad. Sci. USA* **81** 5571–5574.
Miledi, R. (1967) *J. Physiol.* **192** 379–406.
Miwa, A., Kawai, N., Saito, M., Pan-Hou, H. S. and Yoshioka, M. (1987) *J. Neurophysiol.* **58** 319–326.
Piek, T. and Spanjer, W. (1986) In: Piek, T. (ed.) *Venoms of the Hymenoptera.* Academic Press, London, pp. 161–307.
Piek, T., Karst, H., Kruk, C., Lind, A., van Marle, J., Nakajima, T., Nibbering, N. M. M., Shinozaki, H., Spanjer, W. and Tong, Y. C. (1989) In: *Neurotox '88: Nolecular Basis of Drug and Pesticide Action.* Elsevier, Amsterdam, pp. 61–76.
Priestley, T., Woodruff, G. N. and Kemp, J. A. (1989) *Br. J. Pharmacol.* **97** 1315–1323.
Sahara, Y., Miwa, A. and Kawai, N. (1988) *Soc. Neurosci. Abstr.* **14** 97.
Sahara, Y., Robinson, H. P. C., Miwa, A. and Kawai, N. (1991) *Neurosci. Res.* (in press).
Saito, M., Kawai, N., Miwa, A., Yamagishi, S. and Furuya, K. (1985a) *Neurosci. Res.* **2** 297–307.
Saito, M., Kawai, N., Miwa, A., Pan-Hou, H. S. and Yoshioka, M. (1985b) *Brain Res.* **346** 397–399.
Saito, M., Sahara, Y., Miwa, A., Shimizaki, K., Nakajima, T. and Kawai, N. (1989) *Brain Res.* **481** 16–24.
Shimazaki, K., Hagiwara, K., Hirata, Y., Nakajima, T. and Kawai, N. (1988) *Neurosci. Lett.* **84** 173–177.
Shimazaki, K., Hagiwara, K., Kawai, N. and Nakajima, T. (1989s) *Biomed. Res.* **10** 401–403.
Shimazaki, K., Nakajima, T. and Kawai, N. (1989b) *Soc. Neurosci. Abstr.* **15** 531.
Shimazaki, K., Hirata, Y., Nakajima, T. and Kawai, N. (1990) *Neurosci. Lett.* **114** 1–4.
Shinozaki, H. (1989) *Progr. Neurobiol.* **30** 399–435.
Shudo, K., Endo, Y., Hashimoto, Y., Aramaki, Y., Nakajima, T. and Kawai, N. (1987) *Neurosci. Res.* **5** 82–85.
Spanjer, W., May, T. E., Piek, T. and De Hann, N. (1982) *Comp. Biochem. Physiol.* **71C** 149–157.

Tashmukhamedov, B. A., Makhumudova, E. M., Usmanov, P. B., Kazakov, I. (1985) *Gen. Physiol. Biophys.* **4** 625–630.

Teshima, T., Wakamiya, T., Aramaki, Y., Nakajima, T., Kawai, N. and Shiba, T. (1987) *Tetrahedron Lett.* **28** 3509–3510.

Teshima, T., Matsumoto, T., Wakamiya, T., Shiba, T., Nakajima, T. and Kawai, N. (1990) *Tetrahedron* **46** 3813–3818.

Toki, T., Yasuhara, Y., Aramaki, Y., Kawai, N. and Nakajima, T. (1988a) *Biomed. Res.* **9** 75–79.

Toki, T., Yasuhara, T., Aramaki, Y., Osawa, K., Miwa, A., Kawai, N. and Nakajima, T. (1988b) *Biomed. Res.* **9** 421–428.

Tu, A. T. (1984). In: *Handbook of Natural Toxins*, Vol. 2. Dekker, New York.

Usherwood, P. N. R., Duce, I. and Boden, P. (1984) *J. Physiol.* (Paris) **79** 241–245.

Watkins, J. C. and Evans, R. H. (1981) *Annu. Rev. Pharmacol. Toxicol.* **21** 165–204.

Yoshioka, M., Narai, N., Pan-Hou, H. S., Shimazaki, K., Miwa, A. and Kawai, N. (1988) *Toxicon* **26** 414–416.

14

L-AP4 receptor ligands

James F. Koerner
Department of Biochemistry, Medical School, University of Minnesota, Minneapolis, MN 55455, USA
Rodney L. Johnson
Department of Medicinal Chemistry, College of Pharmacy, University of Minnesota, Minneapolis, MN 55455, USA

ABSTRACT

A subclass of excitatory amino acid receptors are characterized by their sensitivity to micromolar concentrations of L-2-amino-4-phosphonobutanoic acid (L-AP4). To date, L-AP4 receptors have been intensively investigated for their inhibitory actions on a few vertebrate neural pathways, where they act presynaptically, and on retinal ON-bipolar cells, where they act postsynaptically. It has recently been shown that the retinal L-AP4 receptors act via a G protein that is coupled to a cyclic GMP second messenger mechanism of signal transduction. In these cells, L-AP4 mediates closure of monovalent cation channels. Current information on presynaptic L-AP4 receptors is also consistent with a second messenger mechanism in which L-AP4 blocks activation of voltage-sensitive calcium ion channels.

Few known structural analogues of excitatory amino acids mimic the activity and potency of L-AP4. A long-known exception, L-serine-*O*-phosphate, has lower potency. Recently, a few conformationally constrained cyclic analogues of L-AP4 have been synthesized, and preliminary results suggest that this approach may yield useful structure-activity relationships. It appears that L-AP4 and its known structural analogues act as agonists. A pharmacology based on antagonists for the L-AP4 receptor does not yet exist.

The compound L-AP4 is also a pharmacological agent for other neural systems with no known relationship to the excitatory amino acid L-AP4 receptor subclass. One of these is a calcium/chloride-dependent glutamic acid uptake system. Another is a novel postsynaptic mechanism that is specifically activated by exposure of neurons to quisqualic acid. The activated neurons show greatly enhanced sensitivity for depolarization by a very limited group of compounds including D- and L-AP4, D- and L-AP5, DL-AP6, and *E*-cyclopentyl-AP4.

INTRODUCTION

L-2-amino-4-phosphonobutanoic acid (L-AP4) is one of dozens of compounds with structural homology to L-glutamic acid (glutamic acid), a neurotransmitter for vertebrate and invertebrate neurons. It, and most other glutamic acid analogues which have been investigated in neural systems, share with their parent compound two centres of negatively charged substituents and one with a positive charge. These structural features are believed to be essential for interaction of glutamic acid as an agonist with its neurotransmitter receptors. The structural homology of L-AP4 with glutamic acid is particularly close; it differs only by replacement of the γ-carboxyl group of glutamic acid with a phosphonic acid moiety. The γ-carboxylic acid substituent of glutamic acid exists under physiological conditions almost entirely as the monoanion ($pK_a = 4.1$) whereas the γ-phosphonic acid substituent of L-AP4 exists as a mixture of mono- and dianionic forms ($pK_{a1} = 1.6$, $pK_{a2} = 6.9$) (Fagg *et al.* 1982a).

The circumstance that focused attention on this particular glutamic acid analogue was the discovery, by several independent groups of investigators in the early 1980s, that L-AP4 is a potent pathway-specific inhibitor of a few very discrete vertebrate neural pathways and that it acts as an agonist for one specific type of retinal neuron. At the time when numerous neuronal systems were being exposed to dozens of acidic amino acid analogues, many of which merely elicited non-specific depolarizations when applied at high micromolar to millimolar concentrations, L-AP4 and D-2-amino-5-phosphonopentanoic acid (D-AP5) seemed spectacular and unique: L-AP4 for its pathway-specific inhibition at low micromolar concentrations; D-AP5 for its highly potent and specific inhibition of *N*-methyl-D-aspartic acid (NMDA)-evoked excitation.

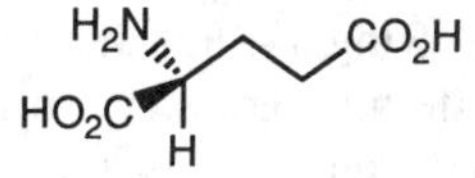

H_2N, CO_2H, HO_2C, H

L-Glutamic Acid

H_2N, $PO(OH)_2$, HO_2C, H

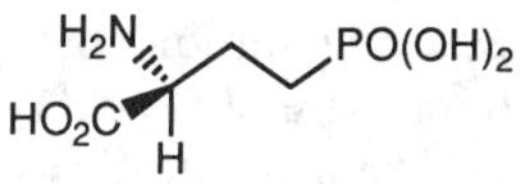

L-AP4

Today, only six neural pathways sensitive to micromolar concentrations of L-AP4 have been defined reasonably well. Five of these are excitatory glutamatergic projection pathways that are specifically inhibited by micromolar concentrations of L-AP4. These are the perforant path–dentate granule cell synapses of the rat hippocampus, the mossy fibre–CA3 pyramidal cell synapses of the guinea-pig hippocampus, the lateral olfactory tract–pyriform cortical neuronal synapses of the rat, the dorsal root–ventral root monosynaptic response of the spinal cord of the rat and cat, and a long latency component of the parallel fibre–Purkinje cell response of the cerebellum of the turtle. L-AP4 is also known to evoke a specific agonist action only for ON-bipolar cells of the amphibian and mammalian retina.

During the past decade, research on mechanisms of action of L-AP4 have disclosed other biological systems for which it shows high potency. One of these is a calcium/chloride-dependent glutamic acid uptake system observed in synaptic plasma membrane preparations. Another is a phenomenon, which we term the QUIS effect, in which neuronal preparations briefly exposed to the agonist quisqualic acid (QUIS) are sensitized to subsequent depolarization by L-AP4 and structurally related phosphonates. Very recently, L-AP4 has also been observed to intervene in certain aspects of the phenomenon of long-term potentiation (LTP). Both calcium/chloride-dependent uptake of L-AP4 and its role as an agonist in the QUIS effect are pharmacologically distinct from its neural pathway specificity.

In this chapter, we will review the current information on known L-AP4-sensitive systems. We will describe their pharmacological similarities and differences and review recent findings on structure-activity relationships for selected systems. We will also present some recently emerging findings on mechanisms of action of L-AP4 and suggest speculative relationships of various receptors and possible approaches for future investigations. Not all of this material has been previously compiled in one article, but four recent general reviews on acidic amino acid agonists and antagonists include descriptions of various aspects of L-AP4-sensitive systems (Collingridge and Lester 1989, Johnson and Koerner 1988, Mayer and Westbrook 1987, Monaghan *et al.* 1989).

L-AP4-SENSITIVE NEURAL PATHWAYS

L-AP4 receptors and pharmacological methods

L-AP4 receptors are uniquely sensitive to activation by L-AP4. The D-isomer of AP4 and higher and lower homologues show little activity. Experimental evidence demonstrates that these receptors are located presynaptically in three of the vertebrate synaptic pathways in which they inhibit the fast excitatory postsynaptic potential (EPSP). They are located postsynaptically on retinal ON-bipolar cells.

The discovery of L-AP4 receptors in neural pathways, and their unique specificity for structural analogues of glutamic acid, have been defined by pharmacological methods. Most of the studies used brain slices, hemisected spinal cords, retinas, or other small tissue preparations bathed in glucose–Ringer solutions that are maintained in an atmosphere of O_2–CO_2. The glutamate analogues and other drugs are applied either via the incubation medium or iontophoretically. In some experiments, the response to an iontophoretically applied agonist is observed in the presence of bath-applied antagonists. However, the fundamental experiments that identified the subclass of glutamate receptors now called L-AP4 receptors measured the inhibition of excitatory synaptic responses evoked by electrical stimuli (or photic stimuli of the visual system); i.e., the responses evoked by release of the endogenous neurotransmitter. These responses have been measured by extracellular or intracellular electrophysiological recordings, but extracelluar recordings have predominated.

Recently, whole-cell patch-clamp techniques have been applied to the study of L-AP4 receptors in cultured neurons and retinal preparations. These methods allow control of both the intracellular and extracellular environment during recording. They are particularly valuable for elucidating the mechanisms of signal transduction mediated by intracellular second messengers.

The pharmacological actions of structural analogues of acidic amino acids must be evaluated in the context that most of these compounds depolarize neurons at a concentration of less than 10 mM. Those that depolarize the preparation in the range of 1–10 mM mediate this effect via unknown mechanisms, although they are often assumed to be interacting with excitatory amino acid receptors. Among those agonists and antagonists with reasonably documented specificities and sites of action, those with IC_{50} values† as low as 0.5 μM are rare indeed. Thus, present pharmacological experiments are constrained to a 'window' of concentrations spanning only 2–3 orders of magnitude. Even within this window, multiple sites of action can be expected. Thus, L-AP4, with IC_{50} values for L-AP4 receptors in the range 1–50 μM, inhibits the stimulus-evoked NMDA receptor response of hippocampal Schaffer collateral–CA1 pyramidal cell synapses with an $IC_{50} = 450$ μM (N. L. Peterson, unpublished data). It depolarizes neurons and/or non-specifically blocks stimulus-evoked responses of CA1 pyramidal neurons at 2.5 mM. (Koerner and Cotman 1982).

Perforant path (rat dentate gyrus)

The perforant path is an excitatory projection pathway from stellate cells of the entorhinal cortex to synapses on apical dendrites of dentate granule cells of the hippocampal formation. Anatomical studies showed that this pathway is mapped topographically; axons from the medial entorhinal cortex terminate on the middle portion of the granule cell dendrites (middle molecular layer of the dentate gyrus) whereas axons from the lateral entorhinal cortex terminate on the most distal portion of the dendritic field (outer molecular layer) (Hjorth-Simonsen 1972, Steward and Scoville 1976).

Koerner and Cotman (1981) discovered that the extracellular excitatory postsynaptic potential (EPSP) of the lateral perforant path is selectively inhibited by micromolar concentrations of L-AP4, in contrast to the medial perforant path which was previously shown to be inhibited by millimolar concentrations (White *et al.* 1977, 1979). Analysis of concentration–response curves for inhibition of the response of the lateral perforant path showed that inhibition of 70–90% of the response could be ascribed to a single component of $IC_{50} = 2.5$ μM and Hill coefficient = 1 (Koerner and Cotman 1981, Koerner *et al.* 1983). The resistant portion of the response may be due to contamination with responses from the medial perforant path.

Although these results are consistent with possible antagonism of a postsynaptic glutamate receptor, the failure of L-AP4 to block excitations by applied glutamic acid, QUIS, or L-serine-*O*-sulphate suggested the alternative possibility of a presynaptic mechanism (Ganong and Cotman 1982). This was strongly supported by experiments by Harris and Cotman (1983), who measured the effect of L-AP4 on paired-pulse potentiation. The phenomenon of paired-pulse potentiation, in which the response to the second of a pair of closely spaced stimuli (40–100 ms) is enhanced, is ascribed to a transient presynaptic increase of intracellular calcium ions elicited by the first stimulus. The percentage increase of the second response is augmented, although its absolute amplitude is inhibited, by lowering extracellular calcium ion concentration

† Data from pharmacological experiments may conform to the Hill equation, which was derived from the law of mass action to describe ligand–receptor interactions:

$$B/B_{max} = [1 + (K_d/c)^h]^{-1}$$

where B/B_{max} = fraction of bound ligand; c = concentration of unbound ligand; K_d = equilibrium dissociation constant; and h = Hill coefficient.

For pharmacological experiments, the term B/B_{max} corresponds to (fraction of inhibited response) or (fraction of uninhibited response)$^{-1}$. Also for pharmacological experiments, the relationship of the calculated K_d to a ligand–receptor dissociation constant is often unknown and may be remote. In articles describing such experiments, the term K_d may be replaced by IC_{50}, EC_{50}, or 'apparent K_d'.

or by presynaptic inhibitors such as L-AP4. In contrast, the percentage of potentiation is not changed by postsynaptic inhibitors such as kynurenic acid.

The medial perforant path is also partially inhibited by micromolar L-AP4, but only 10–40%. The L-AP4-sensitive component gives an $IC_{50} = 45\ \mu M$. It is thus 20-fold less sensitive to L-AP4 than the lateral perforant path (Koerner and Cotman 1981, Koerner *et al.* 1983). Lanthorn and Cotman (1981) discovered that β-(*p*-chlorophenyl)-γ-aminobutanoic acid (baclofen, Lioresal) also inhibits the same component of the medial perforant path but does not inhibit the lateral perforant path. Like the lateral perforant path, the L-AP4- and baclofen-sensitive component of the medial perforant path appears to be inhibited by a presynaptic mechanism (Harris and Cotman 1985). These observations were published when the pharmacology and roles of $GABA_B$ receptors were just beginning to emerge. There have been no further reports on the relationship of L-AP4 receptors to $GABA_B$ receptors.

Lateral olfactory tract (rat)

Inhibition by AP4 of the response of pyriform cortical pyramidal cells to stimulation of the lateral olfactory tract (LOT) was reported ten years ago (Hori *et al.* 1981, 1982). Most of the early studies used a single high concentration of AP4, often as the racemic mixture, and this gave conflicting results for interactions with other agonists and antagonists.

Hearn *et al.* (1986) published detailed concentration–response data for inhibition of the LOT by L-AP4. At 1 mM L-AP4, the stimulus-evoked extracellular EPSP was inhibited only 50%, and the inhibition curve spanned a wide range of concentrations, suggesting receptor heterogeneity. The data could be represented accurately by assuming the presence of three components, with IC_{50} values of 2.5 μM, 45 μM, and much greater than 1 mM, analogous to a mixture of the single component of the lateral perforant path and two components of the medial perforant path (Koerner *et al.* 1983).

There is another analogy to the perforant path: L-AP4 failed to inhibit NMDA, kainic acid (KAIN), and quisqualic acid (QUIS) excitation of pyriform pyramidal cells. This raised the question of whether either glutamic acid or aspartic acid is the endogenous neurotransmitter (Hori *et al.* 1981, 1982). The dipeptide *N*-acetyl-L-aspartyl-L-glutamic acid (NAAG), which occurs in high concentration in the brain, was proposed as a transmitter candidate, and excitations attributed to NAAG were found to be inhibited by L-AP4 (ffrench-Mullen *et al.* 1985). However, this observation proved to be artifactual, since the excitations attributed to NAAG can be mimicked by potassium ions which were present in the early preparations, and these potassium ion-induced excitations are inhibited by L-AP4 (Whittemore and Koerner 1989a).

Experiments measuring paired-pulse potentiation suggest a presynaptic site of action for L-AP4 (Anson and Collins 1987). Recent studies using whole-cell patch-clamps with olfactory bulb mitral/tufted cell cultures showed L-AP4 to act on presynaptic receptors to close voltage-sensitive calcium channels and inhibit neurotransmitter release (Trombley and Westbrook 1990a, 1990b). The only action of NAAG on these preparations was as a moderately potent NMDA receptor agonist ($IC_{50} = 670\ \mu M$).

In summary, the properties of the L-AP4-sensitive pathways of the LOT resemble those of the perforant path in every respect. The action of L-AP4 is presynaptic, which accounts for its failure to inhibit postsynaptic responses to applied agonists. There is a heterogeneous mixture of L-AP4-sensitive and L-AP4-resistant components that are activated by stimulating the LOT. The concentration–response data for the L-AP4-sensitive responses can be modelled adequately by assuming a mixture of the two separate responses observed in the lateral and medial perforant path. However, the presence of multiple components with different IC_{50} values than the perforant path cannot be excluded by this analysis.

Mossy fibre–CA3 pyramidal cells (guinea-pig hippocampus)

Early pharmacological studies on the role of excitatory amino acids on hippocampal mossy fibre–CA3 pyramidal cell synapses yielded conflicting results. The band of mossy fibre axons comprising this pathway is long and narrow. It lies close to the CA3 pyramidal cell body layer, and it forms synapses on the most proximal portion of the apical dendritic tree of these neurons. The massive bundle of commissural fibres and their synapses lies just distal to the mossy fibre bundle. Thus, it is difficult to activate mossy fibres exclusively, and extracellular EPSPs in particular may be contaminated with fibre volleys, antidromic spikes and commissural synaptic responses. These difficulties were often exacerbated by use of 20 mM magnesium ion to selectively inhibit synaptic potentials, but this concentration also partially inhibits antidromic spikes.

First, a species difference in L-AP4 sensitivity was discovered. White *et al.* (1979) reported that the rat mossy fibre–CA3 pyramidal cell response is insensitive to L-AP4, while Yamamoto *et al.* (1983) reported the guinea-pig is sensitive to L-AP4. Lanthorn *et al.* (1984) showed that micromolar L-AP4 inhibits mossy fibre responses in guinea-pig but not in rat hippocampal slices, thus attributing the earlier discrepancies to a species difference†. Mossy fibre responses show a dramatic post-tetanic potentiation, often 3–10-fold, which persists for 1–2 minutes. The potentiated and unpotentiated responses show identical sensitivity to inhibition by L-AP4 (T. H. Lanthorn, personal communication; N. L. Peterson, unpublished data).

Other inhibitors also yielded discrepancies. Thus Robinson *et al.* (1984) reported that kynurenic acid does not inhibit mossy fibre responses in the rat, but others reported that it inhibits these responses in both rat and guinea-pig (Ganong *et al.* 1983, Lanthorn *et al.* 1984, Flatman *et al.* 1986, Harris and Cotman 1986). Using the criteria of Harris and Cotman (1986) for verifying the mossy fibre EPSP, we have demonstrated sensitivity to kynurenic acid, contrary to our earlier conclusion. Likewise, Lanthorn and Cotman (1981) reported that baclofen inhibits mossy fibre responses of the rat, but Ault and Nadler (1982) reported that it does not. Inoue *et al.* (1985) reported that baclofen does not inhibit mossy fibre responses of the guinea-pig. The discrepancy of data for the rat and questions of species variations have not been resolved for baclofen. These are important issues for future research on L-AP4 receptors because of the interaction of L-AP4 and baclofen in the medial perforant path (Lanthorn and Cotman 1981).

† However, we and another investigator sometimes have seen L-AP4 sensitivity in the rat, but not consistently (Robinson *et al.* 1984, T. H. Lanthorn, personal communication).

Cotman *et al.* (1986) demonstrated a presynaptic site of action for L-AP4 by measuring by intracellular recording the effect of this drug on the miniature excitatory postsynaptic potentials evoked by mossy fibre synapses. The amplitude and frequency of these spontaneously released packets of neurotransmitter are not affected by presynaptic inhibitors of stimulus-secretion release, like L-AP4, but their amplitude is decreased by postsynaptic receptor inhibitors, like kynurenic acid.

The phylogenetic difference in distribution of this L-AP4-sensitive response, especially in such closely related species as rat and guinea-pig, is remarkable. Further studies may disclose wide phylogenetic variations for other L-AP4-sensitive pathways. Comparisons of physiological responses in species possessing and lacking L-AP4 sensitivity may lend insight into physiological roles of L-AP4 receptors.

Spinal cord

Much of the classical pharmacology of excitatory amino acids was obtained for spinal neurons. Most of these experiments measured firing patterns of neurons in response to iontophoretically applied agonists and antagonists, or the effects of bath-applied compounds on ventral root outputs in response to dorsal root stimulation.

Most studies on the spinal cord were carried out using incubation media that were free of magnesium ions. In such media, responses mediated by NMDA receptors are prominent. This is in marked contrast to experiments on hippocampal or olfactory cortical slices, which are usually carried out in incubation media containing 1–2.5 mM magnesium ions that inhibit NMDA responses when the frequency of stimulation is low (0.1–0.5 Hz). In a recent perceptive review, Collingridge and Lester (1989) pointed out that the common early belief that the rapid responses of the spinal cord mediated by KAIN and QUIS receptors are monosynaptic whereas the slow responses mediated by NMDA receptors are polysynaptic is probably incorrect. Both the fast and slow responses in this and other systems can now be attributed to simultaneous monosynaptic activation of rapidly responding KAIN/2-amino-3-(3-hydroxy-5-methylisoxazol-4-yl)propanoic acid (KAIN/AMPA) receptors and slowly responding NMDA receptors (Hestrin *et al.* 1990, Lester *et al.* 1990).

In the light of this interpretation, the clearest evidence for the specific spinal L-AP4 receptor was described for the dorsal root-evoked ventral root extracellular EPSP of the rat spinal cord by Evans *et al.* (1982) (see their Fig. 2c and text). They showed that there is an extracellular response of long duration in magnesium ion-free medium. They demonstrated that this response was comprised of two components: a component of long duration that was inhibited by the NMDA antagonist D-AP5 (50 μM) in a medium containing magnesium ions (1 mM), and a component of short duration that was not inhibited by D-AP5 and magnesium ions. The uninhibited component of short duration (which can now be attributed to KAIN/AMPA receptors) was inhibited by L-AP4 but not by D-AP5. Their statement 'The threshold concentration of L-AP4 required to produce this effect was 5 μM, and the effect appeared to be maximal at 250 μM' suggests an $IC_{50} \sim 50$ μM. This is comparable to the L-AP4-sensitive component of the rat medial perforant path and contrasts with the more sensitive component of the lateral perforant path.

Iontophoretic studies in the cat spinal cord preparation demonstrated that L-AP4 does not inhibit KAIN, QUIS, NMDA, glutamic acid, or aspartic acid responses (Davies and Watkins 1982). On the basis of this evidence, and in analogy with evidence that the lateral perforant path L-AP4 receptors are presynaptic (Harris *et al.* 1983), Evans (1986) suggested that L-AP4 also acts in the spinal cord by a presynaptic mechanism. No recent pharmacological studies of L-AP4 analogues have been described for the spinal cord.

Retina

A specific action of AP4 in micromolar concentration on one type of retinal cell, the ON-bipolar cell, was first observed by intracellular recordings from retinal preparations from amphibians and fish (Shiells *et al.* 1981, Slaughter and Miller 1981, 1985).

The electrophysiology of the retina differs from other commonly studied neural systems with regard to its mechanisms of signal transduction (Miller and Slaughter 1986). Unstimulated photoreceptor cells are tonically depolarized and are continually releasing a neurotransmitter, thought to be glutamic acid, at synapses on second-order retinal neurons. When the photoreceptor cells are activated by light, they become hyperpolarized, and glutamate release decreases. Thus, synapses of the second-order ON-bipolar neurons are flooded with neurotransmitter in the dark, but, remarkably, the response triggered by this tonic activation of their postsynaptic glutamate receptors is hyperpolarization, mediated by closure of monovalent cation channels. When the photoreceptor cells are exposed to light, the resulting depolarization of ON-bipolar cells triggers their release of glutamic acid for excitation of third-order neurons.

Application of L-AP4 to ON-bipolar cells mimics the action of the natural transmitter, i.e. closure of monovalent cation channels. Thus, L-AP4 in this system is acting like a potent glutamatergic agonist. Very recently, whole-cell patch-clamp methods have delineated details of the mechanism of signal transduction by the ON-bipolar L-AP4 receptor. As discussed later, this involves a cyclic GMP second messenger system to effect closure of ON-bipolar neuron monovalent cation channels.

Cerebellar Purkinje cells (turtle)

There are two major excitatory inputs to cerebellar Purkinje cells. A single climbing fibre establishes hundreds of synapses on the soma and proximal dendrites of each Purkinje cell. Tens of thousands of parallel fibres establish synapses en passant to the dendritic tree of each Purkinje cell. Very recently, novel slow EPSPs arising from both of these inputs were observed by intracellular recordings of Purkinje cells in preparations of turtle cerebellum (Larson-Prior and Slater 1989, Larson-Prior *et al.* 1990). The slow ESPS from both of these afferents begins to appear after decay of the initial fast EPSP ($\sim$20 ms) and peaks at $\sim$500 ms. The parallel fibre slow EPSP was inhibited 80% by 50 μM L-AP4. Neither the climbing fibre slow EPSP nor the fast EPSP of either pathway was inhibited by 100 μM L-AP4. None of these responses were inhibited by 200 μM L-serine-*O*-phosphate.

OTHER L-AP4-SENSITIVE SYSTEMS

L-AP4 uptake

Early studies that measured displacement of [^{3}H]L-glutamic acid labelling of synaptic plasma membrane preparations disclosed a major component that required calcium and chloride ions for activity (Fagg *et al.* 1982b). Among potent displacers of this labelling was L-AP4. The D-isomer of AP4 was 15-fold less potent. A more direct approach to the AP4-labelling of this system used [^{3}H]AP4 and measured its displacement by other ligands (Butcher *et al.* 1983, Monaghan *et al.* 1983). Although it was hoped that these experiments had identified binding to the L-AP4 receptor demonstrated by electrophysiological experiments, much evidence now suggests that the radiolabelling of these membrane preparations arises from sequestration of the labelled compounds into lipid vesicles formed by the preparation.

This is most directly supported by the physiochemical properties of the labelled entity, including the large temperature coefficient characteristic of uptake processes, an uptake capacity that exceeds the abundance of most known receptors, and increases and decreases of uptake capacity that accompany swelling and shrinking of membrane vesicles in hypo- and hyperosmolar solutions (Pin *et al.* 1984, Zaczek *et al.* 1987). Also, pharmacological studies with glutamic acid analogues showed a different pattern for uptake than the pharmacology of electrophysiologically defined L-AP4 receptors (Fagg and Lanthorn 1985, Robinson *et al.* 1985, Bridges *et al.* 1986, Crooks *et al.* 1986). Although L-AP4 ranks among the compounds with highest known affinity for this uptake system, it is not uniquely potent. The role of this uptake system in excitatory amino acid metabolism is unknown. To date, electrophysiologically defined L-AP4 receptors have not been identified by ligand-receptor binding studies. Their rapid responses to washout of L-AP4 and its analogues suggest that a novel ligand with a much slower rate of dissociation than any known compound will be needed to label them.

QUIS effect

We reported a novel system in which QUIS induces a dramatic and persistent sensitization of neurons to depolarization by D- and L-AP4 and close structural analogues of these compounds (Robinson *et al.* 1986, Whittemore and Koerner 1989b). This phenomenon, termed the QUIS effect, appears to be widely distributed, at least in the forebrain, having been reported for neurons of the CAl region in rat and guinea-pig brain; and the medial perforant path, the lateral olfactory tract, and the cingulate cortex in rat brain (Sheardown 1988, Whittemore and Koerner 1989b). All of these pathways are normally depolarized by millimolar, not micromolar, concentrations of bath-applied phosphonates. In contrast, QUIS does not enhance the pre-existing micromolar sensitivity of the rat lateral perforant path to L-AP4 (Whittemore and Koerner 1989b). The sensitization is dramatic, giving an IC_{50} of 50 μM for depolarization of hippocampal CAl pyramidal cells by L-AP4. This is 40-fold more potent than their sensitivity to depolarization prior to exposure to QUIS.

The QUIS effect is induced in CAl pyramidal cells within the shortest time that can be measured by these experiments, i.e. in less than 4 minutes. The effect persists for at least 4 hours, with less than a threefold decrease in sensitivity to L-AP4 over this time period. The rate of reversal is not dependent on the number of exposures to L-AP4 during this time, nor is it affected by repeated washing of the slice with incubation medium.

Inhibition is accompanied by transient population spiking. This and other extracellular evidence led us to suggest that AP4 depolarizes the CAl pyramidal cells (Robinson *et al.* 1986). This was later confirmed by Harris *et al.* (1987) by intracellular recording. Of particular significance was our early observation that the QUIS effect is not induced by AMPA and therefore likely involves a novel site of action different from the 'classical' QUIS/AMPA-induced depolarization of neurons. To date, no compound other than QUIS itself has been found to induce the QUIS effect. The 'induction site' by which QUIS mediates this effect is insensitive to the non-NMDA antagonist 6-cyano-7-nitroquinoxaline-2,3-dione (CNQX) (Whittemore and Koerner 1989b). Preliminary data suggest that a transient five-fold sensitization to L-AP4 can be induced by ligands specific to the metabotropic excitatory amino acid (EAA) receptors linked to phosphoinositide hydrolysis (Whittemore and Cotman 1990). However, the relationship of this weak, transient sensitization to the phenomenon of the QUIS effect is unknown.

Considerable indirect evidence suggests that cellular uptake of quisqualic acid is a necessary step for induction of the QUIS effect. This was first suggested by Harris *et al.* (1987), who noted a correspondence between potencies of phosphonates for depolarization and uptake by the calcium/chloride-dependent glutamic acid uptake system. Sensitization is attenuated in slices that are exposed to L-α-aminoadipic acid, another potent ligand for this uptake system (Harris 1989). However, other ligands for this system (e.g. L-AP4, glutamic acid) do not attenuate sensitization (Whittemore and Koerner 1989b). Nevertheless, the idea that some kind of uptake is involved is reinforced by the finding that L-α-aminoadipic acid, L-homocysteine sulphinic acid, and L-serine-*O*-sulphate also partially reverse the QUIS effect if they are applied after it is induced (Harris 1989, Whittemore and Koerner 1989b). Even more surprising, these same compounds act as 'pre-blockers', i.e. if they are applied and then removed from the medium before addition of QUIS, they somehow block induction of the QUIS effect by QUIS (Whittemore and Koerner 1991). Pre-blocking also reduces the sensitivity of the neurons to depolarization by QUIS. All of these observations are consistent with a mechanism in which QUIS and its 'pre-blockers' compete for a cellular uptake site. However, other seemingly less likely mechanisms can also be proposed, and at present there are no experimental data which measure uptake and release of these compounds during induction and blockade of the QUIS effect.

Two mechanisms have been proposed for the QUIS-induced enhancement of sensitivity for depolarization by L-AP4. Harris *et al.* (1987) suggested that when L-AP4 is added to the medium there is a heteroexchange reaction between an intracellular pool of QUIS and the extracellular L-AP4. As the QUIS is released, it interacts with extracellular QUIS/AMPA receptors to depolarize the neuron. We regard this mechanism to be unlikely because repeated applications of L-AP4 and repeated washing of the slice do not accelerate the normally slow decay of sensitivity of the cell to L-AP4 (Whittemore and Koerner 1989b). An alternative possibility is that an intracellular pool of sequestered QUIS in some manner triggers events that result in sensitization of a pre-existing cell surface receptor to depolarization by L-AP4.

Long-term potentiation

Long-term potentiation (LTP) is a much studied paradigm for synaptic plasticity (Landfield and Deadwyler 1988). Possible roles of pre- and postsynaptic events for establishing LTP are controversial, but recent evidence supports a combination of pre- and postsynaptic factors and intervention of nitric oxide as a soluble second messenger (Böhme *et al.* 1991, O'Dell *et al.* 1991, Schuman *et al.* 1991, Haley *et al.*

1992). In the hippocampal slice, NMDA receptor blockers inhibit induction of LTP in the Schaffer collateral–CAl pyramidal cell pathway (Harris *et al.* 1984). It has long been known that L-AP4 does not inhibit establishment of LTP in the Schaffer collateral–CAl pyramidal cell pathway. However, it has recently been found that if LTP is induced in the presence of L-AP4 the potentiation of the response is initially established to the normal level but decays to the unpotentiated level within 2 hours (Reymann and Matthies 1989). It is noteworthy that there is no evidence for L-AP4 receptors of the type that mediate inhibition of the fast EPSP in CAl.

Invertebrate systems

Much evidence suggests that the excitatory neurotransmitter of the neuromuscular junction of arthropods is an excitatory amino acid, not acetylcholine. Early pharmacological studies which support this hypothesis included DL-AP4 among the amino acid analogues investigated (Clements and May 1974, Cull-Candy *et al.* 1976, Dudel 1977). Although weak inhibitory activity of DL-AP4 (AP4) supported the idea of glutamatergic transmission, there is presently no evidence that demonstrates the existence of specific, sensitive L-AP4 receptors in these preparations.

MECHANISMS OF SIGNAL TRANSDUCTION

Recently, whole-cell patch-clamp methods have been used to elucidate the mechanism of signal transduction by L-AP4 receptors. These experiments provide detailed voltage–current data for changes in membrane conductance which occur in response to changes of ionic composition of the extracellular medium or application and removal of drugs. They also allow control of the intracellular environment. Soluble intracellular constituents, such as second messenger molecules, are gradually removed by diffusion into the lumen of the patch pipette. Reagents, including second messenger candidates and specific inhibitors, can be placed in the pipette and thus introduced into the cell by diffusion. These experiments suggest that the physiological responses of L-AP4 receptors are mediated by pathways of signal transduction involving second messengers.

The most detailed information is for retinal ON-bipolar cells (Nawy and Jahr 1990, Shiells and Falk, 1990). Patch-clamped preparation of these cells, in the absence of glutamic acid or L-AP4, exhibit an active conductance with a reversal potential near 0 mV, indicating that the current carriers are monovalent cations (predominantly extracellular sodium and intracellular potassium ions). Exposure to glutamic acid or L-AP4 reduces this current, indicating closure of ion channels. Introduction of cyclic GMP via the patch pipette increased the conductance, i.e. opened the channels.

Closure of these ion channels by addition of L-AP4 appears to be mediated by a G protein. Thus, if the L-AP4 response was evoked in the presence of GTP-γ-S, a non-hydrolysable analogue of GTP which maintains G proteins in a permanently activated state, it failed to reverse after removal of L-AP4. Likewise the L-AP4 response was not evoked in the presence of intracellular GTP-β-S, a competitive inhibitor of G protein activation.

These and other observations support a model for signal transduction by L-AP4 receptors in ON-bipolar cells with remarkable similarities to a much-studied system for transduction of photon signals in retinal photoreceptor cells (Yau and Baylor

1989). The proposed transduction of L-AP4 receptor signals is diagrammed in Fig. 1. As shown in this diagram, a likely role for the ON-bipolar cell G protein is activation of a cyclic GMP phosphodiesterase. Activation of this enzyme catalyses hydrolysis of intracellular cyclic GMP, and this results in closure of monovalent cation channels.

There is evidence for similar mechanisms for the L-AP4 receptors of other systems. Thus, Trombley and Westbrook (1990a) obtained whole-cell patch-clamp data from cultures of mitral/tufted cells from the olfactory bulb of rats. (These may include the presynaptic cells of the lateral olfactory tract.) Some pairs of cells which had established synaptic connections in the culture gave an EPSP with a fast component characteristic of KAIN/AMPA receptors and a slow component characteristic of

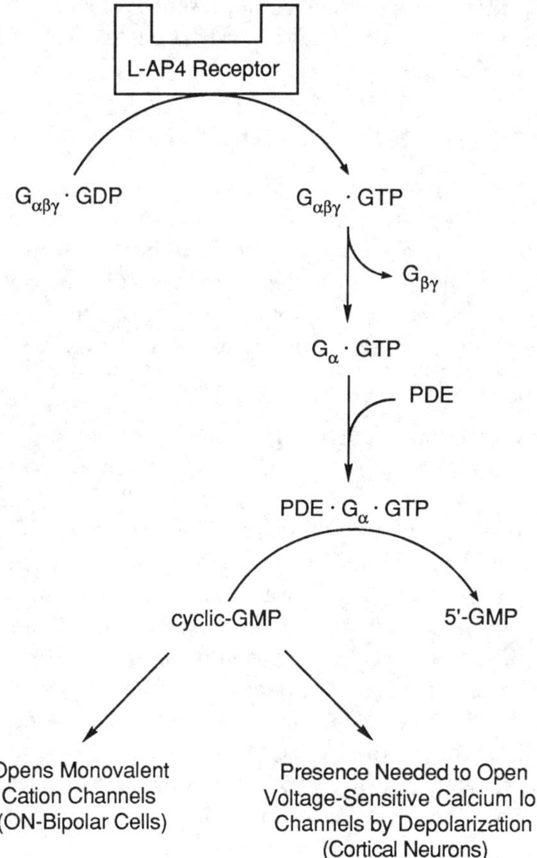

Fig. 1 — Proposed mechanisms for signal transduction by L-AP4 receptors. The pathway illustrated, mediated by a G protein and a cyclic GMP second messenger, is based on evidence from retinal ON-bipolar cells (Nawy and Jahr 1990, Shiells and Falk 1990). An analogous system may modulate presynaptic voltage-sensitive calcium ion channels of cortical neurons (Lester and Jahr 1989, Trombley and Westbrook 1990b). A G protein intermediate has been implicated in the cortical systems, but the identity of the second messenger and details of the transduction process are unknown. (GMP, GDP, GTP, guanosine mono-, di-, and triphosphate; $G_{\alpha\beta\gamma}$, a guanine nucleotide-binding regulatory protein comprised of α, β, and γ subunits; PDE, cyclic GMP phosphodiesterase).

NMDA receptors. The fast response was reduced by 50 μM L-AP4 by a presynaptic mechanism. Presynaptic voltage-sensitive calcium currents in these preparations were reduced by application of L-AP4 (Trombley and Westbrook 1990b). Signal transduction involving a G protein was implicated. It was also reported earlier that L-AP4 reduces the calcium ion currrent in cultured hippocampal cells (Lester and Jahr 1989). As illustrated in Fig. 1, these data are consistent with the same model for L-AP4 receptor signal transduction as in ON-bipolar cells, except the second messenger regulates the response of voltage-sensitive calcium ion channels in cortical neurons whereas it mediates opening of monovalent cation channels in ON-bipolar cells.

There remains the possibility of cellular differences in the second messengers involved. Also, the G protein might in some cases control the activity of a second messenger-synthesizing enzyme rather than a second messenger-degrading enzyme. Whatever differences may emerge in detail, all present evidence supports the hypothesis that the actions of L-AP4 receptors are mediated by G protein-coupled signal transduction mechanisms.

STRUCTURE–ACTIVITY RELATIONSHIPS

Although a number of studies have addressed the structure–activity relationships of AP4, this aspect of the AP4 system is only beginning to be evaluated. The type of AP4 analogues that have been examined to date can be classified into three general categories: (1) lower and higher homologues of AP4; (2) compounds in which the phosphonate group has been modified or replaced with other acidic functionalities; and (3) conformationally constrained analogues of AP4.

Homologues of AP4

Because of the ready availability of the higher and lower homologues of AP4 (Fig. 2), these analogues have been tested in most of the systems shown to be sensitive to AP4. The results of such analyses are summarized in Table 1. The data clearly show that the L-AP4 receptors in excitatory pathways are sensitive to the spacing between the carbon atom to which the amino and carboxyl groups are attached and the side-chain phosphonic acid function. In all the pathways examined, optimum activity is observed with AP4. A significant decrease in activity results when the carbon atom chain linking the α-carbon atom and the phosphonate group is either shortened or lengthened beyond the ethylene chain found in AP4. Stereoselectivity is also observed with AP4, since in the various AP4-sensitive systems the L-isomer is substantially more active than the corresponding D-isomer.

Another system in which AP4 and its homologues have been examined is the Schaffer–CAl pathway. None of these compounds inhibit synaptic transmission in this

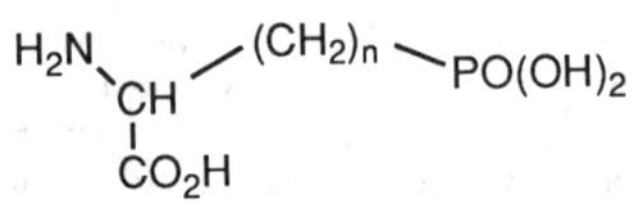

$$H_2N{-}CH(CO_2H){-}(CH_2)_n{-}PO(OH)_2$$

Fig. 2 — Homologues of AP4; $n = 1$, 2-amino-3-phosphonopropanoic acid (AP3); $n = 2$, 2-amino-4-phosphonobutanoic acid (AP4); $n = 3$, 2-amino-5-phosphonopentanoic acid (AP5); $n = 4$, 2-amino-6-phosphonohexanoic acid (AP6); $n = 5$, 2-amino-7-phosphonoheptanoic acid (AP7).

Table 1 — Activity of AP4 and its homologues in various AP4-sensitive systems

Compound	LPP[a] IC_{50} (μM)	MPP[a] IC_{50} (μM)	Inhib CA3[b] % (conc., μM)	ON-Bipolar[c] μM	[^{3}H]AP4 uptake[d] K_i (μM)
L-AP4	2.5	45	59(2)	10	1.45
D-AP4	100	450	2(2)	100	10.1
DL-AP3	5000	1500	ND[e]	*f*	ND
DL-AP5	250	375	50(> 500)	1000	5.46
DL-AP6	ND	ND	> 30(> 500)	ND	2.77
DL-AP7	ND	ND	> 30(> 500)	ND	19.2

[a]Data for the lateral (LPP) and medial (MPP) perforant path were obtained from Koerner and Cotman (1981).
[b]Blockade of mossy fibre–CA3 responses in the guinea-pig (Lanthorn *et al.* 1984).
[c]Hyperpolarization of the retinal ON-bipolar cells (Slaughter and Miller 1985).
[d]AP4 Ca^{2+}/Cl^--dependent uptake. Data from Monaghan *et al.* (1983). Similar relative potencies were observed for some of these homologues by Butcher *et al.* (1986), Butcher *et al.* (1987), and Robinson *et al.* (1985).
[e]ND, no data available.
[f]Inactive at 1000 μM.

pathway at less than 1 mM, and inhibition at higher concentrations is by depolarization of the postsynaptic neurons (Koerner and Cotman 1982). Thus, there is no evidence for specific L-AP4 receptors in this pathway. In this pathway quisqualic acid sensitizes the system to D- and L-AP4, D- and L-AP5 and 2-amino-6-phosphonohexanoic acid (DL-AP6) (QUIS effect). It does not sensitize it to 2-amino-7-phosphonoheptanoic acid (DL-AP7) (Robinson *et al.* 1986, Harris *et al.* 1987). The depolarization of the neurons elicited by L-AP4 in sensitized cells is blocked by the non-NMDA antagonist CNQX (Sheardown 1988). However, the fact that QUIS-sensitized neurons are not significantly sensitized to depolarization by glutamic acid, aspartic acid, KAIN, QUIS, or AMPA suggests that the receptor which is sensitized to L-AP4 and its homologues is not of the known pharmacologically defined KAIN/AMPA receptor subclass.

Among these homologues, L-AP4 is also the most potent inhibitor of [^{3}H]AP4 uptake by the calcium/chloride-dependent uptake system. In contrast to the neural pathways, however, DL-AP5 and DL-AP6, but not DL-AP7, show comparable potencies (Table 1). This ranking of potencies, roughly comparable to their sensitization for the QUIS effect, led to the suggestion that this effect involves uptake of QUIS and AP4–AP6 (Harris *et al.* 1987).

Modifications of the phosphonate moiety

Two general types of modifications of the phosphonate group of AP4 have been made. In one type, the phosphonate moiety has been replaced with non-phosphorus-containing acidic groups such as the tetrazole, sulphonic acid, sulphinic acid, and sulphate functions. The second type of phosphonate modification consists of those modifications in which the phosphonic acid group has been replaced with either a phosphinic acid or phosphate moiety. Structures of the phosphonate-modified AP4 analogues are depicted in Fig. 3 and their activity in several AP4-sensitive systems is summarized in Table 2.

In the calcium/chloride-dependent glutamic acid uptake system, the AP4 analogues containing the non-phosphorus acidic functions show high affinity for this system and are in fact generally more active than L-AP4 itself. These AP4 analogues show little if any enantioselectivity. This is in contrast to the phosphorus-containing acidic functions where good enantioselectivity is seen. Phosphinates generally exhibit about the same activity as L-AP4 in this system, while the phosphates are much weaker than L-AP4.

The nature of the side-chain acidic function in the phosphonate-modified AP4 analogues has been shown to determine whether a compound depolarizes postsynaptic neurons or can manifest the specific presynaptic inhibition of the lateral and medial perforant path mediated by L-AP4 receptors (Koerner *et al.* 1983). When the side-chain acidic function is either a carboxyl, tetrazole, sulphonic acid, or a sulphate group, depolarization is observed. On the other hand, only when the side-chain acidic function is a phosphorus-based acidic group is L-AP4 receptor activity observed in these two pathways. The phosphonic acid group gives rise to a more potent L-AP4 receptor agonist than does the phosphate group, which in turn gives rise to a more potent L-AP4 receptor agonist than does the phosphinic acid moiety.

It was first hypothesized that the ionic state of the side-chain acidic function might be an important determinant in whether a compound would interact with L-AP4 receptors in the lateral and medial perforant path (Koerner *et al.* 1983) since known acidic functions that could only exist as monoanions caused depolarization of the neurons while those that had the potential to exist as dianions mimicked the action of L-AP4. The subsequent finding that the monoanionic phosphonic analogues 2-amino-4-(methylphosphino)butanoic acid and *O*-methylphosphonyl-L-serine also inhibit synaptic transmission without depolarizing the postsynaptic neurons, albeit

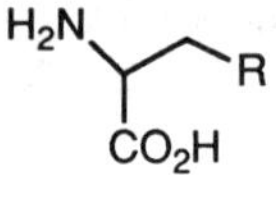

H_2N–CH(CO_2H)–CH_2–R

R	Name
OSO_3H	Serine-O-sulfate
CH_2SO_3H	Homocysteic acid
CH_2SO_2H	Homocysteinesulfinic acid
SSO_3H	S-Sulfocysteine
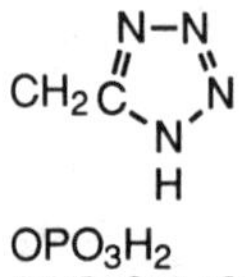 CH_2-(5-tetrazolyl) (N–N, C, N, N–H ring)	L-2-Amino-4-(5-tetrazolyl)butanoic acid
OPO_3H_2	Serine-O-phosphate
$OPO(CH_3)OH$	O-Methylphosphonyl-L-serine
$CH_2PO(H)OH$	2-Amino-4-phosphinobutanoic acid
$CH_2PO(CH_3)OH$	2-Amino-4-(methylphosphino)butanoic acid
$CH_2PO(CH_2CO_2H)OH$	2-Amino-4-(carboxymethylphosphino)butanoic acid

Fig. 3 — Structures of phosphonate-modified AP4 analogues.

Table 2 — Activities of AP4 analogues with modified phosphonate groups

Compound	[³H]AP4 uptake K_i (IC_{50}) (μM)		LPP (MPP)[a] IC_{50} (μM)		QUIS effect blockade[b] ±(conc. mM)
L-AP4	5.3[c]	(1.5)[d]	2.5	(45)[e]	− (2.0)
L-Glutamic acid	0.47[c]	(1.6)[d]	12 000	(8000)*[e]	
L-2-Amino-4-(5-tetrazoyl)butanoic acid	(1.4)[d]		200	(310)*[e]	− (0.1)
L-Serine-*O*-sulphate	0.053[c]	0.37[f]	1 700	(1200)*[e]	
L-Homocysteic acid	0.41[c]	1.3[f]	400	(350)*[e]	+ (0.4)
D-Homocysteic acid	2.4[f]		130	(230)*[e]	+ (2.0)
L-Homocysteine sulphinic acid	1.5[f]				+ (2.0)
D-Homocysteine sulphinic acid	3.3[f]				− (2.0)
L-*S*-Sulphocysteine	0.7[f]				− (2.0)
D-*S*-Sulphocysteine	0.59[f]				− (2.0)
L-Serine-*O*-phosphate	92[c]	72[g]	23	(125)[e]	
D-Serine-*O*-phosphate	303[c]	232[g]	> 3 000	(> 3000)[e]	
O-Methylphosphonyl-L-serine			3 600	(8100)[h]	
2-Amino-4-phosphinobutanoic acid	3.1[i]		> 100[i]		
2-Amino-4-(methylphosphino) butanoic acid	3.4[i]	(3.8)[d]	110	(2200)[h]	
2-Amino-4-(carboxymethyl-phosphino)butanoic acid	7.5[i]		19[i]		

[a]Inhibition or excitation of lateral (LPP) and medial (MPP) perforant path synaptic transmission. An asterisk indicates that the compound acted on the postsynaptic neuron to produce excitation.
[b]Blockade of the quisqualic acid (QUIS)-induced increase in sensitivity to L-AP4. The symbols ' + ' and ' − ' signify that the compound was able to block or not block, respectively, the change in sensitivity. Data from Whittemore and Koerner (1989).
[c]Data from Fagg *et al.* (1982a).
[d]Data from Robinson *et al.* (1985).
[e]Data from Koerner *et al.* (1983).
[f]Data from Pullan *et al.* (1987).
[g]Data from Foster *et al.* (1982).
[h]Data from Freund *et al.* (1984).
[i]Data from Fagg and Lanthorn (1985).

much weaker than AP4 and L-serine-*O*-phosphate, called this hypothesis into question (Freund *et al.* 1984). The observation, however, that 2-amino-4-(carboxy-methylphosphino)butanoic acid is as active an inhibitor of synaptic transmission (i.e. L-AP4 receptor agonist) as L-serine-*O*-phosphate (Fagg and Lanthorn 1985) suggests that the ability of the side-chain acidic function to exist as a dianion is important for maximum potency.

As is true for the perforant path, the presynaptic L-AP4 receptors of the lateral olfactory tract are more sensitive to L-AP4 than D-AP4 (Collins 1982). Likewise, L-serine-*O*-phosphate is a specific but less potent inhibitor than L-AP4, while D-serine-*O*-phosphate has little activity (Hearn *et al.* 1986).

Conformationally constrained AP4 analogues

Analogues of AP4 have been synthesized in an attempt to ascertain the biologically active conformation of AP4 at the L-AP4 receptor. Three sets of analogues have been synthesized to date: *E*- and *Z*-1-amino-3-phosphonocyclohexanecarboxylic acid (*E*- and *Z*-cyclohexyl-AP4), *E*- and *Z*-1-amino-3-phosphonocyclopentanecarboxylic acid

(*E*- and *Z*-cyclopentyl-AP4), and *E*- and *Z*-2-amino-2,3-methano-4-phosphonobutanoic acid (*E*- and *Z*-cyclopropyl-AP4) (Fig. 4). These analogues have been examined in a number of the AP4-sensitive systems (Table 3) and the results clearly demonstrate that the synaptic systems and uptake system recognize different conformations of AP4.

The cyclohexyl-AP4 analogues showed only weak activity in the various pathways and systems investigated. In contrast, the cyclopentyl-AP4 and cyclopropyl-AP4 analogues have been shown to possess activity as well as selectivity in several of the known AP4-sensitive systems. For example, *E*-cyclopentyl-AP4 is quite active in inhibiting the uptake of [^{3}H]AP4 by the calcium/chloride-dependent glutamic acid

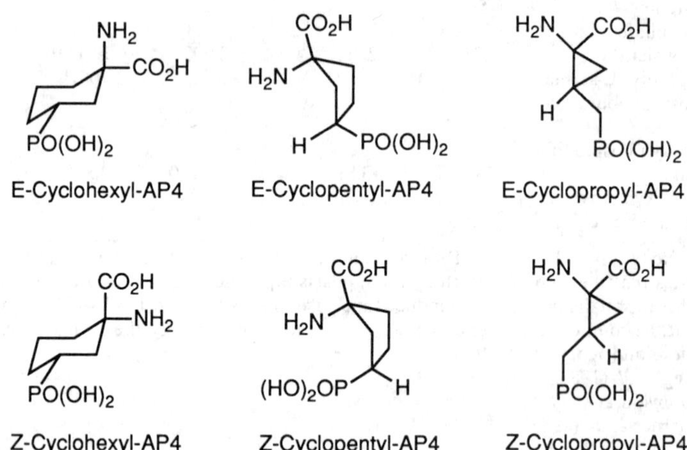

Fig. 4 — Conformationally constrained analogues of AP4. We have adapted the *E,Z* nomenclature to the naming of the cyclohexyl-and cyclopentyl-AP4 analogues in order to avoid the ambiguity produced by the use of the *cis* and *trans* nomenclature that was used by us previously (Crooks *et al.* 1986); *E*- and *Z*-cyclohexyl-AP4 and *E*- and *Z*-cyclopentyl-AP4 correspond to the respective *cis* and *trans* isomers that were reported in that paper.

Table 3 — Pharmacological activity of conformationally constrained AP4 analogues

Compound	IC$_{50}$ (μM)				
	LPP[a]	MPP[a]	[^{3}H]AP4 uptake[a]	CA1 (−QUIS)[b]	CA1 (+QUIS)[b]
E-cyclohexyl-AP4	3100	7300	840	8100	ND[c]
Z-cyclohexyl-AP4	5800	10000	750	9600	ND
E-cyclopentyl-AP4	960	2400	4.7	3800	120
Z-cyclopentyl-AP4	130	1850	80	> 10000	1800
E-cyclopropyl-AP4	18	81	ND	ND	ND
Z-cyclopropyl-AP4	17	1580	ND	ND	ND

[a]Data for lateral (LPP) and medial (MPP) perforant path and for AP4 uptake for the cyclohexyl- and cyclopentyl-AP4 analogues were obtained from Crooks *et al.* (1986), while the data for the cyclopropyl-AP4 analogues were obtained from Kroona *et al.* (1991).
[b]Quisqualic acid (QUIS) induced sensitization of the rat CA1 region. Data were obtained before (−QUIS) (Crooks *et al.* 1986) and after (+QUIS) (Whittemore and Koerner 1989b) exposure to QUIS.
[c]ND, no data available.

uptake system. In addition, after CA1 pyramidal cells are sensitized by QUIS, this conformationally constrained analogue of AP4 is as potent a depolarizing agent as L-AP4 itself (Whittemore and Koerner 1989b).

Based on the relatively similar profiles between *Z*-cyclopentyl-AP4 and L-AP4, it was hypothesized by Crooks *et al.* (1986) that the bioactive conformation of AP4 for L-AP4 receptors in the lateral perforant might be a folded one wherein the *cis*-phosphonate and amino groups could participate in an ionic interaction. The recent results obtained with the *E*- and *Z*-cyclopropyl AP4 analogues indicate that this hypothesis is not a viable one, since in the lateral perforant path *E*- and *Z*-cyclopropyl-AP4 are equally active. In fact, they are the most active analogues synthesized to date. In order to account for their equal potency, the *E*- and *Z*-cyclopropyl-AP4 analogues must be able to assume a conformation that allows the required carboxyl, amino, and phosphonate groups to occupy the same relative place in space. As illustrated in Fig. 5, this can best occur when the αC–βC–γC–P dihedral angle is equal to 150° for *Z*-cyclopropyl-AP4 and − 150° for *E*-cyclopropyl-AP4. These are relatively extended conformations (Kroona *et al.* 1991).

The *E*- and *Z*-cyclopropyl-AP4 analogues show differing and lower potencies in the medial perforant path, in contrast to their equal and higher potencies in the lateral perforant path. Furthermore, the concentration–response curves for their inhibition of the medial perforant path suggests that they are acting on a single component. This is in contrast to L-AP4 which disclosed an L-AP4-sensitive and an L-AP4-resistant component of the medial perforant path (Koerner and Cotman 1981, Koerner *et al.*

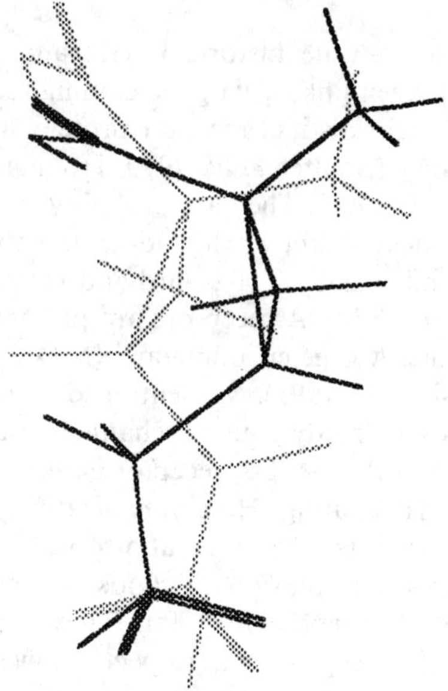

Fig. 5 — Superimposition of *E*-cyclopropyl-AP4 (black) and *Z*-cyclopropyl-AP4 (grey) conformations where the αC–βC–γC–P dihedral angle of *E*-cyclopropyl-AP4 is equal to − 150° and that of *Z*-cyclopropyl-AP4 is equal to + 150°. The structures were built and superimposed using ALCHEMY II, version 1.0, molecular modelling software obtained from Tripos Associates, St Louis, MO.

1983). Perhaps the cyclic AP4 analogues block synaptic transmission to the dentate granule cells at these higher concentrations by a mechanism that does not involve L-AP4 presynaptic receptors. The question of presynaptic or postsynaptic action of *E*- and *Z*-cyclopropyl-AP4 in the medial perforant path has not been addressed. In contrast, their higher and equal potencies for the lateral perforant path was shown to be mediated by a presynaptic mechanism suggesting agonist action on presynaptic L-AP4 receptors in this pathway (Kroona *et al.* 1991).

PERSPECTIVES ON THE L-AP4 RECEPTOR

The past few years have been a watershed in the investigation of excitatory amino acid receptors. Before this time, investigators usually approached these receptors with the methods of classical pharmacology and with electrophysiological assays by conventional extracellular and intracellular recording. Some selective pharmacological agents allowed identification and classification of receptor subtypes, but the data on which these were based were massive and often conflicting and bewildering. Then, in a very short time, two novel approaches were brought to bear: physiological–pharmacological studies using whole-cell patch-clamp techniques, and methods of molecular biology for cloning excitatory amino acid receptors. New discoveries proceed apace, but it is instructive to summarize the present status of L-AP4 receptors in light of these developments.

The summary of current knowledge suggests that there are four known subclasses of excitatory amino acid receptors:

(1) KAIN/AMPA receptors — the historic KAIN and AMPA (QUIS) receptor subtypes have recently been linked via gene cloning techniques to a gene family that codes for at least eight distinct receptor protein subunits with much amino acid sequence homology (Boulter *et al.* 1990, Hollmann *et al.*, 1989, Keinänen *et al.* 1990, Sommer *et al.* 1990). They are probably related remotely to the gene superfamily that includes all known ligand-gated ion channel proteins which mediate fast, millisecond signalling (Barnard and Henley 1990).
(2) NMDA receptors — the NMDA receptors are pharmacologically distinct from KAIN/AMPA receptors. A gene encoding an NMDA receptor has very recently been cloned (Moriyoshi *et al.* 1991). As mentioned earlier, NMDA receptors, like KAIN/AMPA receptors, directly gate ion channels embodied in the structure of the receptor proteins, but the NMDA receptor gating mechanism responds more slowly than KAIN/AMPA gating (Hestrin *et al.* 1990, Lester *et al.* 1990).
(3) Metabotropic EAA receptors — the metabotropic EAA receptors trigger slow physiological responses via phosphatidylinositol second messenger systems (Monaghan *et al.* 1989, Schoepp *et al.* 1990). They appear to be pharmacologically distinct from L-AP4 receptors with respect to low sensitivity to activation by L-AP4, high sensitivity to *trans*-1-aminocyclopentane-1,3-dicarboxylate (*trans*-ACPD), and high sensitivity to inhibition by L-AP3.
(4) L-AP4 receptors — the L-AP4 receptors, like the metabotropic EAA receptors, operate via second messenger systems, but the most likely second messenger for the ON-bipolar cell L-AP4 receptor is cyclic GMP (Nawy and Jahr 1990, Shiells and Falk, 1990). This fact, and the circumstance that known L-AP4 receptors and

metabotropic EAA receptors are pharmacologically and functionally distinct, justifies placing them in separate subclasses.

To date all known receptors that utilize a system for signal transduction involving G proteins are members of a single gene superfamily (Lameh *et al.* 1990). One hallmark of these receptors is a single peptide chain with seven helical domains that span the lipid bilayer membrane. It is possible that both L-AP4 receptors and metabotropic EAA receptors will prove to be members of this superfamily. Future classification of these receptors into a single or separate subclasses must await more detailed information on pharmacology, mechanisms of action, and the structures of the receptor protein molecules.

Whole-cell patch-clamp studies have also provided new insight into the pharmacology of L-AP4 receptors. Thus, it can now be generalized that L-AP4 is a potent and specific agonist for the L-AP4 subclass of glutamatergic receptors, regardless of whether these receptors are located pre- or postsynaptically. Earlier papers often identified L-AP4 as an antagonist, because it inhibits synaptic transmission without depolarizing the postsynaptic cell. This terminology should no longer be used in light of current knowledge of the mechanism of signal transduction by L-AP4 receptors.

It is also likely that the limited number of other glutamic acid analogues known to mimic the specific pharmacological effects of L-AP4 are also agonists for L-AP4 receptors. The pharmacology of L-AP4 receptors is presently restricted to a rank-ordering of potency of these analogues as agonists. It is well known that an agonist may show different IC_{50} values when applied to different biological systems, even though its interaction may be with identical receptor molecules. This is because of the well-documented phenomenon of 'spare receptors' and also because of possible physiological differences of efficacy for signal transduction. There is no evidence on which to decide whether two L-AP4-sensitive systems with different IC_{50} values for L-AP4, such as the lateral and medial perforant path, represent two different L-AP4 receptor subtypes. Potent and specific antagonists, which can provide quantitative data for ligand–receptor interactions at a molecular level, are sorely needed for L-AP4 receptors. At present, such antagonists have not been identified.

Finally, the physiological functions of presynaptic L-AP4 receptors remain unknown. Whether these functions will be elucidated by novel pharmacological agents, site-specific mutagenesis of L-AP4 receptor genes, or by a combination of these or other approaches, must await future developments.

NOTE ADDED IN PROOF

P. Q. Trombley and G. L. Westbrook (J. Neurosci. 1992, in press) report recent evidence pertaining to the mechanism of calcium current inhibition by L-AP4 in cultured olfactory bulb neurons. They demonstrated this inhibition to be mediated by a G protein that appears to interact directly with the calcium channel protein without intervention of a soluble intracellular second messenger (G. L. Westbrook, personal communication).

ACKNOWLEDGEMENT

Research on L-AP4 receptors in our laboratories was supported by NIH grant NS 17944.

REFERENCES

Anson, J. and Collins, G. G. S. (1987) *Br. J. Pharmacol.* **91** 753–761.

Ault, B. and Nadler, J. V. (1982) *J. Pharmacol. Exp. Ther.* **223** 291–297.

Barnard, E. A. and Henley, J. M. (1990) *Trends Pharmacol. Sci.* **11** 500–507.

Böhme, G. A., Bon, C., Stutzmann, J. M., Doble, A. and Bonchard, J. C. (1991) *Eur. J. Pharmacol.* **199** 379–381.

Boulter, J., Hollmann, M., O'Shea-Greenfield, A., Hartley, M., Deneris, E., Maron, C. and Heinemann, S. (1990) *Science* **249** 1033–1037.

Bridges, R. J., Hearn, T. J., Monaghan, D. T. and Cotman, C. W. (1986) *Brain Res.* **375** 204–209.

Butcher, S. P., Collins, J. F. and Roberts, P. J. (1983) *Br. J. Pharmacol.* **80** 355–364.

Butcher, S. P., Roberts, P. J. and Collins, J. F. (1986) *Brain Res.* **381** 305–313.

Butcher, S. P., Roberts, P. J. and Collins, J. F. (1987) *Brain Res.* **419** 294–302.

Clements, A. N. and May, T. E. (1974) *J. Exp. Biol.* **61** 421–442.

Collingridge, G. L. and Lester, R. A. J. (1989) *Pharmacol. Rev.* **40** 143–210.

Collins, G. G. S. (1982) *Brain Res.* **244** 311–318.

Cotman, C. W., Flatman, J. A., Ganong, A. H. and Perkins, M. N. (1986) *J. Physiol.* **378** 403–415.

Crooks, S. L., Robinson, M. B., Koerner, J. F. and Johnson, R. L. (1986) *J. Med. Chem.* **29** 1988–1995.

Cull-Candy, S. G., Donnellan, J. F., James, R. W. and Lunt, G. G. (1976) *Nature* **262** 408–409.

Davies, J. and Watkins, J. C. (1982) *Brain Res.* **235** 378–386.

Dudel, J. (1977) *Pflugers Arch.* **369** 7–16.

Evans, R. H. (1986) *Gen. Pharmacol.* **17** 5–11.

Evans, R. H., Francis, A. A., Jones, A. W., Smith, D. A. S, and Watkins, J. C. (1982) *Br. J. Pharmacol.* **75** 65–75.

Fagg, G. E. and Lanthorn, T. H. (1985) *Br. J. Pharmacol.* **86** 743–751.

Fagg, G. E., Foster, A. C., Harris, E. W., Lanthorn, T. H. and Cotman, C. W. (1982a) *Neurosci. Lett.* **31** 59–64.

Fagg, G. E., Foster, A. C., Mena, E. E. and Cotman, C. W. (1982b) *J. Neurosci.* **2** 958–965.

ffrench-Mullen, J. M. H., Koller, K., Zaczek, R., Coyle, J. T., Hori, N. and Carpenter, D. O. (1985) *Proc. Nat. Acad. Sci. USA* **82** 3897–3900.

Forsythe, I. D. and Clements, J. D. (1990) *J. Physiol.* **429** 1–16.

Foster, A. C., Fagg, G. E., Harris, E. W. and Cotman, C. W. (1982) *Brain Res.* **242** 374–377.

Freund, R. K., Crooks, S. L., Koerner, J. F. and Johnson, R. L. (1984) *Brain Res.* **291** 150–153.

Ganong, A. H. and Cotman, C. W. (1982) *Neurosci. Lett.* **34** 195–200.

Ganong, A. H., Lanthorn, T. H. and Cotman, C. W. (1983) *Brain Res.* **273** 170–174.

Haley, J. E., Wilcox, G. L. and Chapman, P. F. (1992) *Neuron* **8** 211–216.

Harris, E. W. (1989) *Eur. J. Pharmacol.* **161** 107–109.

Harris, E. W. and Cotman, C. W. (1983) *Exp. Brain Res.* **52** 455–460.

Harris, E. W. and Cotman, C. W. (1985) *Brain Res.* **334** 348–353.

Harris, E. W. and Cotman, C. W. (1986) *Neurosci. Lett.* **70** 132–137.

Harris, E. W., Ganong, A. H. and Cotman, C. W. (1984) *Brain Res.* **323** 132–137.

Harris, E. W., Stevens, D. R. and Cotman, C. W. (1987) *Brain Res.* **418** 361–365.
Hearn, T. J., Ganong, A. H. and Cotman, C. W. (1986) *Brain Res.* **379** 372–376.
Hestrin, S., Sah, P. and Nicoll, R. A. (1990) *Neuron* **5** 247–253.
Hjorth-Simonsen, A. (1972) *J. Comp. Neurol.* **146** 219–231.
Hollmann, M., O'Shea-Greenfield, A., Rogers, S. W. and Heinemann, S. (1989) *Nature* **342** 643–648.
Hori, N., Auker, C. R., Braitman, D. J. and Carpenter, D. O. (1981) *Cell. Mol. Neurobiol.* **1** 115–120.
Hori, N., Auker, C. R., Braitman, D. J. and Carpenter, D. O. (1982) *J. Neurophysiol.* **48** 1289–1301.
Inoue, M., Matsuo, T. and Ogata, N. (1985) *Br. J. Pharmacol.* **84** 843–851.
Johnson, R. L. and Koerner, J. F. (1988) *J. Med. Chem.* **31** 2057–2066.
Keinänen, K., Wisden, W., Sommer, B., Werner, P., Herb, A., Verdoorn, T. A., Sakmann, B. and Seeburg, P. H. (1990) *Science* **249** 556–560.
Koerner, J. F. and Cotman, C. W. (1981) *Brain Res.* **216** 192–198.
Koerner, J. F. and Cotman, C. W. (1982) *Brain Res.* **251** 105–115.
Koerner, J. F., Johnson, R. L., Freund, R. K., Robinson, M. B. and Crooks, S. L. (1983) *Brain Res.* **272** 299–309.
Kroona, H. B., Peterson, N. L., Koerner, J. F. and Johnson, R. L. (1991) *J. Med. Chem.* **34** 1692–1699.
Lameh, J., Cone, R. I., Maeda, S., Philip, M., Corbani, M., Nádasdi, L., Ramachandran, J., Smith, G. M. and Sadée, W. (1990) *Pharm. Res.* **7** 1213–1221.
Landfield, P. W. and Deadwyler, S. A. (eds) (1988) *Long Term Potentiation: From Biophysics to Behavior; Neurology and Neuroscience*, Vol. 35. Liss, New York.
Lanthorn, T. H. and Cotman, C. W. (1981) *Brain Res.* **225** 171–178.
Lanthorn, T. H., Ganong, A. H. and Cotman, C. W. (1984) *Brain Res.* **290** 174–178.
Larson-Prior, L. J. and Slater, N. T. (1989) *Neurosci. Lett.* **104** 286–291.
Larsen-Prior, L. J., McCrimmon, D. R. and Slater, N. T. (1990) *J. Neurophysiol.* **63** 637–650.
Lester, R. A. J. and Jahr, C. E. (1989) *Soc. Neurosci. Abstr.* **15** 1159.
Lester, R. A. J., Clements, J. D., Westbrook, G. L. and Jahr, C. E. (1990) *Nature* **346** 565–567.
Mayer, M. L. and Westbrook, G. L. (1987) *Prog. Neurobiol.* **28** 197–276.
Miller, R. F. and Slaughter, M. M. (1986) *Trends Neurosci.* **9** 211–218.
Monaghan, D. T., McMills, M. C., Chamberlin, A. R. and Cotman, C. W. (1983) *Brain Res.* **278** 137–144.
Monaghan, D. T., Bridges, R. J. and Cotman, C. W. (1989) *Annu. Rev. Pharmacol. Toxicol.* **29** 365–402.
Moriyoshi, K., Masu, M., Ishii, T., Shigemoto, R., Mizuno, N. and Nakanishi, S. (1991) *Nature* **354** 31–37.
Nawy, S. and Jahr, C. E. (1990) *Nature* **346** 269–271.
O'Dell, T. J., Hawkins, R. D., Kandel, E. R. and Arancio, O. (1991) *Proc. Nat. Acad. Sci. USA* **88** 11285–11289.
Pin, J.-P., Bockaert, J. and Recasens, M. (1984) *FEBS Lett.* **175** 31–36.
Pullan, L. M., Olney, J. W., Price, M. T., Compton, R. P., Hood, W. F., Michel, J. and Monahan, J. B. (1987) *J. Neurochem.* **49** 1301–1307.
Reymann, K. G. and Matthies, H. (1989) *Neurosci. Lett.* **98** 166–171.
Robinson, M. B., Anderson, K. D. and Koerner, J. F. (1984) *Brain Res.* **309** 119–126.

Robinson, M. B., Crooks, S. L., Johnson, R. L. and Koerner, J. F. (1985) *Biochemistry* **24** 2401–2405.

Robinson, M. B., Whittemore, E. W., Marks, R. L. and Koerner, J. F. (1986) *Brain Res.* **381** 187–190.

Schoepp, D., Bockaert, J. and Sladeczek, F. (1990) *Trends Pharmacol. Sci.* **11** 508–515.

Schuman, E. M. and Madison, D. V. (1991) *Science* **254** 1503–1506.

Sheardown, M. J. (1988) *Eur. J. Pharmacol.* **148** 471–474.

Shiells, R. A. and Falk, G. (1990) *Proc. R. Soc. Lond. B* **242** 91–94.

Shiells, R. A., Falk, G. and Naghshineh, S. (1981) *Nature* **294** 592–594.

Slaughter, M. M. and Miller, R. F. (1981) *Science* **211** 182–185.

Slaughter, M. M. and Miller, R. F. (1985) *J. Neurosci.* **5** 224–233.

Sommer, B., Keinänen, K., Verdoorn, T. A., Wisden, W., Burnashev, N., Herb, A., Kohler, M., Takagi, T., Sakmann, B. and Seeburg, P. H. (1990) *Science* **249** 1580–1585.

Steward, O. and Scoville, S. A. (1976) *J. Comp. Neurol.* **169** 347–370.

Trombley, P. Q. and Westbrook, G. L. (1990a) *J. Neurophysiol.* **64** 598–606.

Trombley, P. and Westbrook, G. (1990b) *Soc. Neurosci. Abstr.* **16** 547.

White, W. F., Nadler, J. V., Hamberger, A., Cotman, C. W. and Cummins, J. T. (1977) *Nature* **270** 356–357.

White, W. F., Nadler, J. V. and Cotman, C. W. (1979) *Brain Res.* **164** 177–194.

Whittemore, E. R. and Cotman, C. W. (1990) *Soc. Neurosci. Abstr.* **16** 1182.

Whittemore, E. R. and Koerner, J. F. (1989a) *Proc. Nat. Acad. Sci. USA* **86** 9602–9605.

Whittemore, E. R. and Koerner, J. F. (1989b) *Brain Res.* **489** 146–156.

Whittemore, E. R. and Koerner, J. F. (1991) *Eur. J. Pharmacol.* **192** 435–438.

Yamamoto, C., Sawada, S. and Takada, S. (1983) *Exp. Brain Res.* **51** 128–134.

Yau, K.-W. and Baylor, D. A. (1989) *Annu. Rev. Neurosci.* **12** 289–327.

Zaczek, R., Arlis, S., Markl, A., Murphy, T., Drucker, H. and Coyle, J. T. (1987) *Neuropharmacology* **26** 281–287.

15

The quisqualate metabotropic receptor coupled to phospholipase C (Qp)

Joël Bockaert, Olivier Manzoni, Michèle Sebben, Jean-Philippe Pin, Aline Dumuis, Laurent Fagni and **Fritz Sladeczek**
Centre CNRS–INSERM de Pharmacologie-Endocrinologie, Rue de la Cardonille, 34094 Montpellier Cedex 5, France

ABSTRACT

The pharmacology of metabotropic Qp receptors has been studied in detail in striatal neurons by measuring inositol phosphate (InsP) production under conditions allowing the blockade of ionotropic glutamate receptors. Quisqualate, ibotenate and glutamate are full agonists, whereas kainate and domoate are weak partial agonists. All these drugs are non-specific metabotropic Qp receptor agonists. *N*-Methyl-D-aspartate (NMDA) and 2-amino-3-(3-hydroxy-5-methylisoxazol-4-yl)propanoate (AMPA) are inactive. *Trans*-1-aminocyclopentane-1,3-dicarboxylate is the only specific metabotropic Qp agonist. There is pharmacological evidence indicating that several metabotropic Qp receptors do exist, although only one has been cloned. Upon stimulation of metabotropic Qp receptors, several transduction events can be observed: (1) production of Ins(1,4,5)P_3, but also Ins(1,3,4,5)P_4; (2) transient translocation and activation of kinase C; (3) phosphorylation of pp80 and GAP 43 proteins; (4) rise in $[Ca^{2+}]_i$; (5) activation or inactivation of K^+ channels; (6) closing of Ca^{2+} channels. One of the most intriguing properties of the metabotropic Qp receptor is its ability to produce, in conjunction with AMPA ionotropic receptors, a message which both receptors, when stimulated separately, are unable to produce, e.g. arachidonic acid production. Such a production is similar to the one produced under physiological conditions (Mg^{2+} present) by the conjunction of AMPA and NMDA receptor stimulations. In cerebellum, it is also likely that the associative stimulation of AMPA and metabotropic Qp receptors is able to produce NO, which can desensitize the AMPA receptors via a cyclic GMP pathway. All these cooperative actions of ionotropic and metabotropic receptors are likely to be involved in synaptic plasticity events.

INTRODUCTION

A few years ago, glutamate (Glu), the main excitatory amino acid in vertebrate brain, was supposed to exert all its physiological effects via ionotropic receptors. These receptors are channels composed of several subunits having four transmembrane domains, the wall of the channel being formed by one transmembrane domain of each subunit (Changeux 1990, Betz 1990). The classical Glu ionotropic receptors were the *N*-methyl-D-aspartate (NMDA) receptors, the kainate (KAIN) and the quisqualate (QUIS) receptors (Watkins and Olverman 1987; see chapter 2).

Pharmacological (Pin *et al.* 1989, Charpentier *et al.* 1990, Rassendren *et al.* 1989) and biochemical data (Henley *et al.* 1989), as well as experiments on cloned KAIN receptors (Boulter *et al.* 1990, Keinänen *et al.* 1990, Bettler *et al.* 1990, Sommer *et al.* 1990; see chapter 3), indicated that KAIN and QUIS receptors are common molecular entities which trigger the ionotropic effects of both QUIS and KAIN. These receptors are now called 2-amino-3-(3-hydroxy-5-methylisoxazol-4-yl)propanoate (AMPA) receptors. However, it is possible that a pure KAIN receptor exists at particular synapses (Watkins *et al.* 1990).

In addition to the ionotropic receptor class, there is another non-structurally related class of neurotransmitter receptors: the G protein-coupled receptors (Strader *et al.* 1989). These receptors are composed of only one subunit spanning seven times the plane of the membrane. Most non-peptidergic neurotransmitters (acetylcholine, serotonin, GABA) act by stimulating both ionotropic and G protein-coupled receptors, this being also the case for Glu. Two G protein-coupled Glu receptor subtypes have already been discovered (Sladeczek *et al.* 1985, Nawy and Jahr 1990). It is likely that several others have yet to be found.

The first G protein-coupled Glu receptor was described in striatal neurons (Sladeczek *et al.* 1985) and hippocampal slices (Nicoletti *et al.* 1986a, 1986b). in striatal neurons, we found that Glu and QUIS potently stimulated phosphoinositide hydrolysis with EC_{50}'s of 4 μM and 0.16 μM, respectively. KAIN, the strong depolarizing agent, and NMDA, which triggers a potent increase in intracellular Ca^{2+}, were much less potent than QUIS in evoking an inositol phosphate response. This indicates that the QUIS-induced phosphoinositide hydrolysis was not a consequence of depolarization or calcium fluxes, but was probably due to stimulation of a new Glu receptor subtype linked to phospholipase C via a G protein. We designated this receptor (Sladeczek *et al.* 1988) as the Qp receptor (QUIS receptor coupled to phospholipase C), but others proposed to call it the metabotropic Glu receptor (mGluR) (Sugiyama *et al.* 1989). In future it would be more appropriate to call it GluG1 (as the first studied Glu receptor coupled to a G protein). Here, we will use the term metabotropic Qp receptor.

The second G protein-coupled Glu receptor was only recently recognized as the (*S*)-2-amino-4-phosphonobutanoate (L-AP4) receptor in retina bipolar cells. This receptor increases the rate of cyclic GMP hydrolysis via a G protein (Nawy and Jahr 1990).

This review is focused on the metabotropic Qp receptor, its pharmacology, mechanism of action, localization and physiological functions.

PHARMACOLOGICAL CHARACTERIZATION AND POSSIBLE DIVERSITY OF METABOTROPIC Qp RECEPTORS

As classically observed in any new field of research, studies on metabotropic Qp receptors are carried out on a wide variety of models, using different species and several parameters to quantify their action. Therefore, the differences in pharmacology already reported by various investigators cannot be interpreted simply in terms of a possible difference in receptor entity.

Models on which Qp receptors have been studied

Stimulation of the inositol phosphates production

The basis of the method is simple. Brain slices, synaptoneurosomes or primary neuronal or glial cultures are incubated with $[^3H]$myoinositol or inorganic ^{32}P to label the phospholipids of the membrane and, in particular, phosphatidylinositol (PI), phosphatidylinositol-4-phosphate (PIP) and phosphatidylinositol-4,5-bisphosphate (PIP2). Following this labelling period, excitatory amino acids (EAAs) are applied for periods which can vary from a few seconds to one hour and generally in the presence of LiCl (10 mM) to inhibit the inositol-4-monophosphate (InsP1) hydrolase (Berridge *et al.* 1983). In most studies, the total content of inositol phosphates was isolated by anion-exchange chromatography.

Only a few studies have been carried out using the high-performance liquid chromatography (HPLC) technique to analyse the inositol phosphates produced and especially the production of inositol-1,4,5-trisphosphate (InsP3), the active second messenger which mobilizes intracellular Ca^{2+} pools (Ambrosini and Meldolesi 1989, Baird and Nahorski 1990, Bockaert *et al.* 1990).

Xenopus oocytes

Sugiyama *et al.* (1987, 1989) were the first to show that stimulation of oocytes (expressing EAA receptors after previous injection with rat brain mRNAs) with QUIS and Glu triggers an increase in Cl^- conductance. This response with an oscillatory pattern can be mimicked by InsP3 injections and is blocked by EGTA as well as by the *Bordetella pertussis* toxin (Sugiyama *et al.* 1987). In this system, such a response is typical of receptors coupled via a G protein to phospholipase C. The sequence of events is: (a) increase in InsP3 production following phospholipase C activation; (b) mobilization of intracellular Ca^{2+} stores; (c) activation of a Ca^{2+}-dependent Cl^- channel. The metabotropic Qp receptors expressed in this system have been studied by other groups (Fong *et al.* 1988, Rassendren *et al.* 1989).

Electrophysiological experiments

Recently, the effects of metabotropic Qp receptors have been studied using electrophysiological experiments. Several ionic conductances have been found to be modified by these receptors (see Table 1) (Stratton *et al.* 1989, Baskys *et al.* 1990, Charpak *et al.* 1990, Lester and Jahr 1990, Fagni *et al.* 1991). A detailed analysis of these electrophysiological responses will be discussed further on in this review.

EAA agonists of metabotropic Qp receptors (see Fig. 1)

In the first studies, several non-specific EAA agonists were used to stimulate metabotropic Qp receptors. QUIS is the most potent agonist for all the models studied and is generally followed by ibotenate (IBO) and Glu (for a review, see Sladeczek *et al.* 1988, Schoepp *et al.*1990). Of course, these agonists are not specific (Fig. 1). QUIS and IBO are also good agonists of AMPA and NMDA receptors,

Table 1 — Electrophysiological consequences of metabotropic Qp receptor stimulation

Model used	Agonist used	Neuronal excitability	Nature of the ionic channel modulated	Reference
• Intracellular recording of CA1 hippocampal neurons in slices	*trans*-ACPD QUIS + CNQX (AMPA inactive)	Increased		Stratton *et al.* (1989)
• Voltage clamp of dentate granule hippocampal neurons in slices	QUIS (AMPA inactive)		• Blockade of a slow calcium-dependent potassium channel (I_{AHP})	Baskys *et al.* (1990)
• Current and voltage clamp of CA3 pyramidal cells in slices	*trans*-ACPD (AMPA inactive)	Increased	• Blockade of I_{AHP} and of a voltage-gated potassium current	Charpak *et al.* (1990)
• Patch-clamp (whole cell recording) of CA1 hippocampal neurons in culture	QUIS (AMPA inactive)		• Inhibition of calcium current	Lester and Jahr (1990)
• Patch-clamp (whole cell and cell attached) of cultured cerebellar granule cells	*trans*-ACPD QUIS + CNQX (AMPA inactive)		• Activation of a calcium-dependent potassium channel (big potassium channels)	Fagni *et al.* (1991)

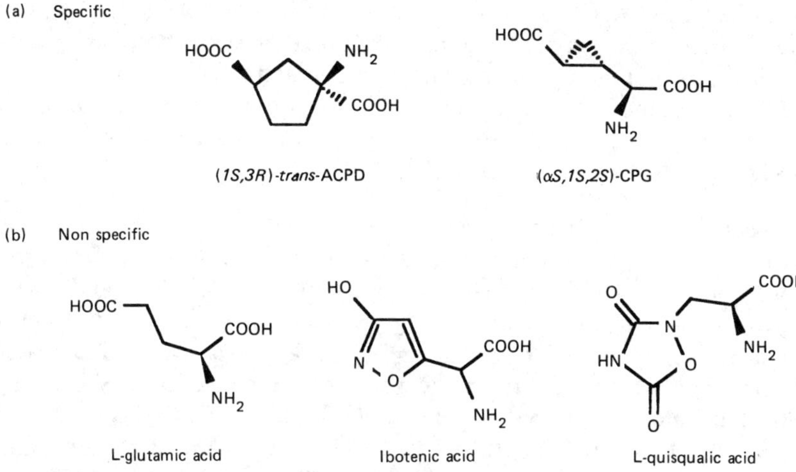

(a) Specific

(1*S*,3*R*)-*trans*-ACPD (α*S*,1*S*,2*S*)-CPG

(b) Non specific

L-glutamic acid Ibotenic acid L-quisqualic acid

Fig. 1 — Metabotropic Qp receptor agonists.

respectively (see chapter 10). Glu is an agonist on all EAA receptors. AMPA at low concentrations is not an agonist of metabotropic Qp receptors (Schoepp and Johnson 1988a, Palmer *et al.* 1988, Récasens *et al.* 1988, Rassendren *et al.* 1989, Patel *et al.* 1990, Manzoni *et al.* 1991). Recently, Manzoni *et al.* (1991) and Patel *et al.* (1990) demonstrated that KAIN and domoate (DOMO), but not NMDA, are weak partial agonists of metabotropic Qp receptors of striatal neurons in culture. The effects of NMDA on the inositol phosphate production are probably indirect via an increase in the intracellular Ca^{2+} concentration. Indeed, it can be blocked by 2-amino-5-phosphonopentanoate (AP5) (Sladeczek *et al.* 1985, Nicoletti *et al.* 1986a).

Two specific agonists of metabotropic Qp receptors have recently been described. The first one is *trans*-1-aminocyclopentane-1,3-dicarboxylate (*trans*-ACPD) (Fig. 1). In brain slices, *trans*-ACPD is less potent than QUIS but has a higher intrinsic activity (Palmer *et al.* 1989, Desai and Conn 1990, Schoepp *et al.* 1991). In contrast, in striatal neurons, *trans*-ACPD has a lower intrinsic activity than QUIS (Manzoni *et al.* 1990a). In this model, a patch-clamp study indicated that *trans*-ACPD does not trigger ionotropic responses up to 1 mM (Manzoni *et al.* 1990a). Other electrophysiological experiments using a more complex model (micro-iontophoretic application into stratum radiatum with intracellular recording from CA1 neurons) suggest that *trans*-ACPD can also activate ionotropic receptors (Curry *et al.* 1988, Magnuson *et al.* 1988). However, in these experiments, as well as a study in which *trans*-ACPD has been shown to produce convulsions after *in vivo* administration (Schoepp *et al.* 1990), one cannot exclude an indirect effect of *trans*-ACPD via an increase in neuronal activity (see Table 1), followed by Glu release. Indeed, the *in vivo* effect of *trans*-ACPD has been shown to be selectively blocked by NMDA antagonists (Schoepp *et al.* 1990).

Trans-ACPD is a racemic mixture of two stereoisomers (1*R*,3*S* and 1*S*,3*R*, see Schoepp and Hillman 1990). In 1990, Irving *et al.* proposed that 1*S*,3*R*-*trans*-ACPD is the only form to activate the metabotropic Qp receptor. They showed that this form was the only one to increase intracellular Ca^{2+} in cerebellar neurons in culture, the

1*R*,*3S* form being inactive when tested at 0.1 mM. Surprisingly, as we showed in a preliminary study, both isomers are able to stimulate inositol phosphate production in striatal neurons and in cerebellar granule cells in culture.

The second compound which has been proposed to be a specific agonist of metabotropic Qp receptors is the 2*S*,*3S*,*4S*-α-(carboxycyclopropyl)glycine (L-CCGI, (α*S*,*1S*,*2S*)-CPG) (Nakagawa *et al.* 1990) (Fig. 1).

EAA antagonists of metabotropic receptors (see Fig. 2)

It is clear that competitive or non-competitive antagonists of NMDA receptors (AP5, 3-(2-carboxypiperazin-4-yl)propyl-1-phosphonate (CPP), MK-801 and phencyclidine (PCP)) as well as competitive antagonists of AMPA receptors (6-cyano-7-nitroquinoxaline-2,3-dione (CNQX) and 6,7-dinitroquinoxaline-2,3-dione (DNQX) do not block metabotropic Qp receptors whatever the model used (Sladeczek *et al.* 1985, 1988, Sugiyama *et al.* 1989, Schoepp and Hillman 1990, Manzoni *et al.* 1991).

L-Serine-*O*-phosphate (L-SOP) and L-AP4 have been found to antagonize the effects of metabotropic Qp receptors but not in all the models tested. Even in the same model, contradictory results have been published (for a review, see Schoepp and Hillman 1990). More consistant results have, however, been published using 2-amino-3-phosphonopropanoate (AP3) as an antagonist. The L-isomer is an antagonist in all the models tested (for a review, see Schoepp and Hillman 1990). It also competes with $[^3H]$glutamate at metabotropic Qp binding sites (Cha *et al.* 1990a), but generally, L-AP3 has been found to be a partial agonist (Schoepp and Johnson 1988a, 1989a, 1989b, Manzoni *et al.* 1991).

Some intriguing properties of the L-AP3 blockade of the metabotropic Qp receptors have been reported: (a) in brain slices (see Schoepp *et al.* 1990) and in striatal neurons in culture (Manzoni *et al.* 1991), the antagonism has been found to be non-competitive and competitive, respectively. However, the non-competitive nature of the inhibition in brain slices could be due to the difficulty met when washing out the drug; (b) as an inhibitor, AP3 is less potent in immature than in mature tissues (Schoepp and Johnson 1989a, 1989b); (c) in *Xenopus* oocytes expressing rat brain metabotropic Qp receptors, AP3 inhibits the *trans*-ACPD, but not the QUIS response (Sugiyama 1990).

This brief review concerning the antagonistic pharmacology of the metabotropic Qp receptors clearly indicates that a potent competitive antagonist of this receptor is essential.

Evidence in favour of several metabotropic receptors

Pharmacologists and molecular biologists have clearly demonstrated that multiplicity of receptor subtypes for a given neurotransmitter is a common rule rather than an

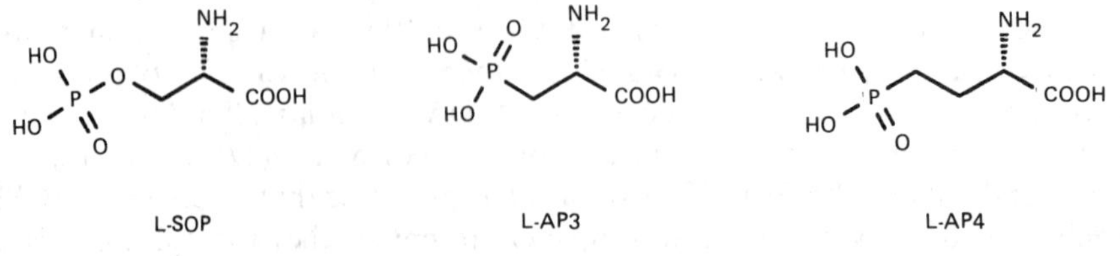

L-SOP L-AP3 L-AP4

Fig. 2 — Putative metabotropic Qp antagonists.

exception. It would not be surprising to find that Glu is able to stimulate several phospholipase C-coupled receptor subtypes. However, in the absence of extensive pharmacological or cloning tools, only indications rather than proof can be accumulated to favour such a heterogeneity:

(1) IBO and QUIS have similar efficacies in triggering inositol phosphate formation in striatal neurons (Manzoni *et al.* 1991) (Fig. 3), whereas IBO is far more efficacious than QUIS in mature hippocampal slices (Nicoletti *et al.* 1986b, Schoepp and Johnson 1988a) (Fig. 3).
(2) Differences in antagonist effectiveness have been found in different preparations (see above and Schoepp and Hillman 1990). In particular, AP4 antagonizes IBO-induced responses more than QUIS-induced responses in the mature rat hippocampus (Schoepp and Johnson 1988a) and has no inhibitory effect in striatal neurons.
(3) As already discussed above, Sugiyama (1990) reported that in *Xenopus* oocytes expressing rat brain mRNA, the Ca^{2+}-sensitive Cl^- current stimulated by *trans*-ACPD, but not by QUIS, is antagonized by AP3. This intriguing result remains to be confirmed.
(4) We have found (Manzoni *et al.* in preparation) that both 1*R*,3*S* and 1*S*,3*R* isomers of *trans*-ACPD increased the inositol phosphate production in striatal neurons and in cerebellar granule cells, whereas Irving *et al.* (1990) found that only the 1*S*,3*R*-*trans*-ACPD increased intracellular Ca^{2+} in cerebellar granule cells.
(5) Regional differences in phosphoinositide responses to *trans*-ACPD versus IBO have been observed (Desai and Conn 1990).
(6) Metabotropic Qp receptors of glial cells may be different from those of neurons, especially when considering the antagonistic effects (for a review, see Schoepp and Hillman 1990).

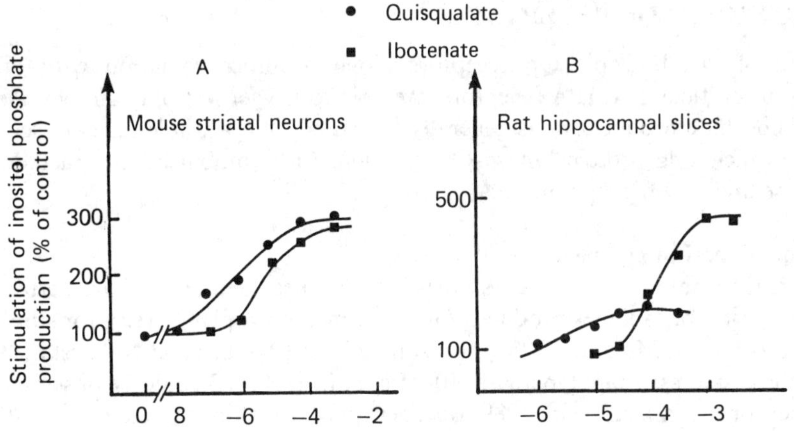

Fig. 3 — Dose-response stimulation of [^{3}H]-labelled phosphoinositide hydrolysis in striatal neurons in culture (A), taken from Manzoni *et al.* (1991), and in brain slices (B), taken from Schoepp and Johnson (1988a).

How can a response, which is mediated by metabotropic receptors, be distinguished from other EAA receptor-mediated responses?
From the above pharmacological and other biochemical data, several simple experiments can be conducted to distinguish a metabotropic Qp-mediated response from other EAA receptor-mediated responses:

(1) AMPA and NMDA should be inactive when tested alone; if they are active, their effects should be blocked by CNQX or MK-801, respectively.
(2) Glu, QUIS and *trans*-ACPD should be active. IBO could be as active as/or more active than QUIS.
(3) CNQX, DNQX and all competitive or non-competitive antagonists of NMDA receptors should be inactive in blocking the *trans*-ACPD or QUIS responses.
(4) L-AP3 should be a partial agonist or an antagonist of the QUIS or *trans*-ACPD-mediated responses.
(5) Phorbol esters (when they can be used) should block the second messenger response (for example, the production of inositol phosphates) mediated by QUIS or *trans*-ACPD. Indeed, it has been shown that phorbol esters inhibit the metabotropic but not the ionotropic-mediated Glu response (Ambrosini and Meldolesi 1989, Manzoni *et al.* 1990b, 1991, Schoepp and Johnson 1988b, Schoepp 1989).

Cloning of the metabotropic Qp receptor
A cDNA of a metabotropic Q receptor has recently been cloned and characterized (Masu *et al.* 1991). This receptor shows no sequence similarity with classical G protein-coupled receptors and has a unique structure with seven transmembrane domains plus a large hydrophilic sequence at both sides. When expressed in *Xenopus* oocytes, its pharmacology is similar to that found in brain, especially in striatal neurons (Manzoni *et al.* 1991).

SECOND MESSENGERS ASSOCIATED WITH THE ACTIVATION OF METABOTROPIC Qp RECEPTORS

Activation of metabotropic Qp receptors alone or in combination with other receptors, in particular AMPA receptors, are able to trigger not only an increase in inositol phosphate production (as generally measured), but a whole series of second messengers including protein kinase C activation, Ca^{2+} mobilization, arachidonic acid release and NO production.

Production of inositol phosphates
Using HPLC techniques, only a few studies have been carried out to characterize the nature of inositol phosphates produced following metabotropic Qp receptor stimulation (Ambrosini and Meldolesi 1989, Manzoni *et al.* 1989, Baird and Nahorski 1990). In all of these studies, a rapid increase (10–15 s) in Ins(1,4,5)P3 could be observed. In striatal neurons (Manzoni *et al.* 1989, Bockaert *et al.* 1991) as well as in rat cortical slices (Baird and Nahorski 1990), stimulation of Ins(1,2,4,5)P4 could also be observed. In rat cortical slices (Baird and Nahorski 1990), AMPA only produced an increase in InsP2, which is probably due to the depolarizing effect of this agent (activation of

phospholipase C by Ca^{2+}). A cooperative action of AMPA receptors and metabotropic Qp receptors has been reported in rat cortical slices (Baird and Nahorski 1990) for the production of inositol phosphates and especially for the production of Ins(1,3,4,5)P4. Such a cooperation between these receptors will be discussed further on in this section.

Mobilization of intra and extracellular Ca^{2+} by metabotropic Qp receptors

Measurement of Ca^{2+} in brain cells

Generally, neuronal or glial cell cultures are used. After loading the cells with the Ca^{2+}-sensitive fluorescent chelator Fura-2 acetoxymethyl ester (Fura-2AM), the Fura-2AM fluorescence from individual cells was monitored after alternative excitation with 340 and 380 nm light, either with a microspectrofluorimeter (Murphy and Miller 1988, 1989) or with an imaging system (Glaum *et al.* 1990, Irving *et al.* 1990, Manzoni *et al.* 1991). For more details on the imaging system technique used in striatal neurons by our group, see legend to Fig. 4. Fagni *et al.* (1991) also estimated (see section below on electrophysiological effects) the increase in Ca^{2+} produced in cerebellar granule cells by measuring the activity of the big K^+ (Ca^{2+}-activated) channel (see Fig. 5).

Mobilization of intracellular Ca^{2+} pools of neurons by metabotropic Qp receptors

Murphy and Miller (1988, 1989) were the first to show that in the absence of extracellular Ca^{2+} these metabotropic receptors are able to induce Ca^{2+} transients (sometimes oscillatory), probably reflecting the release of Ca^{2+} from intracellular stores. In striatal neurons (Fig. 4), the major initial Ca^{2+} increase triggered by *trans*-ACPD seems to be clustered to the proximal parts of neurites and to the soma. No significant *trans*-ACPD-induced Ca^{2+} increase has been seen to occur in remote parts of the neurite. In the same neuron, however, AMPA-induced Ca^{2+} increase is located (at least initially) in the medial and distal parts of the neurites (Fig. 4). The NMDA-induced Ca^{2+} increase is more generalized within the neurons (Fig. 4).

In striatal neurons, we have noted that after a first response to *trans*-ACPD a second response is sometimes difficult to obtain. This may indicate that the refilling of intracellular stores is slow in the absence of stimulatory agents. Similar observations have been made by Murphy and Miller (1988).

Mobilization of intracellular Ca^{2+} pools of glial cells by metabotropic Qp receptors

The metabotropic Qp receptors of glial cells from cortex and hippocampus, but not cerebellum, show an interesting property: the propagation of waves of changes in intracellular free Ca^{2+} concentration through a network of glial cells, possibly via gap junctions (Cornell-Bell *et al.* 1990, Glaum *et al.* 1990). The mechanism and the role of these waves, observed only in glial cells, remain unknown.

Protein kinase C activation and protein phosphorylation induced by metabotropic Qp receptors

As expected for receptors coupled to phospholipase C, the metabotropic Qp receptors should be able to stimulate both the production of inositol phosphates and also protein kinase C. Only two studies have reported protein kinase C activation by

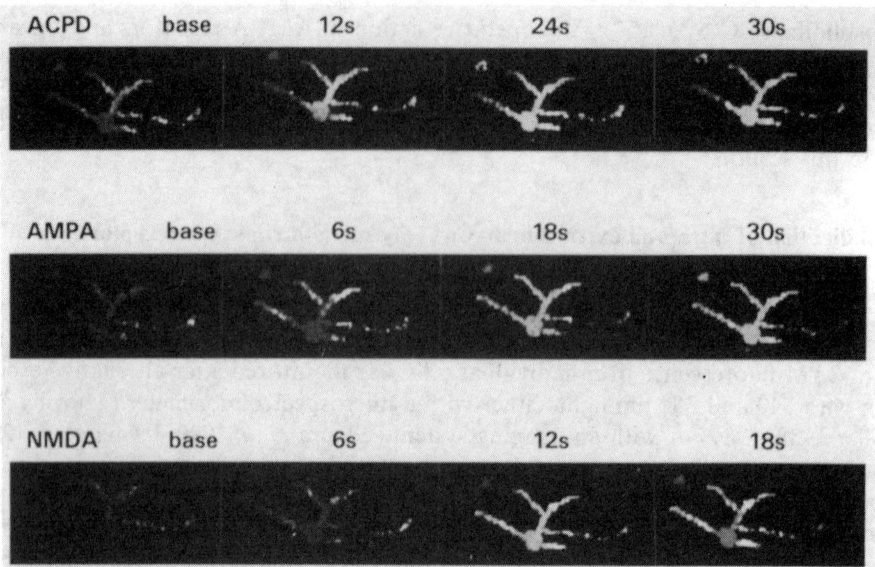

Fig. 4 — (**See colour section.**) Evolution of $[Ca^{2+}]_i$ in striatal neurons after stimulation by ACPD (300 μM) and AMPA (μM). 13 DIV mouse striatal neurons were loaded with the fluorescent calcium chelator Fura-2AM by incubation in 3.3 μM Fura-2AM for 1 hour at 37°C in Hepes buffered saline (HBS) without bovine serum albumin (BSA). After incubation, the cells were washed three times in this buffer. The coverslips containing the cells were mounted for microscopic analysis in HBS without BSA. For $[Ca^{2+}]_i$ measurements by videomicroscopy, the cells were alternatively excited with 340 and 380 nm light and the emission was measured at 500 nm (Grynkiewicz *et al.* 1985). This was done on a Zeiss Axiovert 10 inverse microscope equipped with a photomultiplier and an MSP 20 microprocessor responsible for the filter changes. Images were taken by a low-light-level SIT LH 4036 camera from Lhésa (Paris) and digitized in real time by the Quantel 'Crystal' Imaging System. During the 2-s excitation period at each wavelength, a rolling average of eight videoframes was used and the averaged images saved on a movable Bernoulli disk. The minimum delay between two image couples was 6 s. Ca^{2+} concentrations were calculated by the ratio method described by Grynkiewicz *et al.* (1985). Camera dark noise was subtracted from the recorded crude images. All image treatments were done using home-made software.

metabotropic Qp receptors. Manzoni *et al.* (1990b) demonstrated that in striatal neurons 37.2 ± 2.9% of the total protein kinase C is found in the particulate fraction, whereas after a 30 s stimulation period with 10 μM QUIS 63 ± 7% of this kinase is membrane-bound. Interestingly, the translocation is transient and is not observed when AMPA is used as a stimulator. Similarly, Nicoletti (1990) reported translocation of protein kinase C in cultured glial cells after metabotropic Qp receptor activation. After labelling the striatal neurons with ^{32}P (see legend to Fig. 6 for details), we showed, using a 2D gel separation, that a significant phosphorylation of an 80-kDa and a 50-kDa protein occurred only 5 s after *trans*-ACPD stimulation (300 μM) (Fig. 6) (Manzoni *et al.* 1990c). These proteins are likely to be the major substrates of protein kinase C pp80 and GAP43 — two proteins which are probably implicated in synaptic plasticity (Stumpo *et al.* 1989, Skene 1989).

Production of arachidonic acid and possibly NO by combined activation of AMPA and metabotropic Qp receptors

One of the most exciting results recently obtained with metabotropic Qp receptors is their production of additional intercellular messengers when their stimulation is associated with the stimulation of AMPA receptors. Indeed, such an associative phenomenon is a significant property implicated in synaptic plasticity and especially in the memory process (Hebb 1949).

Production of arachidonic acid in striatal neurons when metabotropic Qp receptor activation is associated with AMPA receptor activation

In striatal neurons in primary cultures after six days *in vitro* (DIV), Glu was able to stimulate basal [^{3}H]arachidonic acid release (Dumuis *et al.* 1988). The effect of Glu was exclusively mediated by NMDA receptors (Dumuis *et al.* 1988). Indeed, NMDA but not QUIS or KAIN was able to produce this [^{3}H]arachidonic acid release. Furthermore, the effects of Glu and NMDA were blocked by competitive and non-competitive NMDA receptor antagonists as well as by high Mg^{2+} concentrations. The effects of NMDA were likely mediated by the Ca^{2+} influx through NMDA receptors followed by activation of phospholipase A2 (PLA2) (Dumuis *et al.* 1988).

We have recently used striatal neurons after 13 DIV (Dumuis *et al.* 1990). These cultures are highly enriched in well-differentiated neurons producing numerous synapses (Weiss *et al.* 1986). These neurons, grown in six-well costar dishes, were incubated with 0.25 μCi [^{3}H]arachidonic acid overnight. They were washed five times with 2 ml per well of Krebs bicarbonate buffer containing 0.05% bovine serum albumin free of fatty acids. Neurons were then stimulated for 15 minutes.

In contrast to what we found in 6 DIV striatal neurons (Dumuis *et al.* 1988), QUIS was as efficacious as NMDA in stimulating [^{3}H]arachidonic acid release in 13 DIV striatal neurons. The potencies of QUIS and NMDA were 1.2 ± 0.2 and 30 ± 7 μM, respectively. We also found that the efficacy of Glu in releasing [^{3}H]arachidonic acid ($620 \pm 78\%$ $n = 10$ increase over the basal release) was at least twice that of NMDA ($245 \pm 49\%$, $n = 7$) or QUIS ($210 \pm 35\%$, $n = 6$) (Dumuis *et al.* 1990).

However, in the presence of MK-801 (1 μM), a specific non-competitive NMDA receptor antagonist: (1) the effect of NMDA (100 mM) was completely suppressed; (2) the effect of QUIS (10 mM) remained unchanged; (3) the effect of Glu (30 mM) was reduced to the level of the QUIS effect (Dumuis *et al.* 1990). Furthermore, the effects of QUIS and NMDA were additive. Taken together these observations demonstrate that the Glu effect results from the activation of both NMDA and QUIS type receptors. This indicates that the NMDA and QUIS receptors involved in [^{3}H]arachidonic acid release were either located in different neurons or that they stimulated different pools of [^{3}H]arachidonic acid precursors, within the same neurons.

HPLC analyses were performed to analyse the nature of the [^{3}H]arachidonic acid metabolites released by these different EAAs. After EAA (QUIS, NMDA or Glu) stimulation, the [^{3}H]arachidonic acid metabolites released from the cells were extracted twice with 5 vol. chloroform. After injection on a reverse-phase HPLC column, the radioactivity was eluted quite exclusively with arachidonic acid. No radioactive compound was detected at the retention time of prostaglandins or hydroxyeicosatetraenoic acids (HETEs). We have previously reported that there were trace amounts co-eluting with some HETEs in Glu-stimulated 6 DIV striatal neurons, (Dumuis *et al.* 1988). Extensive direct and reverse-phase HPLC analyses performed after chloroform extraction of the medium revealed that this radioactivity did not correspond to HETEs (in preparation). Arachidonic acid was found to be the only eicosanoid released from neurons (even when stimulated with ionomycin) whereas cultures of glial cells produced arachidonic acid metabolites (PGF2a; PGE2; PGD2; HHT; 5,11,15-HETE) when stimulated with ionomycin. Such a specificity is intriguing and may be an indication of an important role of arachidonic acid in neuroglial communications.

The nature of the QUIS receptors involved in [^{3}H]arachidonic acid release was investigated. We found that CNQX, a selective antagonist of AMPA receptors (which does not inhibit the metabotropic Qp receptor) (Sladeczek *et al.* 1988) inhibited the QUIS-induced [^{3}H]arachidonic acid release (Dumuis *et al.* 1990) (Ki 3.7 $\pm$ 0.55 mM, $n = 6$). This indicates that AMPA receptors were involved in QUIS-induced [^{3}H]arachidonic acid release. However, the effect of QUIS on [^{3}H]arachidonic acid release was also blocked by the phorbol ester 12,13-dibutyrate (PdBu), a protein kinase C stimulatory agent, and not by 4-α-phorbol 12,13-didecanoate (PdDc), a phorbol ester inactive on C-kinase (Dumuis *et al.* 1990). In the same system, PdBu has been found to block the stimulation of metabotropic Qp receptors (Manzoni *et al.* 1990a) whereas we found this drug inactive on AMPA receptor-mediated [^{3}H]GABA release (Bockaert *et al.* 1990) or AMPA- and NMDA-induced increases in $[Ca^{2+}]_i$ (Manzoni *et al.* 1991). This result indicates that metabotropic Qp was also involved in QUIS-induced [^{3}H]arachidonic acid release. Taken together, these results led us to focus our attention on the possible cooperation between AMPA and metabotropic Qp receptors in stimulating [^{3}H]arachidonic acid release. AMPA, which is the more potent and selective AMPA receptor agonist, elicited only a weak effect in stimulating [^{3}H]arachidonic acid release. *Trans*-ACPD has recently been described as a selective agonist of metabotropic Qp receptors (Manzoni *et al.* 1990a, Palmer *et al.* 1989) and is able to stimulate inositol phosphate production when applied alone on striatal neurons (Manzoni *et al.* 1990a). In the same model, *trans*-ACPD alone has a very weak effect on [^{3}H]arachidonic acid release, whereas it has a strong effect in the presence of 30 mM AMPA (Dumuis *et al.* 1990). Clearly, [^{3}H]arachidonic acid release resulted from the associative stimulation of both AMPA and metabotropic Qp receptors (*trans*-ACPD-sensitive). The parallelism between this associative property of AMPA and Qp receptors and the associative property of AMPA and NMDA receptors to stimulate [^{3}H]arachidonic acid was striking. Indeed, in the presence of physiological concentrations of Mg^{2+}, we could demonstrate that NMDA (100 mM) was able to stimulate [^{3}H]arachidonic acid release only in the presence of AMPA (30 mM) (Dumuis *et al.* 1990).

The mechanisms by which AMPA and NMDA receptors cooperate in stimulating PLA2 are quite obvious. AMPA depolarizes the neurons, the Mg^{2+} blockade of NMDA receptor channels is suppressed, the NMDA receptors are activated and the Ca^{2+} which enters the cell through the NMDA channels activates PLA2. The effect of NMDA is blocked by mepacrine, a PLA2 inhibitor, and by suppression of external Ca^{2+} (Dumuis *et al.* 1988). However, the localization of PLA2 may be closely related to NMDA channels rather than homogeneously distributed in neurons since an increase in intracellular Ca^{2+} due to activation of voltage-sensitive Ca^{2+} channels (VSCC) by K^+ (56 mM) does not stimulate PLA2 (Dumuis *et al.* 1988). Such a specific localization of PLA2 has already been postulated in C62B glioma cells (Brooks *et al.* 1989). In these cells, activation of PLA2 by Ca^{2+} via VSCC was observed, whereas PLA2 was not activated by Ca^{2+} released intracellularly despite the fact that the $[Ca^{2+}]_i$ reached was higher in the latter than in the former situation (Brooks *et al.* 1989).

The mechanisms by which AMPA and metabotropic Qp receptors cooperate in producing [^{3}H]arachidonic acid are unknown. The role of AMPA is likely to depolarize the neurons since we have observed that its role can be mimicked by veratridine. PLA2 activation is involved since mepacrine blocked the QUIS effect whereas RHC 80267, a rather specific inhibitor of diacylglycerol lipase, was without effect (Dumuis *et al.* 1990).

The association between metabotropic Qp and AMPA receptors for the production of arachidonic acid is illustrated in Fig. 7.

Evidence for the production of NO and cyclic GMP when metabotropic Qp receptor activation is associated with AMPA receptor activation

In a very elegant set of experiments, Ito and Karachot (1989, 1990) recently studied the characteristics of long-term desensitization of AMPA responses in Purkinje cells using the wedge recording from rat cerebellar slices.

Long-term (several hours) desensitization of the AMPA response could be obtained following brief exposure to 100 μM QUIS, but not to AMPA alone. Association of the AMPA exposure with *trans*-ACPD exposure did, however, produce desensitization. The roles of NO and cyclic GMP (cGMP) in the desensitization process are suggested by the observation that brief application of AMPA with sodium nitroprusside and 8-bromo-cGMP did in fact produce desensitization. Unfortunately, the experiments reported do not indicate if sodium nitroprusside or 8-bromo-cGMP alone were able to induce desensitization. Furthermore, L-NMMA (an inhibitor of NO synthase), haemoglobin (which strongly binds to NO) and an inhibitor of cGMP protein kinase, were able to block QUIS-induced desensitization.

These experiments suggest that metabotropic Qp receptors alone, or in association with AMPA receptors, produce a cascade of events leading to NO and cGMP production and that the latter desensitizes the AMPA ionotropic receptor. This is schematically represented in Fig. 7.

REGIONAL LOCALIZATION OF METABOTROPIC Qp RECEPTORS

As already stated, metabotropic Qp receptors are localized in glial cells (Usowicz *et al.* 1989, Cornell-Bell *et al.* 1990, Glaum *et al.* 1990). Metabotropic Qp receptors have been described in striatal neurons (Sladeczek *et al.* 1985, Ambrosini and Meldolesi 1989, Weiss *et al.* 1989), hippocampal neurons (Ambrosini and Meldolesi 1989), cerebellar granule cells (Nicoletti *et al.* 1986) and cortical neurons (Patel *et al.* 1990). These receptors have also been found in brain slices prepared from hippocampus, striatum, cerebral and cerebellar cortex (for a review, see Schoepp and Hillman, 1990).

A binding assay on brain slices followed by quantitative autoradiography has been developed using [^{3}H]Glu as a ligand (Cha *et al.* 1990b). This binding was conducted in the presence of 10 μM AMPA and 100 mM NMDA. Non-specific binding was determined in the presence of 2.5 μM QUIS. [^{3}H]Glu binding was displaced by low concentrations of *trans*-ACPD, QUIS and Glu. The regional distribution of these binding sites was different from the distribution of other EAA receptors (Monaghan *et al.* 1989). Binding was highest in lateral septum, cerebellum, striatum, cingulate and entorhinal cortices and lower in hippocampus (CA3 and CA1 regions contained the same binding site densities). Thalamus, brain stem, frontal and parietal cortex had the lowest binding site densities (Cha *et al.* 1990b). [^{3}H]Glu binding on metabotropic Qp receptors was markedly reduced in a mouse mutant lacking the molecular layer (the area in which Purkinje cell dendrites are formed (Cha *et al.* 1990)). Biochemical studies using mice lacking Purkinje cells indicated that the most potent stimulation by EAA of inositol phosphate production in cerebellar slices occurs in these cells (Blackstone *et al.* 1989).

In brain slices, the incorporation of [^{3}H]cytidine into cytidine diphosphate diacylglycerol (a membrane-bound compound) is stimulated when phosphoinositide turnover is increased. In 1990, Hwang *et al.* took advantage of this observation, and using autoradiography techniques they localized the areas in which the incorporation occurred, following *trans*-ACPD stimulation. In cerebellun, the labelling was high in Purkinje cells and in the molecular layer, whereas in the hippocampus it was much higher in CA3 than in CA1. This result was not really expected for binding studies (Cha *et al.* 1990). However, a linear relationship does not necessarily exist between density of sites and their ability to stimulate phosphoinositide metabolism.

ELECTROPHYSIOLOGICAL EFFECTS OF METABOTROPIC Qp RECEPTORS

Activation of Ca^{2+}-sensitive K^+ conductance

Using the cell-attached configuration of the patch-clamp technique to record large-conductance Ca^{2+}-activated K^+ channels (BK channels) in cultured cerebellar granule cells, we found that *trans*-ACPD (100 μM) applied outside the patch pipette, in the absence of external Ca^{2+}, increased the open probability (P_o) of the channel (Fig. 5a) (Fagni *et al.* 1991). A similar effect was observed with QUIS (2 μM) (Fig. 5b), but not with AMPA (2 μM) (Fig. 5c) in the presence of CNQX (30 μM). This pharmacology suggested that the effects were mediated by the metabotropic Qp receptor. Furthermore, because of the tight seal between the recording pipette and the membrane patch, it was likely that BK channel activation involved a second messenger system, probably Ins(1,4,5)P3 and internal Ca^{2+}.

We measured the voltage and Ca^{2+} sensitivities of BK channels in these cells (Fig. 5d and 5e). Hill plots relating P_o to $[Ca^{2+}]_i$ obtained in inside-out patches provided a way of estimating variation in $[Ca^{2+}]_i$ (in the immediate vicinity of the channels) induced by *trans*-ACPD under cell-attached conditions. For instance, P_o of the channel measured in cell-attached configuration at a pipette potential of -60 mV (equivalent to membrane potential of 0 mV, assuming that the cell was at a resting potential of -60 mV) was 0.02 ± 0.01 (mean $\pm$ SD, $n = 11$) in the absence of drug and increased to 0.14 ± 0.06 ($n = 11$) in the presence of *trans*-ACPD. If we consider the linear regression of the Hill plot (Fig. 5e), obtained at 0 mV in inside-out configuration, these P_o values give a pCa_i of 6.58 ($[Ca^{2+}]_i = 260$ nM) and 6.22 ($[Ca^{2+}]_i = 600$ nM), respectively.

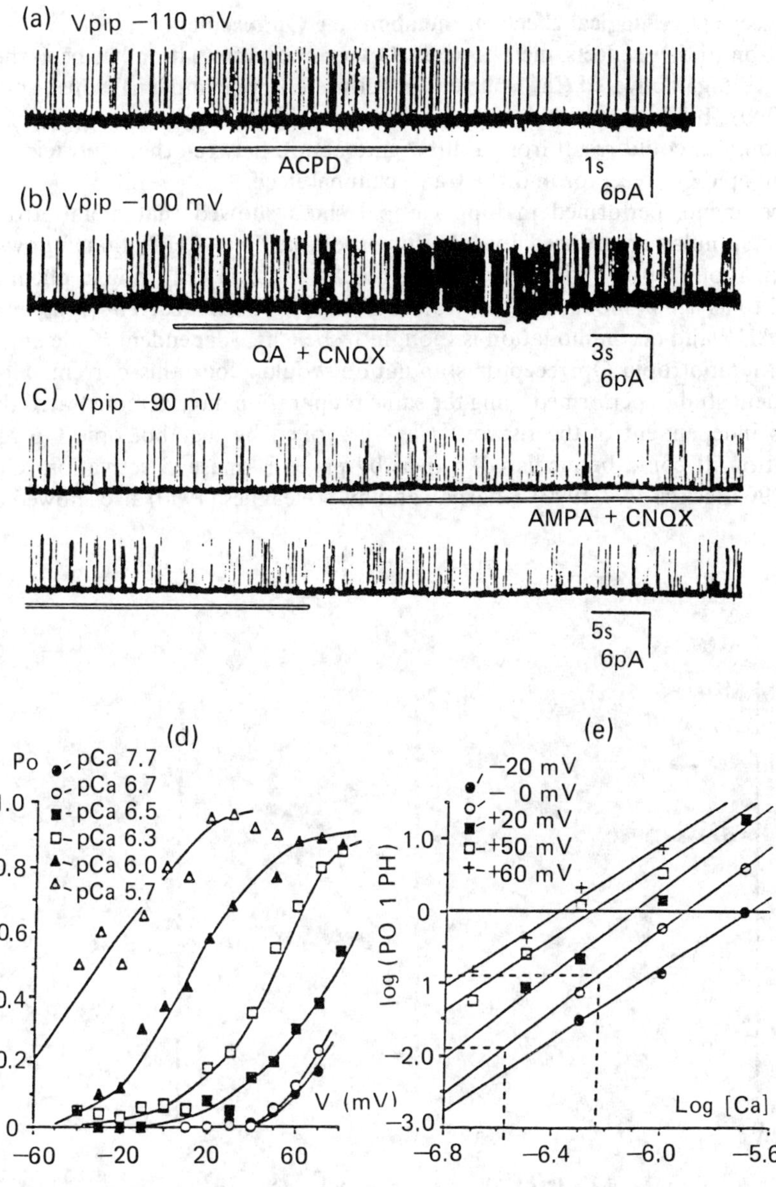

Fig. 5 — Electrophysiological effects of metabotropic Qp receptor stimulation in cultured cerebellar neurons. a, b, c: Effects of bath-applied glutamate agonists (*trans*-ACPD 100 μM, QUIS 2 μM, AMPA 2 μM) and antagonists (CNQX 30 μM) on BK channels recorded under cell-attached conditions. Pipette potential is indicated on each trace. Note the presence of a smaller-amplitude single-channel current different from the studied BK-channel. Recording pipette solution contained (mM): NaCl (130), $CaCl_2$ (2), $MgCl_2$ (2), KCl (5), TTX (0.003), Hepes-Tris (10), pH 7.4. Bath medium had a similar composition except that NaCl was replaced by choline chloride (no TTX), $MgCl_2$ concentration was 5 mM and no Ca^{2+} was added to the solution. D, E: Relationship between P_o and membrane potential obtained at different pCa_i: (d) and related Hill plots (e) obtained from the same inside-out patch. For significance of the dotted lines, see explanation in the text. Internal solution (bath) contained (mM): KCl (130), $MgCl_2$ (2), Hepes-Tris (10) and Ca^{2+} was buffered at the indicated pCa_i with EGTA (11 mM) ($K_D = 10^{-7}$). External solution (pipette) contained (mM): NaCl; (130), $CaCl_2$ (2), $MgCl_2$ (2), KCl (5), Hepes-Tris (10), TTX (0.003), pH 7.4.

Other electrophysiological effects of metabotropic Qp receptors
In addition to these effects, metabotropic Qp receptor stimulation has been shown to depress voltage-activated Ca^{2+} current in cultured hippocampal neurons (Lester and Jahr 1990). Inhibition of this current did not apparently involve protein kinase activation, but could result from a direct interaction between the G protein-coupled metabotropic Qp receptor and the Ca^{2+} channel itself.

Experiments performed in hippocampal slices showed that metabotropic Qp receptor stimulation reduced the afterhyperpolarization (AHP) that follows spike depolarization in pyramidal neurons and blocks pause in cell firing, a phenomenon referred to as accommodation (Stratton *et al.* 1989). The K^+ current (I_{AHP}) responsible for AHP and accommodation is thought to be Ca^{2+}-dependent (Cole and Nicoll 1983). Metabotropic Qp receptor stimulation would block this current. However, subsequent studies performed using the same preparation have demonstrated that this effect is independent of the rise on $[Ca^{2+}]_i$ evoked by metabotropic Qp receptor stimulation. It could be mediated, rather, by protein kinase C activation (Charpak *et al.* 1990, Baskys *et al.* 1990). Charpak and his colleagues (1990) also showed that an

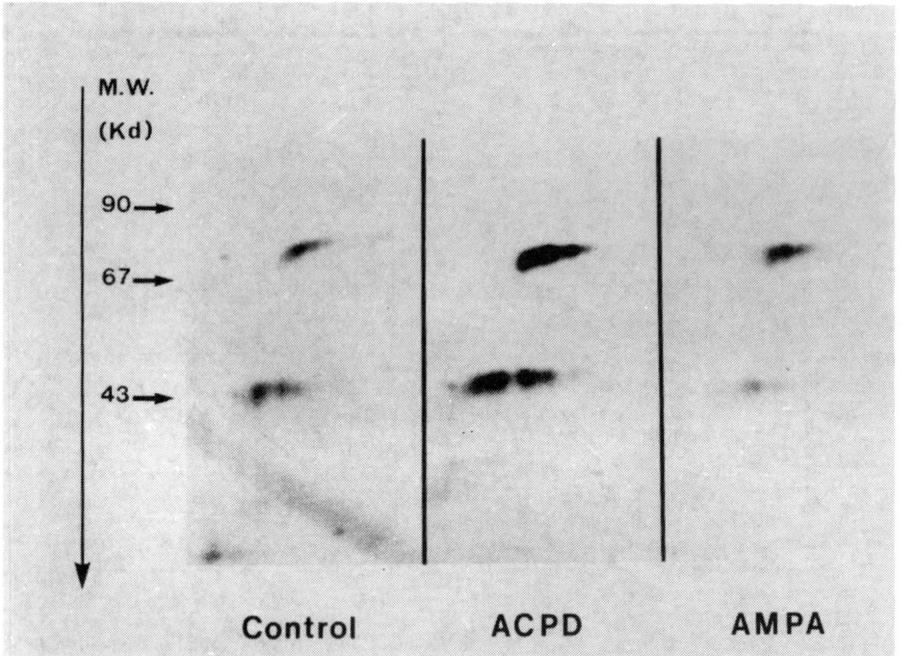

Fig. 6 — Protein phosphorylation after *trans*-ACPD (300 μM) stimulation. Neurons were prelabelled for 45 minutes with 150 μCi/ml of inorganic ^{32}P at room temperature in phosphate-free HBS buffer, pH 7.2. Five seconds before the end of the labelling period, *trans*-ACPD was added to obtain the indicated concentrations. The reaction was stopped by removing the medium and putting the cells on ice. The cells were then rapidly dissolved in 0.5% NP-40, 0.25% SDS and 1.25% BME Tris HCl buffer. After DNAse and RNAse treatment, the cell precipitate was dissolved in urea NP-40 buffer, for two-dimensional gel electrophoresis. Isoelectric focusing was performed using a 30% acrylamide, 2% bis-acrylamide, 10% NP-40, 8% ampholite gel. The specific ampholytes employed had pH 3–10 and 2–4 (80% and 20%, respectively). After focusing at 1000 V for 15 hours, the gels were overlaid on a SDS-acrylamide uniform slab gel (12.5% acrylamide).

additional K^+ current, the M-current, was blocked by metabotropic Qp receptor agonists and suggested that this effect was also Ca^{2+}-independent.

The metabotropic Qp receptor-mediated K^+ channel inhibition reported in the hippocampus is not in discrepancy with the *trans*-ACPD-induced activation of BK channels that we observed in cerebellar cells. Indeed, there are multiple Ca^{2+}-sensitive K^+ channels in central neurons. One class has a small conductancy (SK channels), is sensitive to apamine, is mainly active at negative membrane potentials and is responsible for the slow AHP (Lang and Ritchie 1990, Lancaster *et al.* 1991). A second class of Ca^{2+}-dependent K^+ channels is the BK channel. It is sensitive to charybdotoxin and blocked by low millimolar concentrations of tetraethylammonium ions; it is mainly active at positive membrane potentials (as shown in Fig. 5a–d; see also Lancaster *et al.* 1991) and it contributes to cell repolarization (fast AHP) that immediately follows action potential, especially in the part usually due to voltage-activated Ca^{2+} channel activity. Activation of metabotropic Qp receptors at negative membrane potential would most probably close SK channels via protein kinase C activation. This would explain the small depolarization from resting potential and concomitant cell firing observed in hippocampal cells in the presence of metabotropic Qp receptor agonists (Stratton *et al.* 1989).

PHYSIOLOGICAL FUNCTIONS OF METABOTROPIC Qp RECEPTORS

An increasing number of observations indicate that metabotropic Qp receptors alone or in combination with other receptors, such as AMPA receptors, may have a crucial role in synaptic plasticity events.

General observations in favour of a role in synaptic plasticity

It has long been recognized that the efficacy of metabotropic Qp receptors is greater during the early postnatal period of rat brain development (for reviews, see Sladeczek *et al.* 1988, Schoepp and Hillman 1990). Enhanced EAA-stimulated inositol phosphate production is found in brain slices following amygdala or hippocampal kindling to rats (Iadarola *et al.* 1986, Akiyama *et al.* 1987). In the kitten striate cortex, the age-dependent intensity of the stimulation of inositol phosphate production correlates with postnatal changes in cortical susceptibility to visual deprivation (Dudek and Bear 1989).

Role of the association between metabotropic Qp receptors and AMPA receptors in synaptic plasticity

We have described the unique properties of the association of metabotropic Qp receptors and AMPA receptors in two events: (1) the release of arachidonic acid (Dumuis *et al.* 1988) and (2) the long-term desensitization of AMPA responses in Purkinje cells (Ito and Karachot 1989, 1990) probably mediated by NO and cGMP.

It is clear that the associative stimulation of AMPA and metabotropic Qp receptors produced the same intracellular and intercellular messengers as the associative stimulation of AMPA and NMDA receptors: Na^+ entry, Ins(1,4,5)P3 production, Ca^{2+} mobilization (from intra- and/or extracellular compartments), protein kinase C activation, arachidonic acid release. The role of the associative stimulation of AMPA and NMDA receptors in synaptic plasticity and, in particular, in LTP, is well known

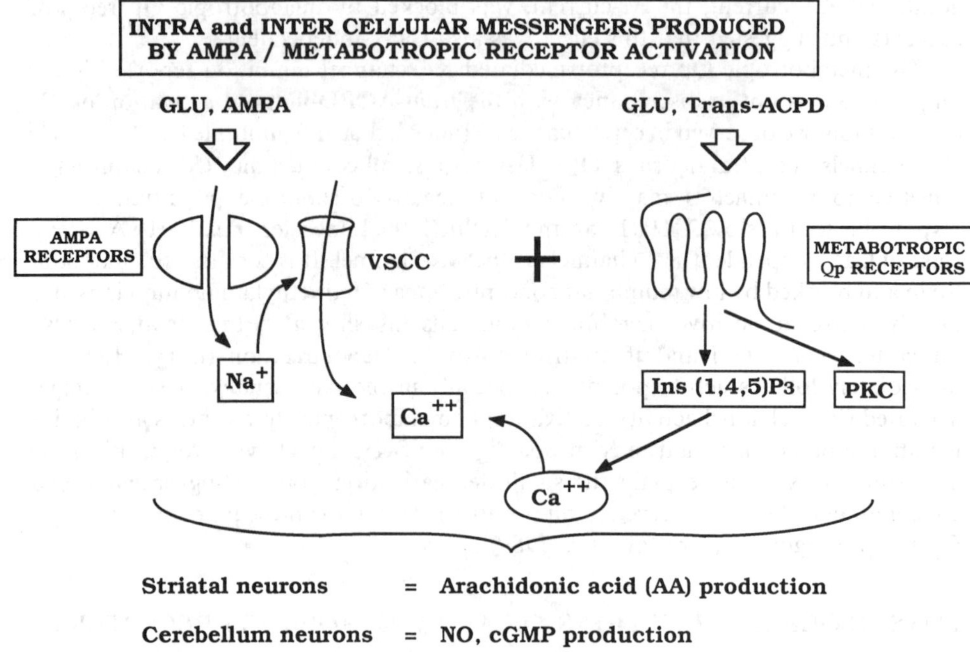

Fig. 7 — Second messengers triggered by the association between AMPA and metabotropic Qp receptor stimulations.

(Nicoll *et al.* 1988). In view of the similarities in the primary effects occurring after activation of AMPA/NMDA and AMPA/metabotropic Qp receptors, some similarities in their final effects might be expected in LTP or LTD, for example.

Indeed, it has been proposed that arachidonic acid could play an important role in synaptic plasticity. Williams and co-workers (1989) showed that arachidonic acid plays a role in the genesis of LTP, and several other groups have reported a blockade of LTP with inhibitors of PLA2 (Lynch *et al.* 1989, Okada *et al.* 1989).

We have already commented on the demonstration made by Ito and Karachot (1989, 1990) that AMPA receptor desensitization in Purkinje cells is induced by the associative stimulation of AMPA and metabotropic Qp receptors and is likely to be mediated by NO and cGMP. LTD is a plasticity event, seen in Purkinje parallel fibre (the axons of cerebellar cortical granule cells) synapses, which may be the key event to the adaptation of the vestibulo-ocular reflex. This LTD occurs when these synapses are repeatedly activated at approximately the same time as the climbing fibre synapse (the axon of an inferior olive neuron) to the same Purkinje cell. The parallel fibre/Purkinje synapse is a Glu synapse, and the receptors involved are of the AMPA type. The climbing fibre neurotransmitter has not yet been identified. It is possible, but has not yet been demonstrated, that metabotropic Qp receptors are involved at climbing fibre/Purkinje cell synapses. It has recently been shown that NO is released after climbing fibre stimulation, that LTD is blocked by haemoglobin (which strongly binds NO) and that exogenous NO or cGMP can substitute for the stimulation of climbing fibres to cause LTD (Shibuki and Okada 1991).

ACKNOWLEDGEMENTS

We would like to thank A. Turner-Madeuf and M. Passama for their help in the preparation of this manuscript. Support was provided by the Centre National de Recherche Scientifique (CNRS), the Institut National de la Santé et de la Recherche Médicale (INSERM), the Direction des Recherches et Etudes Techniques (DRET — contract no. 88/163) and Bayer Troponwerke (France and Germany).

REFERENCES

Akiyama, K., Yamada, N. and Sato, M. (1987) *Exp. Neurol.* **98** 499–508.

Ambrosini, A. and Meldolesi, J. (1989) *J. Neurochem.* **53** 825–833.

Baird, J. G. and Nahorski, S. R. (1990) *Br. J. Pharmacol.* **100** (suppl.) 315.

Baskys, A., Bernstein, N. K., Barolet, A. W. and Carlen, P. L. (1990) *Neurosci. Lett.* **112** 76–81.

Berridge, M. J., Dawson, R. M. C., Downes, C. P., Heslop, J. P. and Irvine, R. F. (1983) *Biochem. J.* **21** 473–482.

Bettler, B., Boulter, J., Hermans-Borgmeyer, I., O'Shea-Greenfield, A., Deneris, E. S., Moll, C., Borgmeyer, U., Hollmann, M. and Heineman, S. (1990) *Neuron* **5** 583–595.

Betz, (1990) *Neuron* **5** 383–392.

Blackstone, C. D., Supattapone, S. and Snyder, S. H. (1989) *Proc. Nat. Acad. Sci. USA* **86** 4316–4320.

Bockaert, J., Dumuis, A., Manzoni, O., Nargeot, J., Oomagari, K., Pin, J. P., Rassendren, F., Sebben, M. and Sladeczek, F. (1990) In: Ben Ari, Y. (ed.) *Excitatory Amino Acids and Neuronal Plasticity.* Plenum Press, New York.

Boulter, J., Hollmann, M., O'Shae-Greenfield, A., Harley, M., Deneris, E., Maron, C. and Heinemann, S. (1990) *Science* **249** 1033–1037.

Brooks, R. C., McCarthy, K. D., Lapetina, E. G. and Morell, P. (1989) *J. Biol. Chem.* **264** 20147–20153.

Cha, J. H. J., Makowiec, R. L., Penney, J. B. and Young, A. B. (1990a) *Soc. Neurosci. Abstr.* **16** 548.

Cha, J. H. J., Makowiec, R. L., Penney, J. B. and Young, A. B. (1990b) *Neurosci. Lett.* **113** 78–83.

Changeux, J. P. (1990) In: Fidia Research Foundation (ed.) *Neuroscience Award Lectures*, Vol. 4. Raven Press, New York, pp. 21–168.

Charpak, S., Gähwiler, B. H., Do, K. Q. and Knöpfel, T. (1990) *Nature* **347** 765–757.

Charpentier, N., Dumuis, A., Sebben, M., Bockaert, J. and Pin, J. P. (1990) *Eur. J. Pharmacol.* **189** 241–251.

Cole, A. E. and Nicoll, R. A. (1983) *Science* **221** 1299–1301.

Cornell-Bell, A. H., Finkbeiner, S. M., Cooper, M. S. and Smith, S. J. (1990) *Science* **247** 470–473.

Curry, K., Peet, M. J., Magnuson, D. S. K. and McLennan, H. (1988) *J. Med. Chem.* **31** 864–867.

Desai, M. A. and Conn, P. J. (1990) *Neurosci. Lett.* **109** 157–162.

Dudek, S. M. and Bear, M. F. (1989) *Science* **246** 673–675.

Dumuis, A., Sebben, M., Haynes, L., Pin, J. P. and Bockaert, J. (1988) *Nature* **336** 68–70.

Dumuis, A., Pin, J. P., Oomagari, K., Sebben, M. and Bockaert, J. (1990) *Nature* **347** 182–184.

Fagni, L., Bossu, J. L. and Bockaert, J. (1991) *Eur. J. Neurosci.* **3** 778–789.

Fong, T. M., Davidson, N. and Lester, H. A. (1988) *Synapse* **2** 657–665.

Glaum, S. R., Holzwarth, J. A. and Miller, R. J. (1990) *Proc. Nat. Acad. Sci. USA* **87** 3454–3458.

Grynkiewicz, G., Poenie, M. and Tsien, R. Y. (1985) *J. Biol. Chem.* **260** 3400–3450.

Hebb, D. O. (1949) *The Organization of Behaviour*. Wiley, New York.

Henley, J. M., Ambrosini, A., Krogsgaard-Larsen, P. and Barnard, E. A. (1989) *New Biologist* **1** 153–158.

Hwang, P. M., Bredt, D. and Snyder, S. H. (1990) *Science* **249** 802–804.

Iadarola, M. J., Nicoletti, F., Naranjo, J. R., Putnam, P. and Costa, E. (1986) *Brain Res.* **347** 174–178.

Irving, A. J., Schofield, J. G., Watkins, J. C., Sunter, D. C. and Collingridge, G. L. (1990) *Eur. J. Pharmacol.* **186** 363–365.

Ito, M. and Karachot, L. (1989) *Neurosci. Res.* **7** 168–171.

Ito, M. and Karachot, L. (1990) *Neuro Report* **1** 129–132.

Keinänen, K., Wisden, W., Sommer, B., Werner, P., Herb, A., Verdoorn, T. A., Sakmann, B. and Seeburg, P. H. (1990) *Science* **249** 556–560.

Lancaster, B., Nicoll, R. A. and Perkel, D. J. (1991) *J. Neurosci.* **11** 23–30.

Lang, D. G. and Ritchie, A. K. (1990) *J. Physiol.* **425** 117–132.

Lester, R. A. J. and Jahr, C. E. (1990) *Neuron* **4** 741–749.

Lynch, M. A., Errington, M. L. and Bliss, T. V. P. (1989) *Neuroscience* **30** 693–701.

Magnuson, D. S. K., Curry, K., Peet, M. J. and McLennan, H. (1988) *Exp. Brain Res.* **73** 541–545.

Manzoni, O., Bockaert, J. and Sladeczek, F. (1989) 3ème Colloque National des Neurosciences, Montpellier, 9–12 May 1989, abstract, p. 297.

Manzoni, O., Fagni, L., Pin, J. P., Rassendren, F., Poulat, F., Sladeczek, F. and Bockaert, J. (1990a) *Mol. Pharmacol.* **38** 1.

Manzoni, O., Finiels-Marlier, F., Sassetti, I., Bockaert, J., Le Peuch, C. and Sladeczek, F. (1990b) *Neurosci. Lett.* **109** 146–151.

Manzoni, O., Finiels-Marlier, F., Poulat, F., Bockaert, J., Le Peuch, C. and Sladeczek, F. (1990c) *Neurochem. Int.* **16** (Suppl. 1) 52.

Manzoni, O. J. J., Poulat, F., Do, E., Sassetti, I., Bockaert, J. and Sladeczek, F. (1991) *Eur. J. Pharmacol.* **207** 231–241.

Masu, M., Tanabe, Y., Tsuchida, K., Shigemoto, R. and Nakanishi, S. (1991) *Nature* **349** 760–765.

Monaghan, D. T., Bridges, R. J. and Cotman, C. W. (1989) *Annu. Rev. Pharmacol. Toxicol.* **29** 365–402.

Murphy, S. N. and Miller, R. J. (1988) *Proc. Nat. Acad. Sci. USA* **85** 8737–8741.

Murphy, S. N. and Miller, R. J. (1989) *Mol. Pharmacol.* **35** 671–680.

Nakagawa, Y., Saitoh, K., Ishihara, T., Ishida, M. and Shinozaki, H. (1990) *Eur. J. Pharmacol.* **184** 205–206.

Nawy, S. and Jahr, C. E. (1990) *Nature* **346** 269–271.

Nicoletti, F., Iadarola, M. J., Wroblewski, J. T. and Costa, E. (1986a) *Proc. Nat. Acad. Sci. USA* **83** 1931–1935.

Nicoletti, F., Wroblewski, J. T., Novelli, A., Alho, H., Guidotti, A. and Costa, E. (1986b) *J. Neurosci.* **6** 1905–1911.

Nicoletti, F. (1990) *J. Neurochem.* **54** 771–777.

Nicoll, R. A., Kauer, J. A. and Malenka, R. C. (1988) *Neuron* **1** 97–103.

Okada, D., Yamagishi, S. and Sugiyama, H. (1989) *Neurosci. Lett.* **100** 141–146.

Palmer, E., Monaghan, D. T. and Cotman, C. W. (1988) *Mol. Brain Res.* **4** 161–165.

Palmer, E., Monaghan, D. T. and Cotman, C. W. (1988) *Eur. J. Pharmacol.* **166** 585–587.

Patel, J., Moore, N. C., Thompson, C., Keith, R. A. and Salama, A. I. (1990) *J. Neurochem.* **54** 1461–1466.

Pin, J. P., Van Vliet, B. J. and Bockaert, J. (1989) *Eur. J. Pharmacol.* **172** 81–91.

Rassendren, F. A., Lory, P., Pin, J. P., Bockaert, J. and Nargeot, J. (1989) *Neurosci. Lett.* **99** 333–339.

Récasens, M., Guiramand, J., Nourigat, A., Sassetti, I. and Devilliers, G. (1988) *Neurochem. Int.* **13** 463–467.

Schoepp, D. D. (1989) *Neurochem. Int.* **15** 131–136.

Schoepp, D. D. and Johnson, B. G. (1988a) *J. Neurochem.* **50** 1605–1613.

Schoepp, D. D. and Johnson, B. G. (1988b) *Biochem. Pharmacol.* **37** 4299–4305.

Schoepp, D. D. and Johnson, B. G. (1989a) *J. Neurochem.* **53** 273–278.

Schoepp, D. D. and Johnson, B. G. (1989b) *J. Neurochem.* **53** 1865–1870.

Schoepp, D. D. and Hillman, C. C. (1990) *Biogenic Amines* **7** 331–340.

Schoepp, D. D., Bockaert, J. and Sladeczek, F. (1990) *Trends Pharmacol. Sci.* **11** 508–515.

Schoepp, D. D., Johnson, B. G., Salhoff, C. R., MacDonald, J. W. and Johnston, M. V. (1991) *J. Neurochem.* **56** (in press).

Shibuku, K. and Okada, D. (1991) *Nature* **349** 326–328.

Skene, J. H. P. (1989) *Annu. Rev. Neurosci.* **12** 127–142.

Sladeczek, F., Pin, J. P., Récasens, M., Bockaert, J. and Weiss, S. (1985) *Nature* **317** 717–719.

Sladeczek, F., Récasens, M. and Bockaert, J. (1988) *Trends Neurosci.* **11** 545–549.

Sommer, B., Keinänen, K., Verdoorn, T. A., Wisden, W., Burnashev, N., Berb, A., Köhler, M., Takagi, T., Sakman, B. and Seeburg, P. H. (1990) *Science* **249** 1580–1585.

Strader, C. O., Sigal, I. S. and Dixon, R. A. F. (1989) *FASEB J.* **3** 1825–1832.

Stratton, K. R., Worley, P. F. and Baraban, J. M. (1989) *Eur. J. Pharmacol.* **173** 235–237.

Stumpo, D. J., Graff, J. M., Albert, K. A., Greengard, P. and Blackshear, P. J. (1989) *Proc. Nat. Acad. Sci. USA* **86** 4012–4015.

Sugiyama, H. (1990) *Neurochem. Int.* **16** (Suppl. 1) 17.

Sugiyama, H., Ito, I. and Hirono, C. (1987) *Nature* **325** 531–533.

Sugiyama, J., Ito, I. and Watanabe, M. (1989) *Neuron* **3** 129–132.

Usowicz, M., Gallo, V. and Cull-Candy, S. G. (1989) *Nature* **339** 380–384.

Watkins, J. C., Krogsgaard-Larsen, P. and Honoré, T., (1990) *Trends Pharmacol. Sci.* **11** 25–33.

Williams, J. H., Errington, M. L., Lynch, M. A. and Bliss, T. V. P. (1989) *Nature* **341** 739–742.

16

Therapeutic opportunities in modulators of excitatory amino acid-mediated neurotransmission

Jørgen Drejer
NeuroSearch A/S, Smedeland 26, DK-2600 Glostrup, Denmark

ABSTRACT

In light of the predominance of excitatory amino acid (EAA)-mediated synapses in the central nervous system (CNS) it is predictable that malfunctions of this system would result in profound pathological changes. EAAs induce hyperexcitation in the brain which may lead to epileptic seizures and it is now recognized that most if not all seizure activity is either initiated or propagated through activation of *N*-methyl-D-aspartate (NMDA) receptors. Under certain pathological conditions a prolonged hyperexcitation mediated via EAA receptors may also lead to neurogeneration in CNS tissue, and it is now the general consensus that glutamate accumulation may be the universal trigger of neurodegeneration in the CNS. The intimate interactions between dopamine and glutamate in the basal ganglia have therapeutic relevance to the treatment of schizophrenia and Parkinson's disease.

During the past ten years a series of NMDA antagonists for different recognition sites of the NMDA receptor complex have become available as well as a few non-NMDA antagonists. By the use of these compounds we have gained much more insight into the involvement of EAA receptors in neurological and psychiatric diseases. So far only a few of these compounds have made their way into humans. However, major efforts have now been put into developing new generations of EAA modulators with properties suitable for future therapeutics.

EAA RECEPTOR PHARMACOLOGY

The excitatory actions of glutamate, aspartate and related acidic amino acids on cat spinal cord neurons were characterized almost thirty years ago (Curtis *et al.* 1959, Curtis and Watkins 1960). However, not until the mid-1970s was the hypothesis put forward with some weight that such acidic amino acids might serve as neurotransmitters in the mammalian central nervous system (CNS) (Curtis and Johnston 1974). At that time the first weak and generally non-selective excitatory amino acid (EAA) antagonists had been found (Haldeman and McLennan 1972, Biscoe *et al.* 1976, Hicks *et al.* 1978). The most important findings with these first-generation EAA antagonists were those with α-aminoapidate (αAA) and glutamate diethylester (GDEE) which led to the classification of EAA receptors into αAA sensitive NMDA receptors, GDEE sensitive quisqualic acid (QUIS) receptors and receptors for kainic acid (KAIN) which were insensitive to both αAA and GDEE (Davies and Watkins 1979, McLennan and Lodge 1979) (for further details see chapter 2). It now appears that QUIS and KAIN in most CNS pathways are mediating fast synaptic excitation via the same receptor complex — the 2-amino-3-(3-hydroxy-5-methylisoxazol-4-yl)propanoic acid (AMPA) receptor (for reviews, see Collingridge and Lester 1989, Honoré 1989).

NMDA receptors

A breakthrough in the development of second-generation EAA antagonists was first seen in the *N*-methyl-D-aspartate (NMDA) receptor area. The phosphono analogues of glutamate (e.g. 2-amino-5-phosphonopentanoic acid (D-AP5)) synthesized by Watkins and colleagues (Davies *et al.* 1981) showed potencies and selectivities sufficiently high for a detailed pharmacological characterization of NMDA responses (for a review see Watkins 1989). During the last decade a series of phosphono-amino acids related to D-AP5 have been synthesized both by Watkins' group and in a number of pharmaceutical companies (for further details see chapter 8). Several of these highly selective competitive NMDA antagonists are of high potencies but only a few have shown acceptable *in vivo* effects after systemic administration (see discussion below). During the last decade a number of other binding sites on the NMDA receptor complex have been characterized and modulators of these sites have become available. Glycine is now known to be a co-transmitter at the NMDA receptor (Johnson and Ascher 1987) and the NMDA responses may also be modulated by Mg^{2+} (Davies and Watkins 1977, Nowak *et al.* 1984), Zn^{2+}, non-competitive channel blockers (e.g. phencylidine (PCP), 5-methyl-10,11-dihydro-5*H*-dibenzo[*a,d*]cyclohepten-5,10-imine (MK-801)), polyamines, ifenprodil, etc. (Fig. 1) (for a review see Lodge and Johnson 1990). The NMDA receptor complex is pharmacologically well characterized and has been the subject of several recent reviews (e.g. Mayer and Westbrook 1987, Watkins 1989, Johnson 1989).

AMPA receptors

Although the agonists AMPA, QUIS and KAIN have been available for almost two decades, a pharmacological characterization of AMPA receptors has only recently become possible with the development of the potent AMPA antagonists 6,7-dinitroquinoxaline-2,3-dione (DNQX), 6-cyano-7-nitroquinonoxaline-2,3-dione (CNQX) and 2,3-dihydroxy-6-nitro-7-sulphamoylbenzo(*F*)quinoxaline (NBQX) (Drejer and Honoré 1988, Honoré *et al.* 1988, Sheardown *et al.* 1990). By use of these highly selective AMPA antagonists it has been verified during the last few years that most excitatory pathways in the CNS mediate their fast synaptic responses via AMPA receptors (e.g. Collingridge and Lester 1989). The AMPA receptor family was recently cloned (Hollmann *et al.* 1989, Wada *et al.* 1989, Wenthold *et al.* 1990) and appear to exist in two subtypes — an AMPA/QUIS-preferring conformation and a KAIN-preferring conformation (Boulter *et al.* 1990, Keinänen *et al.* 1990) (for further details see chapter 3).

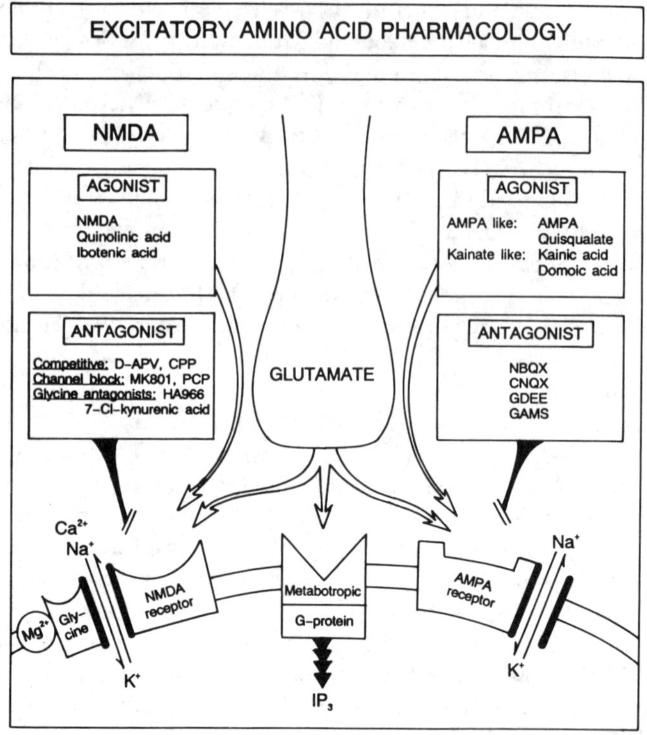

Fig. 1 — EAA pharmacology.

Metabotropic receptors

The so-called metabotropic EAA receptors which lead to stimulation of the inositol phosphate system are distinctly different from the ionotropic NMDA and AMPA receptors (Sladeczek *et al.* 1985, Sugiyama *et al.* 1987). At present no conclusive data exist to assign definitive functions to these receptors. Some evidence for the involvement of metabotropic receptors in excitotoxicity exist (Garthwaite and Garthwaite 1989, Patel *et al.* 1990) but the data are inconclusive, mainly due to the lack of potent and selective agonists and antagonists (for further details see chapter 15). If indeed metabotropic receptors are involved in EAA-induced neurodegeneration, receptor antagonists would appear most interesting from a therapeutic point of view. Such antagonists should not share the potential limitations of ionotropic receptor antagonists, which block fast synaptic neurotransmission.

Glutamate uptake

It has been a puzzle how glutamate, being present in the brain in millimolar concentrations, can act as a specific neurotransmitter. However, the extracellular brain level of glutamate has been shown to be around 1 μM (e.g. Perry and Hansen 1990) and may even be much lower in the synaptic cleft. The extremely efficient inactivation of glutamate is brought about by high-affinity, high-capacity uptake systems both in neurons and in glial cells (Balcar and Johnston 1972, Drejer *et al.* 1982). The maintenance of the 10 000–100 000-fold gradient of glutamate and aspartate between the intracellular and extracellular compartments is essential not only for the transmitter function of glutamate but also in order to avoid exposure of the brain to some of the most potent neurotoxins known: glutamate and aspartate. The uptake system is so efficient that it may be difficult to demonstrate excitation with glutamate in some experimental models in brain slices or *in vivo* even when applying millimolar concentrations, while only low micromolar concentrations are needed in experiments on single cells (e.g. Mayer and Westbrook 1987). For this reason other agonists such as NMDA, quinolinic acid (QUIN), KAIN or QUIS appear to be much more potent excitotoxins *in vivo* than L-glutamate itself (Olney and Gubareff 1978) (see also chapter 1). However, it is possible to induce severe brain damage by local injections of high doses of glutamate. This glutamate neurotoxicity will be potentiated if presynaptic elements are removed (Köhler and Schwarcz 1981). Moreover, injection of glutamate uptake inhibitors such as *threo*-3-hydroxyaspartate (Balcar and Johnston 1972) will evoke similar neurodegeneration (McBean and Roberts 1985). Even minor dysfunctions of the glutamate uptake systems may contribute significantly to the development of neurodegeneration seen both in acute pathological conditions such as cerebral ischaemia, where glutamate uptake has been shown to be altered (Drejer *et al.* 1985) as well as in more slowly developing neurodegenerative diseases such as Huntington's chorea, Alzheimer's disease and Parkinson's disease.

Glutamate release

Inhibition of glutamate release may be an effective means of reducing EAA-mediated neurotransmission. It is noteworthy that the compound lamotrigrine, which is an effective anticonvulsant and neuroprotective compound, has been shown to inhibit KAIN-induced glutamate release (McGeer and Zhu 1990).

EAA ANTAGONISTS

Several of the preceding chapters of this book have dealt with structure–activity relationships within distinct series of NMDA and non-NMDA antagonists. Some of the NMDA antagonists discussed in the present chapter have been listed in Table 1. From the marked structural dissimilarities seen between the different series of EAA antagonists it might appear that the binding sites for NMDA, glycine, QUIS and KAIN have very distinct structural requirements. This may not always be the case. During recent years a series of compounds which may be regarded as analogues of the endogenous EAA agonist QUIN (see below) have been shown to exert antagonistic activities with different pharmacological profiles. The QUIN analogues shown in Table 2 may be derived simply by annelating a benzo ring and modifying the two acidic functions in QUIN. The closest analogue to QUIN with an annelated benzo ring is acridinic acid. This compound has been shown to be a weak and non-selective EAA antagonist (Curry *et al.* 1986). Kynurenic acid, which is another endogenous tryptophan metabolite, is also a weak broad-spectrum antagonist of EAA responses (Perkins and Stone 1982). Kynurenic acid shows some selectivity for antagonizing NMDA responses but will at higher concentrations also block non-NMDA responses (Ganong *et al.* 1983). The 7-chloro derivative of kynurenic acid is 20 times more potent as a NMDA antagonist than the parent compound (Kemp *et al.* 1988). It was demonstrated that the non-competitive NMDA antagonism shown by 7-Cl-kynur-

Table 1 — Chemical structures of selected NMDA antagonists

Competitive antagonists	Channel blockers	Glycine antagonists	Other
αAA	Ketamine	HA 966	Ifenprodil
D-APV	PCP	7-Cl-KYN	SL 820715
CPP	MK 801		
CPPene			
CGS 37849			

Table 2 — EAA antagonists derived from quinolinic acid

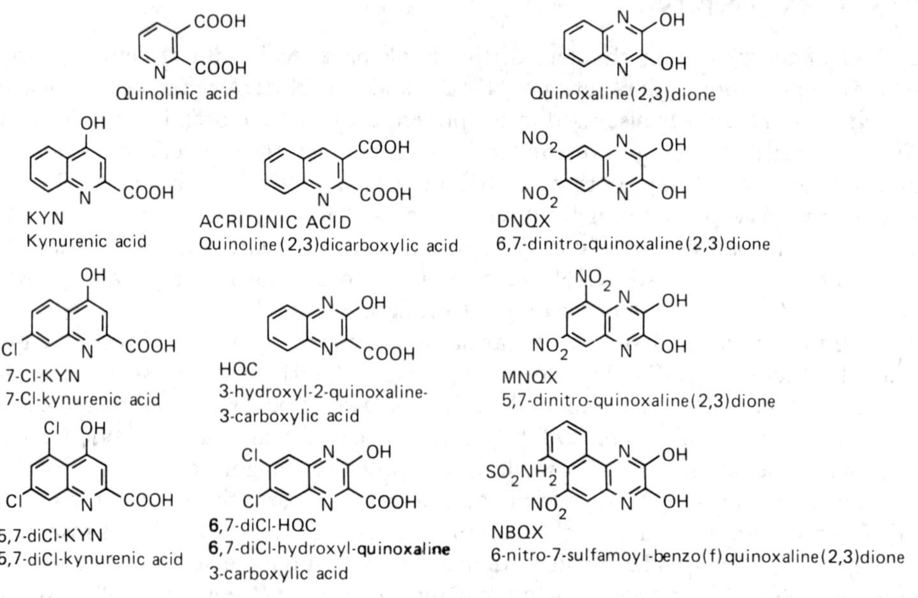

enic acid (7-Cl-KYN) could be reversed by high concentrations of glycine, indicating that this compound is an antagonist at the glycine site of the NMDA receptor complex (Kemp *et al.* 1988). A further increase in potency is obtained with the 5,7-diCl substituted analogue of kynurenic acid (McNamara *et al.* 1990).

The compound 3-hydroxy-2-quinoxaline-3-carboxylic acid (HQC), which is structurally related to kynurenic acid, is also a weak blocker of NMDA and KAIN responses but has essentially no effect on QUIS responses (Erez *et al.* 1985). The 6,7-dichloro derivative of HQC is around ten times more potent as an NMDA antagonist and about 25 times more potent than HQC in blocking KAIN responses (Frey *et al.* 1988). The activity of this compound at QUIS receptors appears to be very weak.

Another series of compounds structurally related to kynurenic acid are the quinoxalinediones (see also chapter 11). The unsubstituted quinoxalinedione is a weak non-selective antagonist. However, the 6,7-dinitro-substituted analogue DNQX was recently shown to be a potent and selective non-NMDA antagonist (Drejer and Honoré 1988, Honoré *et al.* 1988). It has been demonstrated that some of the weak NMDA antagonism shown by DNQX may be reversed by glycine (Birch *et al.* 1988, Drejer *et al.* 1989). The compound MNQX is a potent antagonist at the glycine/NMDA site (Sheardown *et al.* 1989). It is striking that this compound is a very close analogue of DNQX (one nitro group has been moved from position 5 to 6). Another quinoxalinedione, NBQX, on the other hand, is an extremely potent non-NMDA antagonist completely free of NMDA antagonistic activity (Sheardown *et al.* 1990). A general feature of the four series of benzo-annelated QUIN analogues shown in Table 2 is that the simple unsubstituted compounds are rather weak and non-selective antagonists. However, it is also clear that potent antagonists with distinctly different selectivities for the different EAA receptor subtypes including the glycine binding site may be obtained simply by changing the substitution pattern in the benzo group. Although most of the compounds from these chemical series have shown *in vivo* effects (e.g. anticonvulsant effects) when given directly into the brain by intraventricular injection they are all rather polar and pass the blood–brain barrier very poorly (Erez *et al.* 1985, Frey *et al.* 1988). Moreover most of the compounds are excreted rapidly from the bloodstream. Consequently, the compounds have generally shown no or only weak effects when given parenterally or orally. In spite of the poor kinetic properties of the QUIN-related compounds selected substances such as NBQX may be useful clinically in acute conditions where large doses may be administered continuously via the intravenous route.

EAAs IN CNS DISEASES

Excitatory amino acids are now generally accepted as the primary mediators of synaptic excitation in the mammalian CNS (e.g. Collingridge and Lester 1989) (see also chapter 5). As a consequence, possible dysfunctions in EAA-mediated neurotransmission have been postulated to be involved in several CNS diseases. In the following sections some of these pathological conditions will be discussed in relation to a potential therapy using antagonists, agonists or other modulators of the different EAA receptor subtypes. Some key findings in the assignment of a role of EAAs in selected neuropathological conditions are summarized in Table 3.

Table 3 — EAA research and therapeutic prospects 1960–1990

Year	EAA pharmacology		Therapeutic relevance				
			Stroke	Epilepsy	Schizophrenia	Alzheimer	Other
1960	Glutamate activates spinal neurons		Glutamate is neurotoxic (1957)	Glutamate is convulsive (1954)	PCP has psychotomimetic effects		
–	NMDA is an agonist						
–							Ketamine is analgesic
–	QUIS and KAIN are agonists						
1970	NMDA responses vs Non-NMDA responses		Concept of excitotoxicity				
–	αAA blocks NMDA	GDEE blocks QUIS AMPA is an agonist		HA-966 is anticonvulsant			
–							KAIN induces Huntington-like degeneration
–	Mg^{2+} blocks NMDA						
1980	D-AP5 — a competitive antagonist						
–			Glu release in ischaemia	K-801 is anticonvulsant GAMS is anticonvulsant AP5 is anticonvulsant		Selective loss of glutamate in DAT cortex	NMDA induces pain AP5 blocks pain
–			AP5 is anti-ischaemic				
–	PCP — a channel blocker		MK-801 is anti-ischaemic			Selective loss of NMDA receptors in DAT	AP5 is anxiolytic AP5 is muscle-relaxing
–					MK-801 induces dopamine release and stereotypy		AP5 blocks spreading depression (migraine)
–	Glycine site identified	CNQX blocks QUIS/KAIN					
–	Zn^{2+} site postulated		Ifenprodil is anti-ischaemic	7-Cl-KYN is anticonvulsant			
1990	Spermine sitte postulated	AMPA receptor cloned	NBQX is anti-ischaemic		CNQX blocks dopamine release and QUIS-induced stereotypy		
			7-Cl-KYN is anti-ischaemic				

EXCITOTOXICITY AND STROKE

The first demonstration of a neurotoxic effect of glutamate was made more than thirty years ago by Lucas and Newhouse (1957). These findings were followed up by more systematic studies by Olney and coworkers in the late 1960s which led to the suggestions by Olney of the 'excitotoxic' concept in the early 1970s (Olney 1969, 1978) (for a detailed discussion of excitotoxicity see chapter 1). However, less than a decade ago direct *in vivo* evidence became available which indicated an involvement of endogenous brain EAAs in neurodegenerative diseases. Benveniste *et al.* (1984) demonstrated a very significant increase in extracellular glutamate concentration in the rat hippocampus during transient cerebral ischaemia — a condition known to result in an almost complete loss of selective vulnerable CNS neurons, e.g. in the CA1 pyramidal layer of the hippocampus. It was later shown that denervation of the CA1 pyramidal cells by cutting the glutamatergic Schäffer collaterals led to essentially no release of glutamate (Benveniste *et al.* 1989) and no CA1 cell loss after ischaemia (Johansen *et al.* 1987), strongly suggesting that the glutamatergic input via the Schäffer collaterals was essential for the development of ischaemic cell damage to these neurons. Also removal of glutamatergic input from the entorhinal cortex or cutting the angular bundle was shown to limit the ischaemic damage (Wieloch *et al.* 1985, Jørgensen *et al.* 1987).

On the basis of these results the need to evaluate EAA antagonists for potential neuroprotective effect in animal models of global ischaemia, hypoglycaemia and stroke was obvious. The first report of a positive effect of NMDA antagonists was published by Simon *et al.* (1984), who demonstrated protection of CA1 pyramidal cells in the rat hippocampus by local application of D-AP5 prior to transient global ischaemia. Subsequently, neuroprotection by local or intraventricular injection of D-AP5 has been shown by several laboratories in a number of different ischaemia and hypoglycaemia models (for review see Rothman and Olney 1986). For a few competitive antagonists protection has also been seen after systemic administration, whereas the NMDA channel blockers are generally systemically active. 3-(2-Carboxy-piperazin-4-yl)-1-propenyl-1-phosphonic acid (CPPene) and MK-801 have shown remarkable neuroprotection in focal cerebral ischaemia models (stroke) also when the antagonists were given after the insult (McCulloch *et al.* 1990, Ozyurt *et al.* 1988, Park *et al.* 1988, Gotti *et al.* 1988). There have also been several reports of an almost complete protection with MK-801 against neurodegeneration in global ischaemia models. However, several other laboratories have only had negative results in such models (e.g. Buchan and Pulsinelli 1990). Some of the discrepancies may be explained by different experimental set-ups (possible contribution of hypothermia to the effect of MK-801, differences in degree of blood flow reduction, differences in seizure activity in animals after ischaemia) and it presently appears likely that MK-801 has neuroprotective activity in transient complete cerebral ischaemia. Recently it was found that also the glycine antagonists 7-Cl-KYN and HA-966 could prevent ischaemic nerve cell damage (Patel *et al.* 1990). Moreover ifenprodil, an atypical NMDA antagonist, and its analogue SL 820715 have shown remarkable protection in focal ischaemia models (Carter *et al.* 1988, Gotti *et al.* 1988).

The Schäffer collateral synapse on hippocampal CA1 neurons is believed to be mediating primarily non-NMDA responses which are potently blocked by CNQX

(Herreras *et al.* 1989). For this reason it would be rational to evaluate whether non-NMDA antagonists could block ischaemia-induced CA1 cell loss. Recently Sheardown *et al.* (1990) reported that NBQX administered intravenously to gerbils several hours after a 5-minute bilateral artery occlusion could completely protect against hippocampal CA1 cell loss. A similar protection of CA1 neurons was seen in the corresponding rat four-vessel occlusion model (10 minutes occlusion) given NBQX 2 hours after the occlusion (Diemer *et al.* 1990). It is remarkable that protection by NBQX may be obtained by administration several hours after the ischaemia insult. NBQX also seems to be active in focal ischaemia models (M. Sheardown, personal communication).

No conclusive explanations have been proposed for the delayed protective effects of NBQX or MK-801 in ischaemia models. However, the potential of treating stroke patients or cardiac arrest patients hours after their insult with NMDA or AMPA antagonists opens a whole new world of potential cerebrovascular treatments. However, two major complications with NMDA channel blockers like MK-801 have at least for a period of time halted clinical investigations. One concern with this type of compound is the clinical observation that PCP induces severe psychotomimetic effects in humans (Greifenstein *et al.* 1958). Other channel blockers probably share this property with PCP since they all generalize to PCP, e.g. in the PCP-cue test in rats (Willetts *et al.* 1990). The potential psychotomimetic effects of NMDA channel blockers might, however, be tolerable in relation to clinical trials for the stroke indication. Perhaps of more serious concern not only to the pharmaceutical industry but also to the regulatory authorities was the finding of Olney *et al.* (1989) that PCP and MK-801 induce characteristic acute morphological changes in specific neurons in the posterior cingulate cortex. The order of potencies for inducing these morphological changes was the same as their order of affinities for the NMDA receptor complex, indicating that the effect was indeed related to NMDA antagonism. It has now clearly been demonstrated that the vacuolization seen in the specific neurons of the cingulate cortex is completely reversible. It is, however, still unknown whether prolonged exposure to NMDA channel blockers may cause degeneration of these neurons. Although it has been found that a major part of a population of PCP ('angle dust') abusers showed learning and memory impairments (Wilkins 1989) these neuropsychological deficits might be explained by other factors, including poly-drug abuse. However, a systematic histopathological study of post-mortem brains from PCP abusers is highly warranted. The findings of reversible morphological changes after MK-801 administration would call for some caution for the use of NMDA channel blockers in chronic treatment regimes but should not be prohibitive for clinical trials in stroke where administration of compound is only acute and where the alternative to treatment is divesting brain damage. Several pharmaceutical companies have developed NMDA channel blockers with potencies and efficacies comparable to or better than MK-801, but it appears that the industry at the moment is reluctant to move into clinical development with these compounds. However, a few other NMDA channel blockers which have been on the market for several years could be considered for stroke trials. Dextromethorphan is an example of a safe drug with relatively potent NMDA channel blocking activity and significant anticonvulsant and anti-ischaemia activities in animal models. Clinical studies with dextromethorphan in stroke would seem to be relevant (for review see Tortella *et al.* 1989). It has been reported that the competitive NMDA antagonist CGS 19755 as well as the atypical NMDA antagonist SL 820715 derived from ifenprodil are now in phase I clinical trials (company communications). Also CPPene and CGP 37849 are apparently waiting to take the step into clinical development for stroke indication.

ENDOGENOUS EXCITOTOXINS IN RELATION TO NEURODEGENERATIVE DISEASES

Tryptophan is the precursor of several neuroactive compounds, including serotonin, QUIN and kynurenic acid (Stone and Connick 1985). Kynurenic acid, as mentioned above, is a weak, non-selective EAA antagonist (Perkins and Stone 1982) and has been suggested to be an endogenous neuroprotective agent (anti-excitotoxin). QUIN, on the other hand, is a potent and selective NMDA agonist (Stone and Perkins 1981) and like NMDA a potent excitotoxin (Schwarcz *et al.* 1983). It has been speculated that pathologic alterations in tryptophan metabolism leading to elevated levels of QUIN might be responsible for the slow neurodegenerative processes seen, for example, in patients developing Alzheimer's disease, Parkinson's disease or suffering from Huntington's disease. It has also been speculated that changes in brain levels of tryptophan might have neurological consequences (for review see Freese *et al.* 1990). Estimations of brain levels after administration of relatively high doses of tryptophan to experimental animals indicate that QUIN levels of at least 100 times above control may be achieved (During *et al.* 1989). Such brain concentrations of QUIN are equivalent to those shown to be toxic to neurons *in vitro* (Whetsell 1984). Also in the case of hepatic failure significant increase in brain tryptophan levels have been reported (Moroni *et al.* 1986) and it has been found that portacaval shunted rats receiving a tryptophan overload developed significant neuronal degeneration in various brain regions, including the striatum (Bucci *et al.* 1982).

The involvement of an endogenous excitotoxin in the neurodegeneration seen in Huntington's chorea patients was first postulated by McGeer and McGeer (1976) and Coyle and Schwarcz (1976) and has since been supported by several findings. First of all it has been demonstrated that the pattern of axon sparing nerve cell degeneration seen after intrastriatal QUIN injection very closely resembles that seen in patients suffering from Huntington's chorea, with selective preservation of striatal dopaminergic neurons and neuropeptide Y neurons and significant loss of GABAergic interneurons (Beal *et al.* 1986, Ferrante *et al.* 1985). Since QUIN is a selective NMDA agonist the excitotoxic damage seen in the striatum of Huntington's disease patients might very well be mediated via the NMDA receptors. The endogenous causative excitotoxin could be QUIN but it is more likely to be the neurotransmiter glutamate itself (Perry and Hansen 1990), possibly accumulating due to failures in the high-affinity uptake systems for glutamate seen in Huntington's chorea (Cross *et al.* 1986). In any case, Huntington's disease patients and carriers of the Huntington's disease gene should be among the first patient groups to be considered for clinical trials when an NMDA antagonist sufficiently safe for chronical treatment becomes available.

EPILEPSIA

Most of the current approaches to the treatment of epilepsy have focused on interactions with central inhibitory transmitter systems, in particular the GABA system. However, the initiation and propagation of seizures may equally well be regarded as a hyperexcitation. Glutamate has been known for more than thirty years to be convulsive (Hayashi 1954). On this basis a potential treatment of epilepsy involving direct intervention with central excitatory transmitter systems, in particular glutamatergic systems, would appear attractive.

Shortly after the identification of D-AP5 as a potent and selective NMDA antagonist it was demonstrated that administration of this substance locally into rat brain or intraventricularly in DBA/2 mice gave significant protection against cobalt-induced convulsions (Coutinho-Netto *et al.* 1981) and audiogenic seizures (Croucher *et al.* 1982), respectively. At that time MK-801 was already known to be a potent anticonvulsant (Clineschmidt *et al.* 1982) but this was long before its mechanism of action was known. Subsequently numerous studies of the effects of different EAA antagonist, particularly of the NMDA type, have been conducted in animal seizure models (for review see Dingledine *et al.* 1990). It appears that NMDA antagonists show pronounced effects against seizures induced by many different means claimed to be predictive of human epilepsy, ranging from reflex epilepsy (e.g. audiogenic seizures in DBA/2 mice (Croucher *et al.* 1982) and light-induced seizures in *Papio papio* (Meldrum 1983)), over petit mal epilepsy (spontaneous absence seizures in petit mal rats) (Peeters *et al.* 1989), to grand mal epilepsy (electroshock in mice) (Clineschmidt *et al.* 1982). Also in a variety of *in vitro* models showing burst firing or electrographic seizure activity a potent and in many models complete block may be obtained with NMDA antagonists. It appears that NMDA antagonists of different types are more effective in blocking the induction of epileptic events than the expression of an already initiated seizure (Stasheff *et al.* 1989). In agreement with these *in vitro* findings Hönack and Löscher (1990) have shown that MK-801, which potently blocks the development of kindled seizures, only shows marginal protection against seizures in already fully kindled animals.

There is no doubt that NMDA antagonists both of the competitive type and the channel blockers are very potent and effective anticonvulsants. They may have potential clinical advantages over current therapy in that they show protection in a very broad spectrum of animal epilepsy models. This suggests that NMDA antagonists would be effective in certain forms of epilepsy for which at present there is no effective therapy. Despite the fact that NMDA antagonists seem to be excellent broad-spectrum anticonvulsants none of the second-generation NMDA antagonists are presently under clinical development for the epilepsy indication. The shortcomings of the currently available NMDA antagonist are several, some of which have been mentioned above. Essentially all competitive NMDA antagonists are glutamate analogues with a strong acidic group (e.g. phosphono or tetrazole) (see chapter 8) substituted for the amino acid terminal carboxylic acid. These highly polar compounds, which will be almost fully ionized at physiological pH, generally have shown very limited brain penetration (for review see Watkins *et al.* 1990). When given locally or intraventricularly D-AP5, CPP and other phosphono analogues are extremely potent anticonvulsants, but after systemic administration the compounds appear to be either inactive or only active at very high doses in the most sensitive seizure models (Löscher *et al.* 1988). A few of the more recently developed competitive NMDA antagonists of the phosphono type have, however, shown acceptable bioavailability, brain penetration and duration of action. These include CPPene, CGS 19755 and

CGP 37849 (Hönack and Löscher 1990). The generally poor kinetic properties of the competitive antagonists are not shared by the non-competitive channel blockers. These compounds are easily transported into the brain and generally have a very long duration of action (Koek and Colpaert 1990). Also in *in vitro* preparations the channel blockers are generally very difficult to wash out from the preparation whereas the actions of competitive phosphono antagonists are readily reversible.

Another problem limiting the clinical usefulness of the currently available competitive NMDA antagonists and in particular the channel blockers is the relatively poor separation between anticonvulsant doses and doses that will give muscle incoordination, ataxia, sedation or even anaesthesia (e.g. Koek and Colpaert 1990). However, for a few competitive NMDA antagonists (e.g. CGS 37849) (Hönack and Löscher 1990) and NPC 12626 (Ferkany *et al.* 1989) the therapeutic index between anticonvulsant and ataxic doses seems to be acceptable and comparable to what is seen with other broad-spectrum anticonvulsants. Perhaps the major obstacle that has slowed or stopped the development of the more promising competitive NMDA antagonists as anti-epileptics for chronic therapy is the experience from animal studies that a general feature of NMDA antagonists is the potential of disabling learning and memory processes (see below). Moreover, not only the channel blockers, as mentioned above, but also several of the competitive NMDA antagonists have been shown to generalize to PCP in a drug discrimination test and may share the psychotomimetic actions of PCP seen in humans (Compton *et al.* 1987, Koek and Colpaert 1990). This is not seen with the glycine antagonists HA-966 and 7-Cl-KYN or antagonists of the ifenprodil type, although these compounds have been shown to block NMDA-induced convulsions in mice (Perrault *et al.* 1990, Koek and Colpaert 1990).

SCHIZOPHRENIA

NMDA receptor interactions

The suggestion that a hypofunction of the glutamate system might be the primary defect in schizophrenia was first put forward by Kim *et al.* (1980), who showed that the glutamate cerebrospinal fluid (CSF) levels in 20 schizophrenic patients were about half of the normal values. It has been known for almost thirty years that PCP and ketamine induce schizophrenia-like psychotomimetic effects and recently these effects have been associated with their NMDA antagonistic activities. MK-801, which is much more selective for the NMDA receptor complex, as well as D-AP5 show similar animal behavioural responses to PCP, including stereotypes which may be reversed with classical (haloperidol) as well as atypical (clozapine) antipsychotic drugs (Clineschmidt *et al.* 1982, Schmidt 1986, Tiedtke *et al.* 1990). The neuroleptic action of clozapine might actually be mediated via NMDA receptors since this atypical neuroleptic has nanomolar affinity for the binding site and shows a distribution of binding sites in the CNS almost identical to [^{3}H]MK-801 (Janowsky and Berger 1989).

It is clearly premature to suggest a substitution therapy in schizophrenia with activators of the NMDA receptor complex. However, the evidence for psychotic effects of systemic administered NMDA antagonists is now so strong (for review see Carlsson and Carlsson 1990) that further animal studies in psychosis models using systemically active NMDA agonists/partial agonists or glycine agonists are highly warranted for evaluating a possible therapeutic potential of such treatments in schizophrenia.

AMPA receptor interactions

It was shown almost ten years ago that injection of AMPA into the accumbens of the rat results in hypermotility which may be reversed by the dopamine antagonist flupentixol (Arnt 1981). Recently it was confirmed that not only AMPA but also QUIS and KAIN microdialysed into the accumbens induce a significant dopamine release in this nucleus as well as characteristic stereotype behaviour sensitive to the dopamine antagonist haloperidol (Imperato *et al.* 1990). Both non-NMDA induced dopamine release and stereotype behaviour could be antagonized by co-administration of CNQX via the dialysis tube (Imperato *et al.* 1990). CNQX given in the same way could also block the effects of amphetamine on dopamine release and stereotype behaviour (Imperato *et al.* 1990). It is unclear whether the non-NMDA-induced dopamine release in accumbens is a presynaptic or postsynaptic effect but it appears to be mediated via a specific action on AMPA receptors since it could not be induced by relevant concentrations of NMDA and it could not be blocked by NMDA antagonists (Imperato *et al.* 1990). These results suggest a role for AMPA antagonists as potential antipsychotics. However, at present no suitable antagonists of this type are available which could be considered for clinical studies in schizophrenia.

LEARNING AND MEMORY

Induction of a sustained enhancement of synaptic efficacy may be induced by high-frequency stimulation. This phenomenon, termed long-term potentiation (LTP) (Bliss and Lomo 1973), which has been proposed as a neurophysiological substrate of associative learning, is dependent on activation of NMDA receptors (Collingridge and Bliss 1987) and possibly also non-NMDA receptors (Muller *et al.* 1988). The NMDA antagonist D-AP5 injected intraventricularly to rats gives significant impairment of spatial learning and blocks hippocampal LTP at doses which do not interfere with visual discrimination learning (Morris *et al.* 1986, Morris 1989). The impairment of spatial learning by NMDA antagonists was later reproduced by others using different experimental models, different antagonists (e.g. 2-amino-7-phosphonoheptanoic acid (D-AP7) and MK-801) and different routes of administration (Kleinschmidt *et al.* 1987, Mondadori *et al.* 1989, Chiamulera *et al.* 1990, Wozniak *et al.* 1990). Interestingly, however, NMDA antagonists have been reported to facilitate learning in passive avoidance tasks (Mondadori *et al.* 1989). 7-Cl-KYN has been reported to be without effect on learning at doses higher than those effective in anticonvulsant models (Chiamulera *et al.* 1990). Clearly the potential involvement of NMDA antagonists and possibly also non-NMDA antagonists in learning and memory processes is of major concern for the development of such substances as new therapeutics. It is, however, fair to say that much too little is known as yet to warrant a general conclusion on the clinical relevance of the possible negative and positive effects of NMDA antagonists in learning.

PARKINSON'S DISEASE

Recently it was suggested (Klockgether and Turski 1989, Carlsson and Carlsson 1990) that in Parkinson's disease the equilibrium between the glutamatergic system projecting from the cortex via the subthalamic nucleus to the basal ganglia on one hand and the GABAergic system projecting from the striatum to the basal ganglia on the other are shifted to the side of the glutamatergic system. In support of this concept MK-801 has been found to stimulate locomotor activity in monoamine-depleted mice (Carlsson and Carlsson 1989, Carlssson and Svensson 1990) and to reverse neuroleptic-induced catalepsy in rats (Schmidt and Bischoff 1988). Moreover 3-(2-carboxypiperazin-4-yl)-propyl-1-phosphonic acid (CPP) and kynurenate given into the entopeduncular nucleus have been found to alleviate Parkinsonian symptoms in 6-OH-dopamine-treated rats and in reserpinized rats (Brotchie *et al.* 1990). Kynurenate given in the same way may at higher doses even induce hyperkinetic symptoms (Bevan *et al.* 1990). Accordingly NMDA antagonists might have potential as adjuvents of dopamine replacement therapy in Parkinson's disease and might help to save dopaminergic medication and retard the development of resistance to dopaminergic medication.

ALZHEIMER's DISEASE

The pyramidal EAA neurons in cerebral cortex and the hippocampal formation comprise the primary output cells of these brain regions (Lynch and Baudry 1984). These pyramidal neurons are involved in cortical associative processes. In dementia of the Alzheimer type (DAT) presynaptic markers (e.g. [^{3}H]aspartate uptake) of glutamate neurons are reduced 25–80% in cortex and hippocampus (Palmer *et al.* 1986). Postsynaptic NMDA receptors as identified by [^{3}H]glutamate binding and [^{3}H]TCP binding are significantly reduced in cortex and hippocampus (85% loss in hippocampal CA1 pyramidal region) (Greenamyre *et al.* 1985, 1987, Maragos *et al.* 1987). A loss of AMPA receptors of about 50% is also seen in the hippocampus. It is noteworthy that benzodiazepine receptors (GABA receptor complex) and cholinergic receptors are unchanged in the same regions. The loss of presynaptic and postsynaptic components of glutamate pyramidal neurons in the cortex and hippocampus might very well be primary excitotoxin-mediated events since these neurons all receive an intense glutamatergic input. Moreover, NMDA applied to rat cerebral cortex mediates retrograde degeneration of cholinergic neurons in nucleus basalis (Sofroniew and Pearson 1985). In DAT there is a significant degeneration of these cholinergic pathways. This observation forms the basis of several approaches to DAT treatment using cholinergic substitution therapy or activators of endogenous acetylcholine production or release. However, recent studies have shown no correlation between choline acetyltransferase (CAT) activity and any neurophysiological measures in Alzheimer patients but a strong correlation to loss of large cortical pyramidal neurons (Neary *et al.* 1986). Moreover, a significant population of severe Alzheimer patients shows no loss in CAT activity (Palmer *et al.* 1986). On the other hand, degeneration of the cortical input to the hippocampus via the perforant path would functionally disconnect the cortex from the hippocampus and this is thought to occur in Alzheimer's disease (Geddes *et al.* 1985). Consequently, an alternative and probably more effective therapeutic possibility in DAT would be to enhance glutamatergic neurotransmission, as suggested by Deutsch and Morihisa (1988). Glutamate uptake inhibitors would have the potential of effectively elevating extracellular glutamate concentrations and glutamate-mediated events. Such tools would probably be effective independent of degenerated presynaptic terminals since extracellular glutamate levels are controlled not only by reuptake into presynaptic neuronal elements but probably to a greater extent by glial uptake systems (Drejer *et al.* 1982) which should be intact in DAT. It is likely that EAA stimulation via glutamate uptake inhibition would be more easily controllable than direct interaction with postsynaptic EAA receptors. However, also partial agonists for the postsynaptic EAA receptor subtypes without intrinsic excitotoxic properties would be highly relevant as potential therapeutics for the treatment of dementia of the Alzheimer type. Recently a partial agonist for the AMPA receptor subtype was synthesized (Christensen *et al.* 1989).

PAIN

The involvement of EAAs as transmitters in nociceptive pathways is now supported by several lines of evidence. EAAs and their binding sites have been demonstrated in spinal regions associated with nociceptive processes (Greenamyre *et al.* 1984) and glutamate appears to be co-released with substance P from C-fibre terminals (Battaglia and Rustioni 1988). Most of the original electrophysiological characterization of EAA responses was performed on cat spinal cord neurons (for review see Evans 1989) and electrophysiological experiments have shown that NMDA and AMPA activate spinal projection nociceptive neurons (Lei and Wilcox 1990). Intrathecal injection of high doses of NMDA, QUIS or KAIN induce behaviour that mimics responses to noxious stimuli and these effects may be blocked by NMDA antagonists or AMPA antagonists (D-AP5, MK-801, CNQX) (Cahusac *et al.* 1984, Aanonsen and Wilcox 1986, 1987, Yaksh 1989, Lei and Wilcox 1990). NMDA antagonists (e.g. D-AP5, MK-801, 7-Cl-KYN) also block experimentally induced electrical or chemical noxious stimuli (Haley *et al.* 1990, Dickenson 1990). Intrathecal injections of low doses of NMDA or injection of NMDA into the periaquaductal grey matter do not lead to noxious stimuli but on the contrary inhibit noxious responses similarly to morphine (Jacquet 1988). Also this effect of NMDA is sensitive to NMDA antagonists (Siegfried and de Souza 1989).

The Parke-Davis compounds ketamine and PCP have been known for more than twenty years to exert analgesic activity not only in animals but also in humans (Domino *et al.* 1965, Klepstad *et al.* 1990). This effect was originally believed to be mediated via the so-called opiate receptors (Martin *et al.* 1976). However, the establishment of the PCP receptor as a distinct site representing a binding site on the NMDA receptor complex (Anis *et al.* 1983, Largent *et al.* 1986, Kemp *et al.* 1987) led to the suggestion that the antinociceptive effects of ketamine and PCP might be mediated by their NMDA channel blocking activity. Indeed the rank order of potencies for channel blockers, with MK-801 being the most potent analgesic, is in perfect correlation with the affinities for PCP receptors and does not fit with the pharmacology of the σ-receptor (DeLeander and Wahl 1989). Although ketamine has been on the market for decades as an anaesthetic compound its use as an analgesic has been very limited, primarily due to the psychotomimetic side-effects (Domino *et al.* 1965). Since ketamine has an efficiency comparable to pethidine it should also be considered in severe pain when opioids should be avoided or are insufficient (Maurset *et al.* 1989). The glycine antagonist 7-Cl-KYN has also been shown to block noxious responses (Dickenson 1990). Such compounds may prove to have a better potential as pain relievers than the NMDA channel blockers or competitive inhibitors.

SPREADING DEPRESSION AND MIGRAINE

The phenomenon of cortical spreading depression (SD) is characterized by a slow propagating wave of transient EEG depression which, once initiated, will spread over the entire cortical hemisphere (Leao 1944). Spreading depression has been suggested to represent the prodrome of a classical migraine attack in humans (Lauritzen 1987). This hypothesis is mainly based on the finding of a slowly propagating transient increase of cortical blood flow, followed by long-lasting hypoperfusion in patients before a classical migraine attack as well as during SD in animal experiments (Lauritzen 1987). SD may be triggered by mechanical/electrical stimulation of the cortex, by applying depolarizing conditions (e.g. high K^+) or excitatory agents such as EAAs (van Harreveld 1984). Of a long series of agents, representing pharmacological interaction with several transmitter systems, only EAA antagonists have been shown systematically to block the initiation of SD (Mody *et al.* 1987, Lauritzen *et al.* 1988). *In vivo* NMDA antagonists such as D-AP5 and MK-801 have been shown effectively to block SD initiated by different stimuli (Scheller *et al.* 1990). The presently available therapeutic treatment protocols for migraine are far from ideal. However, although NMDA antagonists are clearly active in animal models of SD there is at present no push for clinical studies for the migraine indication with the currently available competitive NMDA antagonists or channel blockers. For reasons discussed previously the side-effect profile of the second-generation NMDA antagonists do not favour a chronic treatment with these compounds for a relative benign indication such as migraine. Other NMDA antagonists acting at different sites in the NMDA receptor complex, e.g. the glycine site or the polyamine site, may as mentioned earlier prove to have a more favourable therapeutic index in relation to, for example, sedation. In this context it should be mentioned that the glycine antagonist HA-966 has been shown to block SD in chick retina by a glycine-sensitive process (Drejer *et al.* 1989). Also the already available drug ifenprodil or its derivative SL 820715 (Carter *et al.* 1988) could probably be considered for clinical trials in migraine.

MUSCLE RELAXATION

Competitive NMDA antagonists such as D-AP5, D-AP7 and CPP have been shown to exert marked effects on spastic muscle contraction in a strain of genetically spastic rats (Turski *et al.* 1985). This would indicate a role for such readily reversible competitive NMDA antagonists as potential tools for acute muscle relaxation during surgery. Moreover, NMDA antagonists with an acceptable side-effect profile might have therapeutic potential in spastic patients.

ANXIETY

The general inhibitory profile of EAA antagonists would suggest that such compounds might show anticonflict activity. D-AP5 has been found to antagonize the suppressive effect of punishment in the four-plate test (Stephens *et al.* 1986) in mice and to enhance rat exploration in the open arms of the plus maze (Bennett and Amrick 1986). Although these findings strongly suggest a potential anxiolytic activity of NMDA antagonists it would at present seem unlikely that an NMDA antagonist could be developed for this indication since the already available anxiolytics are essentially devoid of severe side-effects.

SUMMARY

During the last decade the scientific world has experienced an exponential growth in EAA research. Only ten years ago EAAs were regarded only as putative neurotransmitters and the pharmacology of EAA responses was largely unknown. The igniter of the explosion in EAA research was the chemical input from Watkins and colleagues: D-AP5 and related compounds were found to be potent and selective NMDA antagonists. Shortly after, the clinical relevance of EAA neurotransmission became clear not only to the scientific community but also to government foundations and to the pharmaceutical industry: glutamate was found to be released during cerebral ischaemia, and ischaemia-induced neurodegeneration could be prevented by D-AP5. During the following years an enormous research effort was put into the characterization of anti-ischaemic and anticonvulsant potentials of D-AP5, MK-801 and related compounds in a variety of animal models. The demonstration that MK-801 and later on also CPPene and NBQX showed prominent neuroprotection even when given hours after cerebral ischaemia were extremely encouraging. At the same time evidence of an involvement of EAAs in a variety of other CNS disorders accumulated rapidly. The involvement of EAA neurotransmission in pain, psychotic and depressive disorders, neurodegenerative disorders such as Alzheimer's disease, Parkinson's disease and Huntington's disease were largely speculative a few years ago. The availability of potent and selective EAA antagonists and agonists have now solved many mechanistic questions and suggested a therapeutic potential of EAA modulators in several areas. However, although the current second-generation EAA antagonists and agonists have been excellent pharmacological tools the compounds have generally had very little chance of being developed clinically. The main problems have been poor bioavailability/kinetics and unacceptable side-effects. In Table 4 selected compounds from several classes of first- and second-generation EAA modulators have been compiled and an attempt has been made to review the current prospects of these classes of compounds as potential therapeutics in a variety of CNS disorders. Only very few of these compounds are currently regarded as development candidates (indicated by the symbol [+] for compounds in phase I studies or [!] for compounds ready to go into humans). Most boxes are either empty or given the symbol [()], indicating a rather speculative indication for future therapy with the particular class of compounds.

The ball is now back in the chemists' laboratories. We are now waiting for the third-generation of EAA antagonists and agonists, which should not only be experimental tools but future therapeutics. Several pharmaceutical companies have now devoted major resources into chemical/biological research programmes aimed at developing therapeutically acceptable third-generation modulators of EAA neurotransmission. Much of this effort has shifted from direct-acting competitive antagonists and channel blockers to compounds acting on binding sites at the EAA receptor complexes which allosterically modulate the activation of the macromolecules. If an analogy of the GABA receptor complex and the allosterically coupled benzodiazepine binding site is relevant it is conceivable that interaction with allosterically coupled binding sites of the NMDA receptor complex, e.g. the glycine binding site or the polyamine binding site, may be much easier to control therapeutically than directly activating or blocking the ionophore. The compound SL 82.0517, which is now in clinical phase I studies aimed as a anti-ischaemic compound, may be

Table 4 — Potential clinical applications for EAA modulators

	First generation	Second generation	Third generation	Stroke	Epilepsy	Alzheimer	Huntington	Parkinson	Schizo-phrenia	Spasticity	Pain	Anxiety	Depression
NMDA competitive antagonist	αAA	D-AP5 CPP CPPene CGP 37849 CGS 19755	+[a]	(!)					( )			( )	
channel blockers	Ketamine PCP Dextro-methorph	MK-801		+[b]									
glycine antagonists	HA-966 Kynurenic acid	MNQX 5,7-diCl-KYN		( )	( )		( )	( )		( )	( )		( )
others	Ifenprodil	SL 820715		+[c]	( )		( )	( )		( )	( )		
AMPA antagonists	GDEE GAMS	DNQX CNQX NBQX AMOA[d]		!			( )	( )	( )				
Metabotropic antagonists	AP3												
NMDA agonists/partial agonists/activators	NMDA Quinolinic acid D-serine					( )			( )				
AMPA agonists/partial agonists	QUIS KAIN	AMPA APPA[e]				( )							
Metabotropic agonists/partial agonists	*trans*-ACPD												
Glutamate uptake inhibitors	thr-3-OH-D-Asp					( )			( )				

Symbols used: +, phase I trials initiated; !, ready for clinical trials; ( ), indications for clinical trials.
a CGS 19755 is in clinical phase I studies for stroke indication (company communication, August 1990).
b Development of MK-801 has been halted (company communication, April 1991).
c SL 820715 is in clinical phase I studies for the indication of stroke (company communication, July 1989).
d AMOA: 2-amino-[3-(carboxymethoxy)-5-methylisoxazol-4-yl]propionic acid (Frandsen *et al.* 1990).
e APPA: 2-amino-3-(3-hydroxy-5-phenylisoxazol-4-yl)propanoic acid (Christensen *et al.* 1989).
For other references and abbreviations, see text.

an example of such an allosteric modulator, which apparently lack some of the side-effects shared by competitive NMDA antagonists and channel blockers.

The potential of developing partial agonists which are devoid of neurotoxic actions or partial antagonists which do not have the potential of a complete block of neurotransmission is also interesting. Moreover the evolving knowledge from molecular biology of regional receptor differentiation could lead the way for developing regionally selective componds with fewer side effects. Glutamate uptake inhibition as a means of enhancing EAA neurotransmission is an area of research that has only had little attention from medicinal chemists. However, this approach and other means of modulating endogenous glutamate levels may turn out to be a sensitive and well-tolerated means of modulating EAA function.

While the 1980s could be described as the decade in which EAA neurotransmission was characterized pharmacologically and where the scene of therapeutic opportunities for EAA modulators was set up, the 1990s might very well be the decade in which such EAA modulators will find their way into clinical studies and may revolutionize the treatment of several divesting diseases, including neurodegeneration after stroke.

REFERENCES

Aanonsen, L. M. and Wilcox, G. L. (1986) *Neurosci. Lett.* **67** 191–197.

Aanonsen, L. M. and Wilcox, G. L. (1987) *J. Pharmacol. Exp. Ther.* **243** 9–19.

Anis, N. A., Berry, S. C., Burton, N. R. and Lodge, D. (1983) *Br. J. Pharmacol.* **79** 565–575.

Arnt, J. (1981) *Life Sci.* **28** 1597–1603.

Balcar, V. J. and Johnston, G. A. R. (1972) *J. Neurobiol.* **3** 295–301.

Battaglia, G. and Rustioni, A. (1988) *J. Comp. Neurol.* **277** 302–312.

Beal, M. F., Kowall, N. W., Ellison, D. W., Mazurek, M. F. and Swartz, K. J. (1986) *Nature* **321** 168–171.

Bennett, D. A. and Amrick, C. L. (1986) *Life Sci.* **39** 2455–2461.

Benveniste, H., Drejer, J., Schousboe, A. and Diemer, N. H. (1984) *J. Neurochem.* **43** 1369–1374.

Benveniste, H., Jørgensen, M. B., Sandberg, M., Christensen, T., Hagberg, H. and Diemer, N. H. (1989) *J. Cerebral Blood Flow Metab.* **9** 629–639.

Bevan, M. D., Brotchie, J. M. and Crossman, A. R. (1990) *Eur. J. Pharmacol.* **183** 943.

Birch, P. J., Grossman, C. J. and Hayes, A. G. (1988) *Eur. J. Pharmacol.* **156** 177–180.

Biscoe, T. J., Evans, R. H., Headley, P. M., Martin, M. R. and Watkins, J. C. (1976) *Br. J. Pharmacol.* **58** 373–382.

Blisss, T. V. P. and Lomo, T. (1973) *J. Physiol.* **232** 331–356.

Boulter, J., Hollmann, M., O'Shea-Greenfield, A., Hartley, M., Deneris, E., Maron, C. and Heinemann, S. (1990) *Science* **249** 1033–1037.

Brotchie, J. M., Bevan, M. D., Mitchell, I. J. and Crossman, A. R. (1990) *Eur. J. Pharmacol.* **183** 943–944.

Bucci, L., Ioppolo, A., Chiavarelli, R. and Biogotti, A. (1982) *Br. J. Exp. Pathol.* **63** 235–241.

Buchan, A. and Pulsinelli, W. A. (1990) *J. Neurosci.* **10** 311–316.

Cahusac, P. M. B., Evans, R. H., Mill, R. G., Rodriguez, R. E. and Smith, D. A. (1984) *Neuropharmacology* **23** 719–724.

Carlsson, M. and Carlsson, A. (1989) *J. Neural. Transm.* **77** 65–71.
Carlsson, M. and Carlsson, A. (1990) *Trends Neurosci.* **13** 272–276.
Carlsson, M. and Svensson, A. (1990) *Pharmacol. Biochem. Behav.* **36** 45–50.
Carter, C., Benavides, J., Legendre, P., Vincent, J. D., Noel, F., Thuret, F., Lloyd, K. G., Arbilla, S., Zivkovic, B., MacKenzie, E. T., Scatton, B. and Langer, S. Z. (1988) *J. Pharmacol. Exp. Ther.* **247** 1222–1232.
Chiamulera, C., Costa, S. and Reggiani, A. (1990) *Psychopharmacology* **102** 551–552.
Christensen, I. T., Reinhardt, A., Nielsen, B., Ebert, B., Madsen, U., Nielsen, E. Ø., Brehm, L. and Krogsgaard-Larsen, P. (1989) *Drug Design Delivery* **5** 57–71.
Clineschmidt, B. V., Martin, G. E. and Bunting, P. R. (1982) *Drug Dev. Res.* **2** 123–134.
Collingridge, G. L. and Bliss, T. V. P. (1987) *Trends Neurosci.* **10** 288–293.
Collingridge, G. L. and Lester, R. A. (1989) *Pharmacol. Rev.* **41** 143–210.
Compton, R. P., Contreras, P. C., O'Donohue, T. L. and Monahan, J. B. (1987) *Eur. J. Pharmacol.* **136** 133–134.
Coutinho-Netto, J., Abdul-Ghani, A. S., Collins, J. F. and Bradford, H. F. (1981) *Epilepsia* **22** 289–296.
Coyle, J. T. and Schwarcz, R. (1976) *Nature* **263** 244–246.
Cross, A. J., Slater, P. and Reynolds, G. P. (1986) *Neurosci. Lett.* **67** 198–202.
Croucher, M. J., Collins, J. F. and Meldrum, B. S. (1982) *Science* **216** 899–901.
Curry, K., Magnuson, D. S., McLennan, H. and Peet, M. J. (1986) *Neurosci. Lett.* **66** 101–105.
Curtis, D. R. and Johnston, G. A. R. (1974) *Ergbn. Physiol.* **69** 94–188.
Curtis, D. R. and Watkins, J. C. (1960) *J. Neurochem.* **6** 117–141.
Curtis, D. R., Phillis, J. W. and Watkins, J. C. (1959) *Nature* **183** 611–612.
Davies, J. D. and Watkins, J. C. (1977) *Brain Res.* **130** 364–368.
Davies, J. and Watkins, J. C. (1979) *J. Physiol.* **297** 621–635.
Davies, J., Francis, A. A., Jones, A. W. and Watkins, J. C. (1981) *Neurosci. Lett.* **21** 77–81.
DeLeander, G. E. and Wahl, J. J. (1989) *Eur. J. Pharmacol.* **159** 149–156.
Deutsch, S. I. and Morihisa, J. M. (1988) *Clin. Neuropharmacol.* **11** 18–35.
Dickenson, A. H. (1990) *Trends Pharmacol. Sci.* **11** 307–309.
Diemer, N. H., Johansen, F. F., Sheardown, M., Honoré, T. and Jørgensen, M. B. (1990) In: Krieglstein J. and Oberpichler, H. (eds) *Pharmacology of Cerebral Ischaemia 1990* Wissenschaftliche Verlagsgesellschaft mbH, Stuttgart, pp. 113–116.
Dingledine, R., McBain, C. J. and McNamara, J. O. (1990) *Trends Pharmacol. Sci.* **11** 334–338.
Domino, E. F., Chodoff, P. and Crossen, G. (1965) *Clin. Pharmacol. Ther.* **6** 279–291.
Drejer, J. and Honoré, T. (1988) *Neurosci. Lett.* **87** 104–108.
Drejer, J., Larsson, O. M. and Schousboe, A. (1982) *Exp. Brain Res.* **47** 259–269.
Drejer, J., Benveniste, H., Diemer, N. H. and Schousboe, A. (1985) *J. Neurochem.* **45** 145–151.
Drejer, J., Sheardown, M., Nielsen, E. Ø. and Honoré, T. (1989) *Neurosci. Lett.* **98** 333–338.
During, M. J., Freese, A. and Heyes, M. P. (1989) *FEBS Lett.* **247** 438–444.
Erez, U., Frenk, H., Goldberg, O., Cohen, A. and Teichberg, V. I. (1985) *Eur. J. Pharmacol.* **110** 31–39.
Evans, R. H. (1989) *Prog. Neurobiol.* **33** 255–279.

Ferkany, J. W., Kyle, D. J., Willets, J., Rzeszotarski, W. J., Guzewska, M. E., Ellenberger, S. R., Jones, S. M., Sacaan, A. I., Snell, L. D., Borosky, S., Jones, B. E., Johnson, K. M., Balster, R. L., Burchett, K., Kawasaki, K., Hoch, D. B. and Dingledine, R. (1989) *J. Pharmacol. Exp. Ther.* **250** 100–109.

Ferrante, R. J., Kowall, N. W., Beal, M. F., Richardson, E. P. and Martin, J. B. (1985) *Science* **230** 561–563.

Freese, A., Swartz, K. J., During, M. J. and Martin, J. B. (1990) *Neurology* **40** 691–695.

Frey, P., Berney, D., Herrling, P. L., Mueller, W. and Urwyler, S. (1988) *Neurosci. Lett.* **91** 194–198.

Ganong, A. H., Lanthorn, T. H. and Cotman, C. W. (1983) *Brain Res.* **273** 170–174.

Garthwaite, G. and Garthwaite, J. (1989) *Neurosci. Lett.* **99** 113–118.

Geddes, J. W., Monaghan, D. T., Cotman, C. W., Lott, I. T., Kim, R. C. and Chui, H. C. (1985) *Science* **230** 1179–1181.

Gotti, B., Duverger, D., Bertin, J., Carter, C., Dupont, R., Frost, J., Guadilliere, B., MacKenzie, E. T., Rousseau, J., Scatton, B. and Wick, A. (1988) *J. Pharmacol. Exp. Ther.* **247** 1211–1221.

Greenamyre, J. T., Young, A. B. and Penney, J. B. (1984) *J. Neurosci.* **4** 2133–2144.

Greenamyre, J. T., Penney, J. B., Young, A. B., D'Amato, C. J., Hicks, S. and Shoulson, I. (1985) *Science* **227** 1496–1499.

Greenamyre, J. T., Penney, J. B., D'Amato, C. J. and Young, A. B. (1987) *J. Neurochem.* **48** 543–551.

Greifenstein, F. E., Yoskitake, J., DeVault, M. and Gajewski, J. E. (1958) *Anesth. Anal.* **37** 283–294.

Haldeman, S. and McLennan, H. (1972) *Brain Res.* **45** 393–400.

Haley, J. E., Sullivan, A. F. and Dickenson, A. H. (1990) *Brain Res.* **518** 218–226.

Hayashi, T. (1954) *Keio J. Med.* **3** 183–192.

Herreras, O., Menendez, N., Herranz, A. S., Solis, J. M. and Martin del Rio, R. (1989) *Neurosci. Lett.* **99** 119–124.

Hicks, T. P., Hall, J. G. and McLennan, H. (1978) *Can. J. Physiol. Pharmacol.* **56** 901–907.

Hollmann, M., O'Shea-Greenfield, A., Rogers, S. W. and Heinemann, S. (1989) *Nature* **342** 643–648.

Honoré, T. (1989) *Med. Res. Rev.* **9** 1–23.

Honoré, T., Davies, S. N., Drejer, J., Fletcher, E. J., Jacobsen, P., Lodge, D. and Nielsen, F. E. (1988) *Science* **241** 701–703.

Hönack, D. and Löscher, W. (1990) *Eur. J. Pharmacol.* **183** 950.

Imperato, A., Honoré, T. and Jensen, L. H. (1990) *Brain Res.* **530** 223–228.

Jacquet, Y. F. (1988) *Eur. J. Pharmacol.* **154** 271–276.

Janowsky, A. and Berger, P. (1989) *Schizophrenia Res.* **2** 189.

Johansen, F. F., Jørgensen, M. B. and Diemer, N. H. (1987) In: Hicks, T. P., Lodge, D. and McLennan, H. (eds) *Excitatory Amino Acid Transmission.* Liss, New York, pp. 245–248.

Johnson, G. (1989) *Annu. Rev. Med. Chem.* **24** 41–50.

Johnson, J. W. and Ascher, P. (1987) *Nature* **325** 529–531.

Jørgensen, M. B., Johansen, F. F. and Diemer, N. H. (1987) *Acta Neuropathol.* **73** 189–194.

Keinänen, K., Wisden, W., Sommer, B., Werner, P., Herb, A., Verdoorn, T. A. Sakmann, B. and Seeburg, P. H. (1990) *Science* **249** 556–560.

Kemp, J. A., Foster, A. C. and Wong, E. H. F. (1987) *Trends Neurosci.* **10** 294.

Kemp, J. A., Foster, A. C., Leeson, P. D., Priestley, T., Tridgett, R., Iversen, L. L. and Woodruff, G. N. (1988) *Proc. Nat. Acad. Sci. USA* **85** 6547–6550.

Kim, J. S., Kornhuber, H. H., Schmid-Burgk, W. and Holzmuller, B. (1980) *Neurosci. Lett.* **20** 379–382.

Kleinschmidt, A., Bear, M. F. and Singer, W. (1987) *Science* **238** 355–358.

Klepstad, P., Maurset, A., Moberg, E. R. and Øye, I. (1990) *Eur. J. Pharmacol.* **187** 513–518.

Klockgether, T. and Turski, L. (1989) *Trends Neurosci.* **12** 285–286.

Koek, W. and Colpaert, F. C. (1990) *J. Pharmacol. Exp. Ther.* **252** 349–357.

Köhler, C. and Schwarcz, R. (1981) *Brain Res.* **211** 485–491.

Largent, B. L., Gundlach, A. L. and Snyder, S. H. (1986) *J. Pharmacol. Exp. Ther.* **238** 739–748.

Lauritzen, M. (1987) *Trends Neurosci.* **10** 8–13.

Lauritzen, M., Rice, M., Okada, Y. and Nicholson, C. (1988) *Brain Res.* **475** 317–327.

Leao, A. A. P. (1944) *J. Neurophysiol.* **7** 359–390.

Lei, S. and Wilcox, G. L. (1990) *Eur. J. Pharmacol.* **183** 1438–1439.

Lodge, D. and Johnson, K. M. (1990) *Trends Pharmacol. Sci.* **11** 81–86.

Löscher, W., Nolting, B. and Hönack, D. (1988) *Eur. J. Pharmacol.* **152** 9—17.

Lucas, D. R. and Newhouse, J. P. (1957) *AMA Arch. Opthalmol.* **58** 193–201.

Lynch, G. and Baudry, M. (1984) *Science* **224** 1057–1063.

Maragos, W. F., Chu, D. C. M., Young, A. B., d'Amato, C. J. and Penney, J. B. (1987) *Neurosci. Lett.* **74** 371–376.

Martin, W. R., Eades, C. G., Thompson, J. A., Huppler, R. E. and Gilbert, P. E. (1976) *J. Pharmacol. Exp. Ther.* **197** 517–532.

Maurset, A., Skoglund, L. A., Hustveit, O. and Oye, I. (1989) *Pain* **36** 37–41.

Mayer, M. L. and Westbrook, G. L. (1987) *Prog. Neurobiol.* **28** 197–276.

McBean, G. J. and Roberts, P. J. (1985) *J., Neurochem.* **44** 247–254.

McCulloch, J., Chen, M.-H., Graham, D., Lowe, D. and Bullock, R. (1990) In: Krieglstein, J. and Oberpichler, H. (eds) *Pharmacology of Cerebral Ischaemia 1990.* Wissenschaftliche Verlagsgesellschaft mbH, Stuttgart, pp. 239–244.

McGeer, E. G. and McGeer, P. L. (1976) *Nature* **263** 517–519.

McGeer, E. G. (1989) *Neurobiol. Aging* **10** 614–616.

McGeer, E. G. and Zhu, S. G. (1990) *Neurosci. Lett.* **112** 348–351.

McLennan, H. and Lodge, D. (1979) *Brain Res.* **169** 83-90.

McNamara, D., Smith, E. C. R., Calligaro, D. O., O'Malley, P. J., McQuaid, L. A. and Dingledine, R. (1990) *Neurosci. Lett.* **120** 17–20.

Meldrum, B. S. (1983) *Neurosci. Lett.* **39** 101–104.

Mody, I., Lambert, J. D. C. and Heinemann, U. (1987) *J. Neurophysiol.* **57** 869–888.

Mondadori, C., Weiskrantz, L., Buerki, H., Petschke, F. and Fagg, G. E. (1989) *Exp. Brain Res.* **75** 449–456.

Moroni, F., Lombardi, G., Carla, V., Cal, S., Etienne, P. and Nair, N. P. V. (1986) *J. Neurochem.* **47** 1667–1671.

Morris, R. (1989) *Neurosci. Lett.* **105** 79–85.

Morris, R. G. M., Anderson, E., Lynch, G. S. and Baudry, M. (1986) *Nature* **319** 774–776.

Muller, D., Joly, M. and Lynch, G. (1988) *Science* **242** 1694–1697.

Neary, D., Snowden, J. S. and Mann, D. M. A. (1986) *J. Neurol. Neurosurg. Psychiatry* **49** 229–237.

Nowak, L., Bregestovski, P., Ascher, P., Herbet, A. and Prochiantz, A. (1984) *Nature* **307** 462–465.

Olney, J. W. (1969) *Science* **164** 719–721.

Olney, J. W. (1978) In: McGeer, E., Olney, J. W. and McGeer, P. (eds) *Kainic Acid as a Tool in Neurobiology*. Raven Press, New York, pp. 95–121.

Olney, J. W. and Gubareff, T. (1978) *Nature* **271** 557–559.

Olney, J. W., Labruyere, J. and Price, M. T. (1989) *Science* **244** 1360–1362.

Ozyurt, E., Graham, D. I., Woodruff, G. N. and McCulloch, J. (1988) *J. Cerebral Blood Flow Metab.* **8** 138–143.

Palmer, A. M., Procter, A. W., Stratmann, G. C. and Bowen, D. M. (1986) *Neurosci. Lett.* **66** 199–204.

Park, C. K., Nehls, D. G., Graham, D. I., Teasdale, G. M. and McCulloch, J. (1988) *J. Cerebral Blood Flow Metab.* **8** 757–762.

Patel, J. B., Ross, L. E., Duncan, B., Asiz, M., Salama, A. I. and Valerio, M. (1989) *Soc. Neurosci. Abstr.* **15** 43.

Patel, J., Zinland, W. C., Klika, A. B., Mangano, T. J., Keith, R. A. and Salama, A. I. (1990) *J. Neurochem.* **55** 114–121.

Peeters, B. W. M. M., Van Rijn, C. M., Van Luitelaar, E. L. J. M. and Coenen, A. M. L. (1989) *Epilepsy Res.* **3** 178–181.

Perkins, M. N. and Stone, T. W. (1982) *Brain res.* **247** 184–187.

Perrault, G., Morel, E., Joly, D., Sanger, D. J. and Zivkovic, B. (1990) *Eur. J. Pharmacol.* **183** 942.

Perry, T. L. and Hansen, S. (1990) *Neurology* **40** 20–24.

Rothman, S. M. and Olney, J. W. (1986) *Ann. Neurol.* **19** 105–111.

Scheller, D., Heister, U., Dengler, K. and Peters, T. (1990) In: Krieglstein, J. and Oberpichler, H. (eds) *Pharmacology of Cerebral Ischaemia 1990*. Wissenschaftliche Verlagsgesellschaft mbH, Stuttgart, pp. 205–210.

Schmidt, W. J. (1986) *Psychopharmacology* **90** 123–130.

Schmidt, W. J. and Bischoff, C. (1988) *Psychopharmacology* **96** S51.

Schwarcz, R., Whetsell, W. O. and Mangano, R. M. (1983) *Science* **219** 316–318.

Sheardown, M. J., Drejer, J., Jensen, L. H., Stidsen, C. E. and Honoré, T. (1989) *Eur. J. Pharmacol.* **174** 197–204.

Sheardown, M. J., Nielsen, E. Ø., Hansen, A. J., Jacobsen, P. and Honoré, T. (1990) *Science* **247** 571–574.

Siegfried, B. and de Souza, R. L. (1989) *Eur. J. Pharmacol.* **168** 239–242.

Simon, R. P., Swan, J. H., Griffiths, T. and Meldrum, B. S. (1984) *Science* **226** 850–852.

Sladeczek, F., Pin, J.-P., Recasens, M., Bockaert, J. and Weiss, S. (1985) *Nature* **317** 717–719.

Sofroniew, M. V. and Pearson, R. C. (1985) *Brain Res.* **339** 186–190.

Stasheff, S. F., Anderson, W. W., Clark, S. and Wilson, W. A. (1989) *Science* **245** 648–651.

Stephens, D. N., Meldrum, B. S., Weidmann, R., Schneider, C. and Grutzner, M. (1986) *Psychopharmacology* **90** 166–169.

Stone, T. W. and Connick, J. H. (1985) *Neuroscience* **15** 597–617.

Stone, T. W. and Perkins, M. N. (1981) *Eur. J. Pharmacol.* **72** 411–412.

Sugiyama, H., Ito, I. and Hirono, C. (1987) *Nature* **325** 531–533.

Tiedtke, P. I., Bischoff, C. and Schmidt, W. J. (1990) *J. Neural Transm. Gen. Sect.* **81** 173–182.

Tortella, F. C., Pellicano, M. and Bowery, N. G. (1989) *Trends Pharmacol. Sci.* **10** 501–507.

Turski, L., Schwarz, M., Turski, W. A., Klockgether, T., Sontag, K.-H. and Collins, J. F. (1985) *Neurosci. Lett.* **35** 321–326.

Van Harreveld, A. (1984) *J. Neurobiol.* **15** 333–344.

Wada, K., Dechesne, C. J., Shimasaki, S., King, R. G., Kusano, K., Buonanno, A., Hampson, D. R., Banner, C., Wenthold, R. J. and Nakatani, Y. (1989) *Nature* **342** 684–689.

Watkins, J. C. (1989) In: Watkins, J. C. and Collingridge, G. L. (eds) *The NMDA Receptor.* IRL Press, Oxford, pp. 1–17.

Watkins, J. C., Krosgsgaard-Larsen, P. and Honoré, T. (1990) *Trends Pharmacol. Sci.* **11** 25–33.

Wenthold, R. J., Hampson, D. R., Wada, K., Hunter, C., Oberdorfer, M. D. and Dechesne, C. (1990) *J. Histochem. Cytochem.* **38** 1717–1723.

Whetsell, W. O. (1984) *Clin. Neuropharmacol.* **7** 248–250.

Wieloch, T., Lindvall, O., Blomquist, P. and Gage, F. E. (1985) *Neurol. Res.* **7** 24–26.

Wilkins, J. (1989) *NIDA Res. Monogr.* **95** 275–281.

Willetts, J., Balster, R. L. and Leander, J. D. (1990) *Trends Pharmacol. Sci.* **11** 423–428.

Wozniak, D. F., Olney, J. W., Kettinger, L., Price, M. and Miller, J. P. (1990) *Psychopharmacology* **101** 47–56.

Yaksh, T. L. (1989) *Pain* **37** 111–123.

Index